T0296090

LONDON MATHEMATICAL SOCIETY LECTURE NOTE SERIES

Managing Editor: Professor M. Reid, Mathematics Institute, University of Warwick, Coventry CV4 7AL, United Kingdom

The titles below are available from booksellers, or from Cambridge University Press at www.cambridge.org/mathematics

London Mathematical Society Lecture Note Series: 390

Localization in Periodic Potentials

From Schrödinger Operators to the Gross–Pitaevskii Equation

DMITRY E. PELINOVSKY

McMaster University, Hamilton, Ontario, Canada

CAMBRIDGE
UNIVERSITY PRESS

CAMBRIDGE
UNIVERSITY PRESS

Shaftesbury Road, Cambridge CB2 8EA, United Kingdom

One Liberty Plaza, 20th Floor, New York, NY 10006, USA

477 Williamstown Road, Port Melbourne, VIC 3207, Australia

314–321, 3rd Floor, Plot 3, Splendor Forum, Jasola District Centre, New Delhi – 110025, India

103 Penang Road, #05–06/07, Visioncrest Commercial, Singapore 238467

Cambridge University Press is part of Cambridge University Press & Assessment, a department of the University of Cambridge.

We share the University's mission to contribute to society through the pursuit of education, learning and research at the highest international levels of excellence.

www.cambridge.org
Information on this title: www.cambridge.org/9781107621541

© D.E. Pelinovsky 2011

This publication is in copyright. Subject to statutory exception and to the provisions of relevant collective licensing agreements, no reproduction of any part may take place without the written permission of Cambridge University Press & Assessment.

First published 2011
Reprinted 2012

A catalogue record for this publication is available from the British Library

Library of Congress Cataloging-in-Publication data
Pelinovsky, Dmitry.
Localization in periodic potentials : from Schrödinger operators to the Gross-Pitaevskii equation / Dmitry E. Pelinovsky.
p. cm. – (London Mathematical Society lecture note series ; 390)
Includes bibliographical references and index.
ISBN 978-1-107-62154-1 (pbk.)
1. Schrödinger equation. 2. Gross-Pitaevskii equations. 3. Localization theory.
I. Title. II. Series.
QC174.26.W28P45 2011
530.12´4 – dc23 2011025637

ISBN 978-1-107-62154-1 Paperback

Cambridge University Press & Assessment has no responsibility for the persistence or accuracy of URLs for external or third-party internet websites referred to in this publication and does not guarantee that any content on such websites is, or will remain, accurate or appropriate.

To Anna and my children:
Albert and Edward, Marta and Polina

Contents

Contents

Preface

Bose–Einstein condensation was predicted by S.N. Bose and Albert Einstein in 1925: for a gas of non-interacting particles, below a certain temperature there is a phase transition to a localized (condensed) state of lowest energy. This phenomenon was realized experimentally in 1995 in alkali gases by E. Cornell and C. Wieman in Boulder as well as by W. Ketterle at MIT, who all shared the Nobel Prize in 2001. Since that time, the attention of many mathematicians has turned to the analysis of the mean-field model of this phenomenon, which is known as the Gross–Pitaevskii equation or the nonlinear Schrödinger equation with an external potential.

Various trapping mechanisms of Bose–Einstein condensation were realized experimentally, including a parabolic magnetic confinement and a periodic optical lattice. This book is about the Gross–Pitaevskii equation with a *periodic* potential, in particular about the localized modes supported by the periodic potential. The book is written for young researchers in applied mathematics and so it has the main emphasis on the *mathematical properties* of the Gross–Pitaevskii equation. It can nevertheless serve as a reference for theoretical physicists interested in the phenomenon of localization in periodic potentials.

Compared to recent work by Lieb *et al* on the justification of the Gross–Pitaevskii equation as the mean-field approximation of the linear N-body Schrödinger equation [131], this book takes the Gross–Pitaevskii equation as the starting point of analysis. Hence the validity of the mean-field approximation is not questioned when existence and stability of localized modes are considered. On the other hand, the mean-field model is simplified further to the coupled nonlinear Schrödinger equations, the nonlinear Dirac equations, and the discrete nonlinear Schrödinger equations, which are all justified in the framework of the Gross–Pitaevskii equation with a periodic potential.

Besides optical trapping lattices in the context of Bose–Einstein condensation, periodic structures are very common in many other physical problems, including photonic crystals, arrays of coupled optical waveguides, and optically induced photonic lattices. One of the important features of such systems is the existence of *band gaps* in the wave transmission spectra, which support stationary localized modes known as the *gap solitons* or *discrete breathers*. Similar to other solitons, these localized modes realize a balance between periodicity, dispersion, and nonlinearity of the physical system.

Part of this monograph was used for a one-semester graduate course at McMaster University in 2009–2010. The author is grateful to graduate students of the course, particularly to D. Avalo and A. Sakovich, for thoughtful reading of the text and useful corrections. Many parts of this text would not have been possible without contributions of the coauthors on individual projects, in particular I. Barashenkov, M. Chugunova, S. Cuccagna, G. Iooss, P. Kevrekidis, T. Melvin, G. Schneider, A. Stefanov, V. Vougalter, and J. Yang. The author is extremely grateful to all these researchers for their enthusiasm, support, and many fruitful ideas. Numerical computations of localized modes were performed with the assistance of J. Brown during his B.Sc. project in 2009–2010.

Dmitry Pelinovsky

1

Formalism of the nonlinear Schrödinger equations

Make everything as simple as possible, but not simpler.
– Albert Einstein.

Someone told me that each equation I included in the book would halve the sales.
– Stephen Hawking.

When the author was a graduate student, introductions to texts on nonlinear evolution equations contained a long description of physical applications, numerous references to the works of others, and sparse details of the justification of analytical results. Times have changed, however, and the main interest in the nonlinear evolution equations has moved from modeling to analysis. It is now more typical for applied mathematics texts to start an introduction with the main equations in the first lines, to give no background information on applications, to reduce the list of references to a few relevant mathematical publications, and to focus discussions on technical aspects of analysis.

Since this book is aimed at young mathematicians, we should reduce the background information to a minimum and focus on useful analytical techniques in the context of the nonlinear Schrödinger equation with a periodic potential. It is only in this introduction that we recall the old times and review the list of nonlinear evolution equations that we are going to work with in this book. The few references will provide a quick glance at physical applications, without distracting attention from equations.

The list of nonlinear evolution equations relevant to us begins with the nonlinear Schrödinger equation with an external potential,

$$iu_t = -\Delta u + V(x)u + \sigma|u|^2 u,$$

where $\Delta = \partial_{x_1}^2 + \cdots + \partial_{x_d}^2$ is the Laplacian operator in the space of d dimensions, $u(x, t) : \mathbb{R}^d \times \mathbb{R} \to \mathbb{C}$ is the amplitude function, $V(x) : \mathbb{R}^d \to \mathbb{R}$ is a given potential, and $\sigma \in \{1, -1\}$ is the sign for the cubic nonlinearity. If $\sigma = -1$, the nonlinear Schrödinger equation is usually called focusing or attractive, whereas if $\sigma = +1$, it is called defocusing or repulsive. The names differ depending on the physical applications of the model to nonlinear optics [5], photonic crystals [192], and atomic physics [171].

It is quite common to use the name of the *Gross–Pitaevskii equation* if this equation has a nonzero potential $V(x)$ and the name of the *nonlinear Schrödinger equation* if $V(x) \equiv 0$. Historically, this terminology is not justified as the works of E.P. Gross and L.P. Pitaevskii contained a derivation of the same equation with $V(x) \equiv 0$ as the mean-field model for superfluids [83] and Bose gas [170]. However, given the number of physical applications of this equation with $V(x) \equiv 0$ in nonlinear optics, photonic crystals, plasma physics, and water waves, we shall obey this historical twist in terminology and keep reference to the Gross–Pitaevskii equation if $V(x)$ is *nonzero* and to the nonlinear Schrödinger equation if $V(x)$ is *identically zero*.

Several recent books [2, 30, 133, 199, 201] have been devoted to the nonlinear Schrödinger equation and we shall refer readers to these books for useful information on mathematical properties and physical applications of this equation. As a main difference, this book is devoted to the Gross–Pitaevskii equation with a *periodic* potential $V(x)$.

We shall study localized modes of the Gross–Pitaevskii equation with the periodic potential. These localized modes are also referred to as the *gap solitons* or *discrete breathers* because they are given by time-periodic and space-decaying solutions of the Gross–Pitaevskii equation. In many of our studies, we shall deal with localized modes in one spatial dimension. Readers interested in analysis of vortices in the two-dimensional Gross–Pitaevskii equation with a harmonic potential may be interested to read the recent book of Aftalion [3].

Many other nonlinear evolution equations with similar properties actually arise as asymptotic reductions of the Gross–Pitaevskii equation with a periodic potential, while they merit independent mathematical analysis and have independent physical relevance. One such model is a system of two *coupled nonlinear Schrödinger* equations,

$$\begin{cases} iu_t = -\Delta u + \sigma \left(|u|^2 + \beta |v|^2 \right) u, \\ iv_t = -\alpha \Delta v + \sigma \left(\beta |u|^2 + |v|^2 \right) v, \end{cases}$$

where α and β are real parameters, $\sigma \in \{1, -1\}$, and $(u, v)(x, t) : \mathbb{R}^d \times \mathbb{R} \to \mathbb{C}^2$ stand for two independent amplitude functions. The number of amplitude functions in the coupled nonlinear Schrödinger equations can exceed two in various physical problems. However, two is a good number both from the increased complexity of mathematical analysis compared to the scalar case and from the robustness of the model to the description of practical problems. For instance, two polarization modes in a birefringent fiber are governed by the system of two coupled nonlinear Schrödinger equations and so are the two resonant Bloch modes at the band edge of the photonic spectrum [7].

When the coupled equations for two resonant modes in a periodic potential are derived in the space of one dimension $(d = 1)$, the group velocities of the two modes are opposite to each other and the coupling between the two modes involves linear terms. In this case, the system takes the form of the *nonlinear Dirac* equations,

$$\begin{cases} i(u_t + u_x) = \alpha v + \sigma \left(|u|^2 + \beta |v|^2 \right) u, \\ i(v_t - v_x) = \alpha u + \sigma \left(\beta |u|^2 + |v|^2 \right) v, \end{cases}$$

where α and β are real parameters, $\sigma \in \{1, -1\}$, and $(u, v)(x, t) : \mathbb{R} \times \mathbb{R} \to \mathbb{C}^2$ are the amplitude functions for two resonant modes. Two counter-propagating waves coupled by the Bragg resonance in an optical grating is one of the possible applications of the nonlinear Dirac equations [197]. We recall from quantum mechanics that the Dirac equations represent the relativistic theory compared to the Schrödinger equations that represent the classical theory.

The list of nonlinear evolution equations for this book ends at the spatial discretization of the nonlinear Schrödinger equation, which is referred to as the *discrete nonlinear Schrödinger* equation. This equation represents a system of infinitely many coupled differential equations on a lattice,

$$i\dot{u}_n = -(\Delta u)_n + \sigma |u_n|^2 u_n,$$

where $(\Delta u)_n$ is the discrete Laplacian operator on the d-dimensional lattice, $\sigma \in \{1, -1\}$, and $\{u_n(t)\} : \mathbb{Z}^d \times \mathbb{R} \to \mathbb{C}$ is an infinite set of amplitude functions. The discrete nonlinear Schrödinger equation arises in the context of photonic crystal lattices, Bose–Einstein condensates in optical lattices, Josephson-junction ladders, and the DNA double strand models [110].

Modifications of the nonlinear evolution equations in the aforementioned list with a more general structure, e.g. with additional linear terms and non-cubic nonlinear functions, are straightforward and we adopt these modifications throughout the book if necessary. It is perhaps more informative to mention other nonlinear evolution equations, which are close relatives to the nonlinear Schrödinger equation. Among them, we recall the *Klein–Gordon*, *Boussinesq*, and other nonlinear dispersive wave equations. In the unidirectional approximation, many nonlinear dispersive wave equations reduce to the *Korteweg–de Vries* equation. It would take too long, however, to list all other nonlinear evolution equations, their modifications, and the relationships between them, hence we should stop here and move to the mathematical analysis of the Gross–Pitaevskii equation with a periodic potential. For the sake of clarity, we work with the simplest mathematical models keeping in mind that the application-motivated research in physics leads to more complicated versions of these governing equations.

1.1 Asymptotic multi-scale expansion methods

Our task is to show how the Gross–Pitaevskii equation with a periodic potential can be approximated by the simpler nonlinear evolution equations listed in the beginning of this chapter. To be able to perform such approximations, we shall consider a powerful technique known as the *asymptotic multi-scale expansion method*. This method is applied in a certain asymptotic limit after all important terms of the primary equations are brought to the same order, while all remaining terms are removed from the leading order.

The above strategy does not sound like a rigorous mathematical technique. If power expansions in terms of a small parameter are developed, only a few terms are actually computed in the asymptotic multi-scale expansion method, while the remaining terms are cut down in the hope that they are small in some sense. In the time evolution of a hyperbolic system with energy conservation, it is not unusual to bound the remaining terms at least for a finite time (Sections 2.2–2.4). In the time evolution of a parabolic system with energy dissipation, the theory of invariant manifolds and normal forms is often invoked to obtain global results for all positive times, as we shall see in the same sections in the context of a time-independent elliptic system (which is a degenerate case of a parabolic system).

It is still true that a formal approximation of the asymptotic multi-scale expansion method is half way to the rigorous analysis of the asymptotic reduction. Another way to say this: if you do not know how to solve a problem rigorously, first solve it formally! In many cases, the formal solution may suggest ways to rigorous analysis.

The above slogan explains why, even being unable to analyze asymptotic approximations thirty or even twenty years ago, applied mathematicians tried nevertheless to formalize the asymptotic multi-scale expansion method in many details to avoid misleading computations and failures. The method has been described in several books [143, 199] and the number of original publications in the context of physically relevant equations is truly uncountable!

We shall look at the three different asymptotic limits separately for the reductions of the Gross–Pitaevskii equation with a periodic potential to the nonlinear Dirac equations, the nonlinear Schrödinger equation, and the discrete nonlinear Schrödinger equation. Our approach is based on the classical asymptotic multi-scale expansion method, which has been applied to these three asymptotic reductions in the past [152].

The starting point for our asymptotic analysis is the Gross–Pitaevskii equation in one spatial dimension,

$$iu_t = -u_{xx} + V(x)u + \sigma|u|^2u, \qquad (1.1.1)$$

where $u(x,t) : \mathbb{R} \times \mathbb{R} \to \mathbb{C}$, $\sigma \in \{1, -1\}$, and $V(x + 2\pi) = V(x)$ is a bounded 2π-periodic potential.

Three different asymptotic limits represent different interplays between the strength of the periodic potential $V(x)$ and the strength of the nonlinear potential $\sigma|u|^2$ affecting existence of localized modes in the Gross–Pitaevskii equation (1.1.1). These asymptotic limits are developed for small-amplitude potentials (Section 1.1.1), finite-amplitude potentials (Section 1.1.2), and large-amplitude potentials (Section 1.1.3). In each case, we derive the leading-order asymptotic reduction that belongs to the list of nonlinear evolution equations in the beginning of this chapter.

1.1.1 The nonlinear Dirac equations

In the limit of small-amplitude periodic potentials, the Gross–Pitaevskii equation (1.1.1) can be rewritten in the explicit form,

$$iu_t = -u_{xx} + \epsilon V(x)u + \sigma|u|^2u, \qquad (1.1.2)$$

where ϵ is a small parameter. If the nonlinear term $\sigma|u|^2u$ is crossed over and ϵ is set to 0, equation (1.1.2) becomes the linear Schrödinger equation

$$iu_t = -u_{xx}, \tag{1.1.3}$$

which is solved by the Fourier transform as

$$u(x,t) = \int_{\mathbb{R}} a(k)e^{ikx-i\omega(k)t}dk,$$

where $\omega(k) = k^2$ is the dispersion relation for linear waves and $a(k) : \mathbb{R} \to \mathbb{C}$ is an arbitrary function, which is uniquely specified by the initial condition $u(x,0)$. Since $V(x + 2\pi) = V(x)$ for all $x \in \mathbb{R}$, we can represent $V(x)$ by the Fourier series

$$V(x) = \sum_{m \in \mathbb{Z}} V_m e^{imx}.$$

For any fixed $k \in \mathbb{R}$, the Fourier mode $e^{ikx-i\omega(k)t}$ in $u(x,t)$ generates infinitely many Fourier modes in the term $\epsilon V(x)u(x,t)$ at the Fourier wave numbers $k_m = k + m$ for all $m \in \mathbb{Z}$. These Fourier modes are said to be in *resonance* with the primary Fourier mode $e^{ikx-i\omega(k)t}$ if

$$\omega(k_m) = \omega(k), \quad m \in \mathbb{Z},$$

which gives us an algebraic equation $m(m + 2k) = 0$, $m \in \mathbb{Z}$. Thus, if $2k$ is not an integer, none of the Fourier modes with $m \neq 0$ are in resonance with the primary Fourier mode, while if $2k = n$ for a fixed $n \in \mathbb{N}$, the Fourier mode with $m = -n$ is in resonance with the primary mode with $m = 0$. In terms of the Fourier harmonics, the two resonant modes have $k = \frac{n}{2}$ and $k_{-n} = k - n = -\frac{n}{2}$.

In the asymptotic method, we shall zoom in the two resonant modes at $k = \frac{n}{2}$ and $k_{-n} = -\frac{n}{2}$ by a scaling transformation and obtain a system of nonlinear evolution equations for mode amplitudes by bringing all important terms to the same first order in ϵ, where the resonance is found. The important terms to be included in the leading order are related to the nonlinearity, dispersion, and inter-action of the resonant modes, as well as to their time evolution. To incorporate all these effects, we shall look for an asymptotic multi-scale expansion in powers of ϵ,

$$u(x,t) = \epsilon^p \left[\left(a(X,T)e^{\frac{inx}{2}} + b(X,T)e^{-\frac{inx}{2}} \right) e^{-\frac{in^2t}{4}} + \epsilon u_1(x,t) + \mathcal{O}(\epsilon^2) \right], \tag{1.1.4}$$

where $X = \epsilon^q x$, $T = \epsilon^r t$, (p,q,r) are some parameters to be determined, $n \in \mathbb{N}$ is fixed, $(a,b)(X,T) : \mathbb{R} \times \mathbb{R} \to \mathbb{C}^2$ are some amplitudes to be determined, $\epsilon u_1(x,t)$ is a first-order remainder term, and $\mathcal{O}(\epsilon^2)$ indicates a formal order of truncation of the asymptotic expansion. When the asymptotic expansion (1.1.4) is substituted into the Gross–Pitaevskii equation (1.1.2), the exponents are chosen from the condition that terms coming from derivatives of (a,b) in (X,T) and terms coming from powers of (a,b) enter the same order of the asymptotic expansion. This procedure sets uniquely $p = \frac{1}{2}$ and $q = r = 1$.

The simple choice of the exponents (p, q, r) may fail to bring all terms to the same order. For instance, coefficients in front of the first derivatives of (a, b) or the powers of (a, b) can be zero for some problems. Such failures typically indicate that (p, q, r) must be chosen smaller so that the coefficients in front of the higher order derivatives of (a, b) or the higher-order powers of (a, b) are nonzero. Therefore, similarly to the technique of integration, the asymptotic multi-scale expansion method is a laboratory, in which a researcher plays an active role by employing the strategy of trials and errors until he or she manages to bring all important terms to the same order.

Setting $p = \frac{1}{2}, q = r = 1$ and truncating the terms of the order of $\mathcal{O}(\epsilon^2)$, we write a linear inhomogeneous equation for $u_1(x, t)$,

$$\left(i\partial_t + \partial_x^2\right) u_1 = -i(a_T \mathbf{e}_+ + b_T \mathbf{e}_-) - in(a_X \mathbf{e}_+ - b_X \mathbf{e}_-)$$
$$+ V(a\mathbf{e}_+ + b\mathbf{e}_-) + \sigma |a\mathbf{e}_+ + b\mathbf{e}_-|^2 (a\mathbf{e}_+ + b\mathbf{e}_-), \quad (1.1.5)$$

where $\mathbf{e}_\pm \equiv e^{\pm \frac{in}{2}x - \frac{in^2}{4}t}$ satisfy the linear Schrödinger equation (1.1.3). If terms proportional to either $\mathbf{e}_+$ or $\mathbf{e}_-$ occur on the right-hand side of the linear inhomogeneous equation (1.1.5), the solution $u_1(x, t)$ becomes unbounded in variables (x, t), e.g. $u_1(x, t) \sim t\mathbf{e}_\pm$. Since secular growth is undesired as it destroys the applicability of the asymptotic solution (1.1.4) already at the first order in ϵ, one needs to eliminate the resonant terms on the right-hand side of (1.1.5). This is possible if the amplitudes (a, b) satisfy the system of first-order semi-linear equations,

$$\begin{cases} i(a_T + na_X) = V_0 a + V_n b + \sigma(|a|^2 + 2|b|^2)a, \\ i(b_T - nb_X) = V_{-n} a + V_0 b + \sigma(2|a|^2 + |b|^2)b, \end{cases} \quad (1.1.6)$$

where V_0, V_n, and V_{-n} are coefficients of the Fourier series for $V(x)$. The amplitude equations (1.1.6) are nothing but the nonlinear Dirac equations.

Because the linear Schrödinger equation (1.1.3) is dispersive with $\omega''(k) = 2 \neq 0$, the other Fourier modes on the right-hand side of (1.1.5) which are different from $\mathbf{e}_\pm$ do not produce a secular growth of $u_1(x, t)$. Therefore, the system (1.1.6) gives the necessary and sufficient condition that $u_1(x, t)$ is bounded for all $(x, t) \in \mathbb{R}^2$. Indeed, we can find the explicit bounded solution of the inhomogeneous equation (1.1.5),

$$u_1(x, t) = - \sum_{m \notin \{-n, 0\}} \frac{V_m a}{m(m+n)} e^{i(m + \frac{n}{2})x - \frac{in^2 t}{4}} - \sum_{m \notin \{0, n\}} \frac{V_m b}{m(m-n)} e^{i(m - \frac{n}{2})x - \frac{in^2 t}{4}}$$
$$- \frac{1}{2n^2} \left(a^2 \bar{b} e^{\frac{3inx}{2}} + b^2 \bar{a} e^{-\frac{3inx}{2}} \right) e^{-\frac{in^2 t}{4}}.$$

This expression suggests that the asymptotic solution (1.1.4) to the original equation (1.1.2) may be factored by $e^{-\frac{in^2 t}{4}}$. After the bounded solution is found for the first-order remainder term, one can continue the formal asymptotic solution (1.1.4) to the next order of $\mathcal{O}(\epsilon^2)$.

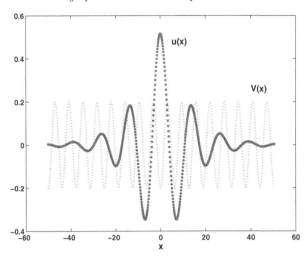

Figure 1.1 Schematic representation of the leading order in the asymptotic solution (1.1.4).

In view of so many restrictive assumptions and so many truncations made in the previous computations, it might be surprising to know that the nonlinear Dirac equations can be rigorously justified for small-amplitude periodic potentials (Section 2.2). Stationary localized modes to the nonlinear Dirac equations (1.1.6) can be constructed in explicit form (Section 3.3.4).

Figure 1.1 shows the leading order of the asymptotic solution (1.1.4) with $p = \frac{1}{2}$ and $q = r = 1$, when (a, b) is the stationary localized mode of the nonlinear Dirac equations (1.1.6) for $V(x) = -2\cos(x)$, $\sigma = -1$, $n = 1$, and $\epsilon = 0.1$.

Exercise 1.1 Consider the nonlinear Klein–Gordon equation,

$$u_{tt} - u_{xx} + u + \sigma u^3 + \epsilon V(x) u = 0,$$

where $u(x, t) : \mathbb{R} \times \mathbb{R} \to \mathbb{R}$, $\sigma \in \{1, -1\}$, and $V(x + 2\pi) = V(x)$ is bounded, and derive the nonlinear Dirac equations in the asymptotic limit $\epsilon \to 0$.

Exercise 1.2 Consider the nonlinear wave–Maxwell equation,

$$E_{xx} - \left(1 + \epsilon V(x) + \sigma |E|^2\right) E_{tt} = 0,$$

where $E(x, t) : \mathbb{R} \times \mathbb{R} \to \mathbb{C}$, $\sigma \in \{1, -1\}$, and $V(x + 2\pi) = V(x)$ is bounded, and derive the nonlinear Dirac equations in the asymptotic limit $\epsilon \to 0$. Show that the first-order correction is not a bounded function for all $(x, t) \in \mathbb{R}^2$ because the linear wave equation $E_{xx} - E_{tt} = 0$ has no dispersion and infinitely many Fourier modes are in resonance with the primary Fourier mode.

Exercise 1.3 Consider the Gross–Pitaevskii equation with periodic coefficients,

$$i u_t = -u_{xx} + \epsilon V(x) u + G(x) |u|^2 u,$$

where $u(x,t) : \mathbb{R} \times \mathbb{R} \to \mathbb{C}$, $V(x + 2\pi) = V(x)$ and $G(x + 2\pi) = G(x)$ are bounded, and derive an extended system of nonlinear Dirac equations in the asymptotic limit $\epsilon \to 0$.

Exercise 1.4   Repeat Exercise 1.3 with even $V(x)$ and odd $G(x)$ and derive the nonlinear Dirac equations with quintic nonlinear terms.

1.1.2 The nonlinear Schrödinger equation

We shall now assume that the 2π-periodic potential $V(x)$ is bounded but make no additional assumptions on the amplitude of $V(x)$. Let us rewrite again the Gross–Pitaevskii equation (1.1.1) in the explicit form,

$$iu_t = -u_{xx} + V(x)u + \sigma|u|^2 u. \tag{1.1.7}$$

If the nonlinear term $\sigma|u|^2 u$ is crossed over, equation (1.1.7) becomes a linear Schrödinger equation with a periodic potential

$$iu_t = -u_{xx} + V(x)u. \tag{1.1.8}$$

The linear modes are now given by the quasi-periodic Bloch waves in the form

$$u(x,t) = \psi_k(x)e^{-i\omega(k)t},$$

where $\psi_k(x)$ is a solution of the boundary-value problem for the second-order differential equation

$$\begin{cases} -\psi_k''(x) + V(x)\psi_k(x) = \omega(k)\psi_k(x), \\ \psi_k(x + 2\pi) = e^{2\pi i k}\psi_k(x), \end{cases} \quad x \in \mathbb{R}.$$

Here k is real and $\omega(k)$ is to be determined (Section 2.1.2).

The asymptotic multi-scale expansion method is now developed in a neighborhood of a particular linear Bloch wave for a given value (k_0, ω_0), where $\omega_0 = \omega(k_0)$. Guided by the asymptotic method from Section 1.1.1, we shall try again the asymptotic multi-scale expansion in powers of ϵ,

$$u(x,t) = \epsilon^p \left[a(X,T)\psi_{k_0}(x)e^{-i\omega_0 t} + \epsilon u_1(x,t) + \mathcal{O}(\epsilon^2) \right], \tag{1.1.9}$$

where $X = \epsilon^q x$, $T = \epsilon^r t$, (p,q,r) are some parameters to be determined, $a(X,T) : \mathbb{R} \times \mathbb{R} \to \mathbb{C}$ is an envelope amplitude to be determined, $\epsilon u_1(x,t)$ is a first-order remainder term, and $\mathcal{O}(\epsilon^2)$ is a formal order of truncation of the asymptotic expansion.

When the asymptotic expansion (1.1.9) is substituted into the Gross–Pitaevskii equation (1.1.7), we can set the exponents as $p = \frac{1}{2}$ and $q = r = 1$, similarly to the previous section. We will see however that this choice is not appropriate and the values for the exponents (p,q,r) will have to be changed.

Setting the exponents $p = \frac{1}{2}$, $q = r = 1$ and truncating the terms of the order of $\mathcal{O}(\epsilon^2)$, we write a linear inhomogeneous equation for $u_1(x,t)$,

$$\left(i\partial_t + \partial_x^2 - V \right) u_1 = \left(-ia_T\psi_{k_0} - 2a_X\psi_{k_0}' + \sigma|a|^2 a|\psi_{k_0}|^2\psi_{k_0} \right) e^{-i\omega_0 t}. \tag{1.1.10}$$

The right-hand side of (1.1.10) produces a secular growth of $u_1(x, t)$ in variables (x, t) because $\psi_{k_0}(x)e^{-i\omega_0 t}$ is a solution of the homogeneous equation (1.1.8). Looking for a solution in the form $u_1(x, t) = w_1(x)e^{-i\omega_0 t}$, we obtain an ordinary differential equation on $w_1(x)$,

$$\left(-\partial_x^2 + V + \omega_0\right) w_1 = ia_T \psi_{k_0} + 2a_X \psi'_{k_0} - \sigma |a|^2 a |\psi_{k_0}|^2 \psi_{k_0}. \tag{1.1.11}$$

Multiplying the left-hand side of (1.1.11) by $\bar{\psi}_{k_0}$ and integrating on $[0, 2\pi]$ we note that if $w_1(x)$ belongs to the same class of functions as $\psi_{k_0}(x)$, then

$$\int_0^{2\pi} \bar{\psi}_{k_0} \left(-\partial_x^2 + V + \omega_0\right) w_1 dx = -\left.\left(\bar{\psi}_{k_0} w_1' - \bar{\psi}'_{k_0} w_1\right)\right|_{x=0}^{x=2\pi} = 0.$$

Multiplying now the right-hand side of (1.1.11) by $\bar{\psi}_{k_0}$ and integrating on $[0, 2\pi]$, we obtain a nonlinear evolution equation on the amplitude $a(X, T)$,

$$ia_T = -\frac{2\langle \psi'_{k_0}, \psi_{k_0}\rangle_{L_{\text{per}}^2}}{\|\psi_{k_0}\|_{L_{\text{per}}^2}^2} a_X + \sigma \frac{\|\psi_{k_0}\|_{L_{\text{per}}^4}^4}{\|\psi_{k_0}\|_{L_{\text{per}}^2}^2} |a|^2 a. \tag{1.1.12}$$

If the nonlinear evolution equation (1.1.12) is violated, then either $w_1(x)$ becomes unbounded on $\mathbb{R}$ because of a linear growth as $|x| \to \infty$ or $u_1(x, t)$ grows secularly in time t as $t \to \infty$.

Equation (1.1.12) is a scalar semi-linear hyperbolic equation, which is easily solvable. Because of the periodic boundary conditions of $|\psi_{k_0}(x)|^2$ on $[0, 2\pi]$, the coefficient $\langle \psi'_{k_0}, \psi_{k_0}\rangle_{L_{\text{per}}^2}$ is purely imaginary, so that the X-derivative term can be removed by the transformation $a(X, T) = a(X - c_g T, T)$, where

$$c_g = \frac{2\langle \psi'_{k_0}, \psi_{k_0}\rangle_{L_{\text{per}}^2}}{i\|\psi_{k_0}\|_{L_{\text{per}}^2}^2} \in \mathbb{R}$$

has the meaning of the group velocity of the Bloch wave $u(x, t) = \psi_{k_0}(x)e^{-i\omega_0 t}$. After the transformation, the nonlinear evolution equation on $a(X - c_g T, T)$ does not have X-derivative terms and can be immediately integrated.

Exercise 1.5 Find the most general solution of the amplitude equation (1.1.12).

The amplitude equation (1.1.12) does not capture effects of dispersion of the Bloch wave at the same order where the effects of nonlinearity, time evolution, and group velocity occur. This outcome of the asymptotic multi-scale expansion method indicates that the leading-order balance misses an important contribution from the wave dispersion which is modeled by the second-order X-derivative terms on $a(X, T)$. Therefore, we have to revise the exponents (p, q, r) of the asymptotic expansion in order to bring all important effects to the same order.

Let us keep the exponent $q = 1$ in the scaling of $X = \epsilon^q x$ for convenience. Since the linear second-order derivative terms are of the order ϵ^{2+p}, while the cubic nonlinear terms are of the order ϵ^{3p}, a non-trivial balance occurs if $p = 1$. The time evolution is of the order ϵ^{q+p} and it matches the balance if $q = 2$. In addition, we need to remove the group-velocity term by the transformation

$$a(X, T) \to a(X - c_g T, T).$$

Combining all at once, we revise the asymptotic expansion (1.1.9) as

$$u(x,t) = \epsilon \left[a(X - c_g T, \tau)\psi_{k_0}(x)e^{-i\omega_0 t} + \epsilon u_1(x,t) + \epsilon^2 u_2(x,t) + \mathcal{O}(\epsilon^3) \right], \quad (1.1.13)$$

where $T = \epsilon t$, $\tau = \epsilon^2 t$, and $\mathcal{O}(\epsilon^3)$ indicates a new order of truncation of the asymptotic expansion. When the asymptotic multi-scale expansion (1.1.13) is substituted into the Gross–Pitaevskii equation (1.1.7), we obtain the first-order correction term in the explicit form

$$u_1(x,t) = \varphi_1(x)a_X(X - c_g T, \tau)e^{-i\omega_0 t},$$

where $\varphi_1(x)$ solves

$$\left(-\partial_x^2 + V + \omega_0 \right)\varphi_1 = -ic_g\psi_{k_0} + 2\psi_{k_0}'.$$

Note that the choice of c_g provides a sufficient condition that $\varphi_1(x)$ belongs to the same class of functions as $\psi_{k_0}(x)$. The second-order remainder term $u_2(x,t)$ satisfies now the linear inhomogeneous equation

$$\left(i\partial_t + \partial_x^2 - V \right)u_2 = \left(-ia_\tau\psi_{k_0} + (ic_g\varphi_1 - 2\varphi_1' - \psi_{k_0})a_{XX} + \sigma|a|^2 a|\psi_{k_0}|^2\psi_{k_0} \right)e^{-i\omega_0 t}.$$

Looking for a solution in the form $u_2(x,t) = w_2(x)e^{-i\omega_0 t}$, we obtain an ordinary differential equation on $w_2(x)$:

$$\left(-\partial_x^2 + V + \omega_0 \right)w_2 = ia_\tau\psi_{k_0} - (ic_g\varphi_1 - 2\varphi_1' - \psi_{k_0})a_{XX} - \sigma|a|^2 a|\psi_{k_0}|^2\psi_{k_0}.$$

Using the same projection algorithm as for equation (1.1.11), we obtain a nonlinear evolution equation on the amplitude $a(X - c_g T, \tau)$,

$$ia_T = \alpha a_{XX} + \sigma\beta|a|^2 a, \quad (1.1.14)$$

where

$$\alpha = \frac{ic_g\langle \varphi_1, \psi_{k_0}\rangle_{L_{\text{per}}^2} - 2\langle \varphi_1', \psi_{k_0}\rangle_{L_{\text{per}}^2}}{\|\psi_{k_0}\|_{L_{\text{per}}^2}^2} - 1, \quad \beta = \frac{\|\psi_{k_0}\|_{L_{\text{per}}^4}^4}{\|\psi_{k_0}\|_{L_{\text{per}}^2}^2}.$$

The amplitude equation (1.1.14) is nothing but the nonlinear Schrödinger (NLS) equation. The asymptotic reduction to the NLS equation is rigorously justified (Section 2.3). The stationary localized mode of the NLS equation (1.1.14) exists in explicit form (Section 1.4.1).

Figure 1.2 shows the leading order of the asymptotic solution (1.1.13), when a is the stationary localized mode of the nonlinear Schrödinger equation (1.1.14) for $V(x) = 0.2(1 - \cos(x))$, $\sigma = -1$, $k_0 = 0$, and $\epsilon = 0.1$. The localized mode corresponds to the lowest Bloch wave.

Exercise 1.6   Consider the nonlinear Klein–Gordon equation,

$$u_{tt} - u_{xx} + u + \sigma u^3 + V(x)u = 0,$$

where $u(x,t) : \mathbb{R} \times \mathbb{R} \to \mathbb{R}$, $\sigma \in \{1, -1\}$, and $V(x + 2\pi) = V(x)$ is bounded, and derive the cubic nonlinear Schrödinger equation for a Bloch wave $u(x,t) = \psi_{k_0}(x)e^{-i\omega_0 t}$.

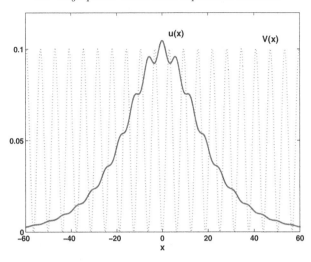

Figure 1.2 Schematic representation of the leading order in the asymptotic solution (1.1.13).

Exercise 1.7 Consider the Gross–Pitaevskii equation with periodic coefficients,

$$iu_t = -u_{xx} + V(x)u + G(x)|u|^2 u,$$

where $u(x,t) : \mathbb{R} \times \mathbb{R} \to \mathbb{C}$, $V(x + 2\pi) = V(x) = V(-x)$ and $G(x + 2\pi) = G(x) = -G(-x)$ are bounded, and derive a quintic nonlinear Schrödinger equation for the Bloch wave $u(x,t) = \psi_{k_0}(x)e^{-i\omega_0 t}$ with $\psi_{k_0}(x) = \psi_{k_0}(-x)$.

1.1.3 The discrete nonlinear Schrödinger equation

In the limit of large-amplitude periodic potentials, which is also referred to as the *semi-classical* limit, the Gross–Pitaevskii equation (1.1.1) can be written as

$$iu_t = -u_{xx} + \epsilon^{-2}V(x)u + \sigma|u|^2 u, \qquad (1.1.15)$$

where ϵ is a small parameter. Let us assume here in addition to the periodicity of V that $V \in C^\infty(\mathbb{R})$, $V(x) = V(-x)$, and 0 is a non-degenerate global minimum of $V(x)$ on $[-\pi, \pi]$. Without loss of generality, we can normalize $V(x)$ by its power series expansion around $x = 0$ with $V(x) = x^2 + \mathcal{O}(x^4)$. When the nonlinearity term $\sigma|u|^2 u$ is crossed out, the linear mode $u(x,t) = \psi(x)e^{-i\omega t}$ satisfies the differential equation

$$-\psi''(x) + \epsilon^{-2}V(x)\psi(x) = \omega\psi(x), \quad x \in \mathbb{R}. \qquad (1.1.16)$$

The truncated linear equation with $V(x) \approx x^2$ has an infinite set of isolated simple eigenvalues $\{\omega_j\}_{j\in\mathbb{N}_0}$ with the L^2-normalized eigenfunctions $\{\varphi_j(x)\}_{j\in\mathbb{N}_0}$ given by *Hermite functions*. More precisely, we have

$$\omega_j = \frac{1+2j}{\epsilon}, \quad \varphi_j(x) = \frac{1}{(\pi\epsilon)^{1/4}(2^j j!)^{1/2}} H_j\left(\frac{x}{\epsilon^{1/2}}\right) e^{-\frac{x^2}{2\epsilon}}, \quad j \in \mathbb{N}_0,$$

where $\{H_j(z)\}_{j \in \mathbb{N}_0}$ is a set of Hermite polynomials. Because the Hermite function $\varphi_j(x)$ decays to zero as $|x| \to \infty$ faster than an exponential function, this function is different from the quasi-periodic Bloch wave $\psi_k(x)$ introduced in Section 1.1.2. However, the small parameter ϵ implies that the local minimum of $V(x)$ near $x = 2\pi n$ is narrow for any $n \in \mathbb{Z}$ so that the Bloch wave $\psi_k(x)$ changes rapidly near the local minima of $V(x)$ and is small between the minima. As a result, we can try approximating $\psi_k(x)$ by a periodic superposition of decaying Hermite functions connected to each other by the small tails and centered at the points $\{2\pi n\}_{n \in \mathbb{Z}}$. In the limit $\epsilon \to 0$, the spectral band function $\omega(k)$ for the selected Bloch wave $\psi_k(x)$ converges to the eigenvalue ω_j for a fixed $j \in \mathbb{N}_0$. This construction is typical in the semi-classical analysis [47, 86]. The relation between the quasi-periodic Bloch waves and the decaying functions (called the *Wannier functions*) will be explained in Section 2.1.3.

Note that the decaying Wannier functions $\{\hat{\varphi}_j\}_{j \in \mathbb{N}_0}$ coincide with the Hermite functions $\{\varphi_j\}_{j \in \mathbb{N}_0}$ only near $x = 0$. Moreover, they are not eigenfunctions of the differential equation (1.1.16) as no decaying solutions exist for periodic potential $V(x)$ (Section 2.1.3). For $x \in [-\pi, \pi]$, the decaying functions $\hat{\varphi}_j(x)$ can be further approximated from the Wentzel–Kramers–Brillouin (WKB) solutions of the differential equation (1.1.16).

Let us pick a particular eigenvalue ω_j for a fixed $j \in \mathbb{N}_0$, for instance, the lowest eigenvalue $\omega_0 = \epsilon^{-1}$ with the ground state $\varphi_0(x) = (\pi\epsilon)^{-1/4} e^{-\frac{x^2}{2\epsilon}}$. The WKB solution for the extension of $\varphi_0(x)$ to $x > 0$ (both φ_0 and $\hat{\varphi}_0$ are even on $\mathbb{R}$) is written by

$$\hat{\varphi}_0(x) \sim A(x) e^{-\frac{1}{\epsilon} \int_0^x S(x')dx'}, \quad x > 0, \tag{1.1.17}$$

where

$$S(x) = \sqrt{V(x)},$$
$$A(x) = \frac{1}{(\pi\epsilon)^{1/4}} \exp\left[\int_0^x \frac{1 - S'(x')}{2S(x')} dx' \right].$$

The WKB solution (1.1.17) is derived by neglecting the term $A''(x)$ on the left-hand side of (1.1.16) and replacing ω by ϵ^{-1}. Note the function $\hat{\varphi}_0(x)$ matches with the function $\varphi_0(x)$ as $x \downarrow 0$. On the other hand, the expression for $A(x)$ diverges as $x \uparrow 2\pi$.

Let us now consider the asymptotic multi-scale expansion in powers of ϵ,

$$u(x,t) = \epsilon^p \left[\sum_{n \in \mathbb{Z}} a_n(T) \hat{\varphi}_0(x - 2\pi n) e^{-i\omega_0 t} + \epsilon u_1(x,t) + \mathcal{O}(\epsilon^2) \right], \tag{1.1.18}$$

where $T = \epsilon^r t$, (p, r) are some parameters to be determined, $\{a_n(T)\}_{n \in \mathbb{Z}} : \mathbb{R} \to \mathbb{C}^{\mathbb{Z}}$ are amplitudes of the Wannier functions to be determined, $\epsilon u_1(x,t)$ is a first-order remainder term, and $\mathcal{O}(\epsilon^2)$ is a formal order of truncation of the asymptotic expansion.

Let us substitute the asymptotic expansion (1.1.18) into the Gross–Pitaevskii equation (1.1.15) and collect all leading-order terms at the resonant frequency $e^{-i\omega_0 t}$. Therefore, we set $u_1(x,t) = w_1(x) e^{-i\omega_0 t}$, where $w_1(x)$ satisfies the ordinary differential equation

$$w_1'' - \epsilon^{-2} V w_1 + \omega_0 w_1 = \epsilon^{-1} F_1, \qquad (1.1.19)$$

where F_1 is given at the leading order by

$$F_1 = \sum_{n \in \mathbb{Z}} a_n \left(-\hat{\varphi}_0''(x - 2\pi n) + \epsilon^{-2} V \hat{\varphi}_0(x - 2\pi n) - \omega_0 \hat{\varphi}_0(x - 2\pi n) \right)$$

$$- i\epsilon^r \sum_{n \in \mathbb{Z}} \dot{a}_n \hat{\varphi}_0(x - 2\pi n) + \sigma \epsilon^{2p} \left| \sum_{n \in \mathbb{Z}} a_n \hat{\varphi}_0(x - 2\pi n) \right|^2 \sum_{n \in \mathbb{Z}} a_n \hat{\varphi}_0(x - 2\pi n).$$

We shall now compute the projection of this equation to $\hat{\varphi}_0$ on $\mathbb{R}$. The right-hand side F_1 has a hierarchic structure with respect to $\hat{\varphi}_0$ since $\hat{\varphi}_0(x)$ gives a projection of the order of $\mathcal{O}(1)$, $\hat{\varphi}_0(x \pm 2\pi)$ gives an exponentially small projection in ϵ, while $\hat{\varphi}_0(x \pm 2\pi m)$ give m powers of the exponentially small projections.

Thanks to the explicit formulas and the symmetry of $\hat{\varphi}_0$ on $\mathbb{R}$, the first off-diagonal linear terms are computed as follows

$$2 \int_{-\infty}^{\pi} \hat{\varphi}_0(x) \left(-\partial_x^2 + \epsilon^{-2} V - \omega_0 \right) \hat{\varphi}_0(x - 2\pi) dx$$

$$= 4\hat{\varphi}_0(\pi)\hat{\varphi}_0'(\pi) + 2 \int_{-\infty}^{\pi} \hat{\varphi}_0(x - 2\pi) \left(-\partial_x^2 + \epsilon^{-2} V - \omega_0 \right) \hat{\varphi}_0(x) dx \equiv -\mu.$$

Neglecting the second integral for the WKB solution (1.1.17) that approximately satisfies equation (1.1.16), we infer that the leading order of μ is given by

$$\mu \sim -4\hat{\varphi}_0(\pi)\hat{\varphi}_0'(\pi)$$

$$\sim \frac{4\sqrt{V(\pi)}}{\pi^{1/2}\epsilon^{3/2}} \exp\left(-\frac{2}{\epsilon} \int_0^{\pi} \sqrt{V(x)} dx + \int_0^{\pi} \frac{1 - S'(x)}{S(x)} dx \right), \qquad (1.1.20)$$

that is, μ is exponentially small as $\epsilon \to 0$. Recall that $V(\pi) \neq 0$ if 0 is a global minimum of V on $[-\pi, \pi]$.

Now the diagonal nonlinear term is computed by

$$\int_{\mathbb{R}} \hat{\varphi}_0^4(x) dx \sim \frac{1}{\pi \epsilon^{1/2}} \int_{\mathbb{R}} e^{-2z^2} dz = \frac{1}{(2\pi\epsilon)^{1/2}},$$

whereas we can use orthogonality of the decaying Wannier functions to write

$$\int_{\mathbb{R}} \hat{\varphi}_0(x)\hat{\varphi}_0(x - 2\pi n) dx = \delta_{0,n}.$$

Because μ is exponentially small in ϵ, the power asymptotic expansion (1.1.18) is not appropriate to balance all terms in F_1 at the leading order. As a result, we have to go back to revise the asymptotic scaling of the asymptotic expansion (1.1.18).

Let us consider a modified asymptotic expansion

$$u(x, t) = \mu^{1/2} (2\pi\epsilon)^{1/4} \left[\sum_{n \in \mathbb{Z}} a_n(T)\varphi_0(x - 2\pi n) e^{-i(\omega_0 + \Omega_0)t} + \mu u_1(x, t) + \mathcal{O}(\mu^2) \right],$$

$$(1.1.21)$$

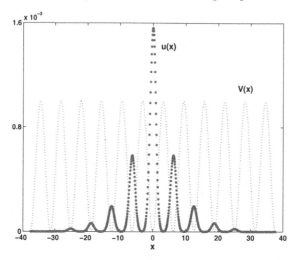

Figure 1.3 Schematic representation of the leading order in the asymptotic solution (1.1.21).

where $T = \mu t$ and

$$\Omega_0 = -\int_{\mathbb{R}} \hat{\varphi}_0(x) \left(-\partial_x^2 + \epsilon^{-2} V - \omega_0\right) \hat{\varphi}_0(x) dx.$$

The correction frequency Ω_0 is used to remove the diagonal linear term in F_1.

Again, using $u_1(x, t) = w_1(x)e^{-i(\omega_0 + \Omega_0)t}$, we obtain an inhomogeneous equation for $w_1(x)$, which is similar to equation (1.1.19) up to the scaling constants. Computing projections of the right-hand side F_1 to $\{\hat{\varphi}_0(x - 2\pi k)\}_{k \in \mathbb{Z}}$, we obtain the differential–difference equations for the amplitudes $\{a_k(T)\}_{k \in \mathbb{Z}}$,

$$i\dot{a}_k + a_{k+1} + a_{k-1} = \sigma|a_k|^2 a_k, \quad k \in \mathbb{Z}. \tag{1.1.22}$$

This amplitude equation is nothing but the discrete nonlinear Schrödinger (DNLS) equation. The asymptotic reduction to the DNLS equation is rigorously justified using the Wannier functions (Section 2.4). Unfortunately, no exact solutions for localized modes of the DNLS equation are available but the proof of existence of localized modes can be developed using different analytical methods (Sections 3.2.5 and 3.3.3).

Figure 1.3 shows the leading order of the asymptotic solution (1.1.21), where $\{a_n\}_{n \in \mathbb{Z}}$ is the stationary localized mode of the discrete nonlinear Schrödinger equation (1.1.22) for $V(x) = 0.5(1 - \cos(x))$, $\sigma = -1$, and $\epsilon = 0.75$. The localized mode corresponds to the ground state φ_0.

Exercise 1.8   Consider the nonlinear Klein–Gordon equation with delta function nonlinear terms,

$$u_{tt} - u_{xx} + u + u^3 + \sum_{n \in \mathbb{Z}} \delta(x - 2\pi n)u = 0,$$

where $u(x, t) : \mathbb{R} \times \mathbb{R} \to \mathbb{R}$, and derive the discrete nonlinear Schrödinger equation using the ground state $\varphi_0(x)$ for the smallest eigenvalue ω_0^2 of the linear problem

$$-\varphi_0''(x) + (1 + \delta(x))\varphi_0(x) = \omega_0^2 \varphi_0(x), \quad x \in \mathbb{R}.$$

Exercise 1.9 Consider the Gross–Pitaevskii equation with periodic coefficients,

$$iu_t = -u_{xx} + \epsilon^{-2} V(x)u + G(x)|u|^2 u,$$

where $u(x,t) : \mathbb{R} \times \mathbb{R} \to \mathbb{C}$, $V(x + 2\pi) = V(x) = V(-x)$ and $G(x + 2\pi) = G(x) = -G(-x)$ are bounded, and derive a quintic discrete nonlinear Schrödinger equation for the ground state $\varphi_0(x)$.

1.2 Hamiltonian structure and conserved quantities

The Gross–Pitaevskii equation with a periodic potential and its asymptotic reductions, the nonlinear Dirac equations, the nonlinear Schrödinger equation, and the discrete nonlinear Schrödinger equation, are particular examples of Hamiltonian dynamical systems. Besides the energy, these systems have additional conserved quantities thanks to symmetries with respect to the spatial and gauge translations. Some of our methods of analysis will rely on the Hamiltonian structure, while the others ignore its presence. Among the former methods, we mention proofs of global well-posedness of the initial-value problem (Section 1.3), the variational theory for existence of stationary solutions (Section 3.1) and analysis of spectral and orbital stability of localized modes (Sections 4.3 and 4.4.2). Among the latter methods, we mention justification of amplitude equations (Sections 2.2, 2.3, and 2.4), direct methods for existence of stationary solutions (Sections 3.2 and 3.3), and analysis of asymptotic stability of localized modes (Section 4.4.3).

Physicists often replace an infinite-dimensional system expressed by a partial differential equation by a reduced finite-dimensional system of ordinary differential equations. This trick is based on a formal Rayleigh–Ritz method, when the integral quantities for the energy and other conserved quantities are evaluated at a particular localized mode with time-varying parameters, after which evolution equations for these time-varying parameters are obtained from the Euler–Lagrange equations. While the accuracy of the qualitative approximations produced by this formal method can be remarkable in some examples, we shall avoid this method in this book, since it does not stand on a rigorous footing. Having issued this warning, the Hamiltonian formulation is often an asset in rigorous analysis, and we shall thus give relevant details on the Hamiltonian structure of the coupled nonlinear Schrödinger equations (Section 1.2.1), the discrete nonlinear Schrödinger equation (Section 1.2.2), and the nonlinear Dirac equations (Section 1.2.3). For simplicity of presentation and without loss of generality, we shall consider these equations in the space of one dimension.

An abstract Hamiltonian system in canonical variables $\mathbf{u} \in \mathbb{R}^{2n}$, $n \in \mathbb{N}$, is formulated as the following evolution problem

$$\frac{d\mathbf{u}}{dt} = J\nabla_{\mathbf{u}} H(\mathbf{u}), \tag{1.2.1}$$

where $J \in \mathbb{M}^{2n \times 2n}$ is a skew-symmetric and invertible matrix associated with the symplectic form,

$$\forall \mathbf{u}, \mathbf{v} \in \mathbb{R}^{2n} : \quad \langle J\mathbf{u}, \mathbf{v} \rangle_{\mathbb{R}^{2n}},$$

and $\nabla_{\mathbf{u}} H(\mathbf{u})$ is a gradient of the Hamiltonian functional $H(\mathbf{u}) : \mathbb{R}^{2n} \to \mathbb{R}$.

The phase space for the one-dimensional Gross–Pitaevskii equation and its reductions is typically a subspace of the Hilbert space $L^2(\mathbb{R}, \mathbb{R}^{2n})$ and $J \in \mathbb{M}^{2n \times 2n}$ is typically given by

$$J = \begin{bmatrix} O_n & I_n \\ -I_n & O_n \end{bmatrix},$$

where O_n and I_n are zero and identity matrices in $\mathbb{M}^{n \times n}$. The gradient operator $\nabla_{\mathbf{u}} H(\mathbf{u})$ is defined by the *Gateaux* derivative of the functional $H(\mathbf{u})$,

$$\forall \mathbf{v} \in L^2(\mathbb{R}, \mathbb{R}^{2n}) : \quad \langle \nabla_{\mathbf{u}} H(\mathbf{u}), \mathbf{v} \rangle_{L^2} := \frac{d}{d\epsilon} H(\mathbf{u} + \epsilon \mathbf{v}) \Big|_{\epsilon=0}. \tag{1.2.2}$$

If $H(\mathbf{u})$ is written as an integral of the density $h(\mathbf{u}, \partial_x \mathbf{u})$ over $x \in \mathbb{R}$, then integration by parts shows that the Gateaux derivative (1.2.2) can be computed using the partial derivatives of the density $h(\mathbf{u}, \mathbf{u}_x)$,

$$\nabla_{u_j} H(\mathbf{u}) := \frac{\partial h}{\partial u_j} - \frac{\partial}{\partial x} \frac{\partial h}{\partial u_{j,x}}, \quad j \in \{1, 2, ..., 2n\}, \tag{1.2.3}$$

where $u_{j,x} := \partial_x u_j$.

Besides the traditional formulation in real-valued canonical coordinates, it is often useful to shorten the formalism by using the complex-valued canonical coordinates $\boldsymbol{\psi} \in \mathbb{C}^n$ and the Hamiltonian form

$$i \frac{d\boldsymbol{\psi}}{dt} = \nabla_{\bar{\boldsymbol{\psi}}} H(\boldsymbol{\psi}, \bar{\boldsymbol{\psi}}). \tag{1.2.4}$$

A similar equation for the time derivative of $\bar{\boldsymbol{\psi}}$ is obtained from equation (1.2.4) after complex conjugation. Note that $H(\mathbf{u})$ in (1.2.1) and $H(\boldsymbol{\psi}, \bar{\boldsymbol{\psi}})$ in (1.2.4) are different by the factor 2 (Exercise 1.10). The symplectic form is now written as the bilinear form

$$\forall \boldsymbol{\psi}, \boldsymbol{\phi} \in \mathbb{C}^n : \quad \langle J_c \boldsymbol{\psi}, \boldsymbol{\phi} \rangle_{\mathbb{C}^n} := \mathrm{Im} \langle \boldsymbol{\psi}, \boldsymbol{\phi} \rangle_{\mathbb{C}^n}, \tag{1.2.5}$$

where J_c is skew-symmetric and invertible. The two Hamiltonian formulations in real-valued and complex-valued coordinates are used equally often in the context of the Gross–Pitaevskii equation.

Exercise 1.10 Consider a planar Hamiltonian system with $H = H(u_1, u_2)$,

$$\frac{du_1}{dt} = \frac{\partial H}{\partial u_2}, \quad \frac{du_2}{dt} = -\frac{\partial H}{\partial u_1}, \tag{1.2.6}$$

associated with the symplectic bilinear form

$$\langle J\mathbf{u}, \mathbf{v} \rangle_{\mathbb{R}^2} = u_2 v_1 - u_1 v_2. \tag{1.2.7}$$

Transform the system after the substitution

$$\psi = u_1 + iu_2, \quad \bar{\psi} = u_1 - iu_2$$

and rewrite the Hamiltonian system (1.2.6) and its symplectic form (1.2.7) in the complex-valued coordinates $(\psi, \bar{\psi})$.

If the Hamiltonian system (1.2.4) is invariant with respect to the translation in time t, then the value of $E = H(\psi, \bar{\psi})$ is conserved in the time evolution of the system. More precisely, we establish the following lemma.

Lemma 1.1 *Let $H(\psi, \bar{\psi})$ be independent of t. Then, the energy $E = H(\psi, \bar{\psi})$ of the Hamiltonian system (1.2.4) is conserved in time $t \in \mathbb{R}$.*

Proof Using (1.2.2) and the complex-valued coordinates, we obtain

$$\frac{d}{dt} H(\psi, \bar{\psi}) = \langle \nabla_\psi H(\psi, \bar{\psi}), \frac{d\bar{\psi}}{dt} \rangle_{L^2} + \langle \nabla_{\bar{\psi}} H(\psi, \bar{\psi}), \frac{d\psi}{dt} \rangle_{L^2}$$

$$= -\mathrm{i}\|\nabla_\psi H(\psi, \bar{\psi})\|_{L^2}^2 + \mathrm{i}\|\nabla_\psi H(\psi, \bar{\psi})\|_{L^2}^2 = 0.$$

Therefore, the value of $E = H(\psi, \bar{\psi})$ is constant in t if ψ is a solution of the Hamiltonian system (1.2.4). □

Exercise 1.11 Prove an analogue of Lemma 1.1 for the Hamiltonian system (1.2.1) in canonical variables $\mathbf{u} \in \mathbb{R}^{2n}$.

Besides symmetry with respect to translation in time t, the Gross–Pitaevskii equation typically has symmetry with respect to gauge transformation. As a result of this symmetry, an additional quantity (called *power* or *charge* or *mass*, depending on the physical context) is also conserved in the time evolution of the Gross–Pitaevskii equation. To classify this symmetry and the conserved quantity, we shall consider *symplectic linear transformations*, which do not change the value of the Hamiltonian functional $H(\psi, \bar{\psi})$ and transform one solution of system (1.2.4) to another solution of the same system.

Lemma 1.2 *Let $G : \mathbb{C}^n \to \mathbb{C}^n$ be an invertible linear transformation. Let $H(\psi, \bar{\psi}) : \mathbb{C}^n \times \mathbb{C}^n \to \mathbb{R}$ be a G-invariant Hamiltonian functional,*

$$H(G\psi, \bar{G}\bar{\psi}) = H(\psi, \bar{\psi}).$$

Transformation G is symplectic, so that if ψ solves the Hamiltonian system (1.2.4), then $\phi = G\psi$ solves the same Hamiltonian system (1.2.4), if and only if

$$G\bar{G}^{\mathrm{T}} = I_n, \tag{1.2.8}$$

where I_n is an identity matrix in $\mathbb{M}^{n \times n}$.

Proof An elementary proof follows from the chain rule

$$\mathrm{i}\frac{d\phi}{dt} = \mathrm{i}G\frac{d\psi}{dt} = G\nabla_{\bar{\psi}}H(\psi, \bar{\psi}) = G\bar{G}^{\mathrm{T}}\nabla_{\bar{\phi}}H(G^{-1}\phi, \bar{G}^{-1}\bar{\phi}) = G\bar{G}^{\mathrm{T}}\nabla_{\bar{\phi}}H(\phi, \bar{\phi})$$

and the fact that G is invertible. □

Among possible symplectic linear transformations G, *near-identity continuously differentiable* transformations have particular importance for the existence of the associated conserved quantities. In other words, we shall assume that the symplectic linear transformation $G(\epsilon) : \mathbb{C}^n \to \mathbb{C}^n$ is parameterized by $\epsilon \in \mathbb{R}$ and can be expanded into power series

$$G(\epsilon) = I_n + \epsilon g + \mathrm{O}(\epsilon^2).$$

The first-order transformation $g : \mathbb{C}^n \to \mathbb{C}^n$ is usually referred to as the *infinitesimal symmetry generator*. If $G(\epsilon)$ is a symplectic transformation, then g satisfies $g = -\bar{g}^{\mathrm{T}}$, which follows immediately by dropping the quadratic term $\epsilon^2 g\bar{g}^{\mathrm{T}}$ in the constraint $G(\epsilon)\bar{G}^{\mathrm{T}}(\epsilon) = I_n$. Now we obtain the conserved quantity, which is related to the near-identity, continuously differentiable, symplectic linear transformation $G(\epsilon)$.

Lemma 1.3 *Let $g : \mathbb{C}^n \to \mathbb{C}^n$ be the symmetry generator for the near-identity, continuously differentiable, symplectic linear transformation $G(\epsilon)$. Then, the power*

$$Q(\boldsymbol{\psi}, \bar{\boldsymbol{\psi}}) = \mathrm{Im}\langle g\boldsymbol{\psi}, \boldsymbol{\psi}\rangle_{L^2}$$

is conserved in time $t \in \mathbb{R}$.

Proof To prove conservation of $Q(\boldsymbol{\psi}, \bar{\boldsymbol{\psi}})$ in time, we take the derivative in ϵ of the equation

$$H(G(\epsilon)\boldsymbol{\psi}, \bar{G}(\epsilon)\bar{\boldsymbol{\psi}}) = H(\boldsymbol{\psi}, \bar{\boldsymbol{\psi}})$$

and set $\epsilon = 0$ to obtain

$$\langle \nabla_{\boldsymbol{\psi}} H(\boldsymbol{\psi}, \bar{\boldsymbol{\psi}}), \bar{g}\bar{\boldsymbol{\psi}}\rangle_{L^2} + \langle \nabla_{\bar{\boldsymbol{\psi}}} H(\boldsymbol{\psi}, \bar{\boldsymbol{\psi}}), g\boldsymbol{\psi}\rangle_{L^2} = 0. \tag{1.2.9}$$

Using this equation, the constraint $g = -\bar{g}^{\mathrm{T}}$, and the Hamiltonian system (1.2.4), we obtain

$$\frac{d}{dt}\langle g\boldsymbol{\psi}, \boldsymbol{\psi}\rangle_{L^2} = -\mathrm{i}\left(\langle g\nabla_{\bar{\boldsymbol{\psi}}} H(\boldsymbol{\psi}, \bar{\boldsymbol{\psi}}), \boldsymbol{\psi}\rangle_{L^2} - \langle g\boldsymbol{\psi}, \nabla_{\bar{\boldsymbol{\psi}}} H(\boldsymbol{\psi}, \bar{\boldsymbol{\psi}})\rangle_{L^2}\right)$$

$$= \mathrm{i}\left(\langle \nabla_{\bar{\boldsymbol{\psi}}} H(\boldsymbol{\psi}, \bar{\boldsymbol{\psi}}), g\boldsymbol{\psi}\rangle_{L^2} + \langle \nabla_{\boldsymbol{\psi}} H(\boldsymbol{\psi}, \bar{\boldsymbol{\psi}}), \bar{g}\bar{\boldsymbol{\psi}}\rangle_{L^2}\right) = 0.$$

Therefore, the value of $\langle g\boldsymbol{\psi}, \boldsymbol{\psi}\rangle_{L^2}$ is constant in t. Since $g = -\bar{g}^{\mathrm{T}}$, then $\langle g\boldsymbol{\psi}, \boldsymbol{\psi}\rangle_{L^2}$ is purely imaginary. Hence $Q(\boldsymbol{\psi}, \bar{\boldsymbol{\psi}}) = \mathrm{Im}\langle g\boldsymbol{\psi}, \boldsymbol{\psi}\rangle_{L^2} \in \mathbb{R}$ is a conserved quantity of the Hamiltonian system (1.2.4). □

Exercise 1.12 Formulate and prove analogues of Lemmas 1.2 and 1.3 for the Hamiltonian system (1.2.1) in canonical variables $\mathbf{u} \in \mathbb{R}^{2n}$.

It is definitely possible to uniquely compute the near-identity, continuously differentiable, symplectic linear transformation $G(\epsilon)$ from the infinitesimal symmetry generator g. This technique, which forms a *Lie group symmetry analysis*, is well described elsewhere [21, 144]. In brief, if g is the symmetry generator such that $g = -\bar{g}^{\mathrm{T}}$ and the Hamiltonian $H(\boldsymbol{\psi}, \bar{\boldsymbol{\psi}})$ satisfies the constraint (1.2.9), then the symplectic transformation $G(\epsilon)$ is uniquely computed by $G(\epsilon) = e^{\epsilon g}$ and the Hamiltonian $H(\boldsymbol{\psi}, \bar{\boldsymbol{\psi}})$ is proved to be invariant under the transformation $G(\epsilon)$. Moreover, if $g = -\bar{g}^{\mathrm{T}}$ and $G(\epsilon) = e^{\epsilon g}$, then $G(\epsilon)\bar{G}^{\mathrm{T}}(\epsilon) = I_n$, so that $G(\epsilon)$ is indeed a symplectic linear transformation.

1.2.1 The coupled nonlinear Schrödinger equations

Let us consider the coupled NLS equations,

$$i\partial_t\psi_k = -\partial_x^2\psi_k + \partial_{\bar\psi_k}W(|\psi_1|^2,|\psi_2|^2,...,|\psi_n|^2), \quad k=1,2,...,n, \tag{1.2.10}$$

where $\boldsymbol\psi := (\psi_1,...,\psi_n) \in \mathbb{C}^n$, $(x,t) \in \mathbb{R}\times\mathbb{R}$, and $W(|\psi_1|^2,|\psi_2|^2,...,|\psi_n|^2) : \mathbb{R}_+^n \to \mathbb{R}$ is assumed to be analytic in its variables. For instance, system (1.2.10) includes the system of two coupled NLS equations

$$\begin{cases} iu_t = -u_{xx} + \sigma\left(|u|^2 + \chi|v|^2\right)u, \\ iv_t = -v_{xx} + \sigma\left(\chi|u|^2 + |v|^2\right)v, \end{cases} \tag{1.2.11}$$

where we have denoted $\psi_1 = u$ and $\psi_2 = v$ and have introduced a coupling constant $\chi \in \mathbb{R}$. As previously, $\sigma \in \{+1,-1\}$. The coupled NLS equations (1.2.10) can be cast as the Hamiltonian system in complex-valued coordinates (1.2.4) with

$$H = \int_{\mathbb{R}}\left[\langle\partial_x\boldsymbol\psi,\partial_x\boldsymbol\psi\rangle_{\mathbb{C}^n} + W(|\psi_1|^2,|\psi_2|^2,...,|\psi_n|^2)\right]dx. \tag{1.2.12}$$

By Lemma 1.1, the value of $E = H(\boldsymbol\psi,\bar{\boldsymbol\psi})$ is constant in t thanks to the translational invariance of the system in time: if $\boldsymbol\psi(x,t)$ is one solution of system (1.2.10), then $\boldsymbol\psi(x,t-t_0)$ is another solution of system (1.2.10) for any $t_0 \in \mathbb{R}$.

By Lemma 1.2, there exists at least n near-identity, continuously differentiable, symplectic linear transformations $G(\epsilon)$ which leave H invariant and satisfy the constraint $G(\epsilon)\bar{G}^{\mathrm{T}}(\epsilon) = I_n$. A particular symplectic transformation $G_k(\epsilon)$ for $k \in \{1,2,...,n\}$ is given by a matrix obtained from the identity matrix I_n after 1 is replaced by $e^{i\epsilon}$ at the kth position. This symplectic transformation is related to the gauge invariance of the kth component of the vector $\boldsymbol\psi$ with respect to the phase rotations: if $(\psi_1,...,\psi_k,...,\psi_n)$ is a solution of system (1.2.10), then $(\psi_1,...,e^{i\alpha_k}\psi_k,...,\psi_n)$ is another solution of system (1.2.10) for any $\alpha_k \in \mathbb{R}$.

The symmetry generator for the symplectic transformation $G_k(\epsilon)$ is given by $g_k = iI_{n,k}$, where $I_{n,k}$ is a diagonal $n \times n$ matrix with 1 at the kth position and 0 at the other positions. By Lemma 1.3, g_k generates the conserved power,

$$Q_k = \int_{\mathbb{R}}|\psi_k|^2 dx. \tag{1.2.13}$$

When some symmetries of the coupled NLS equations (1.2.10) are broken, the number of conserved quantities is reduced.

Exercise 1.13 Assume that the nonlinear function W in the coupled NLS equations (1.2.10) depends on $(\bar\psi_2\psi_1+\bar\psi_1\psi_2)$ in addition to $(|\psi_1|^2,|\psi_2|^2,...,|\psi_n|^2)$. Prove that the values of Q_1 and Q_2 are no longer constant in t, whereas the value of Q_1+Q_2 is still constant in t.

Exercise 1.13 includes, in particular, the system of two coupled NLS equations with linear and nonlinear coupling terms

$$\begin{cases} iu_t = -\partial_x^2 u + Cv + (|u|^2 + A|v|^2)u + Bv^2\bar{u}, \\ iv_t = -\partial_x^2 v + Cu + (A|u|^2 + |v|^2)v + Bu^2\bar{v}, \end{cases} \tag{1.2.14}$$

where A, B, and C are real-valued coefficients and we have set $\psi_1 = u$ and $\psi_2 = v$.

Conversely, when additional symmetries of the coupled NLS equations (1.2.10) are present, the number of conserved quantities is increased. For instance, let us assume that the nonlinear function W actually depends on

$$W = W(|\psi_1|^2 + |\psi_2|^2 + \cdots + |\psi_n|^2). \qquad (1.2.15)$$

In real coordinates, both W and H are invariant with respect to the N-parameter group of rotations in $\mathbb{R}^{2n}$, where

$$N = \binom{2n}{2} = \frac{(2n)!}{2!(2n-2)!} = n(2n-1).$$

However, not all rotations generate conserved quantities because not all rotations are given by the symplectic linear transformations of Lemma 1.2. We can effectively characterize all near-identity, continuously differentiable, symplectic linear transformations $G(\epsilon)$ using the Hamiltonian system in complex coordinates (1.2.4). Moreover, we can start with the symmetry generators g that represent rotations of components of the complex-valued vector $\boldsymbol{\psi} = (\psi_1, \psi_2, ..., \psi_n) \in \mathbb{C}^n$ with respect to each other.

Let $g = A + iB \in \mathbb{M}^{n \times n}$, where A and B have real-valued coefficients. Using the constraint $g = -\bar{g}^{\mathrm{T}}$, we obtain immediately that $A = -A^{\mathrm{T}}$ is skew-symmetric and $B = B^{\mathrm{T}}$ is symmetric. Furthermore, if W satisfies the symmetry (1.2.15) and $g = -\bar{g}^{\mathrm{T}}$, then the constraint (1.2.9) on the Hamiltonian function $H(\boldsymbol{\psi}, \bar{\boldsymbol{\psi}})$ is also satisfied without additional constraints on elements of A and B. Counting the number of independent elements in matrices A and B, we conclude that $N = n(2n-1)$ rotations in $\mathbb{R}^{2n}$ generates only $N_s = n^2$ symplectic rotations in $\mathbb{C}^n$.

By Lemma 1.3, there exist n^2 conserved quantities $\mathrm{Im}\langle A\boldsymbol{\psi}, \boldsymbol{\psi} \rangle_{L^2}$ and $\mathrm{Re}\langle B\boldsymbol{\psi}, \boldsymbol{\psi} \rangle_{L^2}$, which are related to the group of n^2 symplectic rotations. More precisely, these conserved quantities can be reduced to the quadratic functionals

$$Q_{k,m} = \int_{\mathbb{R}} \psi_k \bar{\psi}_m dx, \quad 1 \le k \le m \le n. \qquad (1.2.16)$$

To confirm conservation of $Q_{k,m}$ for the coupled NLS equations (1.2.10) with rotational symmetry (1.2.15), we write the balance equations for densities of $Q_{k,m}$,

$$i\partial_t \left(\psi_k \bar{\psi}_m \right) + \partial_x \left(\bar{\psi}_m \partial_x \psi_k - \bar{\psi}_k \partial_x \psi_m \right) = 0.$$

Integrating this equation in x on $\mathbb{R}$ for solutions with fast decay to zero at infinity as $|x| \to \infty$ gives constant values of $Q_{k,m}$ in t as long as such solutions exist.

Simple accounting shows that there exist n real-valued quantities $Q_{k,k} \equiv Q_k$ which coincide with the conserved quantities (1.2.13). The additional $\frac{1}{2}n(n-1)$ conserved quantities $Q_{k,m}$ with $k \ne m$ are complex-valued. The total number of real-valued conserved quantities $Q_{k,m}$ coincides with the number of symplectic rotational symmetries as $N_s = n^2$.

If the symmetry generators g of the near-identity, continuously differentiable, symplectic linear transformations $G(\epsilon)$ are known, then $G(\epsilon) = e^{\epsilon g}$ for $\epsilon \in \mathbb{R}$. For the two coupled NLS equations (1.2.11) with $n = 2$ and $\chi = 1$ (also known as the *Manakov system*), there exists a group of $N_s = n^2 = 4$ symplectic rotations which are generated by matrices

$$g_1 = \begin{bmatrix} i & 0 \\ 0 & 0 \end{bmatrix}, \quad g_2 = \begin{bmatrix} 0 & 0 \\ 0 & i \end{bmatrix}, \quad g_3 = \begin{bmatrix} 0 & 1 \\ -1 & 0 \end{bmatrix}, \quad g_4 = \begin{bmatrix} 0 & i \\ i & 0 \end{bmatrix}.$$

The four symplectic transformation, which leave H invariant, are given by

$$G_1 = \begin{bmatrix} e^{i\theta_1} & 0 \\ 0 & 1 \end{bmatrix}, \qquad G_2 = \begin{bmatrix} 1 & 0 \\ 0 & e^{i\theta_2} \end{bmatrix},$$

$$G_3 = \begin{bmatrix} \cos\theta_3 & \sin\theta_3 \\ -\sin\theta_3 & \cos\theta_3 \end{bmatrix}, \quad G_4 = \begin{bmatrix} \cos\theta_4 & i\sin\theta_4 \\ i\sin\theta_4 & \cos\theta_4 \end{bmatrix},$$

where θ_1, θ_2, θ_3, and θ_4 are arbitrary real-valued parameters. The parameters θ_1 and θ_2 are related to the gauge invariance of components u and v, while the parameters θ_3 and θ_4 are related to the rotational symmetry of the coupled NLS equations (1.2.11) between components u and v in the case $\chi = 1$.

If $(u, v) \in \mathbb{C}^2$ is a solution of the coupled NLS equations (1.2.11) with $\chi = 1$, then there exists a four-parameter solution $(\tilde{u}, \tilde{v}) \in \mathbb{C}^2$ of the same equation,

$$\begin{cases} \tilde{u} = \alpha_1 e^{i\theta_1} u + \alpha_2 e^{i\theta_2} v, \\ \tilde{v} = -\bar{\alpha}_2 e^{i\theta_1} u + \bar{\alpha}_1 e^{i\theta_2} v, \end{cases} \tag{1.2.17}$$

where $\theta_1, \theta_2 \in \mathbb{R}$ and

$$\begin{cases} \alpha_1 = \cos\theta_3 \cos\theta_4 + i\sin\theta_3 \sin\theta_4, \\ \alpha_2 = \sin\theta_3 \cos\theta_4 + i\cos\theta_3 \sin\theta_4. \end{cases}$$

In a general case $n \geq 2$, there exist n continuations of solutions due to gauge invariance of n individual components and $n(n-1)$ continuations of solutions due to rotational symmetries of the coupled NLS equations (1.2.10) between different components of vector $\boldsymbol{\psi} = (\psi_1, \psi_2, ..., \psi_n) \in \mathbb{C}^n$.

Exercise 1.14 Compute nine transformation matrices $G(\epsilon)$ for the three coupled NLS equations (1.2.10) with $n = 3$ and $W = W(|\psi_1|^2 + |\psi_2|^2 + |\psi_3|^2)$.

Exercise 1.15 Characterize the group of symplectic rotations of the Hamiltonian system (1.2.1) in terms of the matrix

$$g = \begin{bmatrix} A & B \\ C & D \end{bmatrix},$$

where A, B, C, and D are real-valued matrices in $\mathbb{M}^{n \times n}$. Use transformation of real variables in $\mathbb{R}^{2n}$ to complex variables in $\mathbb{C}^n$ and confirm that the result is identical to the computations above.

Because the coupled NLS equations (1.2.10) have no linear potentials, these equations admit additional conserved quantities, which are characterized by the Noether Theorem (Appendix B.9). In particular, the coupled NLS equations (1.2.10) conserve the momentum

$$P = i \int_{\mathbb{R}} \left[\langle \boldsymbol{\psi}, \partial_x \boldsymbol{\psi} \rangle_{\mathbb{C}^n} - \langle \partial_x \boldsymbol{\psi}, \boldsymbol{\psi} \rangle_{\mathbb{C}^n} \right] dx, \tag{1.2.18}$$

thanks to the translational invariance of the system in space: if $\boldsymbol{\psi}(x,t)$ is a solution of system (1.2.10), then $\boldsymbol{\psi}(x - x_0, t)$ is also a solution of system (1.2.10) for any $x_0 \in \mathbb{R}$.

Additionally, the coupled NLS equations (1.2.10) with no linear potentials are invariant with respect to the Galileo transformation: if $\boldsymbol{\psi}(x,t)$ is a solution of system (1.2.10), then

$$e^{\mathrm{i}(vx - v^2 t)} \boldsymbol{\psi}(x - 2vt, t) \tag{1.2.19}$$

is also a solution of system (1.2.10) for any $v \in \mathbb{R}$. Because of the translational and Galileo transformations, any stationary solution

$$\boldsymbol{\psi}(x,t) = e^{-\mathrm{i}\hat{\omega}t} \boldsymbol{\phi}(x),$$

where $\boldsymbol{\phi}(x) : \mathbb{R} \to \mathbb{R}^n$ and $\hat{\omega}$ is a diagonal matrix of parameters $(\omega_1, \omega_2, ..., \omega_n) \in \mathbb{R}^n$, can be translated to any point $x_0 \in \mathbb{R}$ and can be transformed to a traveling wave solution with any speed $v \in \mathbb{R}$.

Traveling wave solutions of the coupled NLS equations (1.2.10) are equivalent to critical points of the energy functional (Section 3.1),

$$\Lambda_{v,\boldsymbol{\omega}} = H + vP + \langle \boldsymbol{\omega}, \mathbf{Q} \rangle_{\mathbb{R}^n},$$

where $\boldsymbol{\omega} = (\omega_1, \omega_2, ..., \omega_n)$ and $\mathbf{Q} = (Q_1, Q_2, ..., Q_n)$. Substituting the Galileo transformation (1.2.19) into $\Lambda_{v,\boldsymbol{\omega}}$, we observe that $\Lambda_{v,\boldsymbol{\omega}}$ transforms to the form

$$\Lambda_{\boldsymbol{\Omega}} = H + \langle \boldsymbol{\Omega}, \mathbf{Q} \rangle_{\mathbb{R}^n}, \tag{1.2.20}$$

where $\boldsymbol{\Omega} = \boldsymbol{\omega} + v^2 \mathbf{1}$ and $\mathbf{1} = (1, 1, ..., 1) \in \mathbb{R}^n$. This transformation enables us to cancel the dependence of the traveling wave solutions on the speed v and set $v = 0$ for many practical computations in the context of the nonlinear Schrödinger equations with no potentials.

When a nonzero bounded vector potential $\mathbf{V}(x) : \mathbb{R} \to \mathbb{R}^n$ is added to the coupled NLS equations (1.2.10), then the translational and Galileo transformations are broken. As a result, the value of P is no longer constant in t and the speed v becomes an important parameter of the traveling wave solutions (Section 5.6).

1.2.2 The discrete nonlinear Schrödinger equation

Let us consider the DNLS equation,

$$\mathrm{i}\dot{\psi}_n = -(\Delta\psi)_n + \partial_{\bar{\psi}_n} W(\boldsymbol{\psi}, \bar{\boldsymbol{\psi}}), \tag{1.2.21}$$

where $\boldsymbol{\psi} = \{\psi_n\} \in \mathbb{C}^{\mathbb{Z}}$, $n \in \mathbb{Z}$, $t \in \mathbb{R}$, $(\Delta\psi)_n = \psi_{n+1} - 2\psi_n + \psi_{n-1}$ is the discrete Laplacian, and $W(\boldsymbol{\psi}, \bar{\boldsymbol{\psi}}) : \mathbb{C}^{\mathbb{Z}} \times \mathbb{C}^{\mathbb{Z}} \to \mathbb{R}$ is assumed to be analytic in its variables. The DNLS equation (1.2.21) can be cast as the Hamiltonian system in complex coordinates (1.2.4) with

$$H = \sum_{n\in\mathbb{Z}} |\psi_{n+1} - \psi_n|^2 + W(\boldsymbol{\psi}, \bar{\boldsymbol{\psi}}).$$

For the simplest case of the power nonlinearity,

$$W = \frac{\sigma}{p+1} \sum_{n \in \mathbb{Z}} |\psi_n|^{2p+2},$$

where $\sigma \in \{1, -1\}$ and $p \in \mathbb{N}$, the DNLS equation (1.2.21) becomes the power DNLS equation

$$i\dot{\psi}_n = -(\Delta \psi)_n + \sigma |\psi_n|^{2p} \psi_n. \tag{1.2.22}$$

Thanks to the translational invariance in time, the value of $E = H(\boldsymbol{\psi}, \bar{\boldsymbol{\psi}})$ is constant in t. Because of the gauge invariance with respect to the phase rotations,

$$Q = \sum_{n \in \mathbb{Z}} |\psi_n|^2 = \|\boldsymbol{\psi}\|_{l^2}^2$$

is another constant in t.

The invariance with respect to the continuous space translations is lost in the discrete systems. As a result, the DNLS equation (1.2.21) has no generally conserved momentum unless the nonlinear function $\partial_{\bar{\psi}_n} W(\boldsymbol{\psi}, \bar{\boldsymbol{\psi}})$ satisfies additional constraints (Section 5.2). There is still a discrete translational invariance in the following sense: if $\{\psi_n(t)\}_{n \in \mathbb{Z}}$ is a solution of the DNLS equation (1.2.21), then $\{\psi_{n-n_0}(t)\}_{n \in \mathbb{Z}}$ is also a solution of the same equation for any $n_0 \in \mathbb{Z}$. However, the discrete translational symmetry cannot be generated by a near-identity, continuously differentiable transformation, and so it does not generate any quantities that are constant in t. In other words, the discrete group of spatial translations allows us to translate any stationary solution to any number of lattice nodes but it does not allow us to translate a stationary solution *relative to* any particular lattice node $n \in \mathbb{Z}$.

Exercise 1.16 Consider the Ablowitz–Ladik lattice

$$i\dot{\psi}_n = -(\Delta \psi)_n + \sigma |\psi_n|^2 (\psi_{n+1} + \psi_{n-1}),$$

where $\sigma \in \{1, -1\}$, and show that the discrete version of the momentum

$$P = \frac{i}{2} \sum_{n \in \mathbb{Z}} (\bar{\psi}_{n+1} \psi_n - \psi_{n+1} \bar{\psi}_n)$$

is constant in t. Show that P in the same form is not constant in t for the power DNLS equation (1.2.22).

1.2.3 The nonlinear Dirac equations

Let us consider the nonlinear Dirac equations,

$$\begin{cases} i(u_t + u_x) + v = \partial_{\bar{u}} W(u, \bar{u}, v, \bar{v}), \\ i(v_t - v_x) + u = \partial_{\bar{v}} W(u, \bar{u}, v, \bar{v}), \end{cases} \tag{1.2.23}$$

where $(u, v) \in \mathbb{C}^2$, $(x, t) \in \mathbb{R}^2$, and $W(u, \bar{u}, v, \bar{v}) : \mathbb{C}^4 \to \mathbb{R}$ is assumed to satisfy the following three conditions:

(W1) W is invariant with respect to the gauge transformation

$$W(e^{i\alpha} u, e^{-i\alpha} \bar{u}, e^{i\alpha} v, e^{-i\alpha} \bar{v}) = W(u, \bar{u}, v, \bar{v}), \quad \alpha \in \mathbb{R}, \tag{1.2.24}$$

(W2) W is symmetric with respect to the interchange of u and v,
(W3) W is analytic in its variables.

The first condition is justified by the derivation of the nonlinear Dirac equations (1.2.23) with the asymptotic multi-scale expansion method (Section 1.1.1). The second condition defines a class of symmetric nonlinear functions, which are commonly met in physical applications. The third condition is related to the theory of normal forms, where the nonlinear terms are approximated by Taylor polynomials. Since the function W contains no quadratic terms that would simply change coefficients of the linear terms in system (1.2.23) and no cubic terms that would violate conditions (W1) and (W3), e.g. the term $u^2\bar{v}$ violates the gauge invariance and the term $|u|u\bar{v}$ violates the analyticity of W, the power expansion of W starts with quartic terms.

The following result characterizes the function W that satisfies conditions (W1)–(W3).

Lemma 1.4 *If condition (W1) is satisfied, then W depends on $|u|^2$, $|v|^2$, and $(\bar{u}v + u\bar{v})$. If conditions (W1)–(W3) are satisfied, then W depends on $(|u|^2 + |v|^2)$, $|u|^2|v|^2$, and $(u\bar{v} + v\bar{u})$.*

Proof We shall assume condition (W1). By differentiating (1.2.24) in α and setting $\alpha = 0$, we obtain

$$DW := \mathrm{i}\,(u\partial_u - \bar{u}\partial_{\bar{u}} + v\partial_v - \bar{v}\partial_{\bar{v}})\,W(u,\bar{u},v,\bar{v}) = 0. \qquad (1.2.25)$$

Consider the set of quadratic variables

$$z_1 = |u|^2, \quad z_2 = |v|^2, \quad z_3 = \bar{u}v + u\bar{v}, \quad z_4 = u^2 + v^2,$$

which is independent of any $u \neq 0$ and $v \neq 0$ in the sense that the Jacobian of (z_1, z_2, z_3, z_4) with respect to $(u, \bar{u}, v, \bar{v})$ is nonzero. It is clear that $Dz_{1,2,3} = 0$ and $Dz_4 = 2\mathrm{i}z_4$. Therefore, $DW = 2\mathrm{i}z_4\partial_{z_4}W = 0$, so that W is independent of z_4.

By conditions (W2) and (W3), the arguments of W can be regrouped as $(|u|^2 + |v|^2)$, $|u|^2|v|^2$, and $(u\bar{v} + v\bar{u})$, whereas W may not depend on $|u|$ and $|v|$. □

The most general quartic function W that satisfies conditions (W1)–(W3) is given by

$$W = \frac{a_1}{2}(|u|^4 + |v|^4) + a_2|u|^2|v|^2 + a_3(|u|^2 + |v|^2)(v\bar{u} + \bar{v}u) + \frac{a_4}{2}(v\bar{u} + \bar{v}u)^2, \quad (1.2.26)$$

where a_1, a_2, a_3, and a_4 are real-valued parameters. It then follows that

$$\begin{cases} \partial_{\bar{u}}W = a_1|u|^2u + a_2u|v|^2 + a_3\left((2|u|^2 + |v|^2)v + u^2\bar{v}\right) + a_4\left(v^2\bar{u} + |v|^2u\right), \\ \partial_{\bar{v}}W = a_1|v|^2v + a_2v|u|^2 + a_3\left((2|v|^2 + |u|^2)u + v^2\bar{u}\right) + a_4\left(u^2\bar{v} + |u|^2v\right). \end{cases}$$

The nonlinear Dirac equations (1.2.23) with W in (1.2.26) is related to solutions of the Gross–Pitaevskii equation (Exercise 1.3),

$$\mathrm{i}u_t = -u_{xx} + \epsilon V(x)u + G(x)|u|^2u,$$

where $\epsilon \in \mathbb{R}$ is a small parameter and

$$V(x) = -2\cos(x), \quad G(x) = g_0 + 2g_1\cos(x) + 2g_2\cos(2x).$$

The correspondence between the numerical coefficients is given by

$$a_1 = g_0, \quad a_2 = 2g_0, \quad a_3 = g_1, \quad a_4 = g_2.$$

Another example of the nonlinear Dirac equations (1.2.23) is given by Porter *et al.* [174] in the context of the Gross–Pitaevskii equation with a time-periodic nonlinearity coefficient.

Exercise 1.17 Write the most general sextic function W that satisfies conditions (W1)–(W3). Find the parameters of W that include the quintic nonlinear Dirac equations [174],

$$\begin{cases} i(u_t + u_x) + v = (2|u|^2 + |v|^2)|v|^2 u, \\ i(v_t - v_x) + u = (|u|^2 + 2|v|^2)|u|^2 v. \end{cases}$$

The nonlinear Dirac equations (1.2.23) can be cast as a Hamiltonian system in complex-valued coordinates (1.2.4) with

$$H = \frac{i}{2} \int_{\mathbb{R}} (u_x \bar{u} - u \bar{u}_x - v_x \bar{v} + v \bar{v}_x) \, dx + \int_{\mathbb{R}} (v\bar{u} + u\bar{v} - W) \, dx. \tag{1.2.27}$$

Conserved quantities of the nonlinear Dirac equations (1.2.23) include the Hamiltonian $E = H(u, \bar{u}, v, \bar{v})$, the momentum

$$P = \frac{i}{2} \int_{\mathbb{R}} (u\bar{u}_x - u_x\bar{u} + v\bar{v}_x - v_x\bar{v}) \, dx, \tag{1.2.28}$$

and the power

$$Q = \int_{\mathbb{R}} \left(|u|^2 + |v|^2 \right) dx. \tag{1.2.29}$$

By the Noether Theorem (Appendix B.9), conservation of H and P follows from the translational invariance of the nonlinear Dirac equations (1.2.23) in time t and space x, whereas conservation of Q follows from the gauge invariance with respect to the phase rotations of u and v. In particular, the conserved quantity Q is found from Lemma 1.3 and the gauge symmetry (1.2.24). On the other hand, conservation of Q in t can be checked directly from the balance equation

$$\frac{\partial}{\partial t}(|u|^2 + |v|^2) + \frac{\partial}{\partial x}(|u|^2 - |v|^2) = DW = 0, \tag{1.2.30}$$

where the last equality follows from identity (1.2.25). Integrating the balance equation in x on $\mathbb{R}$ for solutions with fast decay to zero at infinity as $|x| \to \infty$, we confirm that Q is constant in t as long as such solutions exist.

Exercise 1.18 Write the balance equations for densities of H and P and explain why conservation of H and P in t is related to the time and space translations of the nonlinear Dirac equations (1.2.23).

Exercise 1.19 Consider a multi-component system of nonlinear Dirac equations,

$$i \left(\partial_t u_k + \alpha_k \partial_x u_k \right) + \sum_{l=1}^{n} \beta_{k,l} u_l = \partial_{\bar{u}_k} W(\mathbf{u}, \bar{\mathbf{u}}), \quad k \in \{1, 2, ..., n\},$$

where $\mathbf{u} = (u_1, u_2..., u_n) \in \mathbb{C}^n$, $(x, t) \in \mathbb{R} \times \mathbb{R}$, $(\alpha_1, ..., \alpha_n) \in \mathbb{R}^n$, β is a Hermitian matrix in $\mathbb{M}^{n \times n}$, and $W(\mathbf{u}, \bar{\mathbf{u}}) : \mathbb{C}^n \times \mathbb{C}^n \to \mathbb{R}$. Write this system in the Hamiltonian form (1.2.4) and find the conserved Hamiltonian H, momentum P, and power Q.

1.3 Well-posedness and blow-up

The initial-value problem for a differential equation is said to be *locally well-posed* in a Banach space X if for any initial data from X there exists a unique function in X over a time interval $[0, T_{\max})$ for some $T_{\max} > 0$, which solves the differential equation in some sense and depends continuously on the initial data. If the maximal existence time $T_{\max}$ is infinite, we say that the initial-value problem is *globally well-posed*.

If no solution exists, or the solution is not unique, or the solution is not continuous with respect to the initial data, the initial-value problem is said to be *ill-posed*. Although there can be different reasons why the initial-value problem for a particular differential equation is ill-posed and understanding these reasons can be challenging, it is a common practice to revise the derivation of the governing equation in a particular physical problem and to account for additional terms such as viscosity, higher-order dispersion or higher-order nonlinearity, which would make the initial-value problem well-posed.

Physicists have typically a very good intuition on whether the initial-value problem for a given equation is well-posed or ill-posed. This intuition is based on a number of numerical simulations, which show various scenarios of the time evolution, as well as on understanding the basic physical laws behind the derivation of a given model. As a result, physicists are typically sceptical about rigorous proofs of well-posedness since they think that the corresponding results are obvious and need no proofs.

Without making long arguments with physicists on the issue, we would like to explain a simple and rather universal method of how to prove local and, sometimes, global well-posedness of the Gross–Pitaevskii equation and its asymptotic reductions. This method combines Picard's iteration for systems of ordinary differential equations and Kato's theory for semi-linear partial differential equations. As a result, the analytical technique is often referred to as the *Picard–Kato* method.

As a first step of the Picard–Kato method, the differential equation is written in the integral form, e.g. by decomposing the vector field into linear and nonlinear parts, introducing the fundamental solution operator for the linear part, and using Duhamel's principle for the nonlinear term. The integral equation is then shown to be a Lipschitz continuous map from a local neighborhood of an initial point in a suitably chosen Banach space X to itself, which is also a contraction if a non-empty time interval $[0, T]$ is sufficiently small. Existence of a unique fixed point of the integral equation in a complete metric space $C([0, T], X)$ follows by the Banach Fixed-Point Theorem (Appendix B.2). This gives also a continuous dependence on initial data in X and the existence of a continuously differentiable solution in a larger Banach space Y that contains $X \subset Y$. To extend the maximal existence time to infinity, if possible, one needs to use other properties of the nonlinear evolution equation such as the conserved quantities (Section 1.2).

There are many locally well-posed differential equations, solutions of which cannot be extended to an infinite maximal existence time, because they become singular by reaching the boundary of X in a finite time. When it happens, we say that the differential equation admits a *finite-time blow-up* in the chosen vector space X. Specialists in harmonic analysis would then look for other choices of X to exclude the finite-time blow-up while preserving local well-posedness. From a practical point of view, however, the choice of X is often determined by the finite energy constraints and it would make less sense to look for exotic spaces, where the energy is not defined.

We shall apply the general algorithm of the well-posedness analysis to the nonlinear Schrödinger equation (Section 1.3.1), the discrete nonlinear Schrödinger equation (Section 1.3.2), and the nonlinear Dirac equations (Section 1.3.3). While the proof of local well-posedness follows for these semi-linear equations by repeating the same arguments, analysis of global well-posedness differs between different classes of equations. Readers interested in other equations such as the Korteweg–de Vries equation and various modifications of the nonlinear Schrödinger equation can look at recent texts [11, 133, 201] on foundations of the well-posedness theory for nonlinear evolution equations.

1.3.1 The nonlinear Schrödinger equation

Let us consider the initial-value problem for the NLS equation,

$$\begin{cases} iu_t = -\Delta u + V(x)u + f(|u|^2)u, \\ u(0) = u_0, \end{cases} \qquad (1.3.1)$$

where $u(x,t) : \mathbb{R}^d \times \mathbb{R}_+ \to \mathbb{C}$ is the wave function, $\Delta = \partial_{x_1}^2 + ... + \partial_{x_d}^2$ is the Laplacian in d dimensions, $V(x) : \mathbb{R}^d \to \mathbb{R}$ is a bounded potential, and $f(|u|^2) : \mathbb{R}_+ \to \mathbb{R}$ is a real analytic function.

The main question is: if $u_0 \in H^s(\mathbb{R}^d)$ for some $s \geq 0$, does there exist a unique local solution $u(t)$ to the initial-value problem (1.3.1) that remains in $H^s(\mathbb{R}^d)$ for all $t \in [0,T]$ and, if there is, does it extend globally for all times $t \in \mathbb{R}_+$? The classical theory of partial differential equations considers smooth solutions for sufficiently large values of s. It was only recently when this theory was extended to solutions of low regularity (for smaller values of s). The following theorem gives local well-posedness of the initial-value problem (1.3.1) in $H^s(\mathbb{R}^d)$ for $s > \frac{d}{2}$.

Theorem 1.1   *Fix $s > \frac{d}{2}$ and let $u_0 \in H^s(\mathbb{R}^d)$. Assume that $V \in C_b^s(\mathbb{R}^d)$ and f is analytic in its variable with $f(0) = 0$. There exist a $T > 0$ and a unique solution $u(t)$ of the initial-value problem (1.3.1) such that*

$$u(t) \in C([0,T], H^s(\mathbb{R}^d)) \cap C^1([0,T], H^{s-2}(\mathbb{R}^d)),$$

and $u(t)$ depends continuously on initial data u_0.

Proof   The proof consists of the following four steps.

Step 1: Linear estimates. We shall interpret the linear potential term $V(x)u$ and the nonlinear term $f(|u|^2)u$ as a perturbation to the linear evolution problem

$$\begin{cases} iS_t = -\Delta S, \\ S(0) = S_0, \end{cases} \qquad (1.3.2)$$

where $S_0 \in H^s(\mathbb{R}^d)$ for some $s \geq 0$. Since the linear Schrödinger equation has constant coefficients and is defined on an unbounded domain, it can be solved uniquely in $H^s(\mathbb{R}^d)$ by the Fourier transform.

Let $S_0 \in H^s(\mathbb{R}^d)$ be represented by the Fourier transform

$$\hat{S}_0(k) = \frac{1}{(2\pi)^{d/2}} \int_{\mathbb{R}^d} S_0(x)e^{-ik \cdot x}dx, \quad k \in \mathbb{R}^d.$$

The Fourier transform of the initial-value problem (1.3.2) gives

$$\begin{cases} i\hat{S}_t = |k|^2 \hat{S}, \\ \hat{S}(0) = \hat{S}_0, \end{cases}$$

and the unique solution is written in the Fourier form

$$S(x,t) = \frac{1}{(2\pi)^{d/2}} \int_{\mathbb{R}^d} \hat{S}_0(k)e^{ik \cdot x - it|k|^2} dk.$$

Let us denote the solution operator $H^s(\mathbb{R}^d) \ni S_0 \mapsto S \in C(\mathbb{R}, H^s(\mathbb{R}^d))$ by $e^{it\Delta}$. By Parseval's equality for the Fourier transform, the solution operator is unitary (norm-preserving) in the sense

$$\|S(t)\|_{H^s} = \|e^{it\Delta}S_0\|_{H^s} = \|e^{-it|k|^2}\hat{S}_0\|_{L^2_s} = \|\hat{S}_0\|_{L^2_s} = \|S_0\|_{H^s}, \quad s \geq 0. \qquad (1.3.3)$$

Step 2: Duhamel's principle. Let us now write the initial-value problem (1.3.1) as the inhomogeneous equation

$$\begin{cases} iu_t = -\Delta u + F(t), \\ u(0) = u_0, \end{cases}$$

where $F(t) = Vu(t) + f(|u(t)|^2)u(t)$. Solving the inhomogeneous equation by the variation of constant formula, we substitute $u(t) = e^{it\Delta}v(t)$ and obtain

$$\begin{cases} iv_t = e^{-it\Delta}F(t), \\ v(0) = u_0. \end{cases}$$

Note that $e^{it\Delta}$ is invertible and commutes with Δ for all $v \in H^s(\mathbb{R}^d)$ thanks to the norm preservation (1.3.3). As a result, we obtain

$$v(t) = u_0 - i \int_0^t e^{-it'\Delta}F(u(t'))dt'.$$

Thus, the initial-value problem (1.3.1) can be written as the fixed-point problem for the integral equation

$$u(t) = e^{it\Delta}u_0 - i \int_0^t e^{i(t-t')\Delta}\left(Vu(t') + f(|u(t')|^2)u(t')\right)dt'. \qquad (1.3.4)$$

Step 3: Fixed-point iterations. Let us recall the Banach Fixed-Point Theorem (Appendix B.2). We shall choose the Banach space $X = C([0, T], H^s(\mathbb{R}^d))$ for some $T > 0$ and $s \geq 0$ and the closed non-empty set $M = \bar{B}_\delta(X) \subset X$, the ball of radius $\delta > 0$ in X centered at $0 \in X$ together with its boundary.

Let $A(u)$ denote the right-hand side of the integral equation (1.3.4). We need to prove that A maps an element in M to an element in M and that A is a contraction operator. In other words, we need to prove that

$$\forall u(t) \in C([0, T], H^s(\mathbb{R}^d)) : \quad \sup_{t \in [0,T]} \|u(t)\|_{H^s} \leq \delta \quad \Rightarrow \quad \sup_{t \in [0,T]} \|A(u(t))\|_{H^s} \leq \delta$$

and there is $q \in (0, 1)$ such that

$$\forall u(t), v(t) \in C([0, T], H^s(\mathbb{R}^d)) : \quad \sup_{t \in [0,T]} \|u(t)\|_{H^s}, \ \sup_{t \in [0,T]} \|v(t)\|_{H^s} \leq \delta$$

$$\Rightarrow \sup_{t \in [0,T]} \|A(u(t) - v(t))\|_{H^s} \leq q \sup_{t \in [0,T]} \|u(t) - v(t)\|_{H^s}.$$

To be able to prove these two properties, we may need to adjust the choice of $T > 0$, $\delta > 0$, and $s \geq 0$.

By the triangle inequality and the norm preservation (1.3.3), we obtain

$$\|A(u(t))\|_{H^s} \leq \|e^{it\Delta}u_0\|_{H^s} + \int_0^t \left\| e^{i(t-t')\Delta} \left(Vu(t') + f(|u(t')|^2)u(t') \right) \right\|_{H^s} dt'$$

$$\leq \|u_0\|_{H^s} + \int_0^t \left(\|Vu(t')\|_{H^s} + \|f(|u(t')|^2)u(t')\|_{H^s} \right) dt'.$$

If $V \in C_b^s(\mathbb{R}^d)$ for the same $s \geq 0$, there is a constant $C_V > 0$ that depends on $\|V\|_{C_b^s}$ such that

$$\|Vu\|_{H^s} \leq C_V \|u\|_{H^s}.$$

If f is analytic in variable $|u|^2$ and $f(0) = 0$, then $f(|u|^2)u$ can be expressed by the convergent Taylor series $f(|u|^2)u = \sum_{k=1}^\infty f_k |u|^{2k} u$ for some coefficients $\{f_k\}_{k \in \mathbb{N}}$. To bound the powers of the Taylor series, it would be useful to have an estimate like

$$\exists C_s > 0 : \quad \forall u, v \in H^s(\mathbb{R}^d) : \quad \|uv\|_{H^s} \leq C_s \|u\|_{H^s} \|v\|_{H^s}.$$

This bound above holds for $s > \frac{d}{2}$ because $H^s(\mathbb{R}^d)$ for $s > \frac{d}{2}$ is a Banach algebra with respect to multiplication (Appendix B.1). Using this bound, we obtain

$$\|f(|u|^2)u\|_{H^s} \leq \sum_{k=1}^\infty |f_k| C_s^{2k} \|u\|_{H^s}^{2k+1}.$$

We also recall that space $C_b^0([0, T])$ also forms a Banach algebra with respect to multiplication. Therefore, assuming that $s > \frac{d}{2}$, we can close the bound on $\|A(u(t))\|_{H^s}$ by

$$\sup_{t \in [0,T]} \|A(u(t))\|_{H^s} \leq \|u_0\|_{H^s} + T(C_V + C_u(\delta)) \sup_{t \in [0,T]} \|u(t)\|_{H^s},$$

where

$$C_u(\delta) = \sum_{k=1}^{\infty} |f_k| C_s^{2k} \delta^{2k} < \infty$$

and $\delta = \sup_{t \in [0,T]} \|u(t)\|_{H^s}$ is sufficiently small. If we choose $\delta \geq 2\|u_0\|_{H^s}$, then the integral operator $A(u)$ in the integral equation (1.3.4) maps $\bar{B}_\delta(C([0,T], H^s(\mathbb{R}^d)))$ to itself if $T > 0$ satisfies the inequality

$$T(C_V + C_u(\delta)) \leq \frac{1}{2}. \tag{1.3.5}$$

To achieve a contraction of the integral operator $A(u)$, we note that for any $u, v \in C([0,T], H^s(\mathbb{R}^d))$,

$$\sup_{t \in [0,T]} \|V(u-v)\|_{H^s} \leq C_V \sup_{t \in [0,T]} \|u-v\|_{H^s},$$

$$\sup_{t \in [0,T]} \|f(|u|^2)u - f(|v|^2)v\|_{H^s} \leq C_{uv}(\delta) \sup_{t \in [0,T]} \|u-v\|_{H^s},$$

where C_V is the same constant and

$$C_{uv}(\delta) = \sum_{k=1}^{\infty} (2k+1)|f_k| C_s^{2k} \delta^{2k}.$$

For instance, if $f(|u|^2) = |u|^2$, then $C_u(\delta) = \delta^2$ and $C_{uv}(\delta) = 3\delta^2$. The contraction of $A(u)$ in $\bar{B}_\delta(C([0,T], H^s(\mathbb{R}^d)))$ is achieved if $T > 0$ satisfies another inequality

$$T(C_V + C_{uv}(\delta)) < 1. \tag{1.3.6}$$

By the Banach Fixed-Point Theorem, there exists a unique solution in the ball of radius $\delta > 0$ in $C([0,T], H^s(\mathbb{R}^d))$ with $s > \frac{d}{2}$ and $\delta \geq 2\|u_0\|_{H^s}$, where $T > 0$ is chosen to satisfy the constraints (1.3.5) and (1.3.6) simultaneously. The continuous dependence from $u_0 \in H^s(\mathbb{R}^d)$ also follows from the Banach Fixed-Point Theorem.

Step 4: Differentiability in time. We can iterate back the obtained solution of the integral equation (1.3.4) to a solution of the partial differential equation (1.3.1). Recall that $H^s(\mathbb{R}^d)$ is the Banach algebra with respect to multiplication for $s > \frac{d}{2}$ and if $u \in H^s$, then $\Delta u \in H^{s-2}$ for any $s \geq 0$. As a result, if u is a solution of the integral equation (1.3.4), then

$$-\Delta u + Vu + f(|u|^2)u \in C([0,T], H^{s-2}(\mathbb{R}^d)).$$

Integrating the evolution equation (1.3.1) in time, we obtain

$$u(t) \in C^1((0,T), H^{s-2}(\mathbb{R}^d)).$$

The end points of $[0,T]$ can be included since $[0,T]$ constructed in Step 3 is not the maximal existence interval. The proof of Theorem 1.1 is now complete. $\qquad \square$

Exercise 1.20 Consider a system of coupled NLS equations,

$$i\mathbf{u}_t = -A\Delta\mathbf{u} + \nabla_{\bar{\mathbf{u}}} W(\mathbf{u}, \bar{\mathbf{u}}),$$

where $\mathbf{u} \in \mathbb{C}^n$, A is a Hermitian matrix in $\mathbb{M}^{n \times n}$, and $W(\mathbf{u}, \bar{\mathbf{u}}) : \mathbb{C}^n \times \mathbb{C}^n \to \mathbb{R}$ is a homogeneous quartic polynomial in variables $\mathbf{u}$ and $\bar{\mathbf{u}}$. Prove local well-posedness of the system of coupled NLS equations in $C([0, T], H^s(\mathbb{R}^d, \mathbb{C}^n))$ for some $T > 0$ and $s > \frac{d}{2}$.

It is tempting to iterate the arguments of fixed-point iterations to a proof of the global existence of the solution $u(t)$ for all $t \in \mathbb{R}_+$ using $u(T)$ as a new initial data for another time step on $[T, T + T']$ with $T' > 0$. However, we will immediately get into trouble because the existence time T in Theorem 1.1 is inversely proportional to the radius δ of the ball in function space $C([0, T], H^s(\mathbb{R}^d))$ as follows from constraints (1.3.5) and (1.3.6) with the nonlinear terms in $C_u(\delta)$ and $C_{uv}(\delta)$. As a result, the sequence of the time steps $\{T_n\}_{n \geq 1}$ may converge to zero, while the sequence of initial data $\{u(t_n)\}_{n \geq 1}$ with $t_n = T_1 + \dots + T_n$ may diverge to infinity in H^s norm with $t_n < \infty$ as $n \to \infty$. Therefore, to prove global well-posedness, we need to control the H^s norm of the solution $u(t)$ on an elementary time interval $[0, T]$. To do so, we can use the conserved quantities for the nonlinear Schrödinger equation (Section 1.2.1).

For simplicity, we shall consider $d = 1$ and the power nonlinear function

$$f(|u|^2) = \sigma(p+1)|u|^{2p}$$

for an integer $p \geq 1$ and $\sigma \in \{1, -1\}$. The following theorem gives global existence for some values of p and σ.

Theorem 1.2 *Let $u_0 \in H^1(\mathbb{R})$, $V \in C_b^1(\mathbb{R})$, and $f(|u|^2) = \sigma(p+1)|u|^{2p}$ for an integer $p \geq 1$. If $\sigma = 1$ or $\sigma = -1$ and $p = 1$ or $\sigma = -1$, $p = 2$, and $\|u_0\|_{L^2}$ is sufficiently small, then the initial-value problem (1.3.1) admits a unique global solution $u(t) \in C(\mathbb{R}_+, H^1(\mathbb{R}))$ satisfying $u(0) = u_0$.*

Proof Local well-posedness holds in $H^1(\mathbb{R})$ since $s = 1$ exceeds $\frac{d}{2} = \frac{1}{2}$. Therefore, we only need to control $\|u(T)\|_{H^1}$ in terms of $\|u_0\|_{H^1}$ for the local solution $u(t) \in C([0, T], H^1(\mathbb{R}))$. Because $V \in C_b^1(\mathbb{R})$, we can assume that $V(x) \geq 0$ for all $x \in \mathbb{R}$. This assumption does not limit the generality of our approach since a transformation $u(x, t) = \tilde{u}(x, t)e^{-i\omega_0 t}$ with $\omega_0 = \min_{x \in \mathbb{R}} V(x)$ gives the same nonlinear Schrödinger equation for $\tilde{u}(x, t)$ but with $V(x)$ replaced by

$$\tilde{V}(x) = V(x) - \min_{x \in \mathbb{R}} V(x) \geq 0.$$

If we can show that $\tilde{u}(x, t)$ exists for all $t \in \mathbb{R}_+$ in $H^1(\mathbb{R})$, then $u(x, t)$ is also defined for all $t \in \mathbb{R}_+$ in $H^1(\mathbb{R})$.

Recall the conserved quantity for the energy and power of the nonlinear Schrödinger equation (Section 1.2.1):

$$E = \|\nabla u\|_{L^2}^2 + \|V^{1/2}u\|_{L^2}^2 + \sigma\|u\|_{L^{2(p+1)}}^{2(p+1)}, \quad Q = \|u\|_{L^2}^2.$$

By the Sobolev Embedding Theorem (Appendix B.10), $H^1(\mathbb{R})$ is continuously embedded into $L^{2(p+1)}(\mathbb{R})$ for any $p \geq 0$. If $\sigma = 1$ and $u_0 \in H^1(\mathbb{R})$, then $E, Q < \infty$ and by conservation of these quantities, we obtain that

$$\|u(t)\|_{H^1} \leq \sqrt{Q + E}, \quad t \in [0, T].$$

This is enough to guarantee global existence, since the local solution in $C([0,T], H^1(\mathbb{R}))$ can be uniquely continued to $C(\mathbb{R}_+, H^1(\mathbb{R}))$ with the same time step $T > 0$.

If $\sigma = -1$, we need an additional inequality to control the balance between norms in $H^1(\mathbb{R})$ and $L^{2(p+1)}(\mathbb{R})$. This inequality is known as the Gagliardo–Nirenberg inequality (Appendix B.5):

$$\exists C_p > 0: \quad \forall u \in H^1(\mathbb{R}): \quad \|u\|_{L^{2(p+1)}}^{2(p+1)} \leq C_p \|\nabla u\|_{L^2}^p \|u\|_{L^2}^{2+p}.$$

Using the conserved quantities and the Gagliardo–Nirenberg inequality, we obtain

$$E \geq \|\nabla u\|_{L^2}^2 - C_p Q^{1+p/2} \|\nabla u\|_{L^2}^p.$$

If $p = 1$, we obtain the bound

$$\|u(t)\|_{H^1} \leq Q^{1/2} + \frac{1}{2}C_1 Q^{3/2} + \sqrt{E + \frac{1}{4}C_1^2 Q^3}, \quad t \in [0,T],$$

which gives global existence by extending the local solution $u(t)$ from $[0,T]$ to $\mathbb{R}_+$.

If $p = 2$, the norm $\|\nabla u\|_{L^2}^2$ is bounded by E only if $u_0 \in H^1(\mathbb{R})$ is sufficiently small such that $C_2 Q^2 < 1$. If $p > 2$, the above method fails to control the H^1 norm of $u(t)$ for any initial data with finite E and Q. □

We mention that no global well-posedness in $H^1(\mathbb{R})$ can be proved for $p > 2$ and for $p = 2$ with large initial data $u_0 \in H^1(\mathbb{R})$. Indeed, the finite-time blow-up in H^1 norm is known to occur in these cases [133, 167, 199].

Exercise 1.21 Consider the NLS equation with power nonlinearity in one dimension

$$iu_t + u_{xx} + (p+1)|u|^{2p}u = 0 \tag{1.3.7}$$

and derive the evolution equation for the second moment

$$\frac{d^2}{dt^2} \int_{\mathbb{R}} x^2 |u(x,t)|^2 dx = 4pE + 8\left(1 - \frac{p}{2}\right)\|\nabla u\|_{L^2}^2,$$

for the solution $u(t) \in C([0,T], H^2(\mathbb{R}))$. Prove that the finite-time blow-up always occurs if $E < 0$ and $p \geq 2$.

Exercise 1.22 Consider the quintic NLS equation in one dimension

$$iu_t + u_{xx} + 3|u|^4 u = 0 \tag{1.3.8}$$

and show that it admits an exact solution in the form

$$u(x,t) = a(t)\,\text{sech}^{1/2}(2a^2(t)x)e^{i\theta(t)+ix^2 b(t)},$$

where a, b, and θ are some functions of t. Find equations for a, b, and θ and construct an exact solution describing a finite-time blow-up at the critical power $Q = \frac{\pi}{2}$.

If we try to extend the global well-posedness analysis for $d \geq 2$, the first problem we will be facing is the lack of the local existence in space $H^1(\mathbb{R}^d)$. Thanks to the Gagliardo–Nirenberg inequality (Appendix B.5), we are able to control the norm in $L^{2p+2}(\mathbb{R}^d)$ in terms of the norm in $H^1(\mathbb{R}^d)$ for $p \in [0, \frac{2}{d-2})$. As a result, we can still control the H^1 norm of the solution $u(t)$ in the subcritical case $p < \frac{2}{d}$ by using conserved quantities E and Q. However, we are not able to control the higher-order Sobolev norm H^s for $s > \frac{d}{2}$, where local well-posedness holds. This obstacle has been removed when the local well-posedness theory was pushed down below the constraint $s > \frac{d}{2}$ and down below $s = 1$ [30, 133, 201]. As a result, global well-posedness of the nonlinear Schrödinger equation for $d \in \{1, 2, 3\}$ has been proved for $\sigma = -1$ or $\sigma = 1$ and $p < \frac{2}{d}$ or $\sigma = 1$, $p = \frac{2}{d}$, and $\|u_0\|_{L^2}$ is sufficiently small [199].

1.3.2 The discrete nonlinear Schrödinger equation

Let us consider the initial-value problem for the DNLS equation,

$$\begin{cases} i\dot{u}_n(t) = -\Delta u_n + V_n u_n + f(|u_n|^2)u_n, \\ u_n(0) = u_{n,0}, \end{cases} \tag{1.3.9}$$

where $\{u_n(t)\}_{n \in \mathbb{Z}^d} : \mathbb{R}_+ \to \mathbb{R}^{\mathbb{Z}^d}$, Δ is the discrete Laplacian on the d-dimensional cubic lattice

$$\Delta u_n = \sum_{j=1}^{d} \left(u_{n+e_j} + u_{n-e_j} - 2u_n \right),$$

$\{V_n\}_{n \in \mathbb{Z}^d} \in \mathbb{R}^{\mathbb{Z}^d}$ is a bounded potential, and $f(|u_n|^2) : \mathbb{R}_+ \to \mathbb{R}$ is a real analytic function. We shall prove that the initial-value problem (1.3.9) enjoys local and global well-posedness in the discrete space $l_s^2(\mathbb{Z}^d)$ for any $s \geq 0$, independently of the dimension of the lattice $d \geq 1$ and the power in the nonlinear function f. Moreover, we will show that the unique solution is continuously differentiable in time in the same space $l_s^2(\mathbb{Z}^d)$. These nice properties are explained partly by the boundedness of the discrete Laplacian operator and partly by the cancelation of the nonlinear term in the time evolution of the l_s^2 norm. In what follows, the sequence $\{u_n(t)\}_{n \in \mathbb{Z}^d}$ in a discrete vector space is denoted by $\mathbf{u}(t)$.

Theorem 1.3 *Fix $s \geq 0$ and let $\mathbf{u}_0 \in l_s^2(\mathbb{Z}^d)$. Assume that $V \in l^\infty(\mathbb{Z}^d)$ and f is analytic in its variable. There exists a unique solution $\mathbf{u}(t) \in C^1(\mathbb{R}, l_s^2(\mathbb{Z}^d))$ of the initial-value problem (1.3.9) that depends continuously on initial data $\mathbf{u}_0$.*

Proof To prove local existence, it is sufficient to write the initial-value problem (1.3.9) in the integral form

$$\mathbf{u}(t) = \mathbf{u}_0 - i \int_0^t \left(-\hat{\Delta}\mathbf{u}(t') + \hat{V}\mathbf{u}(t') + \hat{\mathbf{f}}(|\mathbf{u}(t')|^2)\mathbf{u}(t') \right) dt', \tag{1.3.10}$$

where $\hat{\Delta}$, $\hat{V}$, and $\hat{\mathbf{f}}(|\mathbf{u}|^2)$ are operator extensions of Δ, V, and $\mathbf{f}(|\mathbf{u}|^2)$ acting on $\mathbf{u} \in l_s^2(\mathbb{Z}^d)$.

We do not need to worry about linear estimates and Duhamel's principle (Steps 1 and 2 in the proof of Theorem 1.1) because the discrete Laplacian is a positive bounded operator. Indeed, we have

$$\langle \mathbf{u}, (-\Delta)\mathbf{u} \rangle_{l^2} = \sum_{n \in \mathbb{Z}^d} \sum_{j=1}^{d} |u_{n+e_j} - u_n|^2 \geq 0$$

and

$$\langle \mathbf{u}, (-\Delta)\mathbf{u}\rangle_{l^2} \leq 2d\|\mathbf{u}\|_{l^2}^2 + \sum_{n\in\mathbb{Z}^d} |u_n| \sum_{j=1}^{d} (|u_{n+e_j}| + |u_{n-e_j}|) \leq 4d\|\mathbf{u}\|_{l^2}^2,$$

where the Cauchy–Schwarz inequality is used in the last inequality.

Using the fact that $l_s^2(\mathbb{Z}^d)$ forms a Banach algebra with respect to multiplication for any $s \geq 0$ and $d \geq 1$ (Appendix B.1), we see that the right-hand side of the integral equation (1.3.10) defines a Lipschitz continuous map in a ball B_δ of radius $\delta > 0$ in $C([0,T], l_s^2(\mathbb{Z}^d))$ for any $T > 0$. Furthermore, choosing $\mathbf{u}_0 \in l_s^2(\mathbb{Z}^d)$ and a sufficiently small $T > 0$, we can also make the Lipschitz constant strictly smaller than unity. We thus have a contraction in $B_\delta(C([0,T], l_s^2(\mathbb{Z}^d)))$, and a unique fixed point of the integral equation (1.3.10) exists by the Banach Fixed-Point Theorem (Appendix B.2). Since $(-\Delta)$ is a bounded positive operator, the integral of a continuous function gives a continuously differentiable function, hence $\mathbf{u}(t) \in C^1([0,T], l_s^2(\mathbb{Z}^d))$.

To extend T to infinity, it is sufficient to derive a global bound on the l_s^2 norm of the solution. Multiplying (1.3.9) by $\bar{u}_n$ and subtracting the complex conjugate equation, we eliminate the nonlinear term in the equation

$$i\frac{d}{dt}|u_n|^2 = u_n\Delta\bar{u}_n - \bar{u}_n\Delta u_n, \quad n \in \mathbb{Z}^d.$$

By the Cauchy–Schwarz inequality, there is a constant $C_{s,d} > 0$ such that

$$\sum_{n\in\mathbb{Z}^d} (1+n^2)^s |u_n||u_{n+e_j}| \leq C_{s,d}\|\mathbf{u}\|_{l_s^2}^2$$

for any $j \in \{1, 2, ..., d\}$. Therefore, it follows from the integral equation

$$|u_n(t)|^2 = |u_n(0)|^2 + i\int_0^t [\bar{u}_n(t')\Delta u_n(t') - u_n(t')\Delta\bar{u}_n(t')]\,dt'$$

that

$$\|\mathbf{u}(t)\|_{l_s^2}^2 \leq \|\mathbf{u}(0)\|_{l_s^2}^2 + 4dC_{s,d}\int_0^t \|\mathbf{u}(t')\|_{l_s^2}^2\,dt'.$$

By Gronwall's inequality (Appendix B.6), we have

$$\|\mathbf{u}(t)\|_{l_s^2}^2 \leq \|\mathbf{u}_0\|_{l_s^2}^2 e^{4dC_{s,d}|t|}, \quad \forall t \in \mathbb{R},$$

which proves that $\mathbf{u}(t) \in C^1(\mathbb{R}, l_s^2(\mathbb{Z}^d))$. This bound also gives the continuous dependence of $\mathbf{u}(t)$ from $\mathbf{u}_0$. $\square$

Remark 1.1   Embedding of $l_s^2(\mathbb{Z}^d)$ into $l^1(\mathbb{Z}^d)$ for $s > \frac{d}{2}$ implies that the unique solution of the initial-value problem (1.3.9) in $C^1(\mathbb{R}, l_s^2(\mathbb{Z}^d))$ for $s > \frac{d}{2}$ also belongs to space $C^1(\mathbb{R}, l^1(\mathbb{Z}^d))$.

Remark 1.2   Because of the global conservation of the l^2 norm, we know that $\|\mathbf{u}(t)\|_{l^2} = \|\mathbf{u}_0\|_{l^2}$. However, we have no control of $\limsup_{t\to\infty} \|\mathbf{u}(t)\|_{l_s^2}$ for $s > 0$, which may generally diverge.

Exercise 1.23 Prove local well-posedness in $l_s^2(\mathbb{Z})$ for any $s \geq 0$ of the discrete nonlinear Klein–Gordon equation,

$$\ddot{u}_n - \Delta u_n + u_n^3 = 0,$$

where $\{u_n(t)\}_{n \in \mathbb{Z}} : \mathbb{R} \to \mathbb{R}^{\mathbb{Z}}$.

1.3.3 The nonlinear Dirac equations

Let us consider the initial-value problem for the nonlinear Dirac equations,

$$\begin{cases} \mathrm{i}(u_t + u_x) + v = \partial_{\bar{u}} W(u, \bar{u}, v, \bar{v}), & u(0) = u_0, \\ \mathrm{i}(v_t - v_x) + u = \partial_{\bar{v}} W(u, \bar{u}, v, \bar{v}), & v(0) = v_0, \end{cases} \tag{1.3.11}$$

where $(u, v) \in \mathbb{C}^2$, $(x, t) \in \mathbb{R}^2$, and $W(u, \bar{u}, v, \bar{v}) : \mathbb{C}^4 \to \mathbb{R}$ is given by Lemma 1.4,

$$W(u, \bar{u}, v, \bar{v}) \equiv W(|u|^2 + |v|^2, |u|^2|v|^2, \bar{u}v + u\bar{v}). \tag{1.3.12}$$

System (1.3.11) can be written as a semi-linear equation,

$$\frac{d}{dt}\mathbf{u} = \mathrm{i}A\mathbf{u} - \mathrm{i}\mathbf{f}(\mathbf{u}), \tag{1.3.13}$$

where $\mathbf{u} = (u, v) \in \mathbb{C}^2$, $A = \sigma_1 + \mathrm{i}\sigma_3 \partial_x$ is a differential operator defined by the Pauli matrices, and $\mathbf{f}(\mathbf{u}) = (\partial_{\bar{u}} W, \partial_{\bar{v}} W)$ is a nonlinear vector field.

Since A is a differential operator with constant coefficients, its spectrum $\sigma(A)$ in $L^2(\mathbb{R})$ is purely continuous. The location of $\sigma(A)$ is found using the Fourier transform from the range of $\sigma(A(k))$, where $A(k) = \sigma_1 - k\sigma_3$ is a continuous family of 2×2 symmetric matrices for all $k \in \mathbb{R}$. The two eigenvalues of $A(k)$ are $\lambda_{\pm}(k) = \pm\sqrt{1 + k^2}$, so that

$$\sigma(A) = (-\infty, -1] \cup [1, \infty).$$

Because $e^{\mathrm{i}tA(k)}$ is a unitary matrix operator for all $k \in \mathbb{R}$, the linear evolution operator enjoys the norm-preservation property

$$\forall \mathbf{u} \in H^s(\mathbb{R}) : \quad \|e^{\mathrm{i}tA}\mathbf{u}\|_{H^s} = \|\mathbf{u}\|_{H^s}, \quad s \geq 0. \tag{1.3.14}$$

Local well-posedness in $C([0, T], H^s(\mathbb{R}))$ for any $s > \frac{1}{2}$ follows from the integral formulation of the semi-linear equation (1.3.13) and the norm-preserving properties of the group $e^{\mathrm{i}tA}$ in $H^s(\mathbb{R})$.

Theorem 1.4 *Assume that W is an analytic function (1.3.12) in its variables. Fix $s > \frac{1}{2}$ and let $\mathbf{u}_0 \in H^s(\mathbb{R})$. There exist a $T > 0$ such that the nonlinear Dirac equations (1.3.11) admit a unique solution*

$$\mathbf{u}(t) \in C([0, T], H^s(\mathbb{R})) \cap C^1([0, T], H^{s-1}(\mathbb{R})),$$

where $\mathbf{u}(t)$ depends continuously on the initial data $\mathbf{u}_0$.

Proof Since A is a self-adjoint operator from $H^1(\mathbb{R})$ to $L^2(\mathbb{R})$, then $e^{\mathrm{i}tA}$ is a unitary norm-preserving operator in H^s for any $s \geq 0$. By Duhamel's principle, we write the semi-linear equation (1.3.13) with the initial data $\mathbf{u}(0) = \mathbf{u}_0$ in the integral form,

$$\mathbf{u}(t) = e^{\mathrm{i}tA}\mathbf{u}_0 - \mathrm{i}\int_0^t e^{\mathrm{i}(t-t')A}\mathbf{f}(\mathbf{u}(t'))dt'. \qquad (1.3.15)$$

Because W is an analytic function of its arguments and $H^s(\mathbb{R})$ is a Banach algebra with respect to multiplication for any $s > \frac{1}{2}$, $\mathbf{f}$ is a C^∞ map from H^s to itself. By the norm-preservation property (1.3.14), the integral operator in equation (1.3.15) is a Lipschitz map in a ball of radius $\delta > 0$ in $C([0,T], H^s(\mathbb{R}))$ provided $T > 0$ is sufficiently small. Moreover, it is a contraction for a sufficiently small $T > 0$, which gives existence of a unique fixed point $\mathbf{u}(t) \in C([0,T], H^s(\mathbb{R}))$ and a continuous dependence of $\mathbf{u}(t)$ from $\mathbf{u}_0$.

Finally, $\mathbf{u}(t) \in C^1([0,T], H^{s-1}(\mathbb{R}))$ follows from $A : H^s(\mathbb{R}) \to H^{s-1}(\mathbb{R})$ and bootstrapping arguments for system (1.3.13). $\qquad \square$

Global well-posedness of the nonlinear Dirac equations (1.3.11) in $C(\mathbb{R}, H^1(\mathbb{R}))$ follows from the energy estimates if $W \equiv W(|u|^2 + |v|^2, |u|^2|v|^2)$. These arguments are given by Goodman *et al.* [72] for the particular example $W = (|u|^2 + |v|^2)^2 + 2|u|^2|v|^2$.

Theorem 1.5 *Assume that W is a polynomial in variables $|u|^2$ and $|v|^2$. The local solution of Theorem 1.4 is extended to the global solution in $C(\mathbb{R}, H^1(\mathbb{R}))$.*

Proof To extend the solution $\mathbf{u}(t)$ of Theorem 1.4 for all $t \in \mathbb{R}$, it is sufficient to prove that if T_0 is an arbitrary time, then the H^s norm of the solution satisfies the estimate

$$\sup_{t \in [0,T_0]} \|\mathbf{u}(t)\|_{H^s} \leq C(T_0), \qquad (1.3.16)$$

where the constant $C(T_0)$ is finite for $T_0 < \infty$ but may grow as $T_0 \to \infty$. By the power conservation, we have

$$\|\mathbf{u}(t)\|_{L^2} = \|\mathbf{u}_0\|_{L^2}, \quad t \in \mathbb{R}_+. \qquad (1.3.17)$$

To consider the H^1 norm, we multiply the first equation of system (1.3.11) by $|u|^{2p}\bar{u}$ and the second equation by $|v|^{2p}\bar{v}$ for a fixed $p > 0$, add the two equations, and take the imaginary part. If W depends only on $|u|^2$ and $|v|^2$, the nonlinear function is canceled out and we obtain

$$\frac{1}{p+1}\partial_t \left(|u|^{2p+2} + |v|^{2p+2}\right) + \frac{1}{p+1}\partial_x \left(|u|^{2p+2} - |v|^{2p+2}\right) = \mathrm{i}(v\bar{u} - \bar{v}u)(|u|^{2p} - |v|^{2p}).$$

Integrating this balance equation on $x \in \mathbb{R}$ for a local solution in $C([0,T], H^s(\mathbb{R}))$ and using inequality $|u||v| \leq \frac{1}{2}(|u|^2 + |v|^2)$, we obtain the energy estimate

$$\frac{d}{dt}\|\mathbf{u}(t)\|_{L^{2p+2}}^{2p+2} \leq 4(p+1)\|\mathbf{u}(t)\|_{L^{2p+2}}^{2p+2}. \qquad (1.3.18)$$

By Gronwall's inequality (Appendix B.6), we have

$$\|\mathbf{u}(t)\|_{L^{2p+2}} \leq e^{2|t|}\|\mathbf{u}_0\|_{L^{2p+2}}, \quad t \in \mathbb{R}. \qquad (1.3.19)$$

Since the estimate holds for any $p > 0$, it holds for $p \to \infty$ and gives a direct *a priori* estimate on the L^∞ norm of the local solution. The bound on the L^∞ norm is needed to control the growth rate of the L^2 norm of the x-derivative of $\mathbf{u}(t)$.

Taking x-derivatives and performing a similar computation, we obtain the balance equation

$$\partial_t \left(|\partial_x u|^2 + |\partial_x v|^2 \right) + \partial_x \left(|\partial_x u|^2 - |\partial_x v|^2 \right)$$
$$= \mathrm{i} \left(\partial_x u \partial_x \partial_{\bar{u}} + \partial_x v \partial_x \partial_{\bar{v}} - \partial_x \bar{u} \partial_x \partial_u - \partial_x \bar{v} \partial_x \partial_v \right) W.$$

Let W be a polynomial of degree $(N+1)$ in variables $|u|^2$ and $|v|^2$, where $N \geq 1$. Integrating over $x \in \mathbb{R}$ for a polynomial W and using bound (1.3.19), we obtain the estimate

$$\frac{d}{dt} \|\partial_x \mathbf{u}(t)\|_{L^2}^2 \leq C_W e^{4N|t|} \|\partial_x \mathbf{u}(t)\|_{L^2}^2,$$

where the constant $C_W > 0$ depends on the coefficients of W. By Gronwall's inequality again, we obtain

$$\|\partial_x \mathbf{u}(t)\|_{L^2}^2 \leq e^{\frac{C_W}{4N}(e^{4N|t|}-1)} \|\partial_x \mathbf{u}_0\|_{L^2}^2, \quad t \in \mathbb{R}. \tag{1.3.20}$$

The exponential factor remains bounded for any finite time $T_0 > 0$ in (1.3.16). This gives global well-posedness in the H^1 norm. $\square$

Remark 1.3 Unlike the NLS equation (1.3.1), the conserved quantities H and P of the nonlinear Dirac equations (1.3.11) (Section 1.2.3) are not useful for analysis of global well-posedness because the quadratic parts of H and P are sign-indefinite. Nevertheless, conservation of power Q is still used in the proof of Theorem 1.5.

If W also depends on $(\bar{u}v + u\bar{v})$, the energy estimates of the L^p norm include nonlinear terms which may lead to the finite-time blow-up of solutions in L^∞ and H^1 norms. However, no results on blow-up in the nonlinear Dirac equations have been reported to date.

Exercise 1.24 Prove local well-posedness in $H^s(\mathbb{R})$ for $s > \frac{1}{2}$ of the multi-component system of nonlinear Dirac equations,

$$\mathrm{i} \left(\partial_t u_k + \alpha_k \partial_x u_k \right) + \sum_{l=1}^n \beta_{k,l} u_l = f_k(\mathbf{u}, \bar{\mathbf{u}}), \quad k \in \{1, 2, ..., n\},$$

where $\mathbf{u} = (u_1, u_2 ..., u_n) \in \mathbb{C}^n$, $(x, t) \in \mathbb{R} \times \mathbb{R}$, $(\alpha_1, ..., \alpha_n) \in \mathbb{R}^n$, β is a Hermitian matrix in $\mathbb{M}^{n \times n}$, and $\mathbf{f}(\mathbf{u}, \bar{\mathbf{u}}) : \mathbb{C}^n \times \mathbb{C}^n \to \mathbb{R}^n$ is a homogeneous cubic polynomial in variables $\mathbf{u}$ and $\bar{\mathbf{u}}$.

1.4 Integrable equations and solitons

Among the class of the Gross–Pitaevskii equation and its asymptotic reductions, there exist remarkable *integrable* equations, which can be solved analytically by a method called the *inverse scattering transform*. This method is used when the nonlinear evolution equation in variables (x, t) is associated with a pair of linear operators, one of which determines the spectral problem for a fixed t and the other one determines evolution of the spectral data in t. The potentials of the linear operators are expressed by solutions of the nonlinear evolution equation under the condition that the spectral problem has a t-independent spectrum. The inverse scattering transform is described elsewhere [1, 2].

Since asymptotic reductions of the Gross–Pitaevskii equations lead often to integrable equations, this book will not be complete without a brief reference to these equations and the exact analytical solutions describing localized modes. In the context of integrable equations, these localized modes are referred to as *solitons*. Unfortunately, the Gross–Pitaevskii equation with a periodic potential does not reduce to an integrable equation (Section 1.4.2). Nevertheless, the stationary Gross–Pitaevskii equation with periodic linear and nonlinear coefficients can be integrable and can admit exact localized modes (Section 3.3.1).

The inverse scattering transform is thought to be a nonlinear version of the Fourier transform method for solutions of linear partial differential equations with constant coefficients. It consists of the following three steps.

- Spectral data are computed for the spectral problem, when the potential is given by the initial data at $t = 0$.
- Time dependence of the spectral data is found from the linear evolution problem for all $t > 0$.
- The inverse problem is solved to recover the potential for $t > 0$ from the t-dependent spectral data.

In addition to solving the Cauchy problem for integrable nonlinear evolution equations, the inverse scattering transform method offers a universal algorithm for construction of exact stationary and traveling solutions, exact solutions of the linearized stability problems associated with the stationary and traveling solutions, and exact solutions describing the long-term nonlinear dynamics, instabilities and interactions of localized modes.

It turns out that every class of nonlinear evolution equations described in this chapter has a representative among the class of integrable equations. It will be sufficient for our purpose to consider only the focusing version of these equations, so we will set $\sigma = -1$ in all equations below. The list of integrable equations opens with the cubic nonlinear Schrödinger equation,

$$iu_t + u_{xx} + |u|^2 u = 0, \tag{1.4.1}$$

where $u(x,t) : \mathbb{R} \times \mathbb{R} \to \mathbb{C}$.

The coupled nonlinear Schrödinger equations are integrable for the rotation-invariant cubic nonlinear function,

$$i\mathbf{u}_t + \mathbf{u}_{xx} + \|\mathbf{u}\|^2_{\mathbb{C}^n} \mathbf{u} = 0, \tag{1.4.2}$$

where $\mathbf{u}(x,t) : \mathbb{R} \times \mathbb{R} \to \mathbb{C}^n$. For $n = 2$, the vector extension of the nonlinear Schrödinger equation is referred to as the *Manakov system*,

$$\begin{cases} iu_t + u_{xx} + (|u|^2 + |v|^2)u = 0, \\ iv_t + v_{xx} + (|u|^2 + |v|^2)v = 0, \end{cases} \tag{1.4.3}$$

where $(u, v)(x, t) : \mathbb{R} \times \mathbb{R} \to \mathbb{C}^2$.

Among the nonlinear Dirac equations, the integrable system is referred to as the *massive Thirring model*,

$$\begin{cases} \mathrm{i}(u_t + u_x) + v + |v|^2 u = 0, \\ \mathrm{i}(v_t - v_x) + u + |u|^2 v = 0, \end{cases} \tag{1.4.4}$$

where $(u, v)(x, t) : \mathbb{R} \times \mathbb{R} \to \mathbb{C}^2$.

Finally, the integrable discrete nonlinear Schrödinger equation is referred to as the *Ablowitz–Ladik lattice*,

$$\mathrm{i}\dot{u}_n + u_{n+1} - 2u_n + u_{n-1} + |u_n|^2(u_{n+1} + u_{n-1}) = 0, \tag{1.4.5}$$

where $\{u_n(t)\}_{n \in \mathbb{Z}} : \mathbb{R} \to \mathbb{C}^{\mathbb{Z}}$.

For the sake of briefness, we shall only describe the inverse scattering transform method for the nonlinear Schrödinger equation (1.4.1) (Section 1.4.1). The other integrable equations (1.4.2)–(1.4.5) can be treated similarly. In particular, the inverse scattering method for the Ablowitz–Ladik lattice (1.4.5) can be considered to be a semi-discretization of the inverse scattering method for the nonlinear Schrödinger equation (1.4.1), where differential operators are replaced by difference operators. The inverse scattering transform for the coupled NLS equations (1.4.2) and the Ablowitz–Ladik lattice (1.4.5) is described in full details by Ablowitz *et al.* [2].

We will also show that a general nonlinear Schrödinger equation with variable coefficients in space x and time t can be treated similarly to the cubic nonlinear Schrödinger equation (1.4.1) if the variable coefficients satisfy some constraints (Section 1.4.2). While finding these constraints becomes popular in physics literature from time to time, the Gross–Pitaevskii equation with a periodic potential does not belong to the list of integrable nonlinear evolution equations.

1.4.1 The NLS equation with constant coefficients

Let us consider the following pair of linear equations

$$\partial_x \boldsymbol{\psi} = (Q(u) + \lambda \sigma_3)\, \boldsymbol{\psi}, \quad \partial_t \boldsymbol{\psi} = A(\lambda, u) \boldsymbol{\psi}, \tag{1.4.6}$$

where $\lambda \in \mathbb{C}$ is the spectral parameter, $\boldsymbol{\psi}(x, t) : \mathbb{R} \times \mathbb{R} \to \mathbb{C}^2$ is the wave function, σ_3 is the Pauli matrix, $Q(u)$ and $A(\lambda, u)$ are 2×2 matrices in the form

$$Q(u) = \begin{bmatrix} 0 & -u \\ \bar{u} & 0 \end{bmatrix}, \quad A(\lambda, u) = 2\mathrm{i}\lambda^2 \sigma_3 + \mathrm{i}|u|^2 \sigma_3 + 2\mathrm{i}\lambda Q(u) + \mathrm{i}\sigma_3 Q(u_x).$$

Let $\boldsymbol{\psi} \in \mathcal{C}^2(\mathbb{R} \times \mathbb{R})$ be a solution of system (1.4.6) such that $\partial_x \partial_t \boldsymbol{\psi} = \partial_t \partial_x \boldsymbol{\psi}$. The spectral parameter λ is independent of t and x if the linear system (1.4.6) satisfies the compatability condition also known as the *Lax equation*,

$$\partial_t Q(u) - \partial_x A(\lambda, u) + (Q(u) + \lambda \sigma_3)A(\lambda, u) - A(\lambda, u)(Q(u) + \lambda \sigma_3) = 0. \tag{1.4.7}$$

Explicit differentiation and matrix multiplications show that the Lax equation (1.4.7) is equivalent to the cubic nonlinear Schrödinger (NLS) equation,

$$\mathrm{i}u_t + u_{xx} + 2|u|^2 u = 0. \tag{1.4.8}$$

The spectral analysis of the first equation in the linear system (1.4.6) depends on the localization of the bounded potential $u(x)$. When the potential $u(x)$ decays exponentially at infinity, the spectral data are determined by the union of the continuous and discrete spectra. The continuous spectrum is located along a continuous curve on the complex λ plane. The discrete spectrum consists of a finite number of isolated eigenvalues of finite multiplicity. When the potential $u(x)$ decays algebraically at infinity, the spectral data may have additional singularities such as embedded eigenvalues in the continuous spectrum [115].

Let us assume that the bounded potential $u(x)$ decays exponentially to zero as $|x| \to \infty$ so that

$$\exists C > 0, \ \exists \kappa > 0: \quad |u(x)| \le Ce^{-\kappa|x|}, \quad x \in \mathbb{R}. \tag{1.4.9}$$

By Weyl's Theorem (Appendix B.15), the location of the continuous spectrum of the spectral problem

$$\partial_x \psi = (Q(u) + \lambda\sigma_3)\,\psi \tag{1.4.10}$$

is found from bounded solutions of the same problem (1.4.10) with

$$\lim_{x\to\infty} Q(u(x)) = Q(0) = O \in \mathbb{M}^{2\times 2},$$

that is, for any $\lambda \in i\mathbb{R}$. Let us define two sets

$$\{\phi^+(x,k), \phi^-(x,k)\} \quad \text{and} \quad \{\psi^+(x,k), \psi^-(x,k)\}$$

of linearly independent solutions of system (1.4.10) for $\lambda = -ik$, $k \in \mathbb{R}$ from the boundary conditions

$$\lim_{x\to-\infty} e^{\pm ikx}\phi^\pm(x,k) = e_\pm, \quad \lim_{x\to+\infty} e^{\pm ikx}\psi^\pm(x,k) = e_\pm, \tag{1.4.11}$$

where $e_+ = (1,0)$ and $e_- = (0,1)$.

Interpreting the term $Q(u)$ as the source term and solving the first-order linear system (1.4.10) by variations of parameters, we obtain integral equations for the first set of fundamental solutions,

$$\phi^+(x,k) = e^{-ikx}e_+ + \int_{-\infty}^x \left[\begin{array}{cc} e^{-ik(x-x')} & 0 \\ 0 & e^{ik(x-x')} \end{array} \right] Q(u(x'))\phi^+(x',k)dx'$$

$$\tag{1.4.12}$$

and

$$\phi^-(x,k) = e^{ikx}e_- + \int_{-\infty}^x \left[\begin{array}{cc} e^{-ik(x-x')} & 0 \\ 0 & e^{ik(x-x')} \end{array} \right] Q(u(x'))\phi^-(x',k)dx',$$

$$\tag{1.4.13}$$

where $x \in \mathbb{R}$ and $k \in \mathbb{R}$. Let us consider the homogeneous integral equation

$$\phi(x,k) = \int_{-\infty}^x \left[\begin{array}{cc} e^{-ik(x-x')} & 0 \\ 0 & e^{ik(x-x')} \end{array} \right] Q(u(x'))\phi(x',k)dx', \quad k \in \mathbb{R}. \tag{1.4.14}$$

Because of the exponential decay (1.4.9), equation (1.4.14) implies for any $x_0 < 0$ that

$$\sup_{x \leq x_0} \|\phi(x,k)\|_{\mathbb{C}^2} \leq Ce^{\kappa x_0} \sup_{x \leq x_0} \|\phi(x,k)\|_{\mathbb{C}^2}.$$

For large negative x_0, $Ce^{\kappa x_0} < 1$ and the only bounded solution of the homogeneous integral equation (1.4.14) for $x \in (-\infty, x_0]$ is the zero solution, which is hence extended to the zero solution for all $x \in \mathbb{R}$. By the Fredholm Alternative Theorem (Appendix B.4), bounded solutions of the integral equation (1.4.12) for fixed $k \in \mathbb{R}$ are uniquely defined for all $x \in \mathbb{R}$ including the limit $x \to \infty$, where we recover the *scattering problem*

$$\phi^+(x,k) = a(k)\psi^+(x,k) + b(k)\psi^-(x,k), \quad k \in \mathbb{R}, \tag{1.4.15}$$

with the *scattering coefficients*

$$a(k) = 1 + \int_{\mathbb{R}} e^{ikx} \left(Q(u(x))\phi^+(x,k)\right)_1 dx, \quad k \in \mathbb{R} \tag{1.4.16}$$

and

$$b(k) = \int_{\mathbb{R}} e^{-ikx} \left(Q(u(x))\phi^+(x,k)\right)_2 dx, \quad k \in \mathbb{R}. \tag{1.4.17}$$

Although the scattering problem (1.4.15) is derived in the limit $x \to \infty$, it is valid for any $x \in \mathbb{R}$, because one set of fundamental solutions $\{\phi^+(x,k), \phi^-(x,k)\}$ is linearly dependent on the other set of fundamental solutions $\{\psi^+(x,k), \psi^-(x,k)\}$ for all $x \in \mathbb{R}$.

Since system (1.4.10) is nothing but

$$\begin{cases} (\partial_x + ik)\psi_1 = -u\psi_2, \\ (\partial_x - ik)\psi_2 = \bar{u}\psi_1, \end{cases} \tag{1.4.18}$$

it admits the following symmetry reduction: if (ψ_1, ψ_2) is a solution of system (1.4.18) for $k \in \mathbb{R}$, then $(-\bar{\psi}_2, \bar{\psi}_1)$ is also a solution of the same system (1.4.18). By uniqueness of solutions with boundary conditions (1.4.11), we conclude that

$$\phi^-(x,k) = \sigma_1\sigma_3\bar{\phi}^+(x,k), \quad \psi^-(x,k) = \sigma_1\sigma_3\bar{\psi}^+(x,k), \tag{1.4.19}$$

where σ_1 is another Pauli matrix. Therefore, the scattering problem for $\phi^-(x,k)$ can be expressed from the scattering problem (1.4.15) by

$$\phi^-(x,k) = -\bar{b}(k)\psi^+(x,k) + \bar{a}(k)\psi^-(x,k), \tag{1.4.20}$$

for any $x \in \mathbb{R}$ and any $k \in \mathbb{R}$.

Because the Wronskian of any two solutions of system (1.4.18) is independent of x, the Wronskian of linearly independent solutions $\{\phi^+(x,k), \phi^-(x,k)\}$ can be computed in the limits $x \to -\infty$ and $x \to +\infty$. Using the scattering relations (1.4.15) and (1.4.20), we obtain the relation between the scattering coefficients

$$|a(k)|^2 + |b(k)|^2 = 1, \quad k \in \mathbb{R}. \tag{1.4.21}$$

Besides the continuous spectrum for $\lambda \in i\mathbb{R}$ that corresponds to the wave functions $\phi^\pm(\cdot, k) \in L^\infty(\mathbb{R})$, the spectral problem (1.4.10) may have isolated eigenvalues

$\lambda \notin i\mathbb{R}$ that correspond to the eigenvectors in $H^1(\mathbb{R})$. Because the scattering coefficients $a(k)$ and $b(k)$ as well as the fundamental solutions $\phi^{\pm}(\cdot, k) \notin L^2(\mathbb{R})$ are not singular for any $k \in \mathbb{R}$, no eigenvalues may be located for $k \in \mathbb{R}$, that is, for $\lambda \in i\mathbb{R}$. However, for a particular value $k = \xi_j + i\eta_j$ with $\eta_j > 0$, there may exist an eigenvector $\phi_j^+ \in H^1(\mathbb{R})$ of the spectral problem (1.4.10) that decays to zero exponentially fast according to the boundary conditions

$$\lim_{x \to -\infty} e^{(i\xi_j - \eta_j)x}\phi_j^+(x) = e_+, \quad \lim_{x \to +\infty} e^{(-i\xi_j + \eta_j)x}\phi_j^+(x) = c_j e_-, \qquad (1.4.22)$$

where c_j is referred to as the *norming constant*. Compared with the definition of the fundamental solution $\phi^+(x, k)$ for $k \in \mathbb{R}$, the eigenvector $\phi_j^+(x)$ can be thought to be identical to $\phi^+(x, \xi_j + i\eta_j)$ if there is a way to extend $\phi^+(x, k)$ analytically in the upper half of the complex k plane. The following lemma shows that this is indeed the case.

Lemma 1.5 *Let $u \in L^{\infty}(\mathbb{R})$ satisfy the exponential decay (1.4.9). Then, for any $\kappa_0 \in (0, \frac{1}{2}\kappa)$, $e^{ikx}\phi^+(x, k)$ and $e^{-ikx}\psi^-(x, k)$ are analytic functions of k for $\mathrm{Im}(k) > -\kappa_0$ whereas $e^{-ikx}\phi^-(x, k)$ and $e^{ikx}\psi^+(x, k)$ are analytic functions of k for $\mathrm{Im}(k) < \kappa_0$.*

Proof   It is sufficient to prove the statement for the function

$$M(x, k) = e^{ikx}\phi^+(x, k).$$

The proof for the other functions is developed similarly. The integral equation (1.4.12) is rewritten in the equivalent form

$$\begin{cases} M_1(x, k) = 1 - \int_{-\infty}^x u(x')M_2(x', k)dx', \\ M_2(x, k) = \int_{-\infty}^x e^{2ik(x-x')}\bar{u}(x')M_1(x', k)dx', \end{cases} \qquad x \in \mathbb{R}, \quad k \in \mathbb{R}. \;\; (1.4.23)$$

Consider the Neumann series,

$$M_1(x, k) = 1 + \sum_{n \in \mathbb{N}} m^{(2n)}(x, k), \quad M_2(x, k) = \sum_{n \in \mathbb{N}} m^{(2n+1)}(x, k),$$

where the terms are defined recurrently by

$$\begin{cases} m^{(2n-1)}(x, k) = \int_{-\infty}^x e^{2ik(x-x')}\bar{u}(x')m^{(2n-2)}(x', k)dx', \\ m^{(2n)}(x, k) = -\int_{-\infty}^x u(x')m^{(2n-1)}(x', k)dx', \end{cases}$$

for $x \in \mathbb{R}$, $k \in \mathbb{R}$, and $n \in \mathbb{N}$. For any $\kappa_0 \in [0, \frac{1}{2}\kappa)$, $k \in \{\mathbb{C} : \mathrm{Im}(k) > -\kappa_0\}$, and $x < 0$, we use the exponential decay (1.4.9) and obtain

$$|m^{(1)}(x, k)| \leq \int_{-\infty}^x \left| e^{2ik(x-x')}\bar{u}(x') \right| dx' = \int_0^{\infty} e^{2\kappa_0 y}|u(x-y)|dy$$

$$\leq C \int_0^{\infty} e^{2\kappa_0 y - \kappa|x-y|}dy = \frac{C}{\kappa - 2\kappa_0}e^{\kappa x}$$

and

$$|m^{(2)}(x,k)| \le \int_{-\infty}^{x} \left| u(x')m^{(1)}(x',k) \right| dx'$$

$$\le \frac{C^2}{\kappa - 2\kappa_0} \int_{-\infty}^{x} e^{2\kappa x'} dx' = \frac{C^2}{2\kappa(\kappa - 2\kappa_0)} e^{2\kappa x}.$$

By the induction method, for any $x < 0$ and $k \in \{\mathbb{C} : \text{Im}(k) > -\kappa_0\}$ we obtain

$$|m^{(2n-1)}(x,k)| \le \frac{C^{2n-1}}{(2\kappa)^{n-1}(\kappa - 2\kappa_0)^n} e^{\kappa(2n-1)x},$$

$$|m^{(2n)}(x,k)| \le \frac{C^{2n}}{(2\kappa)^n(\kappa - 2\kappa_0)^n} e^{\kappa(2n)x}, \quad n \in \mathbb{N}.$$

Therefore, the Neumann series for any $x < 0$ is majorized by the uniformly convergent series for any $k \in \mathbb{C}$ such that $\text{Im}(k) > -\kappa_0$. Since each function of the Neumann series is analytic in k (because it does not depend on $\bar{k}$), the uniformly convergent series of the analytic function (function $M(x,k)$ for $x < 0$) is analytic in k in the same domain. The analyticity is then extended to $x \ge 0$. $\qquad \square$

Because of Lemma 1.5, it follows from expressions (1.4.16) and (1.4.17) that

$$a(k) = 1 - \int_{\mathbb{R}} u(x)M_2(x,k)dx, \quad k \in \mathbb{R}$$

is analytic for any $\text{Im}(k) > -\kappa_0$, whereas

$$b(k) = \int_{\mathbb{R}} e^{-2ikx}\bar{u}(x)M_1(x,k)dx, \quad k \in \mathbb{R}$$

cannot be extended off the real k-axis in the general case. Comparison of the scattering problem (1.4.15) with the boundary conditions (1.4.22) shows that $a(\xi_j + i\eta_j) = 0$, whereas the norming constant c_j serves as an equivalent of $b(k)$ at $k = \xi_j + i\eta_j$.

Note again the symmetry: if (ψ_1, ψ_2) is a solution of system (1.4.18) for $k \in \mathbb{C}$, then $(-\bar{\psi}_2, \bar{\psi}_1)$ is also a solution of the same system (1.4.18) for $\bar{k}$. Using this symmetry, we can see that if $k = \xi_j + i\eta_j$ is the eigenvalue with the eigenvector $\phi_j^+(x)$ for $\eta_j > 0$, then $k = \xi_j - i\eta_j$ is also an eigenvalue with the eigenvector

$$\phi_j^-(x) = \sigma_1\sigma_3\bar{\phi}_j^+(x). \tag{1.4.24}$$

We note that there is at most one eigenvector of the spectral problem (1.4.10) for each value $k = \xi_j + i\eta_j$ since the other linearly independent solution is exponentially growing as $x \to -\infty$. Therefore, the geometric multiplicity of the eigenvalue $k = \xi_j + i\eta_j$ is one. If $k = \xi_j + i\eta_j$ with $\eta_j > 0$ is the m_jth root of $a(k) = 0$, we say that m is the *algebraic multiplicity* of the eigenvalue $k = \xi_j + i\eta_j$.

Although we are not precise on the index j and on the algebraic multiplicity of eigenvalue at $k = \xi_j + i\eta_j$, we can show that the total number of eigenvalues taking account of their algebraic multiplicity is finite thanks to the exponential decay (1.4.9).

Let us denote

$$D_+ := \{k \in \mathbb{C}: \quad \text{Im}(k) > 0\}, \quad D_- := \{k \in \mathbb{C}: \quad \text{Im}(k) < 0\}.$$

The closed half-planes are denoted by $\bar{D}_+$ and $\bar{D}_-$.

Lemma 1.6 *Let $u \in L^\infty(\mathbb{R})$ satisfy the exponential decay (1.4.9). There exists at most a finite number of eigenvalues of the spectral problem (1.4.10) taking account of their algebraic multiplicity.*

Proof By Lemma 1.5, functions $M(x,k) = e^{ikx}\phi^+(x,k)$ and $a(k)$ are analytic in $\bar{D}_+$. Integrating equation (1.4.23) by parts for large k in D_+ yields the asymptotic expansion

$$\begin{cases} M_1(x,k) = 1 + \frac{1}{2ik}\int_{-\infty}^x |u(x')|^2 dx' + \mathcal{O}(k^{-2}), \\ M_2(x,k) = -\frac{1}{2ik}\bar{u}(x) + \mathcal{O}(k^{-2}), \end{cases} \quad x \in \mathbb{R}, \quad (1.4.25)$$

which hence induces the asymptotic expansion

$$a(k) = 1 + \frac{1}{2ik}\int_{\mathbb{R}} |u(x)|^2 dx + \mathcal{O}(k^{-2}). \quad (1.4.26)$$

Therefore, $a(k)$ has no zeros for large values of k in D_+ because $\lim_{|k|\to\infty} a(k) = 1$ in D_+. On the other hand, $a(k)$ is analytic in $\bar{D}_+$ and hence it has only finitely many zeros of finite multiplicities in $\bar{D}_+$. $\square$

Having identified both the continuous spectrum and the eigenvalues of the spectral problem (1.4.10) for a given $u(x)$, we should now address the *inverse problem*: recover the potential $u(x)$ from the given continuous spectrum and eigenvalues of the spectral problem (1.4.10). Because the time evolution of spectral data is going to be somewhat trivial (Exercise 1.25), the inverse problem will enable us to recover solutions of the NLS equation (1.4.8) for any $t > 0$.

Exercise 1.25 Consider the time evolution of the second equation in system (1.4.6) in the limit $|x| \to \infty$, that is

$$\partial_t \psi = 2i\lambda^2 \sigma_3 \psi,$$

and prove that the scattering data evolve in $t \in \mathbb{R}$ as follows:

$$a(k,t) = a(k,0), \quad b(k,t) = b(k,0)e^{-4ik^2 t}, \quad c_j(t) = c_j(0)e^{-4i(\xi_j+i\eta_j)^2 t}, \quad (1.4.27)$$

where (ξ_j, η_j) are t-independent.

We shall now consider solutions of the inverse problem for the spectral problem (1.4.10). Let us rewrite the scattering relations (1.4.15) and (1.4.20) in the equivalent form,

$$\begin{cases} \mu^+(x,k) = \nu^-(x,k) + \rho(k)e^{2ikx}\nu^+(x,k), \\ \mu^-(x,k) = \nu^+(x,k) - \bar{\rho}(k)e^{-2ikx}\nu^-(x,k), \end{cases} \quad k \in \mathbb{R}, \quad (1.4.28)$$

where

$$\mu^+(x,k) = \frac{1}{a(k)}e^{ikx}\phi^+(x,k), \quad \mu^-(x,k) = \frac{1}{\bar{a}(k)}e^{-ikx}\phi^-(x,k),$$

$$\nu^+(x,k) = e^{-ikx}\psi^-(x,k), \quad \nu^-(x,k) = e^{ikx}\psi^+(x,k), \quad \text{and} \quad \rho(k) = \frac{b(k)}{a(k)}.$$

We know from Lemmas 1.5 and 1.6 that $\mu^\pm(x, k)$ are meromorphic functions in $\bar{D}_\pm$ and $\nu^\pm(x, k)$ are analytic functions in $\bar{D}_\pm$. The boundary conditions

$$\lim_{|k|\to\infty} \mu^\pm(x, k) = e_\pm \quad \text{and} \quad \lim_{|k|\to\infty} \nu^\pm(x, k) = e_\mp$$

hold in the corresponding domains $D_\pm$. Using these data, we set up the *Riemann–Hilbert problem*: find meromorphic extensions of 2×2 matrix functions

$$m^+(x, k) = \left[\mu^+(x, k), \nu^+(x, k)\right], \quad m^-(x, k) = \left[\nu^-(x, k), \mu^-(x, k)\right],$$

such that $m^\pm(x, k) \to I \in \mathbb{M}^{2\times 2}$ as $\mathrm{Im}(k) \to \pm\infty$ from the jump condition on the real axis

$$m^+(x, k) - m^-(x, k) = m^-(x, k)W(x, k), \quad k \in \mathbb{R}, \tag{1.4.29}$$

where

$$W(x, k) = \begin{bmatrix} |\rho(k)|^2 & \bar{\rho}(k)e^{-2ikx} \\ \rho(k)e^{2ikx} & 0 \end{bmatrix}.$$

Let us consider the two simplest solutions of the Riemann–Hilbert problem (1.4.29).

Case I: Assume that $a(k)$ has no zeros in $\bar{D}_+$, which implies that $m^\pm(x, k)$ are analytic in $\bar{D}_\pm$.

Let $f(z) : \mathbb{R} \to \mathbb{C}$ be a continuous function such that $\lim_{|z|\to\infty} f(z) = 0$. Let us define Plemelej's projection operators

$$(P^\pm f)(z) = \frac{1}{2\pi i} \int_\mathbb{R} \frac{f(\zeta)}{\zeta - (z \pm i0)} d\zeta, \tag{1.4.30}$$

where $\pm i0$ indicates that if $\mathrm{Im}(z) \to 0$, the contour of integration in ζ should pass z from below for $(P^+ f)(z)$ and from above for $(P^- f)(z)$. From the Cauchy Integral Theorem, we know that $f^\pm(z) := (P^\pm f)(z)$ is analytic in $\bar{D}_\pm$ and $f^\pm(z) \to 0$ as $|z| \to \infty$ in $\bar{D}_\pm$. Complex integration shows that

$$\lim_{\mathrm{Im}(z)\to 0} f^\pm(z) = \frac{1}{2\pi i}\text{p.v.} \int_\mathbb{R} \frac{f(\zeta)}{\zeta - z} d\zeta \pm \frac{1}{2}f(z), \quad z \in \mathbb{R},$$

where p.v. denotes the principal value integral, and the limits are taken inside the corresponding domains $\bar{D}_\pm$. Therefore, $f^+(z)$ and $f^-(z)$ solve the Riemann–Hilbert problem,

$$f^+(z) - f^-(z) = f(z), \quad z \in \mathbb{R},$$

for a given $f \in C_b^0(\mathbb{R})$ subject to the boundary conditions $f^\pm(z) \to 0$ as $|z| \to \infty$ in $\bar{D}_\pm$.

Let $f^\pm(k) = m^\pm(x, k) - I$, $k \in \bar{D}_\pm$ and $f(k) = m_-(x, k)W(x, k)$, $k \in \mathbb{R}$. To ensure the decay of $f(k)$ at infinity, we need the decay of scattering coefficient $\rho(k)$ as $|k| \to \infty$. The following lemma is obtained from Plemelej's projection operators (1.4.30).

Lemma 1.7 *Assume that $a(k)$ has no zeros in $\bar{D}_+$ and $\rho(k) \to 0$ as $|k| \to \infty$. There exists a unique solution of the Riemann–Hilbert problem (1.4.29) in the form*

$$m_+(x, k) = I + \frac{1}{2\pi i} \int_\mathbb{R} \frac{m_-(x, \xi)W(x, \xi)}{\xi - (k + i0)} d\xi \tag{1.4.31}$$

and

$$m_-(x,k) = I + \frac{1}{2\pi i} \int_{\mathbb{R}} \frac{m_-(x,\xi)W(x,\xi)}{\xi - (k-i0)} d\xi. \qquad (1.4.32)$$

From the solution (1.4.31)–(1.4.32) of the Riemann–Hilbert problem (1.4.29), we can recover the solution of the inverse problem using the asymptotic expansions (1.4.25) and (1.4.26) in inverse powers of k. It is clear that

$$m_-(x,k)W(x,k) = \left[\rho(k)\psi^-(x,k)e^{ikx}, \bar{\rho}(k)\psi^+(x,k)e^{-ikx}\right].$$

On the other hand, we have the asymptotic expansion in $\bar{D}_+$,

$$m_+(x,k) = I - \frac{1}{2\pi i k} \int_{\mathbb{R}} m_-(x,\xi)W(x,\xi)d\xi + \mathcal{O}(k^{-2}) \quad \text{as} \quad |k| \to \infty. \qquad (1.4.33)$$

Comparing (1.4.33) with (1.4.25) and (1.4.26), we obtain

$$\bar{u}(x) = \frac{1}{\pi} \int_{\mathbb{R}} \rho(k)e^{ikx}\psi_2^-(x,k)dk. \qquad (1.4.34)$$

Thanks to symmetry (1.4.19), equation (1.4.34) is equivalent to

$$u(x) = \frac{1}{\pi} \int_{\mathbb{R}} \bar{\rho}(k)e^{-ikx}\psi_1^+(x,k)dk. \qquad (1.4.35)$$

Case II: Assume that $a(k)$ has only simple zeros for $\text{Im}(k) > 0$ and $\rho(k) = 0$ for all $k \in \mathbb{R}$.

The Riemann–Hilbert problem (1.4.29) tells us that $m^+(x,k) = m^-(x,k)$ represents a globally meromorphic function for all $k \in \mathbb{R}$ with a finite number of simple poles. Therefore, we write

$$m_\pm(x,k) = I + \sum_j \frac{m_j^+(x)}{k - k_j} + \sum_j \frac{m_j^-(x)}{k - \bar{k}_j},$$

where the range of summation in j is finite and $k_j = \xi_j + i\eta_j$ with $\eta_j > 0$. It is more convenient to rewrite this solution of the Riemann–Hilbert problem in the component form

$$\frac{1}{a(k)}e^{ikx}\phi^+(x,k) = e^{ikx}\psi^+(x,k) = e_+ + \sum_j \frac{e^{ik_j x}\phi_j^+(x)}{a'(k_j)(k - k_j)} \qquad (1.4.36)$$

and

$$e^{-ikx}\psi^-(x,k) = \frac{1}{\bar{a}(k)}e^{-ikx}\phi^-(x,k) = e_- + \sum_j \frac{e^{-i\bar{k}_j x}\phi_j^-(x)}{\bar{a}'(k_j)(k - \bar{k}_j)}, \qquad (1.4.37)$$

where eigenvectors $\phi_j^\pm(x)$ are defined by the boundary conditions (1.4.22) and symmetry (1.4.24). Using the norming constants, we set $\phi_j^+(x) = c_j\psi_j^-(x)$ and rewrite the second equation (1.4.37) in the equivalent form

$$e^{-ikx}\psi^-(x,k) = e_- + \sum_j \frac{e^{-i\bar{k}_j x}c_j\sigma_1\sigma_3\psi_j^-(x)}{\bar{a}'(k_j)(k - \bar{k}_j)}. \qquad (1.4.38)$$

By Lemma 1.5, function $e^{-ikx}\psi^-(x,k)$ is analytic in $\bar{D}^+$ and, hence, we can associate $\psi_j^-(x) = \psi_j^-(x, k_j)$. As a result, the eigenvectors $\psi_j^-(x)$ are determined by the system of algebraic equations

$$e^{-ik_jx}\psi_j^-(x) = e_- + \sum_j \frac{e^{-i\bar{k}_jx}C_j\sigma_1\sigma_3\psi_j^-(x)}{(k_j - \bar{k}_j)}, \qquad (1.4.39)$$

where $C_j = c_j/\bar{a}'(k_j)$. The inverse problem is solved from the asymptotic expansion in inverse powers of k by

$$\bar{u}(x) = -2i\sum_j C_j e^{ik_jx}(\psi_j^-)_2(x). \qquad (1.4.40)$$

Exercise 1.26 Show that if $\rho(k) = 0$ for all $k \in \mathbb{R}$ and $a(k)$ has only simple zeros in D_+, then

$$a(k) = \prod_j \frac{k - k_j}{k - \bar{k}_j}.$$

From solution (1.4.40) with only one zero of $a(k)$, we obtain a soliton of the NLS equation (1.4.8). Let $C_j = 2i\eta_j e^{2\eta_j s_j - i\theta_j}$ for convenience, where $s_j \in \mathbb{R}$ and $\theta_j \in \mathbb{R}$. Solving the system of two equations (1.4.39) for only one j, we obtain

$$\psi_j^-(x) = \frac{e^{ik_jx}}{1 + e^{-4\eta_j(x - s_j)}} \left[\begin{array}{c} -e^{-2\eta_j(x - s_j) - i\theta_j} \\ 1 \end{array} \right].$$

Using the representation (1.4.40), we obtain the soliton at $t = 0$:

$$\bar{u}(x) = 2\eta_j \operatorname{sech}(2\eta_j(x - s_j))e^{2i\xi_jx - i\theta_j}.$$

The time evolution of the scattering coefficients follows from equation (1.4.27) as

$$C_j(t) = 2i\eta_j e^{2\eta_j s_j} e^{-4i(\xi_j + i\eta_j)^2 t - i\theta_j}.$$

As a result, we obtain the time-dependent soliton of the NLS equation (1.4.8),

$$\bar{u}(x,t) = 2\eta_j \operatorname{sech}(2\eta_j(x - 4\xi_j t - s_j))e^{2i\xi_jx - 4i(\xi_j^2 - \eta_j^2)t - i\theta_j},$$

or explicitly,

$$u(x,t) = 2\eta_j \operatorname{sech}(2\eta_j(x - 4\xi_j t - s_j))e^{-2i\xi_jx + 4i(\xi_j^2 - \eta_j^2)t + i\theta_j}. \qquad (1.4.41)$$

The exact soliton (1.4.41) is a stationary and traveling localized mode of the NLS equation (1.4.8) with arbitrary parameters $(s_j, \theta_j, \xi_j, \eta_j) \in \mathbb{R}^4$.

Exercise 1.27 Construct a solution of the Riemann–Hilbert problem (1.4.29) with one double zero of $a(k)$ in D_+ and $\rho(k) = 0$ for all $k \in \mathbb{R}$ and find the corresponding solution of the NLS equation (1.4.8).

Zeros of $a(k)$ in D_+ may bifurcate from zeros of $a(k)$ on $\mathbb{R}$ when the potential $u(x)$ is continued along a parameter deformation. As a result of this bifurcation, the number of eigenvalues of the discrete spectrum in the spectral problem (1.4.10) jumps. These bifurcations in the context of the NLS equation have been studied by Klaus & Shaw [116].

Exercise 1.28   Consider system (1.4.18) for a real-valued potential $u \in L^1(\mathbb{R})$ and prove that $a(0) = 0$ if and only if $\int_{\mathbb{R}} u(x)dx = \frac{\pi}{2}(2n - 1)$, $n \in \mathbb{N}$.

The inverse scattering transform is also useful to obtain exact eigenvectors and eigenvalues of the spectral stability problem associated with the soliton of the NLS equation (1.4.8). The corresponding expressions can be found in the paper of Kaup [109], where some earlier results are also mentioned.

1.4.2 The NLS equation with variable coefficients

The inverse scattering transform can also be applied to the nonlinear Schrödinger equation with variable coefficients in space x and time t. Besides the x-periodic potentials, which are met in this book, the t-periodic coefficients in front of the linear dispersive and cubic nonlinear terms are often used to model effects of dispersion and diffraction management, and for nonlinearity management. It is common to refer to the equations with time-dependent coefficients as *non-autonomous* systems and to the equations with space-dependent coefficients as *inhomogeneous* systems.

Let us consider the cubic NLS equation with variable coefficients

$$iu_t + D(x,t)u_{xx} + V(x,t)u + G(x,t)|u|^2u = 0, \qquad (1.4.42)$$

where $D(x,t), V(x,t), G(x,t) : \mathbb{R} \times \mathbb{R} \to \mathbb{R}$ are given coefficients. We are looking for the change of variables under which the NLS equation with variable coefficients (1.4.42) reduces to the cubic NLS equation,

$$iU_T + D_0 U_{XX} + G_0|U|^2U = 0, \qquad (1.4.43)$$

where D_0 and G_0 are constant coefficients. The change of variables that transforms (1.4.42) into (1.4.43) is known as the *point transformation*. Recall that the cubic NLS equation (1.4.43) is integrable by the inverse scattering transform method (Section 1.4.1).

The point transformation of (1.4.42) to (1.4.43) exists under certain constraints on $D(x,t)$, $V(x,t)$, and $G(x,t)$. An algorithmic search for the most general constraints on the coefficients of the equation can be performed with the method of Lie group symmetries [21] because the existence of non-trivial commuting continuous symmetries implies the possibility of a transformation of a given equation with variable coefficients to an equation with constant coefficients. The transformation of a general NLS equation with variable coefficients (1.4.42) was studied by Gagnon & Winternitz [61]. Some particular cases of this transformation were recently rediscovered in the physics literature [8, 22, 188].

Consider the change of variables

$$T = T(x,t), \quad X = X(x,t), \quad u = u(X, T, U(X, T)).$$

The chain rule gives

$$u_t = u_X X_t + u_T T_t + u_U(U_X X_t + U_T T_t)$$

and

$$u_{xx} = u_X X_{xx} + u_T T_{xx}$$
$$+ u_U (U_X X_{xx} + U_T T_{xx} + U_{XX}(X_x)^2 + 2U_{XT}X_x T_x + U_{TT}(T_x)^2)$$
$$+ X_x(u_{XX}X_x + u_{XT}T_x + u_{XU}(U_X X_x + U_T T_x))$$
$$+ T_x(u_{TX}X_x + u_{TT}T_x + u_{TU}(U_X X_x + U_T T_x))$$
$$+ (U_X X_x + U_T T_x)(u_{UX}X_x + u_{UT}T_x + u_{UU}(U_X X_x + U_T T_x)).$$

Removing the terms U_X^2 and U_{TT}, which are absent in the NLS equation (1.4.43), we have constraints $u_{UU} = 0$ and $T_x = 0$, so that

$$u(X,T,U) = Q(X,T)U + R(X,T), \quad T = T(t).$$

Removing terms U^2, we set $R(X,T) \equiv 0$. It is easier technically to write Q in old variables (x,t), so that we write $Q = Q(x,t)$. Expressing iU_T from the NLS equation (1.4.43) and removing terms U_X, U_{XX}, $|U|^2 U$, and U, we obtain the remaining constraints

$$0 = iQX_t + D(x,t)QX_{xx} + 2D(x,t)Q_x X_x, \qquad (1.4.44)$$

$$0 = D(x,t)\frac{(X_x)^2}{T'(t)} - D_0, \qquad (1.4.45)$$

$$0 = G(x,t)\frac{|Q|^2}{T'(t)} - G_0, \qquad (1.4.46)$$

$$0 = iQ_t + V(x,t)Q + D(x,t)Q_{xx}. \qquad (1.4.47)$$

Constraints (1.4.44)–(1.4.47) on parameters of the transformation $X(x,t)$, $T(t)$, and $Q(x,t)$ on the one hand, and variable coefficients $D(x,t)$, $G(x,t)$, and $V(x,t)$ on the other hand, give non-trivial conditions for integrability of the NLS equation with variable coefficients (1.4.42). Of course, there are too many possibilities in the general transformation. Therefore, it is useful to consider a number of particular examples.

Case I: Assume that $D(x,t) \equiv D_0$ and $G(x,t) \equiv G_0$.
Equations (1.4.44)–(1.4.46) admit solutions for $X(x,t)$ and $Q(x,t)$:

$$X(x,t) = x\left(T'(t)\right)^{1/2}, \quad Q(x,t) = \left(T'(t)\right)^{1/2} \exp\left(-\frac{ix^2 T''(t)}{8D_0 T'(t)}\right).$$

Substituting these expressions into the last equation (1.4.47), we obtain

$$V(x,t) = \frac{x^2(3(T'')^2 - 2T'T''')}{16D_0(T')^2} - \frac{iT''}{4T'}.$$

The requirement of $V \in \mathbb{R}$ sets up the constraint $T''(t) = 0$, after which $V(x,t) \equiv 0$. As a result, the Gross–Pitaevskii equation with a potential $V(x,t)$ is not an integrable model for any choice of the potential.

Case II: Assume that $D = D(t)$ and $G = G(t)$.

Equations (1.4.44)–(1.4.46) admit solutions for $X(x,t)$ and $Q(x,t)$:

$$X(x,t) = x \left(\frac{D_0 T'(t)}{D(t)} \right)^{1/2},$$

$$Q(x,t) = \left(\frac{G_0 T'(t)}{G(t)} \right)^{1/2} \exp\left(-\frac{ix^2}{8} \left(\frac{T''(t)}{T'(t)D(t)} - \frac{D'(t)}{D^2(t)} \right) \right).$$

Substituting these expressions into the last equation (1.4.47), we obtain

$$V(x,t) = \frac{x^2}{8} \left(\frac{T''}{DT'} - \frac{D'}{D^2} \right)' - \frac{Dx^2}{16} \left(\frac{T''}{DT'} - \frac{D'}{D^2} \right)^2 + \frac{i}{4} \left(\frac{T''}{T'} - \frac{2G'}{G} + \frac{D'}{D} \right).$$

The requirement of $V \in \mathbb{R}$ is satisfied if

$$T'(t) = \frac{G^2(t)}{D(t)},$$

after which we find the potential $V(x,t) = \Omega^2(t)x^2$ with

$$\Omega^2(t) = -\frac{D(t)D''(t) - (D'(t))^2}{4D^3(t)} + \frac{G(t)G''(t) - 2(G'(t))^2}{4D(t)G^2(t)} + \frac{D'(t)G'(t)}{4D^2(t)G(t)}.$$

Serkin *et al.* [188] obtain the same constraint using a direct search of the pair of Lax operators for integrability of the NLS equation with variable coefficients (1.4.42).

Exercise 1.29   Assume that $D = D(x)$, $V = V(x)$, and $G = G(x)$, and obtain a constraint on coefficients D, V, and G for integrability of the NLS equation with variable coefficients (1.4.42). Can this constraint be satisfied for real periodic functions D, V, and G?

Exercise 1.30   Consider a linear wave–Maxwell equation with the variable speed,

$$u_{tt} - c^2(x)u_{xx} = 0,$$

where $u(x,t) : \mathbb{R} \times \mathbb{R} \to \mathbb{R}$ and $c^2(x) : \mathbb{R} \to \mathbb{R}$. Find the constraint on $c(x)$ and the change of variables $(x,t,u) \mapsto (X,T,U)$, so that the linear wave–Maxwell equation can be reduced to the Klein–Gordon equation,

$$U_{TT} - U_{XX} + \kappa U = 0,$$

where $U(X,T) : \mathbb{R} \times \mathbb{R} \to \mathbb{R}$ and $\kappa \in \mathbb{R}$ is a constant coefficient.

2

Justification of the nonlinear Schrödinger equations

As far as the laws of mathematics refer to reality, they are not certain, as far as they are certain, they do not refer to reality.
– Albert Einstein.

The 20th century was the century of least squares, and the 21st may very well be that of ell-one magic.
– B.A. Cipra, "Ell 1-magic", SIAM News (November 2006).

Show us the equations! – a common wail of mathematicians when they hear physicists speak. Physicists argue that even the basic physical laws may incorrectly model nature. Physicists like to discuss various observations of real-world effects and thus admit a freedom of juggling with model equations and their simplifications.

This approach is however unacceptable in mathematics. Even if one model can be simplified to another model in a certain limit, mathematicians would like to understand the exact meaning of this simplification, proving convergence in some sense, studying bounds on the distance between solutions of the two models, and spending years in searches for sharper bounds.

Since this book is written for young mathematicians, we shall explain elements of analysis needed for rigorous justification of the nonlinear Dirac equations, the nonlinear Schrödinger equation, and the discrete nonlinear Schrödinger equation, in the context of the Gross–Pitaevskii equation with a periodic potential. Justification of nonlinear evolution equations in other contexts has been a subject of intense research in the past twenty years (see [186, 187] on the Korteweg–de Vries equations for water waves, [66, 68] on the Boussinesq equations for Fermi–Pasta–Ulam lattices, and [182, 183] on the Ginzburg–Landau equations for reaction–diffusion systems).

The models we are interested in are formally derived with the asymptotic multi-scale expansion method (Section 1.1). Good news is that the scaling of the asymptotic multi-scale method and computations of the first terms of the asymptotic expansion can be incorporated into a rigorous treatment of the corresponding solution, hence our previous results will be used for the work in this chapter. Nevertheless, this time we shall go deeper into analysis of the Gross–Pitaevskii equation with a periodic potential. In particular, we shall review elements of the Floquet–Bloch–Wannier theory for Schrödinger operators with periodic potentials and develop nonlinear analysis involving decompositions in Sobolev and Wiener spaces, fixed-point iteration schemes, implicit function arguments, and Gronwall's inequality.

We shall separate justification of the time-dependent and stationary equations for localized modes in a periodic potential. Our reasoning is that the time-dependent problem is governed by a hyperbolic system, while the stationary problem is given by an elliptic system. Asymptotic reductions of hyperbolic systems hold generally for large but finite time intervals, whereas reductions of parabolic and elliptic systems can be extended to all positive times with the invariant manifold theory. In other words, time-dependent localized modes in periodic potentials may become delocalized beyond finite time intervals, whereas the stationary localized modes can be controlled for all times in an appropriate function space.

A general algorithm of the justification analysis for time-dependent problems consists of the following four steps.

(1) Decompose the solution into the leading-order term, which solves a given non-linear evolution equation in a formal limit $\epsilon \to 0$, and a remainder term.
(2) Remove the non-vanishing part of the residual as $\epsilon \to 0$ by a normal form transformation of the remainder term.
(3) Choosing an appropriate function space, prove that the residual of the time-evolution problem maps elements of this function space to elements of the same function space under some constraints on the leading-order term.
(4) Prove the local well-posedness of the time evolution problem for the remainder term and control its norm by Gronwall's inequality.

A general algorithm of the justification analysis for stationary problems reminds us of the method of Lyapunov–Schmidt reductions (Section 3.2) and consists of the following three steps.

(1) Decompose the solution into two components, one of which solves an ϵ-dependent equation for a singular operator and the other one solves another ϵ-dependent equation for an invertible operator.
(2) Prove the existence of a unique smooth map from the first (large) component of the solution to the second (small) component of the solution uniformly as $\epsilon \to 0$.
(3) Approximate solutions of the ϵ-dependent equation for the first component by solutions of the limiting equation as $\epsilon \to 0$ and estimate the distance between the two solutions.

In both problems, we shall use similar function spaces, which generalize the space of absolutely convergent infinite series. An alternative choice would be the space of squared summable infinite series and its generalizations. There is a similarity between analysis in the two different spaces, but we believe that our choice is better suited for the problem of justification of asymptotic multi-scale expansions for localized modes in periodic potentials.

2.1 Schrödinger operators with periodic potentials

Since we are working with the nonlinear evolution equations for localized modes in periodic potentials, we shall start with a simpler linear problem. It is true here and in many other instances in the book that linear problems provide useful information for solving nonlinear problems associated with the localized modes.

Let us start with the one-dimensional Schrödinger operator $L = -\partial_x^2 + V(x)$, where V is a real-valued 2π-periodic potential on $\mathbb{R}$. If $V \in L^\infty(\mathbb{R})$, then the operator L maps continuously $H^2(\mathbb{R})$ to $L^2(\mathbb{R})$ in the following sense. There exists $C > 0$ such that

$$\forall f \in H^2(\mathbb{R}): \quad \|Lf\|_{L^2} \leq C\|f\|_{H^2}. \tag{2.1.1}$$

To see this bound, we use the Cauchy–Schwarz inequality and obtain

$$\begin{aligned}
\|Lf\|_{L^2}^2 &= \int_{\mathbb{R}} \left((f'')^2 - 2Vff'' + V^2f^2 \right) dx \\
&\leq \int_{\mathbb{R}} (f'')^2 dx + 2\|V\|_{L^\infty} \int_{\mathbb{R}} |f||f''| dx + \|V\|_{L^\infty}^2 \int_{\mathbb{R}} f^2 dx \\
&\leq \left(\|f''\|_{L^2} + \|V\|_{L^\infty}\|f\|_{L^2} \right)^2 \\
&\leq (1 + \|V\|_{L^\infty})^2 \|f\|_{H^2}^2.
\end{aligned}$$

As a result of bound (2.1.1), we conclude that $H^2(\mathbb{R})$ is continuously embedded into the domain of operator L,

$$\text{Dom}(L) = \left\{ f \in L^2(\mathbb{R}): \quad Lf \in L^2(\mathbb{R}) \right\},$$

for which operator L is closed in $L^2(\mathbb{R})$.

Moreover, L is extended to a self-adjoint operator in $L^2(\mathbb{R})$, since

$$\begin{aligned}
\forall f, g \in H^2(\mathbb{R}): \quad \langle Lf, g \rangle_{L^2} &= \int_{\mathbb{R}} (-f'' + Vf)g\, dx \\
&= (g'f - f'g)\Big|_{x \to -\infty}^{x \to \infty} + \int_{\mathbb{R}} f(-g'' + Vg)\, dx \\
&= \langle f, Lg \rangle_{L^2},
\end{aligned}$$

where $f(x), f'(x) \to 0$ as $|x| \to \infty$ if $f \in H^2(\mathbb{R})$ (Appendix B.10).

Because L is self-adjoint in $L^2(\mathbb{R})$, the spectrum of L is a subset of the real axis and the eigenvectors of L are orthogonal. Although these two properties follow from the Spectral Theorem for any self-adjoint operator (Appendix B.11), we shall clarify different representations which exist for solutions of the spectral problem $Lu = \lambda u$. In particular, we shall work with the Floquet theory [59] and the series of periodic eigenfunctions (Section 2.1.1), the Bloch theory [20] and the integral transform of quasi-periodic eigenfunctions (Section 2.1.2), and the Wannier theory and the series of decaying functions (Section 2.1.3).

The first representation is described in the easy-reading text of Eastham [52]. The second representation is covered in Chapter XIII of Volume IV of the mathematical physics bible by Reed and Simon [176]. The third representation originates from the classical paper of Kohn [117]. Although this theory is widely known by specialists (in particular, in solid-state physics), we shall present a coherent introduction suitable for young mathematicians.

2.1.1 Floquet theory and the series of periodic eigenfunctions

When we deal with the spectrum of an operator L in $L^2(\mathbb{R})$, we set up the spectral problem

$$Lu - \lambda u, \tag{2.1.2}$$

for an eigenfunction $u \in \mathrm{Dom}(L) \subset L^2(\mathbb{R})$ and an eigenvalue $\lambda \in \mathbb{C}$. Since $L = -\partial_x^2 + V(x)$ is self-adjoint in $L^2(\mathbb{R})$, then $\lambda \in \mathbb{R}$, and since $Lu = \lambda u$ becomes a differential equation with real coefficients, we may choose a real-valued eigenfunction $u \in L^2(\mathbb{R})$. However, a striking feature of the spectral problem (2.1.2) for $L = -\partial_x^2 + V(x)$ with 2π-periodic $V(x)$ is the fact that it admits *no* solutions in $L^2(\mathbb{R})$!

Therefore, it may be strategically simpler to consider all possible solutions of the differential equation (2.1.2) for $x \in [0, 2\pi]$, thanks to the 2π-periodicity of the potential V. Thus, we set

$$-u''(x) + V(x)u(x) = \lambda u(x), \quad x \in (0, 2\pi), \tag{2.1.3}$$

subject to some boundary conditions at $x = 0$ and $x = 2\pi$. To understand the boundary conditions, let us view the second-order differential equation (2.1.2) in the matrix–vector notation

$$\frac{d}{dx} \begin{bmatrix} u \\ v \end{bmatrix} = \begin{bmatrix} 0 & 1 \\ V(x) - \lambda & 0 \end{bmatrix} \begin{bmatrix} u \\ v \end{bmatrix} \tag{2.1.4}$$

or simply as

$$\dot{y} = A(x)y, \quad y = \begin{bmatrix} u \\ v \end{bmatrix}, \quad A(x) = \begin{bmatrix} 0 & 1 \\ V(x) - \lambda & 0 \end{bmatrix}. \tag{2.1.5}$$

This formulation is common in dynamical system theory (since we work in the space of one dimension). Because the dynamical system (2.1.5) is defined by a 2π-periodic matrix $A(x)$, the Floquet Theorem (Appendix B.3) states that there exists a fundamental matrix of two linearly independent solutions $\Phi(x) \in \mathbb{M}^{2\times 2}$ of the matrix system $\dot{\Phi} = A(x)\Phi$ in the form

$$\Phi(x) = P(x)e^{xQ},$$

where $P(x) \in \mathbb{M}^{2\times 2}$ is 2π-periodic and $Q \in \mathbb{M}^{2\times 2}$ is x-independent.

If $\Phi(x)$ is normalized by the initial condition $\Phi(0) = I$, then $P(0) = I$. The matrix $M = \Phi(2\pi) = e^{2\pi Q}$ is referred to as the *monodromy* matrix. A unique solution of the vector system (2.1.5) starting with the initial condition $y(0) = y_0$ is given by

$$y(x) = \Phi(x)y_0.$$

Repeated evolution of $y(x)$ after several periods $[0, 2\pi]$, $[2\pi, 4\pi]$, and so on, depends on the eigenvalues of M called *Floquet multipliers.*

Because the trace of matrix $A(x)$ is zero for any $x \in \mathbb{R}$, the Wronskian of two solutions of the linear system (2.1.5) is constant in x, so that

$$\det(M) = \det(I) = 1.$$

Let μ_1 and μ_2 be the two Floquet multipliers of M. Multipliers μ_1 and μ_2 are defined by the roots of the quadratic equation

$$\mu^2 - \mu \operatorname{tr}(M) + 1 = 0. \tag{2.1.6}$$

In particular, we note that $\mu_1 + \mu_2 = \operatorname{tr}(M)$ and $\mu_1 \mu_2 = 1$.

By the Spectral Mapping Theorem (Appendix B.3), we have

$$\sigma(M) = e^{2\pi(\sigma(Q) + in)}, \tag{2.1.7}$$

for any integer n, where $\sigma(M)$ denotes the spectrum of eigenvalues of M. Eigenvalues of Q are referred to as the *characteristic exponents*. Because of the above representation with an arbitrary integer n, characteristic exponents are uniquely defined in the strip

$$-\frac{1}{2} < \operatorname{Im}(\nu) \le \frac{1}{2}.$$

Let ν_1 and ν_2 be the two characteristic exponents. There are only three possibilities for the values of Floquet multipliers and the corresponding characteristic exponents.

Case I: $|\operatorname{tr}(M)| < 2$.

It follows from the quadratic equation (2.1.6) that $\mu_1, \mu_2 \in \mathbb{C}$ and since $\mu_2 = \bar{\mu}_1$, then $|\mu_1| = |\mu_2| = 1$. Let $k \in [0, \frac{1}{2}]$ and define

$$\mu_1 = e^{2\pi i k}, \quad \mu_2 = e^{-2\pi i k}.$$

From the spectral mapping (2.1.7), we recognize that $\nu_1 = ik$ and $\nu_2 = -ik$. Since the eigenvalues of Q are distinct, matrix Q is diagonalizable with a similarity transformation

$$S^{-1}QS = \begin{bmatrix} ik & 0 \\ 0 & -ik \end{bmatrix}$$

involving an invertible matrix S. The fundamental matrix $\Phi(x)$ is self-similar to

$$S^{-1}\Phi(x)S = S^{-1}P(x)Se^{xS^{-1}QS} = \begin{bmatrix} e^{ikx}u_{11}(x) & e^{-ikx}u_{12}(x) \\ e^{ikx}u_{21}(x) & e^{-ikx}u_{22}(x) \end{bmatrix},$$

where $u_{jl}(x)$ for $1 \le j, l \le 2$ are complex-valued 2π-periodic functions. Since components of the first row of matrix $\Phi(x)S$ give two particular solutions of the scalar equation (2.1.3) and since the equation has real-valued coefficients, the two solutions have the form

$$u(x) = e^{ikx}w(x), \quad \bar{u}(x) = e^{-ikx}\bar{w}(x),$$

where $w(x) = w(x + 2\pi)$ for all $x \in \mathbb{R}$. We note that $u \in L^{\infty}(\mathbb{R})$ but $u \notin L^2(\mathbb{R})$ because $u(x)$ is a quasi-periodic function of $x \in \mathbb{R}$.

Case 2: $|\mathrm{tr}(M)| > 2$.

It follows from the quadratic equation (2.1.6) that $\mu_1, \mu_2 \in \mathbb{R}$ and since $\mu_1\mu_2 = 1$ we may have two situations: either $\mu_1, \mu_2 > 0$ or $\mu_1, \mu_2 < 0$. Let $\kappa > 0$ and define

$$\mu_1, \mu_2 > 0: \quad \mu_1 = e^{2\pi\kappa}, \quad \mu_2 = e^{-2\pi\kappa},$$

$$\mu_1, \mu_2 < 0: \quad \mu_1 = e^{2\pi\kappa + i\pi}, \quad \mu_2 = e^{-2\pi\kappa + i\pi}.$$

From the spectral mapping (2.1.7), we recognize that $\nu_1 = \kappa$ and $\nu_2 = -\kappa$ for $\mu_1, \mu_2 > 0$ or $\nu_1 = \kappa + \frac{i}{2}$ and $\nu_2 = -\kappa + \frac{i}{2}$ for $\mu_1, \mu_2 < 0$. In what follows, let us only consider the case $\mu_1, \mu_2 > 0$.

Since the eigenvalues of Q are distinct, matrix Q is diagonalizable with a similarity transformation

$$S^{-1}QS = \begin{bmatrix} \kappa & 0 \\ 0 & -\kappa \end{bmatrix}$$

involving an invertible matrix S. The fundamental matrix $\Phi(x)$ is self-similar to

$$S^{-1}\Phi(x)S = S^{-1}P(x)Se^{xS^{-1}QS} = \begin{bmatrix} e^{\kappa x}u_{11}(x) & e^{-\kappa x}u_{12}(x) \\ e^{\kappa x}u_{21}(x) & e^{-\kappa x}u_{22}(x) \end{bmatrix},$$

where $u_{jl}(x)$ for $1 \leq j, l \leq 2$ are real-valued 2π-periodic functions. Two solutions of the scalar equation (2.1.3) now take the form

$$u_+(x) = e^{\kappa x}w_+(x), \quad u_-(x) = e^{-\kappa x}w_-(x),$$

where $w_\pm(x)$ are real-valued 2π-periodic functions. Note that $u_\pm \notin L^\infty(\mathbb{R})$ because $e^{\pm \kappa x}$ grows exponentially either as $x \to \infty$ or as $x \to -\infty$.

If $\mu_1, \mu_2 < 0$ needs to be considered, we can use the previous construction with the 2π-antiperiodic functions $u_{jl}(x)$ and $w_\pm(x)$.

Case 3: $|\mathrm{tr}(M)| = 2$.

It follows from the quadratic equation (2.1.6) that either $\mu_1 = \mu_2 = 1$ if $\mathrm{tr}(M) = 2$ or $\mu_1 = \mu_2 = -1$ if $\mathrm{tr}(M) = -2$. Therefore, we have a double Floquet multiplier, which can be obtained after a coalescence of two simple multipliers of Case 2 as $\kappa \to 0$. The two characteristic exponents are repeated too, either at $\nu_1 = \nu_2 = 0$ for $\mu_1 = \mu_2 = 1$ or at $\nu_1 = \nu_2 = \frac{i}{2}$ for $\mu_1 = \mu_2 = -1$.

Since the eigenvalues of Q are now repeated, matrix Q may not be diagonalizable. This leads to the branching of Case 3 in the following two subcases. Again, we restrict our attention to the case $\mu_1 = \mu_2 = 1$.

Case 3a: Q is diagonalizable.

There exist two 2π-periodic linearly independent solutions $u_1(x)$ and $u_2(x)$ of the scalar equation (2.1.3).

Case 3b: Q is not diagonalizable.

There exists only one 2π-periodic solution $u_1(x)$ of the scalar equation (2.1.3).

Exercise 2.1 For Case 3b, assume that there exists an invertible matrix S, which transforms Q to the Jordan block

$$S^{-1}QS = \begin{bmatrix} 0 & 1 \\ 0 & 0 \end{bmatrix}.$$

Show that the second solution of equation (2.1.3) grows linearly in x on $\mathbb{R}$.

Exercise 2.2 Assume that $V(x)$ is a smooth confining potential near $x = 0$ and consider a smooth T-periodic solution $x(t; E)$ of the nonlinear oscillator equation

$$\ddot{x} + V'(x) = 0 \quad \Rightarrow \quad E = \frac{1}{2}\dot{x}^2 + V(x) = \text{const in } t \in \mathbb{R},$$

such that $x(0; E) = a(E) > 0$ and $\dot{x}(0; E) = 0$, where $a(E)$ is found from $V(a) = E$ with $V'(a) > 0$. Compute the monodromy matrix M associated with the Schrödinger operator with T-periodic potential

$$L = -\partial_t^2 - V''(x(t))$$

for $\lambda = 0$ and show that Case 3a occurs if $T'(E) = 0$ and Case 3b occurs if $T'(E) \neq 0$.

The three cases above occur for different values of the spectral parameter λ in equation (2.1.3). It is now time to give a precise meaning of the spectrum of operator L in the case when the 2π-periodic potential V is restricted on $[0, 2\pi]$.

Let $u(x) = e^{ikx}w(x)$ for any fixed $k \in \mathbb{T}_+ := \left[0, \frac{1}{2}\right]$ and rewrite the differential equation (2.1.3) as a regular Sturm–Liouville spectral problem,

$$\begin{cases} -w''(x) - 2ikw'(x) + k^2 w(x) + V(x)w(x) = \lambda w(x), & x \in (0, 2\pi), \\ w(2\pi) = w(0), \quad w'(2\pi) = w'(0). \end{cases} \tag{2.1.8}$$

The operator

$$L_k := e^{-ikx} L e^{ikx} \equiv -\partial_x^2 - 2ik\partial_x + k^2 + V(x)$$

is self-adjoint in $L^2(\mathbb{R})$ since a multiplication by e^{ikx} is a unitary operator. When L_k is restricted on the compact interval $[0, 2\pi]$ subject to the periodic boundary conditions, its spectrum is purely discrete. In other words, there exists a countable infinite set of eigenvalues $\{E_n(k)\}_{n\in\mathbb{N}}$ for each fixed $k \in \mathbb{T}_+$, which can be ordered as

$$E_1(k) \leq E_2(k) \leq E_3(k) \leq \dots.$$

If $w(x)$ is a solution of the second-order differential equation (2.1.8), the second solution of the same equation is $e^{-2ikx}\bar{w}(x)$ and it does not satisfy the periodic boundary conditions unless $k = 0$ or $k = \frac{1}{2}$. Therefore, the equality in the ordering of the eigenvalues may only happen if $k = 0$ or $k = \frac{1}{2}$, while the strict inequality holds for $k \in \left(0, \frac{1}{2}\right)$.

If $V \in L^\infty(\mathbb{R})$, then there exist constants $C_+ \geq C_- > 0$ such that eigenvalues $\{E_n(k)\}_{n\in\mathbb{N}}$ satisfy a uniform asymptotic distribution (Theorem 4.2.3 in [52])

$$C_- n^2 \leq |E_n(k)| \leq C_+ n^2, \quad n \in \mathbb{N}, \quad k \in \mathbb{T}_+. \tag{2.1.9}$$

Eigenvalues $E_n(k)$ for $k = 0$ and $k = \frac{1}{2}$ play an important role in the spectral analysis of operator L in $L^2(\mathbb{R})$. Let us denote the corresponding eigenvalues by $E_n^+ = E_n(0)$ and $E_n^- = E_n\left(\frac{1}{2}\right)$. The corresponding eigenfunctions of operator L are 2π-periodic for E_n^+ and 2π-antiperiodic for E_n^-. Both E_n^+ and E_n^- are eigenvalues of the same operator L restricted on $[-2\pi, 2\pi]$ with the 4π-periodic boundary conditions. Since there are at most two eigenfunctions of a second-order self-adjoint operator for the same eigenvalue and the 2π-periodic and 2π-antiperiodic eigenfunctions may not coexist for the same eigenvalue, we conclude that the eigenvalues $\{E_n^+, E_n^-\}_{n\in\mathbb{N}}$ can be sorted as follows:

$$E_1^+ < E_1^- \leq E_2^- < E_2^+ \leq E_3^+ < E_3^- \leq E_4^- < E_4^+ \leq E_5^+ < \dots. \qquad (2.1.10)$$

The equality sign corresponds to Case 3a, when Q is diagonalizable, while the strict inequality corresponds to Case 3b, when Q is not diagonalizable.

Exercise 2.3 Set $V \equiv 0$ and show that $E_1^+ = 0$ and

$$E_{2n}^+ = E_{2n+1}^+ = n^2, \quad E_{2n-1}^- = E_{2n}^- = \frac{(2n-1)^2}{4}, \quad n \in \mathbb{N}.$$

In other words, only Case 3a occurs if $V \equiv 0$.

By the Spectral Theorem (Appendix B.11), the set of eigenfunctions of L_k forms an orthogonal basis in $L^2_{\text{per}}([0, 2\pi])$. Let us denote the set of orthogonal eigenfunctions of L_k for a fixed $k \in \mathbb{T}_+$ by $\{w_n(x; k)\}_{n\in\mathbb{N}}$. If the amplitude factors of the eigenfunctions are normalized by their L^2 norms, the orthogonal eigenfunctions satisfy

$$\langle w_n(\cdot; k), w_{n'}(\cdot; k)\rangle_{L^2_{\text{per}}} := \int_0^{2\pi} w_n(x; k)\bar{w}_{n'}(x; k)dx = \delta_{n,n'}, \qquad (2.1.11)$$

where $n, n' \in \mathbb{N}$ and $\delta_{n,n'}$ is the Kronecker symbol. Since the set $\{w_n(\cdot; k)\}_{n\in\mathbb{N}}$ is a basis in $L^2_{\text{per}}([0, 2\pi])$, there exists a unique set of coefficients $\{\phi_n\}_{n\in\mathbb{N}}$ in the decomposition

$$\forall \phi \in L^2_{\text{per}}([0, 2\pi]) : \qquad \phi(x) = \sum_{n\in\mathbb{N}} \phi_n w_n(x; k), \qquad (2.1.12)$$

given by

$$\phi_n = \langle \phi, w_n(\cdot; k)\rangle_{L^2_{\text{per}}}, \quad n \in \mathbb{N}.$$

Substituting ϕ_n back into (2.1.12), we find that the set of eigenfunctions $\{w_n(x; k)\}_{n\in\mathbb{N}}$ must satisfy the completeness relation in $L^2_{\text{per}}([0, 2\pi])$:

$$\sum_{n\in\mathbb{N}} w_n(x; k)\bar{w}_n(y; k) = \delta(x - y), \quad x, y \in [0, 2\pi], \qquad (2.1.13)$$

where $\delta(x - y)$ is the Dirac delta function in the distribution sense. Since $w_n(x; k)$ is periodic in x, the completeness relation (2.1.13) can be extended on the real axis by

$$\sum_{n\in\mathbb{N}} w_n(x;k)\bar{w}_n(y;k) = \sum_{m\in\mathbb{Z}} \delta(x+2\pi m - y), \quad x,y\in\mathbb{R}. \tag{2.1.14}$$

We shall use symbol $\boldsymbol{\phi}$ to denote the vector of elements of the sequence $\{\phi_n\}_{n\in\mathbb{N}}$. The decomposition (2.1.12) with the orthogonality relations (2.1.11) immediately gives the Parseval equality

$$\|\phi\|_{L^2_{\mathrm{per}}([0,2\pi])} = \|\boldsymbol{\phi}\|_{l^2(\mathbb{N})}.$$

The series of periodic eigenfunctions (2.1.12) can be useful to replace an element of the function space (e.g. $H^k_{\mathrm{per}}([0,2\pi]) \subset L^2_{\mathrm{per}}([0,2\pi])$, $k\in\mathbb{N}$) with an element of another function space (e.g. $l^2_k(\mathbb{N}) \subset l^2(\mathbb{N})$, $k\in\mathbb{N}$) and to reformulate the problem from one coordinate representation to another coordinate representation. In other words, if $\phi\in L^2_{\mathrm{per}}([0,2\pi])$ satisfies additional restrictions, we can still work with the series of eigenfunctions (2.1.12) but add some constraints on $\boldsymbol{\phi}\in l^2(\mathbb{N})$. As a result, an underlying problem for $\phi(x)$ can be reformulated as a new problem for $\{\phi_n\}_{n\in\mathbb{N}}$. For applications to differential equations, we need to work with $\phi\in C^r_{\mathrm{per}}([0,2\pi])$ for an integer $r\geq 0$.

The vector $\boldsymbol{\phi}$ can be considered alternatively in two weighted spaces $l^1_s(\mathbb{N})$ or $l^2_s(\mathbb{N})$ for any $s\geq 0$. Although l^1 spaces are used throughout this chapter, some results extend easily to l^2 spaces and will be treated in exercises.

Lemma 2.1 *Fix an integer $r\geq 0$ and define $\phi(x)$ by the decomposition (2.1.12). If $\boldsymbol{\phi}\in l^1_s(\mathbb{N})$ for any integer $s > r + \frac{1}{2}$, then $\phi\in C^r_{\mathrm{per}}([0,2\pi])$.*

Proof By the triangle inequality, we obtain

$$\|\phi\|_{C^r_{\mathrm{per}}} \leq \sum_{n\in\mathbb{N}} |\phi_n| \|w_n(\cdot;k)\|_{C^r_{\mathrm{per}}}.$$

By the Sobolev Embedding Theorem (Appendix B.10), there exists a constant $C_{s,r} > 0$ such that

$$\|w_n(\cdot;k)\|_{C^r_{\mathrm{per}}} \leq C_{s,r}\|w_n(\cdot;k)\|_{H^s_{\mathrm{per}}}, \quad s > r + \frac{1}{2}, \quad n\in\mathbb{N}, \quad k\in\mathbb{T}_+.$$

Since V is bounded in the supremum norm, fix a constant C_V such that

$$C_V \geq 1 + \|V\|_{L^\infty([0,2\pi])}.$$

For any $f\in L^2_{\mathrm{per}}([0,2\pi])$, there exists a constant $C > 0$ such that

$$\langle (C_V + L_k)f, f\rangle_{L^2_{\mathrm{per}}} \geq \left(C_V - \|V\|_{L^\infty_{\mathrm{per}}}\right)\|g\|^2_{L^2_{\mathrm{per}}} + \|\partial_x g\|^2_{L^2_{\mathrm{per}}} \geq \|g\|^2_{H^1_{\mathrm{per}}},$$

where $g(x) = e^{ikx}f(x)$. Continuing this estimate for an integer $s\geq 0$, we obtain

$$\|u_n(\cdot;k)\|^2_{H^s_{\mathrm{per}}} \leq \langle (C_V + L_k)^s w_n(\cdot;k), w_n(\cdot;k)\rangle_{L^2_{\mathrm{per}}} = (C_V + E_n(k))^s,$$

where $u_n(x;k) = e^{ikx}w_n(x;k)$ and the normalization condition (2.1.11) has been used. By the asymptotic distribution of eigenvalues (2.1.9), we finally obtain

$$\|\phi\|_{C^r_{\text{per}}} \leq C \sum_{n \in \mathbb{N}} |\phi_n| \|w_n(\cdot;k)\|_{H^s_{\text{per}}}$$

$$\leq C' \sum_{n \in \mathbb{N}} |\phi_n| \|u_n(\cdot;k)\|_{H^s_{\text{per}}}$$

$$\leq C'' \sum_{n \in \mathbb{N}} (1 + n^2)^{s/2} |\phi_n|$$

for some constants $C, C', C'' > 0$ and any integer $s > r + \frac{1}{2}$. $\qquad\square$

Remark 2.1   The statements of Lemma 2.1 can be extended for non-integer s and r if the definitions of the norms in H^s and C^r with fractional derivatives are used and a spectral family of operators $(C_V + L_k)^s$ is considered.

Exercise 2.4   Prove that if $\phi \in l^2_s(\mathbb{N})$ for an integer $s > r + \frac{1}{2}$, then $\phi \in C^r_{\text{per}}([0, 2\pi])$ for any given integer $r \geq 0$.

We finish this section with two explicit examples of the periodic potential $V(x)$. In the first example, $V(x)$ is a piecewise-constant 2π-periodic function

$$V(x) = \begin{cases} b, & x \in (0, \pi), \\ 0, & x \in (\pi, 2\pi). \end{cases} \qquad (2.1.15)$$

The scalar equation (2.1.3) can be solved explicitly by

$$u(x) = \begin{cases} u(0) \cosh\sqrt{b - \lambda}\,x + \frac{u'(0)}{\sqrt{b-\lambda}} \sinh\sqrt{b - \lambda}\,x, & x \in [0, \pi], \\ u(2\pi) \cos\sqrt{\lambda}(x - 2\pi) + \frac{u'(2\pi)}{\sqrt{\lambda}} \sin\sqrt{\lambda}(x - 2\pi), & x \in [\pi, 2\pi]. \end{cases}$$

for any $0 < \lambda < b$. Continuity of $u(x)$ and $u'(x)$ across the jump point $x = \pi$ leads to the monodromy matrix with the trace

$$\text{tr}(M) = 2\cosh(\pi\sqrt{b - \lambda})\cos(\pi\sqrt{\lambda}) + \frac{b - 2\lambda}{\sqrt{\lambda(b - \lambda)}} \sinh(\pi\sqrt{b - \lambda})\sin(\pi\sqrt{\lambda}).$$

This equation is valid for $0 < \lambda < b$ and it is analytically extended for $\lambda > b$ to the equation

$$\text{tr}(M) = 2\cos(\pi\sqrt{\lambda - b})\cos(\pi\sqrt{\lambda}) + \frac{b - 2\lambda}{\sqrt{\lambda(\lambda - b)}} \sin(\pi\sqrt{\lambda - b})\sin(\pi\sqrt{\lambda}).$$

Figure 2.1 shows a typical behavior of $\text{tr}(M)$ versus λ (top) and the eigenvalues $\{E_n(k)\}_{n \in \mathbb{N}}$ versus k (bottom) for the piecewise-constant potential (2.1.15) with $b = 2$.

Figure 2.2 shows the first three eigenfunctions $\{u_n^+(x)\}_{n \in \mathbb{N}}$ for $k = 0$ (top) and $\{u_n^-(x)\}_{n \in \mathbb{N}}$ for $k = \frac{1}{2}$ (bottom) for the shifted potential $V(x - \frac{3\pi}{2})$, for which 0 is a center of symmetry in the middle of the potential well.

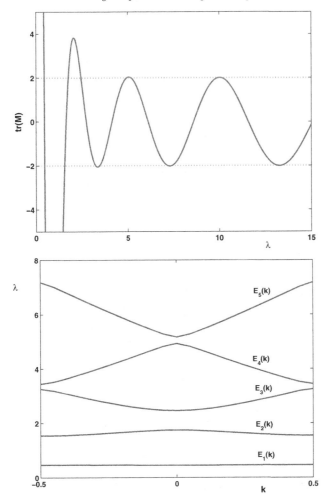

Figure 2.1 Top: schematic representation of the behavior of $\mathrm{tr}(M)$ versus λ. Bottom: typical behavior of the eigenvalues $\{E_n(k)\}_{n\in\mathbb{N}}$ versus k.

In the second example, $V(x)$ is a smooth 2π-periodic function

$$V(x) = V_0 \sin^2\left(\frac{x}{2}\right). \tag{2.1.16}$$

Figure 2.3 shows the dependence of the first five spectral bands between $\{E_n^+\}_{n\in\mathbb{N}}$ and $\{E_n^-\}_{n\in\mathbb{N}}$ versus V_0 for the smooth potential (2.1.16).

Exercise 2.5   Let $V = 2\epsilon V_n \cos(nx)$ for a fixed $n \in \mathbb{N}$ and a small $\epsilon > 0$. Compute the asymptotic dependence of (E_n^+, E_{n+1}^+) for even n and (E_n^-, E_{n+1}^-) for odd n on ϵ up to the terms of $\mathcal{O}(\epsilon^2)$.

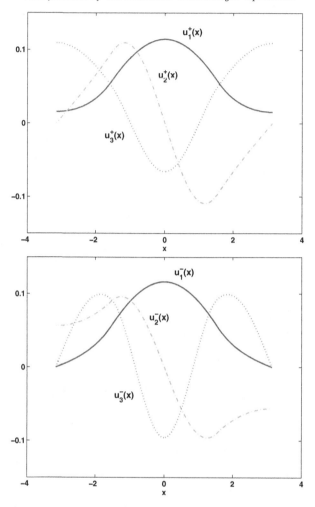

Figure 2.2 The first three eigenfunctions $\{u_n^+(x)\}_{n\in\mathbb{N}}$ (top) and $\{u_n^-(x)\}_{n\in\mathbb{N}}$ (bottom).

2.1.2 Bloch theory and the integral transform of quasi-periodic eigenfunctions

We can now reiterate on the question about the spectrum of operator $L = -\partial_x^2 + V(x)$ in $L^2(\mathbb{R})$. To answer a similar question in $L^2_{\mathrm{per}}([0, 2\pi])$, we have introduced the set of orthogonal 2π-periodic eigenfunctions $\{w_n(x; k)\}_{n\in\mathbb{N}}$ for the set of eigenvalues $\{E_n(k)\}_{n\in\mathbb{N}}$ of the self-adjoint operator $L_k = e^{-ikx}Le^{ikx}$ for a fixed $k \in \mathbb{T}_+ \equiv [0, \frac{1}{2}]$. Let us now continue these eigenfunctions on the closed interval $\mathbb{T}_+$.

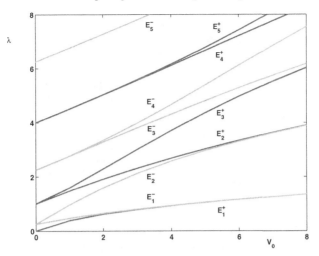

Figure 2.3 The boundaries of the first five spectral bands versus V_0.

Setting $u_n(x;k) = e^{ikx}w_n(x;k)$, we obtain a quasi-periodic solution of $Lu = \lambda u$ for $\lambda = E_n(k)$ that satisfies

$$u_n(x + 2\pi; k) = e^{2\pi ik}u_n(x;k), \quad x \in \mathbb{R}, \ k \in \mathbb{T}_+, \ n \in \mathbb{N}.$$

In many applications of solid-state physics, eigenfunctions $u_n(x;k)$ are referred to as the *Bloch wave functions*, while the parameter k is referred to as the *quasi-momentum*.

For any fixed $k \in \mathbb{T}_+$, $u_n(\cdot;k)$ belongs to $L^\infty(\mathbb{R})$. On the other hand, solutions $u_\pm(x)$ of $Lu = \lambda u$ for $\lambda \notin \cup_{n \in \mathbb{N}} \text{range}_{k \in \mathbb{T}}[E_n(k)]$ are unbounded on $\mathbb{R}$ and thus $u_\pm \notin L^\infty(\mathbb{R})$. Therefore, it should be no surprise to know that the spectrum of $L = -\partial_x^2 + V(x)$ in $L^2(\mathbb{R})$ is purely continuous and consists of the union of the intervals in the range of functions $\{E_n(k)\}_{n \in \mathbb{N}}$.

Definition 2.1 Let M be the monodromy matrix associated with the system

$$\frac{d}{dx}\begin{bmatrix} u \\ v \end{bmatrix} = \begin{bmatrix} 0 & 1 \\ V(x) - \lambda & 0 \end{bmatrix}\begin{bmatrix} u \\ v \end{bmatrix}.$$

We say that λ belongs to the *spectral band* of operator $L = -\partial_x^2 + V(x)$ if $|\text{tr}(M)| \leq 2$ and to the *spectral gap* if $|\text{tr}(M)| > 2$. The point of λ for which $|\text{tr}(M)| = 2$ is called a *band edge*.

The nth spectral band of operator L in $L^2(\mathbb{R})$ is hence defined by

$$\text{range}_{k \in \mathbb{T}_+}[E_n(k)] \subset \mathbb{R},$$

whereas the interval

$$\left(\sup_{k\in\mathbb{T}_+}[E_n(k)], \inf_{k\in\mathbb{T}_+}[E_{n+1}(k)]\right),$$

if it is non-empty, defines the spectral gap between the nth and $(n+1)$th spectral bands. The spectral gap is non-empty in Case 3b, when $E_n^\pm < E_{n+1}^\pm$, and it is empty in Case 3a, when $E_n^\pm = E_{n+1}^\pm$, where $n \in \mathbb{N}$. In other words, the spectral bands are *disjoint* in Case 3b and *overlap* at the point $E_n^\pm = E_{n+1}^\pm$ in Case 3a.

Exercise 2.6 Show that $\text{tr}(M)$ is a C^1 function of λ that crosses with a nonzero slope the values ± 2 in Case 3b and touches with a zero slope the values ± 2 in Case 3a.

All eigenfunctions of L can be defined for values of k taken in the interval $\mathbb{T}_+ \equiv \left[0, \frac{1}{2}\right]$, but we would then need two eigenfunctions $u_n(x;k)$ and $\bar{u}_n(x;k)$. To use the same eigenfunction $u_n(x;k)$ in both cases, we adopt a technical trick and extend the interval for k to $\mathbb{T} := \left[-\frac{1}{2}, \frac{1}{2}\right]$. The interval $\mathbb{T}$ is referred to as the *Brillouin zone* in the Bloch theory. Since $u_n(x;k)$, $\bar{u}_n(x;k)$, and $u_n(x;-k)$ solve the same equation $Lu = \lambda u$ for $\lambda = E_n(k)$, it is clear that one solution is a linear combination of the other two solutions. Therefore, it would make sense to normalize uniquely the phase factors of the eigenfunctions by the constraint

$$\bar{u}_n(x;k) = u_n(x;-k), \quad n \in \mathbb{N}, \ k \in \mathbb{T}, \ x \in \mathbb{R}, \qquad (2.1.17)$$

which implies the reflection

$$E_n(-k) = E_n(k), \quad n \in \mathbb{N}, \ k \in \mathbb{T}. \qquad (2.1.18)$$

We shall now list useful properties of functions $E_n(k)$ in $\mathbb{T}$ for a fixed $n \in \mathbb{N}$.

- By Theorem XIII.89 in [176], $E_n(k)$ is analytic in k on $\mathbb{T}\setminus\left\{-\frac{1}{2}, 0, \frac{1}{2}\right\}$ and continuous at the points $k \in \left\{-\frac{1}{2}, 0, \frac{1}{2}\right\}$.
- By Theorem XIII.90 in [176], the extremal values of $E_n(k)$ may only occur at the points $k = 0$ and $k = \pm\frac{1}{2}$, that is at the values E_n^+ and E_n^- respectively.
- By Theorem XIII.95 in [176], $E_n(k)$ is extended to an analytic function of k on $\mathbb{T}$ if the nth spectral band is disjoint from the adjacent spectral bands.

The reason why we describe properties of functions $E_n(k)$ and $u_n(x;k)$ in so much detail lies in our ultimate goal to use the eigenfunctions of L as an orthogonal basis in $L^2(\mathbb{R})$. By Theorems XIII.97 and XIII.98 in [176], the orthogonal eigenfunctions satisfy

$$\langle u_n(\cdot;k), u_{n'}(\cdot;k')\rangle_{L^2} := \int_\mathbb{R} u_n(x;k)\bar{u}_{n'}(x;k')dx = \delta_{n,n'}\delta(k-k'), \qquad (2.1.19)$$

where $n, n' \in \mathbb{N}$ and $k, k' \in \mathbb{T}$. Completeness of the Bloch functions follows from the completeness relation (2.1.14) by explicit computations

$$\int_{\mathbb{T}} \sum_{n \in \mathbb{N}} u_n(x; k) \bar{u}_n(y; k) dk = \int_{\mathbb{T}} e^{ik(x-y)} \left(\sum_{n \in \mathbb{N}} w_n(x; k) \bar{w}_n(y; k) \right) dk$$

$$= \int_{\mathbb{T}} e^{ik(x-y)} \left(\sum_{m \in \mathbb{Z}} \delta(x + 2\pi m - y) \right) dk$$

$$= \sum_{m \in \mathbb{Z}} \delta(x + 2\pi m - y) \int_{\mathbb{T}} e^{-i2\pi mk} dk$$

$$= \delta(x - y), \quad x, y \in \mathbb{R}. \tag{2.1.20}$$

Let us define a unitary transformation from $L^2(\mathbb{R})$ to $L^2(\mathbb{T}, l^2(\mathbb{N}))$ by

$$\hat{\phi}_n(k) = \int_{\mathbb{R}} \phi(y) \bar{u}_n(y; k) dy, \quad n \in \mathbb{N}, \quad k \in \mathbb{T}. \tag{2.1.21}$$

The inverse transformation referred to as the *Bloch decomposition* is

$$\forall \phi \in L^2(\mathbb{R}): \quad \phi(x) = \int_{\mathbb{T}} \sum_{n \in \mathbb{N}} \hat{\phi}_n(k) u_n(x; k) dk, \quad x \in \mathbb{R}. \tag{2.1.22}$$

By explicit computations, we obtain the Parseval equality,

$$\|\phi\|_{L^2}^2 = \int_{\mathbb{T}} \int_{\mathbb{T}} \sum_{n \in \mathbb{N}} \sum_{n' \in \mathbb{N}} \hat{\phi}_n(k) \bar{\hat{\phi}}_{n'}(k') \langle u_n(\cdot; k), u_{n'}(\cdot; k') \rangle_{L^2} dk dk'$$

$$= \int_{\mathbb{T}} \sum_{n \in \mathbb{N}} |\hat{\phi}_n(k)|^2 dk =: \|\hat{\phi}\|_{L^2(\mathbb{T}, l^2(\mathbb{N}))}^2.$$

According to the Bloch decomposition (2.1.22), space $L^2(\mathbb{R})$ is decomposed into a direct sum of invariant closed bounded subspaces associated with the spectral bands in the spectrum of operator L. Let $\mathcal{E}_n$ for a fixed $n \in \mathbb{N}$ be an invariant closed subspace of $L^2(\mathbb{R})$ associated with the nth spectral band. Then,

$$\forall \phi \in \mathcal{E}_n \subset L^2(\mathbb{R}): \quad \phi(x) = \int_{\mathbb{T}} \hat{\phi}_n(k) u_n(x; k) dk, \tag{2.1.23}$$

where $\hat{\phi}_n(k)$ is defined by (2.1.21).

Similarly to what we have done in the case of $L^2_{\text{per}}([0, 2\pi])$, we can now work with the Bloch decomposition in some restrictions of $L^2(\mathbb{R})$. Since $\mathbb{T}$ is compact, we do not need a weight under the integration sign and can work either in space $L^1(\mathbb{T}, l^1_s(\mathbb{N}))$ or in space $L^2(\mathbb{T}, l^2_s(\mathbb{N}))$. A natural definition of the norm in $L^1(\mathbb{T}, l^1_s(\mathbb{N}))$ is

$$\|\hat{\phi}\|_{L^1(\mathbb{T}, l^1_s(\mathbb{N}))} := \int_{\mathbb{T}} \sum_{n \in \mathbb{N}} (1 + n^2)^{s/2} |\hat{\phi}_n(k)| dk, \quad s \geq 0.$$

The following lemma gives an analogue of Lemma 2.1.

Lemma 2.2 *Fix an integer $r \geq 0$ and define $\phi(x)$ by the Bloch decomposition (2.1.22). If $\hat{\phi} \in L^1(\mathbb{T}, l^1_s(\mathbb{N}))$ for an integer $s > r + \frac{1}{2}$, then $\phi \in C^r_b(\mathbb{R})$ and $\phi(x), ..., \phi^{(r)}(x) \to 0$ as $|x| \to \infty$.*

Proof   The proof is similar to that of Lemma 2.1 since the asymptotic bound (2.1.9) is uniform for all $k \in \mathbb{T}$. Therefore,

$$\|\phi\|_{C_b^r} \leq \int_{\mathbb{T}} \sum_{n \in \mathbb{N}} |\hat{\phi}_n(k)| \|u_n(\cdot; k)\|_{C_{\text{per}}^r} dk$$

$$\leq C \int_{\mathbb{T}} \sum_{n \in \mathbb{N}} |\hat{\phi}_n(k)| \|u_n(\cdot; k)\|_{H_{\text{per}}^s} dk$$

$$\leq C' \int_{\mathbb{T}} \sum_{n \in \mathbb{N}} |\hat{\phi}_n(k)| (C_V + E_n(k))^{s/2} dk$$

$$\leq C'' \|\hat{\phi}\|_{L^1(\mathbb{T}, l_s^1(\mathbb{N}))},$$

for some constants $C, C', C'' > 0$, $C_V \geq 1 + \|V\|_{L_{\text{per}}^\infty}$, and any integer $s > r + \frac{1}{2}$. The decay of $\phi(x), ..., \phi^{(r)}(x)$ to zero as $|x| \to \infty$ follows from the Riemann–Lebesgue Lemma applied to the Bloch decomposition, after the integrals on $k \in \mathbb{T}$ and the summation on $n \in \mathbb{N}$ are rewritten as a Fourier-type integral

$$\phi(x) = \int_{\mathbb{R}} \hat{\phi}(p) u(x; p) dp. \tag{2.1.24}$$

Here $\hat{\phi}(p)$ and $u(x; p)$ for $p \in \left[-\frac{n}{2}, -\frac{n-1}{2}\right] \cup \left[\frac{n-1}{2}, \frac{n}{2}\right]$ are equal to $\hat{\phi}_n(k)$ and $\hat{u}_n(x; k)$ for $k \in \mathbb{T}$. Since $u(x; p)$ is uniformly bounded in $C_b^0(\mathbb{R} \times \mathbb{R})$ with respect to both x and p and $\hat{\phi}(p) \in L_s^1(\mathbb{R})$ with $s > r + \frac{1}{2}$, the Riemann–Lebesgue Lemma (Appendix B.10) applies to the Fourier-type integral (2.1.24) and gives the decay of $\phi(x), ..., \phi^{(r)}(x)$ to zero as $|x| \to \infty$. □

Exercise 2.7   Prove that if $\hat{\phi} \in L^2(\mathbb{T}, l_s^2(\mathbb{N}))$ for an integer $s > r + \frac{1}{2}$, then $\phi \in C_b^r(\mathbb{R})$ and $\phi(x), ..., \phi^{(r)}(x) \to 0$ as $|x| \to \infty$ for any given integer $r \geq 0$. Use the property

$$\forall \phi \in \text{Dom}(L): \quad \|\phi\|_{H^s} \leq \|(C_V + L)^{s/2} \phi\|_{L^2}$$

for some $C_V \geq 1 + \|V\|_{L^\infty}$ and any integer $s \geq 0$.

The Bloch decomposition can be used for the representation of solutions of partial differential equations with space–periodic coefficients. In the same context, a different representation that also involves the Fourier-type integrals has been used in the literature, e.g. by Eckmann & Schneider [53] and Busch *et al.* [24]. For the sake of completeness, we shall finish this section by formulating the main ingredients of this approach. Setting $u_n(x; k) = e^{ikx} w_n(x; k)$ and denoting

$$\tilde{\phi}(x; k) = \sum_{n \in \mathbb{N}} \hat{\phi}_n(k) w_n(x; k), \quad x \in \mathbb{R}, \quad k \in \mathbb{T}, \tag{2.1.25}$$

we rewrite the decomposition formula (2.1.22) in a compact form,

$$\forall \phi \in L^2(\mathbb{R}): \quad \phi(x) = \int_{\mathbb{T}} \tilde{\phi}(x; k) e^{ikx} dk, \quad x \in \mathbb{R}. \tag{2.1.26}$$

We recognize that $\tilde{\phi}(x; k)$ is represented by the series of 2π-periodic eigenfunctions $\{w_n(x; k)\}_{n \in \mathbb{N}}$ for any fixed $k \in \mathbb{T}$, which provide an orthogonal basis in

$L^2_{\mathrm{per}}([0, 2\pi])$. Therefore,

$$\tilde{\phi}(x + 2\pi; k) = \tilde{\phi}(x; k), \quad x \in \mathbb{R}, \quad k \in \mathbb{T}.$$

Using the orthogonality relation (2.1.11), we can invert the representation as

$$\hat{\phi}_n(k) = \langle \tilde{\phi}(\cdot; k), w_n(\cdot; k) \rangle_{2\pi} = \int_0^{2\pi} \tilde{\phi}(y; k) \bar{w}_n(y; k) dy.$$

On the other hand, using the definition of $\hat{\phi}_n(k)$ in (2.1.21) and the completeness relation (2.1.14), we obtain

$$\begin{aligned} \tilde{\phi}(x; k) &= \sum_{n \in \mathbb{N}} \int_{\mathbb{R}} \phi(y) e^{-iky} w_n(x; k) \bar{w}_n(y; k) dy \\ &= \int_{\mathbb{R}} \phi(y) e^{-iky} \left(\sum_{n \in \mathbb{N}} w_n(x; k) \bar{w}_n(y; k) \right) dy \\ &= \int_{\mathbb{R}} \phi(y) e^{-iky} \left(\sum_{m \in \mathbb{Z}} \delta(x + 2\pi m - y) \right) dy \\ &= e^{-ikx} \sum_{m \in \mathbb{Z}} \phi(x + 2\pi m) e^{-2\pi i m k}, \quad x \in \mathbb{R}. \end{aligned} \tag{2.1.27}$$

Substitution of (2.1.27) back into (2.1.26) recovers $\phi(x)$. Transformations between $\phi(x)$ and $\tilde{\phi}(x; k)$ are not standard, so it is important to establish the precise meaning of the Bloch transform (2.1.26).

Lemma 2.3 *The Bloch transform (2.1.26) is an isomorphism between $L^2(\mathbb{R})$ for $\phi(x)$ and $L^2_{\mathrm{per}}([0, 2\pi] \times \mathbb{T})$ for $\tilde{\phi}(x; k)$ with the Parseval equality*

$$\|\phi\|^2_{L^2} = \int_{\mathbb{T}} \int_0^{2\pi} |\tilde{\phi}(x; k)|^2 dx dk =: \|\tilde{\phi}\|^2_{L^2_{\mathrm{per}}([0, 2\pi] \times \mathbb{T})}. \tag{2.1.28}$$

Proof This relation is proved with explicit computations since

$$\|\phi\|^2_{L^2} = \|\hat{\phi}\|^2_{L^2(\mathbb{T}, l^2(\mathbb{N}))} = \int_{\mathbb{T}} \sum_{n \in \mathbb{N}} |\hat{\phi}_n(k)|^2 dk = \int_{\mathbb{T}} \int_0^{2\pi} |\tilde{\phi}(x; k)|^2 dx dk,$$

where we have used the decomposition (2.1.25) and the orthogonality relations (2.1.11). □

Exercise 2.8 Prove the Parseval equality (2.1.28) in the opposite direction starting with direct representation (2.1.27) of $\tilde{\phi}(x; k)$ in terms of $\phi(x)$.

The Bloch transform (2.1.26) is directly related to the Fourier transform for functions in $L^2(\mathbb{R})$. With a slight abuse of notation, let us introduce here the standard Fourier transform by

$$\phi(x) = \int_{\mathbb{R}} \hat{\phi}(p) e^{ipx} dp, \quad \hat{\phi}(p) = \frac{1}{2\pi} \int_{\mathbb{R}} \phi(x) e^{-ipx} dx, \quad x \in \mathbb{R}, \quad p \in \mathbb{R}.$$

Using this definition and representation (2.1.27), we obtain

$$\tilde{\phi}(x;k) = \sum_{m\in\mathbb{N}} \int_{\mathbb{R}} \hat{\phi}(p) e^{\mathrm{i}(p-k)(x+2\pi m)} dp$$

$$= \int_{\mathbb{R}} \hat{\phi}(p) e^{\mathrm{i}(p-k)x} \left(\sum_{m\in\mathbb{N}} e^{\mathrm{i}2\pi m(p-k)} \right) dp$$

$$= \int_{\mathbb{R}} \hat{\phi}(p) e^{\mathrm{i}(p-k)x} \left(\sum_{n\in\mathbb{Z}} \delta(p-k-n) \right) dp$$

$$= \sum_{n\in\mathbb{Z}} \hat{\phi}(k+n) e^{\mathrm{i}kn}, \quad x\in\mathbb{R}, \ k\in\mathbb{T}.$$

Properties of the Bloch transform (2.1.26) have similarities with those of the standard Fourier transform. In particular, multiplication of two functions in x-space corresponds to the convolution operator in k-space

$$(\widetilde{uv})(x;k) := (\tilde{u}\star\tilde{v})(x;k) = \int_{\mathbb{T}} \tilde{u}(x;k-k')\tilde{v}(x;k')dk'.$$

However, the convolution integral involves values of $\tilde{u}(x;k)$ beyond the interval $\mathbb{T}$ and, therefore, it is necessary to extend definitions of the Bloch functions $u_n(x;k)$ as 1-periodic functions in k-space.

2.1.3 Wannier theory and the series of decaying functions

Recall that the characteristic exponents ν associated with the monodromy matrix M are unique in the strip $\mathrm{Im}(\nu) \in (-\frac{1}{2}, \frac{1}{2}]$. Thanks to relation (2.1.7), they are 1-periodic along the imaginary axis. Therefore, we can now extend functions $E_n(k)$ and $u_n(x;k)$ in k beyond the interval $\mathbb{T} \equiv [-\frac{1}{2}, \frac{1}{2}]$ using 1-periodic continuations of these functions. The best tool for a periodic extension of functions on a real axis is the Fourier series. Therefore, we fix $n\in\mathbb{N}$ and represent these functions by the Fourier series

$$E_n(k) = \sum_{m\in\mathbb{Z}} \hat{E}_{n,m} e^{\mathrm{i}2\pi mk}, \quad k\in\mathbb{R}, \tag{2.1.29}$$

and

$$u_n(x;k) = \sum_{m\in\mathbb{Z}} \hat{u}_{n,m}(x) e^{\mathrm{i}2\pi mk}, \quad k\in\mathbb{R}, \ x\in\mathbb{R}. \tag{2.1.30}$$

Inverting the Fourier series, we find Fourier coefficients in the form

$$\hat{E}_{n,m} = \int_{\mathbb{T}} E_n(k) e^{-\mathrm{i}2\pi mk} dk, \quad m\in\mathbb{Z}, \tag{2.1.31}$$

and

$$\hat{u}_{n,m}(x) = \int_{\mathbb{T}} u_n(x;k) e^{-\mathrm{i}2\pi mk} dk, \quad m\in\mathbb{Z}, \ x\in\mathbb{R}. \tag{2.1.32}$$

Since

$$E_n(k) = \bar{E}_n(k) = E_n(-k), \quad u_n(x;k) = \bar{u}_n(x;-k),$$

the coefficients of the Fourier series (2.1.29) satisfy properties

$$\hat{E}_{n,m} = \bar{\hat{E}}_{n,-m} = \hat{E}_{n,-m}, \quad \hat{u}_{n,m}(x) = \bar{\hat{u}}_{n,m}(x). \tag{2.1.33}$$

In particular, the functions $\hat{u}_{n,m}(x)$ are real-valued. Because of the quasi-periodicity

$$u_n(x + 2\pi; k) = u_n(x; k)e^{i2\pi k}, \quad x \in \mathbb{R}, \quad k \in \mathbb{R},$$

we obtain another property of the functions $\hat{u}_{n,m}(x)$:

$$\hat{u}_{n,m}(x) = \hat{u}_{n,m-1}(x - 2\pi) = \dots = \hat{u}_{n,0}(x - 2\pi m), \quad n \in \mathbb{N}, \quad m \in \mathbb{Z}, \tag{2.1.34}$$

which shows that all functions $\hat{u}_{n,m}(x)$ are generated from the same "mother" function $\hat{u}_{n,0}(x)$ by a spatial translation to the m-multiple of the period of $V(x)$.

Functions $\{\hat{u}_{n,m}\}_{n\in\mathbb{N},m\in\mathbb{Z}}$ satisfying properties (2.1.33) and (2.1.34) are called the *Wannier functions*. These functions are not eigenfunctions of operator $L = -\partial_x^2 + V(x)$. Nevertheless, for a fixed $n \in \mathbb{N}$, Wannier functions $\{\hat{u}_{n,m}\}_{m\in\mathbb{Z}}$ lie in the invariant subspace $\mathcal{E}_n$ associated with the nth spectral band and they satisfy a closed system of second-order differential equations

$$L\hat{u}_{n,m} = \sum_{m'\in\mathbb{Z}} \hat{E}_{n,m-m'}\hat{u}_{n,m'}, \quad m \in \mathbb{Z}. \tag{2.1.35}$$

System (2.1.35) is obtained after the Fourier series (2.1.29) and (2.1.30) are substituted into the linear problem $Lu_n(x; k) = E_n(k)u_n(x; k)$. Therefore, the set $\{\hat{u}_{n,m}\}_{n\in\mathbb{Z},m\in\mathbb{Z}}$ can be used for the same purpose of building an orthogonal basis in $L^2(\mathbb{R})$ as the set $\{u_n(x; k)\}_{n\in\mathbb{N},k\in\mathbb{T}}$.

Exercise 2.9 Assume that $u_n(0; k) > 0$ for all $k \in \mathbb{T}$ and a fixed $n \in \mathbb{N}$ and consider functions

$$g_m(x) = \int_{\mathbb{T}} \frac{u_n(x; k)}{u_n(0; k)} e^{-i2\pi km} dk, \quad m \in \mathbb{Z}, \quad x \in \mathbb{R}.$$

Prove that $g_m(2\pi m') = \delta_{m,m'}$ for all $m, m' \in \mathbb{Z}$, so that $\phi(x) = \sum_{m\in\mathbb{Z}} \phi_m g_m(x)$ solves the interpolation problem

$$\phi(2\pi m) = \phi_m, \quad m \in \mathbb{Z}.$$

Orthogonality and normalization of the Wannier functions follow from the orthogonality relation (2.1.19) for the Bloch functions by explicit computations

$$\int_{\mathbb{R}} \hat{u}_{n,m}(x)\hat{u}_{n',m'}(x)dx = \int_{\mathbb{R}}\int_{\mathbb{T}}\int_{\mathbb{T}} u_n(x; k)\bar{u}_{n'}(x; k')e^{i2\pi(k'm'-km)} dk dk' dx$$

$$= \delta_{n,n'} \int_{\mathbb{T}}\int_{\mathbb{T}} \delta(k - k')e^{i2\pi k(m'-m)} dk dk'$$

$$= \delta_{n,n'} \int_{\mathbb{T}} e^{i2\pi k(m'-m)} dk$$

$$= \delta_{n,n'}\delta_{m,m'}, \quad n, n' \in \mathbb{N}, \quad m, m' \in \mathbb{Z}.$$

Similarly, completeness of the Wannier functions follows from the completeness relation (2.1.20) for the Bloch functions by explicit computations

$$\sum_{n\in\mathbb{N}}\sum_{m\in\mathbb{Z}}\hat{u}_{n,m}(x)\hat{u}_{n,m}(y) = \sum_{n\in\mathbb{N}}\sum_{m\in\mathbb{Z}}\int_{\mathbb{T}}\int_{\mathbb{T}}u_n(x;k)\bar{u}_n(y;k)e^{i2\pi(k'-k)m}dkdk'$$

$$= \sum_{n\in\mathbb{N}}\int_{\mathbb{T}}\int_{\mathbb{T}}u_n(x;k)\bar{u}_n(y;k')\left(\sum_{m\in\mathbb{Z}}e^{i2\pi(k'-k)m}\right)dkdk'$$

$$= \sum_{n\in\mathbb{N}}\int_{\mathbb{T}}u_n(x;k)\bar{u}_n(y;k)dk$$

$$= \delta(x-y),\quad x,y\in\mathbb{R},$$

where we have used the orthogonality relation for discrete Fourier modes

$$\sum_{m\in\mathbb{Z}}e^{i2\pi(k-k')m} = \delta(k-k'),\quad k,k'\in\mathbb{T}.$$

A unitary transformation from $L^2(\mathbb{R})$ for $\phi(x)$ to $l^2(\mathbb{N}\times\mathbb{Z})$ for $\{\phi_{n,m}\}_{n\in\mathbb{N},m\in\mathbb{Z}}$ is given by

$$\phi_{n,m} = \int_{\mathbb{R}}\phi(x)\hat{u}_{n,m}(x)dx,\quad n\in\mathbb{N},\ m\in\mathbb{Z}.\tag{2.1.36}$$

The inverse transformation referred to as the *Wannier decomposition* is

$$\forall\phi\in L^2(\mathbb{R}):\quad \phi(x) = \sum_{n\in\mathbb{N}}\sum_{m\in\mathbb{Z}}\phi_{n,m}\hat{u}_{n,m}(x),\quad x\in\mathbb{R}.\tag{2.1.37}$$

The Parseval equality now reads

$$\|\phi\|_{L^2}^2 = \sum_{n\in\mathbb{N}}\sum_{m\in\mathbb{Z}}|\phi_{n,m}|^2 =: \|\phi\|_{l^2(\mathbb{N}\times\mathbb{Z})}^2,$$

where ϕ denotes a vector with elements in the double summable sequence $\{\phi_{n,m}\}_{n\in\mathbb{N},m\in\mathbb{Z}}$.

Exercise 2.10 Assume that the nth spectral band of L is disjoint from the adjacent spectral bands, so that $E_n(k)$ and $u_n(x;k)$ are analytic in k on $\mathbb{T}$. Show that there exist $\eta_n > 0$ and $C_n > 0$ for any fixed $n\in\mathbb{N}$ such that

$$|\hat{u}_{n,0}(x)| \le C_n e^{-\eta_n|x|},\quad x\in\mathbb{R}.\tag{2.1.38}$$

Figure 2.4 shows the first two Wannier functions $\hat{u}_{1,0}(x)$ (top) and $\hat{u}_{2,0}(x)$ (bottom) for the piecewise-constant potential (2.1.15) (dotted line).

Recall that $\mathcal{E}_n$ for a fixed $n\in\mathbb{N}$ denotes the invariant closed subspace of $L^2(\mathbb{R})$ associated with the nth spectral band. If $\phi\in\mathcal{E}_n$, it can be represented by the integral (2.1.23) involving Bloch functions. Although Wannier decomposition can be used to represent any element of $L^2(\mathbb{R})$ by the double sum (2.1.37), it is more practical to use the Wannier decomposition for a particular $\mathcal{E}_n\subset L^2(\mathbb{R})$. The relevant representation involves a single summation

$$\forall\phi\in\mathcal{E}_n\subset L^2(\mathbb{R}):\quad \phi(x) = \sum_{m\in\mathbb{Z}}\phi_m\hat{u}_{n,m}(x).\tag{2.1.39}$$

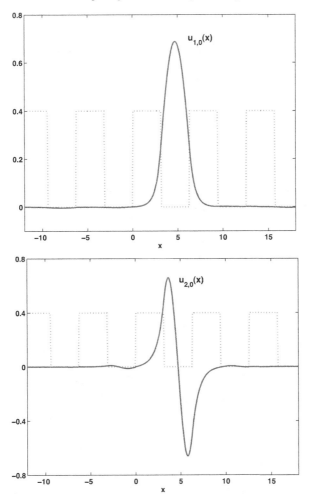

Figure 2.4 A typical example of Wannier functions $\hat{u}_{1,0}(x)$ (top) and $\hat{u}_{2,0}(x)$ (bottom) relative to the periodic potential $V(x)$ (dotted line).

With a slight abuse of notation, we shall now use $\boldsymbol{\phi}$ to denote a vector with elements in the set $\{\phi_m\}_{m\in\mathbb{Z}}$ for a representation of $\phi \in \mathcal{E}_n$. Similarly to other decompositions of this section, we shall study constraints on $\boldsymbol{\phi}$ to ensure that ϕ belongs to a subspace of $\mathcal{E}_n \subset L^2(\mathbb{R})$.

Lemma 2.4   *Let $\phi(x)$ be defined by the decomposition (2.1.39). If $\boldsymbol{\phi} \in l^1(\mathbb{Z})$, then $\phi \in H^s(\mathbb{R})$ for any integer $s \geq 0$, so that $\phi \in C_b^r(\mathbb{R})$ for any integer $r \geq 0$ and $\phi(x), ..., \phi^{(r)}(x) \to 0$ as $|x| \to \infty$.*

Proof   Since $\|\boldsymbol{\phi}\|_{l^2} \leq \|\boldsymbol{\phi}\|_{l^1}$, we have $\boldsymbol{\phi} \in l^2(\mathbb{Z})$ and, hence, if ϕ is given by (2.1.39) and $\boldsymbol{\phi} \in l^1(\mathbb{Z})$, then $\phi \in \mathcal{E}_n \subset L^2(\mathbb{R})$ for a fixed $n \in \mathbb{N}$. We use the triangle inequality and obtain

$$\|\phi\|_{H^s} \leq \sum_{m \in \mathbb{Z}} |\phi_m| \|\hat{u}_{n,m}\|_{H^s} = \|\hat{u}_{n,0}\|_{H^s} \|\phi\|_{l^1},$$

thanks to property (2.1.34). Therefore, we only need to check that $\hat{u}_{n,0} \in H^s(\mathbb{R})$ for any integer $s \geq 0$. Since $V \in L^\infty(\mathbb{R})$, we obtain

$$\forall f \in \text{Dom}(L): \quad \|f\|_{H^s}^2 \leq \|(C_V + L)^{s/2} f\|_{L^2}^2, \quad s \geq 0,$$

for any fixed $C_V \geq 1 + \|V\|_{L^\infty}$. It follows from system (2.1.35) that $\hat{u}_{n,0} \in \text{Dom}(L)$. Using the integral representation (2.1.31), the orthogonality relations (2.1.19), and the asymptotic distribution (2.1.9), we hence obtain

$$\|\hat{u}_{n,0}\|_{H^s}^2 \leq \|(C_V + L)^{s/2} \hat{u}_{n,0}\|_{L^2}^2$$
$$= \int_{\mathbb{R}} \int_{\mathbb{T}} \int_{\mathbb{T}} [(C_V + L)^s u_n(x;k)] \, \bar{u}_n(x;k') dk dk' dx$$
$$= \int_{\mathbb{T}} (C_V + E_n(k))^s \, dk$$
$$\leq C_s (1 + n^2)^s,$$

where the positive constant C_s depends on s for any $s \geq 0$. Therefore, for a fixed $n \in \mathbb{N}$, $\phi \in H^s(\mathbb{R})$ for any integer $s \geq 0$ if $\phi \in l^1(\mathbb{Z})$. By the Sobolev Embedding Theorem (Appendix B.10), we obtain that $\phi \in C_b^r(\mathbb{R})$ for any integer $r < s - \frac{1}{2}$ and $\phi(x), ..., \phi^{(r)}(x)$ decay to zero as $|x| \to \infty$. $\square$

Exercise 2.11 Prove that if $\phi \in l^2(\mathbb{Z})$ and there is $C > 0$ such that $\sum_{m \in \mathbb{Z}} |\hat{E}_{n,m}| \leq C$ for a fixed $n \in \mathbb{N}$, then $\phi \in H^s(\mathbb{R})$ for any integer $s \geq 0$.

Because of our interest in the exponentially decaying solutions of nonlinear evolution equations, we shall also study the conditions under which the Wannier decomposition (2.1.39) recovers exponentially decaying functions $\phi(x)$ as $|x| \to \infty$.

Lemma 2.5 *Fix $n \in \mathbb{N}$ and assume that $\hat{u}_{n,m}(x)$ satisfy the exponential decay (2.1.38). Let $\phi(x) = \sum_{m \in \mathbb{Z}} \phi_m \hat{u}_{n,m}(x)$. If there exist $C > 0$ and $r \in (0,1)$ such that $|\phi_m| \leq Cr^{|m|}$ for all $m \in \mathbb{Z}$, then there exist $C^\pm > 0$ and $\eta^\pm > 0$ such that*

$$\lim_{x \to \pm\infty} \phi(x) e^{\pm \eta_\pm x} = C_\pm.$$

Proof By continuity of $\phi(x)$, it is sufficient to prove that there exist $C > 0$ and $q \in (0,1)$ such that $|\phi(2\pi k)| \leq Cq^{|k|}$ for all $k \in \mathbb{Z}$. Let $k \geq 1$ (the proof for $k \leq -1$ is similar). Using the exponential decay (2.1.38), we obtain

$$|\phi(2\pi k)| \leq C_n(I_1 + I_2 + I_3), \tag{2.1.40}$$

where

$$I_1 = \sum_{m=1}^{\infty} |\phi_{k+m}| e^{-2\pi\eta_n m},$$

$$I_2 = \sum_{m=0}^{k} |\phi_{k-m}| e^{-2\pi\eta_n m},$$

$$I_3 = e^{-2\pi\eta_n k} \sum_{m=1}^{\infty} |\phi_{-m}| e^{-2\pi\eta_n m}.$$

Since $r < 1$, and $q := e^{-2\pi\eta_n} < 1$, I_1 is bounded by $C_1 r^k$ for some $C_1 > 0$, while I_3 is bounded by $C_3 q^k$ for some $C_3 > 0$. On the other hand, I_2 is bounded by

$$r^k + r^{k-1} e^{-2\pi\eta_n} + \dots + e^{-2\pi\eta_n k} = \begin{cases} r^k \frac{1-p^{k+1}}{1-p}, & p < 1, \\ r^k p^k \frac{1-p^{-k-1}}{1-p^{-1}}, & p > 1, \end{cases}$$

where $pr = e^{-2\pi\eta_n}$. If $p \leq 1$, I_2 is bounded by $C_2 r^k$, while if $p \geq 1$, I_2 is bounded by $C_2' q^k$ for $q = e^{-2\pi\eta_n} < 1$, where $C_2, C_2' > 0$ are some constants. Thus, all three terms of (2.1.40) decay to zero exponentially fast as $k \to \infty$. □

Exercise 2.12 Using only the exponential decay (2.1.38) for Wannier functions, show that if $\phi \in l^1(\mathbb{Z})$, then $\phi \in C_b^0(\mathbb{R})$.

Results of Lemmas 2.4 and 2.5 specify constraints on ϕ to ensure that ϕ in the Wannier decomposition (2.1.39) belongs to a subspace of $\mathcal{E}_n \subset L^2(\mathbb{R})$. On the other hand, the decomposition formula (2.1.39) can be used for representation of other functions as well. For instance, the Fourier series (2.1.30) for the Bloch function $u_n(x; k)$ is an example of the Wannier decomposition (2.1.39) with the explicit representation

$$\phi(x) := u_n(x; k), \qquad \phi_m := e^{i2\pi mk}.$$

This decomposition corresponds to the case when $\phi \notin l^1(\mathbb{Z})$ and $\phi \notin L^2(\mathbb{R})$.

2.2 Justification of the nonlinear Dirac equations

The first example of the justification analysis deals with the nonlinear Dirac equations,

$$\begin{cases} i(a_T + na_X) = V_0 a + V_n b + (|a|^2 + 2|b|^2)a, \\ i(b_T - nb_X) = \bar{V}_n a + V_0 b + (2|a|^2 + |b|^2)b, \end{cases} \tag{2.2.1}$$

where $(a, b) \in \mathbb{C}^2$ are amplitude functions in new (slow) variables $(X, T) \in \mathbb{R}^2$, $n \in \mathbb{N}$ is fixed, V_0 and V_n are the zeroth and nth Fourier coefficients of the Fourier series for $V(x)$. The nonlinear Dirac equations (2.2.1) appear from the Gross–Pitaevskii equation

$$iu_t = -u_{xx} + \epsilon V(x)u + |u|^2 u \tag{2.2.2}$$

in the limit of *small* potentials, measured by the small parameter $\epsilon > 0$ (Section 1.1.1). The functions (a, b) represent amplitudes of two resonant counter-propagating Fourier modes $e^{\pm i k_n x}$ with $k_n = \frac{n}{2}$ at the bifurcation point $E_n = \frac{n^2}{4}$, from which the band gap between two adjacent band edges $E_n^{\pm}$ and $E_{n+1}^{\pm}$ diverges if $V_n \neq 0$. Here the plus sign occurs if n is even and the minus sign occurs if n is odd (Exercise 2.5).

The limit of small potentials changes the construction of the Bloch functions. Operator $L = -\partial_x^2 + \epsilon V(x)$ becomes $L_0 = -\partial_x^2$ at $\epsilon = 0$ and the Fourier transform replaces the Bloch decomposition (2.1.22).

Exercise 2.13 Construct eigenfunctions of $L_0 = -\partial_x^2$ in $L_{\text{per}}^2([0, 2\pi])$ and $L^2(\mathbb{R})$ and write Fourier analogues of the Bloch decompositions (2.1.12) and (2.1.22).

As an alternative to the direct method in Exercise 2.13, the relevant Fourier decompositions can be deduced from the Bloch functions (Section 2.1.1). First, solving the regular Sturm–Liouville spectral problem (2.1.8) with $V \equiv 0$ for a fixed $k \in \mathbb{T}$, we find one eigenfunction

$$w_1(x; k) = \frac{1}{\sqrt{2\pi}}$$

for the first eigenvalue

$$E_1(k) = k^2$$

and two linearly independent eigenfunctions

$$w_{2n}(x; k) = \frac{1}{\sqrt{2\pi}} e^{inx}, \quad w_{2n+1}(x; k) = \frac{1}{\sqrt{2\pi}} e^{-inx}, \quad n \in \mathbb{N}$$

for all subsequent eigenvalues

$$E_{2n}(k) = E_{2n+1}(k) = (k + n)^2, \quad n \in \mathbb{N},$$

where the normalization condition (2.1.11) has been used.

Let us list two eigenfunctions for $n \in \mathbb{N}$ by extending n from $\mathbb{N}$ to $\mathbb{Z}$. In this way, the decomposition (2.1.12) becomes the complex Fourier series

$$\forall \phi \in L_{\text{per}}^2([0, 2\pi]) : \quad \phi(x) = \sum_{n \in \mathbb{Z}} c_n e^{inx}, \tag{2.2.3}$$

where

$$c_n = \frac{1}{2\pi} \int_0^{2\pi} \phi(x) e^{-inx} dx, \quad n \in \mathbb{Z}.$$

We note that the complex Fourier series (2.2.3) is independent of parameter k. Recall that $u_n(x; k) = e^{ikx} w_n(x; k)$ satisfies

$$u_n(x + 2\pi; k) = e^{i2\pi k} u_n(x; k), \quad x \in \mathbb{R}.$$

If k is set to 0, then the complex Fourier series (2.2.3) represents a 2π-periodic function $\phi(x)$. If k is set to $\frac{1}{2}$, then the complex Fourier series represents a 2π-antiperiodic function $\phi(x)$ satisfying $\phi(x + 2\pi) = -\phi(x)$, or explicitly,

$$\forall \phi \in L_{\text{antiper}}^2([0, 2\pi]) : \quad \phi(x) = \sum_{n \in \mathbb{Z}} d_n e^{i\left(n + \frac{1}{2}\right)x},$$

where

$$d_n = \frac{1}{2\pi} \int_0^{2\pi} \phi(x) e^{-i\left(n+\frac{1}{2}\right)x} dx, \quad n \in \mathbb{Z}.$$

As an immediate use of the complex Fourier series, we write

$$V(x) = \sum_{n \in \mathbb{Z}} V_n e^{inx}. \qquad (2.2.4)$$

We note that if V is real, then $V_{-n} = \bar{V}_n$ for all $n \in \mathbb{N}$.

The complex Fourier series (2.2.4) is an isomorphism between $H^s_{\text{per}}([0, 2\pi])$ for functions $V(x)$ and $l^2_s(\mathbb{Z})$ for vectors $\mathbf{V}$ representing sequences $\{V_n\}_{n \in \mathbb{Z}}$ for any integer $s \geq 0$. This discretization of $H^s_{\text{per}}([0, 2\pi])$ can be exploited to study bifurcations of stationary 2π-periodic or 2π-antiperiodic solutions of the Gross–Pitaevskii equation (2.2.2) [65].

Exercise 2.14 Consider a stationary periodic solution $u(x,t) = \phi(x)e^{-i\omega t}$ of the Gross–Pitaevskii equation (2.2.2), where $\phi(x + 2\pi) = \phi(x)$ and $\omega = \frac{n^2}{4} + \epsilon\Omega$ for an even integer $n \geq 2$ and a fixed $\Omega \in \mathbb{R}$. Assume that $V(x)$ is given by (2.2.4), $\mathbf{V} \in l^2_s(\mathbb{Z})$ for a fixed $s > \frac{1}{2}$, and $V_n \neq 0$. Prove that there exists a non-trivial periodic solution $\phi(x)$ given by Fourier series (2.2.3) with $\mathbf{c} \in l^2_s(\mathbb{Z})$ if and only if there exists a non-trivial solution for $(a, b) \in \mathbb{C}^2$ of the bifurcation equations

$$\begin{cases} \Omega a = V_0 a + V_n b + (|a|^2 + 2|b|^2)a + \epsilon A_\epsilon(a,b), \\ \Omega b = \bar{V}_n a + V_0 b + (2|a|^2 + |b|^2)b + \epsilon B_\epsilon(a,b), \end{cases} \qquad (2.2.5)$$

where $A_\epsilon(a, b)$ and $B_\epsilon(a, b)$ are analytic functions of ϵ satisfying the bounds

$$\forall \epsilon \in (0, \epsilon_0): \quad \exists C_0 > 0: \quad |A_\epsilon(a,b)| + |B_\epsilon(a,b)| \leq C_0(|a| + |b|), \qquad (2.2.6)$$

for some $\epsilon_0 > 0$. Moreover, prove that

$$\forall \epsilon \in (0, \epsilon_0): \quad \exists C > 0: \quad \left\| \phi - \sqrt{\epsilon}\left(a e^{ik_n \cdot} + b e^{-ik_n \cdot}\right) \right\|_{C^0_{\text{per}}} \leq C\epsilon^{3/2}, \qquad (2.2.7)$$

where $k_n = \frac{n}{2}$.

Exercise 2.15 Repeat Exercise 2.14 for a stationary antiperiodic solution of the Gross–Pitaevskii equation (2.2.2) with $\omega = \frac{n^2}{4} + \epsilon\Omega$ for an odd integer $n \geq 1$ and a fixed $\Omega \in \mathbb{R}$.

To work with the nonlinear Dirac equations (2.2.1) in the time–space domain, we shall look for a representation of an element of $L^2(\mathbb{R})$ rather than that of $L^2_{\text{per}}([0, 2\pi])$ in terms of eigenfunctions of $L_0 = -\partial_x^2$. To do so, let us define the Bloch functions in the form

$$u_1(x;k) = \frac{1}{\sqrt{2\pi}} e^{ikx}$$

and

$$u_{2n}(x;k) = \frac{1}{\sqrt{2\pi}} e^{i(k+n)x}, \quad u_{2n+1}(x;k) = \frac{1}{\sqrt{2\pi}} e^{i(k-n)x}, \quad n \in \mathbb{N},$$

where $k \in \mathbb{T} := [-\frac{1}{2}, \frac{1}{2}]$. The Bloch decomposition (2.1.22) involves integration over $k \in \mathbb{T}$ and summation over $n \in \mathbb{N}$. Extending the enumeration of n to $\mathbb{Z}$ as before, we can incorporate the summation into an integral over $\mathbb{R}$ since the interval $\mathbb{T}$ for k translates to the interval $\left[n - \frac{1}{2}, n + \frac{1}{2}\right]$ for $k + n$, where $n \in \mathbb{Z}$. In this way, the Bloch decomposition (2.1.22) becomes the inverse Fourier transform

$$\forall \phi \subset L^2(\mathbb{R}): \quad \phi(x) = \frac{1}{2\pi} \int_{\mathbb{R}} \hat{\phi}(k) e^{ikx} dk, \quad x \in \mathbb{R}, \tag{2.2.8}$$

where $\hat{\phi}(k)$ is the Fourier transform given by

$$\hat{\phi}(k) = \int_{\mathbb{R}} \phi(x) e^{-ikx} dx, \quad k \in \mathbb{R}.$$

The Fourier integral (2.2.8) reminds us of the Fourier-type integral (2.1.24) introduced in the proof of Lemma 2.2 by unfolding $\mathbb{T}$ into $\left[-\frac{n}{2}, -\frac{n-1}{2}\right] \cup \left[\frac{n-1}{2}, \frac{n}{2}\right]$. Although the method of concatenations of $\mathbb{T} \times \mathbb{Z}$ into $\mathbb{R}$ is different from what is used in Lemma 2.2, the resulting Fourier integral is the same if $V \equiv 0$.

The inverse Fourier transform (2.2.8) is an isomorphism between $H^s(\mathbb{R})$ for $\phi(x)$ and $L_s^2(\mathbb{R})$ for $\hat{\phi}(k)$ for any integer $s \geq 0$. However, the Sobolev space $H^s(\mathbb{R})$ is not convenient for our analysis in the sense that if $\phi(x) = a(X) e^{ik_0 x}$, where $X = \epsilon x$ and $k_0 \in \mathbb{R}$ is fixed, then

$$\hat{\phi}(k) = \frac{1}{\epsilon} \hat{a}\left(\frac{k - k_0}{\epsilon}\right), \quad k \in \mathbb{R}. \tag{2.2.9}$$

As a result, for any sufficiently small $0 < \epsilon \ll 1$, there is $C_s > 0$ such that

$$\|\phi\|_{H^s} \leq C_s \epsilon^{-1/2} \|a\|_{H^s}.$$

Therefore, the H^s norm of ϕ may diverge as $\epsilon \to 0$ even if $a \in H^s(\mathbb{R})$. On the other hand, let us consider the Wiener space $W^s(\mathbb{R})$ of functions with the Fourier transform in $L_s^1(\mathbb{R})$ equipped with the norm $\|\phi\|_{W^s} := \|\hat{\phi}\|_{L_s^1}$. If $\phi(x) = a(X) e^{ik_0 x}$, where $X = \epsilon x$ and $k_0 \in \mathbb{R}$ is fixed, for any sufficiently small $0 < \epsilon \ll 1$, there is $C_s > 0$ such that

$$\|\hat{\phi}\|_{L_s^1} \leq C_s \|\hat{a}\|_{L_s^1}.$$

Moreover, if $s = 0$, then $L^1(\mathbb{R})$ is invariant with respect to the scaling transformation (2.2.9) and

$$\|\hat{\phi}\|_{L^1} = \|\hat{a}\|_{L^1}.$$

As a result, the Wiener space $W^s(\mathbb{R})$ is more convenient for the justification analysis of the nonlinear Dirac equations (2.2.1). By the Sobolev Embedding Theorem (Appendix B.10), there is $C_{s,q} > 0$ such that

$$\|\hat{\phi}\|_{L_s^1} \leq C_{s,q} \|\hat{\phi}\|_{L_{s+q}^2}, \quad s \geq 0, \quad q > \frac{1}{2}. \tag{2.2.10}$$

Hence $H^{s+q}(\mathbb{R})$ is embedded into $W^s(\mathbb{R})$ and the L_{s+q}^2 norm of $\hat{\phi}(k)$ may diverge as $\epsilon \to 0$, while the L_s^1 norm of $\hat{\phi}(k)$ may remain bounded.

Justifications of the time-dependent nonlinear Dirac equations (2.2.1) were developed by Goodman *et al.* [72] in the context of the Maxwell–Lorentz equations and by Schneider & Uecker [185] in the context of the Klein–Fock equation with quadratic nonlinear terms. The error terms were bounded in space $H^1(\mathbb{R})$ in [72] and in Wiener space $W(\mathbb{R}) \equiv W^0(\mathbb{R})$ in [185]. A similar use of the Wiener spaces for justification of the envelope equations for description of modulated short pulses in hyperbolic systems was developed by Colin & Lannes [39]. Stationary reduction of the nonlinear Dirac equations for gap solitons was justified by Pelinovsky & Schneider [162]. Persistence of traveling wave solutions of the nonlinear Dirac equations (2.2.1) was considered by Pelinovsky & Schneider [163].

According to the strategy of this chapter, we shall separate justification of the time-dependent and stationary equations. The error bounds for the time-dependent solutions are valid on a finite time interval, whereas the error bounds for the stationary solutions can be controlled globally in time in the corresponding norm. In the latter case, the justification analysis also proves persistence of stationary solutions of the nonlinear Dirac equations (2.2.1) as stationary solutions of the Gross–Pitaevskii equation (2.2.2).

2.2.1 Justification of the time-dependent equation

Let us fix $n \in \mathbb{N}$ and assume that $V_n \neq 0$ in the Fourier series (2.2.4). A solution of the Gross–Pitaevskii equation (2.2.2) is represented by the asymptotic multi-scale expansion

$$u(x,t) = \epsilon^{1/2}\left(a(X,T)e^{ik_n x} + b(X,T)e^{-ik_n x} + \epsilon\varphi_\epsilon(x,t)\right)e^{-iE_n t},$$

where $X = \epsilon x$, $T = \epsilon t$, and parameters (k_n, E_n) are fixed at the bifurcation values $k_n = \frac{n}{2}$ and $E_n = \frac{n^2}{4}$. We let (a,b) satisfy the nonlinear Dirac equations (2.2.1) in variables (X,T) and write the time evolution problem for the *remainder* term $\varphi_\epsilon(x,t)$ in the form

$$i\partial_t\varphi_\epsilon + E_n\varphi_\epsilon = -\partial_x^2\varphi_\epsilon + F_\epsilon(a,b,\varphi_\epsilon), \tag{2.2.11}$$

where the *residual* term F_ϵ is written explicitly by

$$F_\epsilon = -\epsilon\left(a_{XX}e_+ + b_{XX}e_-\right) + \tilde{V}_a ae_+ + \tilde{V}_b be_- + V\varphi_\epsilon$$
$$+ |ae_+ + be_- + \epsilon\varphi_\epsilon|^2(ae_+ + be_- + \epsilon\varphi_\epsilon) - (|a|^2 + 2|b|^2)ae_+ - (2|a|^2 + |b|^2)be_-,$$

with $e_\pm = e^{\pm ik_n x}$ and

$$\tilde{V}_a = V - (V_0 + V_{-n}e^{-inx}), \quad \tilde{V}_b = V - (V_0 + V_n e^{inx}).$$

We have performed the first step in the general algorithm of the justification analysis. It remains to perform three more steps.

- Remove the non-vanishing terms of the residual F_ϵ as $\epsilon \to 0$ by a normal form transformation of the remainder term φ_ϵ.
- Choosing the Wiener space $W(\mathbb{R})$ for φ_ϵ at a fixed $t \in \mathbb{R}_+$, prove that the residual term F_ϵ of the time evolution problem (2.2.11) maps an element of $W(\mathbb{R})$ to an element of $W(\mathbb{R})$ under some constraints on the solution (a,b) of the nonlinear Dirac equations (2.2.1).
- Prove the local well-posedness of the time evolution problem (2.2.11) in $W(\mathbb{R})$ and control the size of $\|\varphi_\epsilon\|_W$ using Gronwall's inequality.

We will show that all three goals can be achieved if there is a time $T_0 > 0$ such that

$$(a,b)(T) \in C([0,T_0], W^2(\mathbb{R}) \times W^2(\mathbb{R})) \cap C^1([0,T_0], W^1(\mathbb{R}) \times W^1(\mathbb{R})). \quad (2.2.12)$$

By Theorem 1.4 (Section 1.3.3), the nonlinear Dirac equations (2.2.1) are locally well-posed in Sobolev space $H^r(\mathbb{R}) \times H^r(\mathbb{R})$ for any $r > \frac{1}{2}$. Inequality (2.2.10) tells us that if $s = 2$, then any $r > \frac{5}{2}$ gives a good solution (2.2.12). For instance, $r = 3$ does the job. The following theorem formulates the main result of the justification analysis.

Theorem 2.1   *Assume that V is given by (2.2.4) and $\mathbf{V} \in l^1(\mathbb{Z})$. Fix $n \in \mathbb{N}$ such that $V_n \neq 0$. Fix $T_0 > 0$ such that*

$$(a,b)(T) \in C([0,T_0], H^3(\mathbb{R}) \times H^3(\mathbb{R})) \cap C^1([0,T_0], H^2(\mathbb{R}) \times H^2(\mathbb{R}))$$

is a solution of the nonlinear Dirac equations (2.2.1) with initial data $(a_0,b_0) \in H^3(\mathbb{R}) \times H^3(\mathbb{R})$. Fix $C_0 > 0$ such that $u_0 \in W(\mathbb{R})$ satisfy the bound

$$\left\| u_0 - \epsilon^{1/2} \left(a_0(\epsilon \cdot)e^{ik_n \cdot} + b_0(\epsilon \cdot)e^{-ik_n \cdot} \right) \right\|_{C_b^0} \leq C_0 \epsilon^{3/2}.$$

There exist $\epsilon_0 > 0$ and $C > 0$ such that for all $\epsilon \in (0, \epsilon_0)$, the Gross–Pitaevskii equation (2.2.2) admits a solution $u(t) \in C([0, \epsilon^{-1}T_0], W(\mathbb{R}))$ such that $u(0) = u_0$ and for any $t \in [0, \epsilon^{-1}T_0]$,

$$\left\| u(\cdot,t) - \epsilon^{1/2} \left(a(\epsilon\cdot, \epsilon t)e^{ik_n \cdot} + b(\epsilon\cdot, \epsilon t)e^{-ik_n \cdot} \right) e^{-iE_n t} \right\|_{C_b^0} \leq C\epsilon^{3/2}, \quad (2.2.13)$$

where $k_n = \frac{n}{2}$ and $E_n = \frac{n^2}{4}$.

Remark 2.2   Solution $u(x,t)$ constructed in Theorem 2.1 is bounded, continuous in x on $\mathbb{R}$, and decaying to zero at infinity as $|x| \to \infty$ for any $t \in [0, \epsilon^{-1}T_0]$ thanks to the Riemann–Lebesgue Lemma (Appendix B.10).

Our first task is to show that the non-vanishing terms in F_ϵ as $\epsilon \to 0$ can be removed by a normal form transformation of φ_ϵ. The non-vanishing terms in F_ϵ as $\epsilon \to 0$ are written explicitly by

$$F_0 = \tilde{V}_a a e_+ + \tilde{V}_b b e_- + a^2 \bar{b} e_+^3 + b^2 \bar{a} e_-^3.$$

Explicitly, F_0 can be rewritten as the Fourier series

$$F_0 = \sum_{n' \in \mathbb{Z} \setminus \{0, -n\}} f_{n'}(X,T) e^{i\left(n' + \frac{n}{2}\right)x},$$

which contains no resonant terms with $n' = 0$ and $n' = -n$, where

$$f_n = a^2\bar{b} + V_n a + V_{2n} b, \quad f_{-2n} = b^2\bar{a} + V_{-2n}a + V_{-n}b,$$

and

$$f_m = V_m a + V_{m+n} b, \quad m \notin \{n, 0, -n, -2n\}.$$

By writing $\varphi_\epsilon = \varphi_1 + \psi_\epsilon$ with

$$\varphi_1 = -\sum_{n' \in \mathbb{Z}\backslash\{0,-n\}} \frac{f_{n'}(X,T)}{n'(n+n')} e^{i(n'+\frac{n}{2})x}, \qquad (2.2.14)$$

we rewrite the time evolution problem for ψ_ϵ in the abstract form

$$i\partial_t \psi_\epsilon + E_n \psi_\epsilon = -\partial_x^2 \psi_\epsilon + \epsilon G_\epsilon(a,b,\psi_\epsilon), \qquad (2.2.15)$$

where

$$G_\epsilon = \frac{F_\epsilon(a,b,\varphi_1+\psi_\epsilon) - F_0(a,b)}{\epsilon} - \left(i\partial_T + 2\partial_{xX}^2 + \epsilon\partial_{XX}^2\right)\varphi_1. \qquad (2.2.16)$$

Our second task is to show that the residual term F_ϵ of the time evolution problem (2.2.11) maps an element of $W(\mathbb{R})$ to an element of $W(\mathbb{R})$ provided that $(a,b) \in W^2(\mathbb{R}) \times W^2(\mathbb{R})$.

Lemma 2.6 *Assume that* $\mathbf{V} \in l^1(\mathbb{Z})$. *Then,*

$$F_\epsilon(a,b,\varphi_\epsilon) : W^2(\mathbb{R}) \times W^2(\mathbb{R}) \times W(\mathbb{R}) \to W(\mathbb{R})$$

is a locally Lipschitz map for any (a,b,φ_ϵ).

Proof We shall convert $F_\epsilon(a,b,\varphi_\epsilon)$ to Fourier space and show that

$$\hat{F}_\epsilon(\hat{a},\hat{b},\hat{\varphi}_\epsilon) : L_2^1(\mathbb{R}) \times L_2^1(\mathbb{R}) \times L^1(\mathbb{R}) \to L^1(\mathbb{R})$$

is a locally Lipschitz map for any $(\hat{a},\hat{b},\hat{\varphi}_\epsilon)$. Multiplication of two functions $u(x)$ and $v(x)$ in physical space corresponds to the convolution integral

$$(\hat{u} \star \hat{v})(k) := \int_{\mathbb{R}} \hat{u}(k')\hat{v}(k-k')dk'$$

in Fourier space. Thanks to the bound

$$\forall \hat{u}, \hat{v} \in L^1(\mathbb{R}) : \quad \|\hat{u} \star \hat{v}\|_{L^1} \leq \|\hat{u}\|_{L^1}\|\hat{v}\|_{L^1},$$

space $L^1(\mathbb{R})$ forms the Banach algebra with respect to convolution integrals (Appendix B.1). Thanks to the scaling invariance of $W(\mathbb{R})$ with respect to the scaling transformation, we have $\|ae_+\|_W = \|\hat{a}\|_{L^1}$. Similarly,

$$\|a_{XX}e_+\|_W = \int_{\mathbb{R}} p^2|\hat{a}(p)|dp \leq \|\hat{a}\|_{L_2^1}.$$

Therefore, the V-independent terms of $\hat{F}_\epsilon(\hat{a},\hat{b},\hat{\varphi}_\epsilon)$ give a Lipschitz map from $L_2^1(\mathbb{R}) \times L_2^1(\mathbb{R}) \times L^1(\mathbb{R})$ to $L^1(\mathbb{R})$. It remains to bound multiplications of $V(x)$ given by the Fourier series (2.2.4) and $u(x)$ given by the Fourier integral (2.2.8). In Fourier space, this multiplication corresponds to the convolution sum

$$(\hat{u} \star \mathbf{V})(k) := \sum_{m \in \mathbb{Z}} V_m \hat{u}(k-m),$$

which can be bounded by

$$\forall \hat{u} \in L^1(\mathbb{R}), \forall \mathbf{V} \in l^1(\mathbb{Z}) : \quad \|\hat{u} \star \mathbf{V}\|_{L^1} \leq \|\hat{u}\|_{L^1}\|\mathbf{V}\|_{l^1}.$$

Therefore, the V-dependent terms of $\hat{F}_\epsilon(\hat{a},\hat{b},\hat{\varphi}_\epsilon)$ give also a Lipschitz map from $L_2^1(\mathbb{R}) \times L_2^1(\mathbb{R}) \times L^1(\mathbb{R})$ to $L^1(\mathbb{R})$. $\qquad\square$

Using Lemma 2.6, we show that $G_\epsilon(a, b, \psi_\epsilon)$ is a Lipschitz map in a ball $B_\delta(W(\mathbb{R}))$ of radius $\delta > 0$ provided that (a, b) are solutions of the nonlinear Dirac equations (2.2.1) in a ball $B_{\delta_0}(W^2(\mathbb{R}) \times W^2(\mathbb{R}))$ of radius $\delta_0 > 0$. Again, we shall formulate and prove the statement in Fourier space for $\hat{\psi}_\epsilon$ and $(\hat{a}, \hat{b})$.

Lemma 2.7   *Assume that* $\mathbf{V} \in l^1(\mathbb{Z})$ *and* (a, b) *are solutions of the nonlinear Dirac equations (2.2.1). Assume that there exists* $\delta_0 > 0$ *such that* $(\hat{a}, \hat{b}) \in B_{\delta_0}(L_2^1(\mathbb{R}) \times L_2^1(\mathbb{R}))$. *For any fixed* $\delta > 0$ *and any small* $\epsilon > 0$, *there exist* $C(\delta_0, \delta) > 0$ *and* $K(\delta_0, \delta) > 0$ *such that if* $\hat{\psi}_\epsilon, \hat{\tilde{\psi}}_\epsilon \in B_\delta(L^1(\mathbb{R}))$, *then*

$$\|\hat{G}_\epsilon(\hat{a}, \hat{b}, \hat{\psi}_\epsilon)\|_{L^1} \leq C(\delta_0, \delta) \left(\|\hat{a}\|_{L_2^1} + \|\hat{b}\|_{L_2^1} + \|\hat{\psi}_\epsilon\|_{L^1} \right), \qquad (2.2.17)$$

$$\|\hat{G}_\epsilon(\hat{a}, \hat{b}, \hat{\psi}_\epsilon) - \hat{G}_\epsilon(\hat{a}, \hat{b}, \hat{\tilde{\psi}}_\epsilon)\|_{L^1} \leq K(\delta_0, \delta)\|\hat{\psi}_\epsilon - \hat{\tilde{\psi}}_\epsilon\|_{L^1}. \qquad (2.2.18)$$

Proof   Although the definition of G_ϵ in (2.2.15) involves T-derivatives of (a, b), these terms are eliminated by the nonlinear Dirac equations (2.2.1). From the exact expression (2.2.14), there is a constant $C(\delta_0) > 0$ such that

$$\|i\partial_T \varphi_1\|_W + \|\partial_{xX}^2 \varphi_1\|_W + \|\partial_{XX}^2 \varphi_1\|_W \leq C(\delta_0) \left(\|\hat{a}\|_{L_2^1} + \|\hat{b}\|_{L_2^1} \right).$$

Combining this analysis with Lemma 2.6, we know that $\hat{G}_\epsilon(\hat{a}, \hat{b}, \hat{\psi}_\epsilon)$ is a locally Lipschitz map from $L_2^1(\mathbb{R}) \times L_2^1(\mathbb{R}) \times L^1(\mathbb{R})$ to $L^1(\mathbb{R})$. Bounds (2.2.17) and (2.2.18) follow from expression (2.2.16). $\qquad \square$

Our third task is to prove local well-posedness of the time evolution problem (2.2.15) in the same Wiener space $W(\mathbb{R})$ for ψ_ϵ. This is not a difficult task, thanks to Lemma 2.7.

Lemma 2.8   *Assume that* $\mathbf{V} \in l^1(\mathbb{Z})$. *Fix* $T_0 > 0$ *such that*

$$(\hat{a}, \hat{b})(T) \in C([0, T_0], L_2^1(\mathbb{R}) \times L_2^1(\mathbb{R})) \cap C^1([0, T_0], L_1^1(\mathbb{R}) \times L_1^1(\mathbb{R})).$$

Fix $\delta > 0$ *such that* $\hat{\psi}_\epsilon(0) \in B_\delta(L^1(\mathbb{R}))$. *There exist a* $t_0 > 0$ *and a* $\epsilon_0 > 0$ *such that, for any* $\epsilon \in (0, \epsilon_0)$, *the time evolution problem (2.2.15) admits a unique solution* $\psi_\epsilon(t) \in C([0, t_0], W(\mathbb{R}))$.

Proof   Transforming (2.2.15) to Fourier space and using Duhamel's principle, we write the time evolution problem in the integral form

$$\hat{\psi}_\epsilon(t) = e^{it(E_n - k^2)}\hat{\psi}_\epsilon(0) + \epsilon \int_0^t e^{i(t-s)(E_n - k^2)}\hat{G}_\epsilon(\hat{a}(\epsilon s), \hat{b}(\epsilon s), \hat{\psi}_\epsilon(s))ds. \qquad (2.2.19)$$

Existence of a unique solution $\hat{\psi}_\epsilon(t) \in C([0, t_0], W(\mathbb{R}))$ for a fixed $t_0 > 0$ and any small $\epsilon > 0$ follows from bounds (2.2.17)–(2.2.18) by the Banach Fixed-Point Theorem (Appendix B.2). $\qquad \square$

It remains now to control the remainder term $\hat{\psi}_\epsilon$ in the L^1 norm, after which the proof of Theorem 2.1 appears to be complete.

Proof of Theorem 2.1   Using bound (2.2.17) and the integral equation (2.2.19), we obtain

$$\|\hat{\psi}_\epsilon(t)\|_{L^1} \leq \|\hat{\psi}_\epsilon(0)\|_{L^1} + \epsilon C(\delta_0, \delta) \int_0^t \left(\|\hat{a}(\epsilon s)\|_{L_2^1} + \|\hat{b}(\epsilon s)\|_{L_2^1} + \|\hat{\psi}_\epsilon(s)\|_{L^1} \right) ds.$$

By Gronwall's inequality (Appendix B.6), it follows that

$$\sup_{t\in[0,\epsilon^{-1}T_0]} \|\hat{\psi}_\epsilon(t)\|_{L^1}$$

$$\leq \left(\|\hat{\psi}_\epsilon(0)\|_{L^1} + C(\delta_0,\delta) \int_0^{T_0} \left(\|\hat{a}(T)\|_{L_{\frac{1}{2}}} + \|\hat{b}(T)\|_{L_{\frac{1}{2}}} \right) dT \right) e^{T_0 C(\delta_0,\delta)}$$

$$\leq \delta.$$

This bound is satisfied by a choice for $\delta_0 > 0$ and $\delta > 0$. $\qquad\square$

Exercise 2.16 Use two subsequent normal form transformations and justify nonlinear Dirac equations (2.2.1) for the system of Gross–Pitaevskii equations with quadratic nonlinear terms and a bounded 2π-periodic potential $V(x)$

$$\begin{cases} iu_t = -u_{xx} + \epsilon V(x)u + uw, \\ iw_t = -w_{xx} + u^2, \end{cases}$$

where ϵ is a small positive parameter.

2.2.2 Justification of the stationary equation

Consider the stationary reduction of the nonlinear Dirac equations (2.2.1),

$$a(X,T) = A(X)e^{-i(\Omega+V_0)T}, \quad b(X,T) = B(X)e^{-i(\Omega+V_0)T}$$

where Ω is a real-valued parameter and (A, B) satisfy the stationary nonlinear Dirac equations,

$$\begin{cases} inA'(X) + \Omega A = V_n B + (|A|^2 + 2|B|^2)A, \\ -inB'(X) + \Omega B = \bar{V}_n A + (2|A|^2 + |B|^2)B. \end{cases} \tag{2.2.20}$$

We are looking for localized modes of the stationary system (2.2.20) that may exist for any $\Omega \in (-|V_n|, |V_n|)$. The interval $\epsilon(V_0 - |V_n|, V_0 + |V_n|)$ corresponds to the narrow gap between two spectral bands of $L = -\partial_x^2 + V(x)$ that bifurcates from the point $E_n = \frac{n^2}{4}$ as $\epsilon > 0$ (Exercise 2.5). Localized modes of the stationary nonlinear Dirac equations (2.2.20) correspond to localized modes of the stationary Gross–Pitaevskii equation

$$-\Phi''(x) + \epsilon V(x)\Phi(x) + |\Phi(x)|^2\Phi(x) = \omega\Phi(x), \tag{2.2.21}$$

which arises from the Gross–Pitaevskii equation (2.2.2) after the substitution

$$u(x,t) = \Phi(x)e^{-i\omega t}.$$

Decaying solutions of system (2.2.20) are found explicitly (Section 3.3)

$$A(X - X_0)e^{i\Theta_0} = \frac{\sqrt{2}}{\sqrt{3}} \frac{\sqrt{V_n^2 - \Omega^2}}{\sqrt{V_n - \Omega}\cosh(KX) + i\sqrt{V_n + \Omega}\sinh(KX)}$$

$$= \bar{B}(X - X_0)e^{i\Theta_0}, \tag{2.2.22}$$

where $K = \frac{1}{n}\sqrt{V_n^2 - \Omega^2}$, (Θ_0, X_0) are arbitrary parameters and $V_n > 0$. (Normalization of V_n to a real positive number can be achieved by a shift $x \to x + x_0$ in the stationary Gross–Pitaevskii equation (2.2.21).)

Before we estimate the distance between decaying solutions of systems (2.2.20) and (2.2.21), we should realize that the nonlinear Dirac equations (2.2.20) are invariant with respect to gauge transformation and spatial translation

$$(A, B) \mapsto (A, B)e^{i\Theta_0}, \quad (A, B)(X) \mapsto (A, B)(X - X_0), \quad (\Theta_0, X_0) \subset \mathbb{R}^2,$$

while the Gross–Pitaevskii equation (2.2.21) with a nonzero $V(x)$ is only invariant with respect to the gauge transformation

$$\Phi \mapsto \Phi e^{i\theta_0}, \quad \theta_0 \in \mathbb{R}.$$

Without loss of generality, we can set $\Theta_0 = 0$ in the exact solution (2.2.22). On the other hand, we do not have freedom to remove parameter X_0 from consideration.

To simplify the justification analysis, we shall add an assumption that

$$V(-x) = V(x) \quad \text{for all} \quad x \in \mathbb{R}.$$

In this case, the stationary Gross–Pitaevskii equation (2.2.21) admits two solutions $\Phi(x)$ and $\Phi(-x)$, which are identical if $\Phi(x) = \Phi(-x)$. This construction allows us to consider even solutions (2.2.22) and set $X_0 = 0$. We note that

$$V(-x) = V(x), \ x \in \mathbb{R} \quad \Rightarrow \quad V_{-n} = V_n, \ n \in \mathbb{Z} \tag{2.2.23}$$

and

$$\Theta_0 = X_0 = 0 \quad \Rightarrow \quad A(X) = \bar{A}(-X) = \bar{B}(X), \ X \in \mathbb{R}. \tag{2.2.24}$$

The following theorem formulates the main result of the justification analysis.

Theorem 2.2   *Let $V(-x) = V(x)$ be given by the Fourier series (2.2.4) and $\mathbf{V} \in l^1(\mathbb{Z})$. Fix $n \in \mathbb{N}$ such that $V_n \neq 0$. Let $A(X) = \bar{A}(-X) = \bar{B}(X)$ be the localized mode (2.2.22) for $\Omega \in (-|V_n|, |V_n|)$. There exist $\epsilon_0 > 0$ and $C_0 > 0$ such that for all $\epsilon \in (0, \epsilon_0)$, the stationary Gross–Pitaevskii equation (2.2.21) with $\omega = E_n + \epsilon(V_0 + \Omega)$ admits a real-valued solution $\Phi \in W(\mathbb{R})$, even in x, decaying to zero as $|x| \to \infty$, satisfying the bound*

$$\left\| \Phi - \epsilon^{1/2} \left(A(\epsilon \cdot)e^{ik_n \cdot} + B(\epsilon \cdot)e^{-ik_n \cdot} \right) \right\|_{C_b^0} \leq C_0 \epsilon^{5/6}, \tag{2.2.25}$$

where $k_n = \frac{n}{2}$ and $E_n = \frac{n^2}{4}$.

Remark 2.3   If $V_n > 0$ and $\Omega = V_n$, the exact solution (2.2.22) with $\Theta_0 = X_0 = 0$ degenerates into the algebraically decaying solution

$$A(X) = \frac{2n\sqrt{V_n}}{\sqrt{3}(n + 2iV_n X)}, \quad B(X) = \frac{2n\sqrt{V_n}}{\sqrt{3}(n - 2iV_n X)}. \tag{2.2.26}$$

Although $A \in W(\mathbb{R})$, the persistence of algebraically decaying solutions cannot be proved with our analysis. The reason lies in Lemma 2.10 below, where the linearized operator (2.2.35) associated with the algebraically decaying solution (2.2.26) admits a continuous spectrum that touches the origin.

To prove Theorem 2.2, we substitute the Fourier transform (2.2.8) for $\Phi(x)$ and the Fourier series (2.2.4) for $V(x)$ into the stationary Gross–Pitaevskii equation

(2.2.21) with $\omega = E_n + \epsilon(V_0 + \Omega)$. If we use the rescaling $\hat{\Phi}(k) \mapsto \epsilon^{1/2}\hat{\Phi}(k)$ for convenience, the resulting integral equation becomes

$$\left(E_n + \epsilon\Omega - k^2\right)\hat{\Phi}(k) + \epsilon \sum_{n' \in \mathbb{Z}\backslash\{0\}} V_{n'}\hat{\Phi}(k - n')$$

$$= \epsilon \int_{\mathbb{R}}\int_{\mathbb{R}} \hat{\Phi}(k_1)\bar{\hat{\Phi}}(k_2)\hat{\Phi}(k - k_1 + k_2)dk_1 dk_2. \qquad (2.2.27)$$

Since $E_n = \frac{n^2}{4}$, the first term in brackets vanishes near $k = \pm k_n = \pm\frac{n}{2}$ and $\epsilon = 0$. This fact results in singularities of $\hat{\Phi}(k)$ near $k = \pm k_n$ if we are to invert the first term. To single out these singular contributions, we decompose $\hat{\Phi}(k)$ into three parts

$$\hat{\Phi}(k) = \hat{\Phi}_+(k)\chi_{\mathbb{R}'_+}(k) + \hat{\Phi}_-(k)\chi_{\mathbb{R}'_-}(k) + \hat{\Phi}_0(k)\chi_{\mathbb{R}'_0}(k), \qquad (2.2.28)$$

where $\chi_S(k)$ is a characteristic function and the sets $\mathbb{R}'_+$, $\mathbb{R}'_-$ and $\mathbb{R}'_0$ are

$$\mathbb{R}'_\pm = \left[\pm\omega_n - \epsilon^{2/3}, \pm\omega_n + \epsilon^{2/3}\right], \quad \mathbb{R}'_0 = \mathbb{R}\backslash(\mathbb{R}'_+ \cup \mathbb{R}'_-).$$

The components $\hat{\Phi}_\pm(k)$ represent the largest part of the solution $\hat{\Phi}(k)$ near the resonant values $k = \pm k_n$, whereas the component $\hat{\Phi}_0(k)$ represents a small remainder term. The largest terms are approximated by solutions of the nonlinear Dirac equations (2.2.20) in Fourier space, whereas the small remainder $\hat{\Phi}_0(k)$ is determined uniquely from $\hat{\Phi}_+(k)$ and $\hat{\Phi}_-(k)$.

We have performed the first step in the general algorithm of the justification analysis. It remains to perform two more steps.

- Prove the existence of a unique smooth map from $(\hat{\Phi}_+, \hat{\Phi}_-) \in L^1(\mathbb{R}'_+) \times L^1(\mathbb{R}'_-)$ to $\hat{\Phi}_0 \in L^1(\mathbb{R}'_0)$.
- Approximate $(\hat{\Phi}_+, \hat{\Phi}_-)$ by solutions $(\hat{A}, \hat{B})$ of the nonlinear Dirac equations (2.2.20) rewritten in Fourier space and prove the persistence of solutions for $(\hat{\Phi}_+, \hat{\Phi}_-)$ in $L^1(\mathbb{R}'_+) \times L^1(\mathbb{R}'_-)$.

These steps are performed in the following two lemmas. Note that we can take $L^1(\mathbb{R}'_\pm)$ instead of $L^1_{\frac{1}{2}}(\mathbb{R}'_\pm)$ compared to Section 2.2.1 because $\mathbb{R}'_\pm$ are compact intervals.

Lemma 2.9 *Let V be given by (2.2.4) with $\mathbf{V} \in l^1(\mathbb{Z})$. There exist $\epsilon_0 > 0$, $\delta_0 > 0$, $C_0 > 0$, and a unique continuous map*

$$\Psi(\hat{\Phi}_+, \hat{\Phi}_-, \epsilon) : L^1(\mathbb{R}'_+) \times L^1(\mathbb{R}'_-) \times \mathbb{R} \to L^1(\mathbb{R}'_0), \qquad (2.2.29)$$

such that $\hat{\Phi}_0(k) = \Psi(\hat{\Phi}_+, \hat{\Phi}_-, \epsilon)(k)$ solves the integral equation (2.2.27) on $\mathbb{R}'_0$ for any $\epsilon \in (0, \epsilon_0)$ and $\hat{\Phi}_\pm \in B_{\delta_0}(L^1(\mathbb{R}'_\pm))$. Moreover, the map satisfies

$$\|\Psi(\hat{\Phi}_+, \hat{\Phi}_-, \epsilon)\|_{L^1(\mathbb{R}'_0)} \leq \epsilon^{1/3}C_0\left(\|\hat{\Phi}_+\|_{L^1(\mathbb{R}'_+)} + \|\hat{\Phi}_-\|_{L^1(\mathbb{R}'_-)}\right). \qquad (2.2.30)$$

Proof We project the integral equation (2.2.27) onto $\mathbb{R}'_0$:

$$\left(E_n + \epsilon\Omega - k^2\right)\hat{\Phi}_0(k) + \epsilon\sum_{n'\in\mathbb{Z}\setminus\{0\}}V_{n'}\chi_{\mathbb{R}'_0}(k)\hat{\Phi}(k - n')$$

$$= \epsilon\chi_{\mathbb{R}'_0}(k)\int_{\mathbb{R}}\int_{\mathbb{R}}\hat{\Phi}(k_1)\bar{\hat{\Phi}}(k_2)\hat{\Phi}(k - k_1 + k_2)dk_1 dk_2, \qquad (2.2.31)$$

where $\hat{\Phi}(k)$ is represented by (2.2.28). Because of the definition of $\mathbb{R}'_0$, for any $\epsilon \in (0, \epsilon_0)$, there is $C_n > 0$ such that

$$\min_{k\in\mathbb{R}'_0}\left|E_n - k^2\right| \geq C_n\epsilon^{2/3}.$$

Since $\mathbf{V} \in l^1(\mathbb{Z})$, the convolution sum in the integral equation represents a bounded perturbation of the linear operator $(E_n + \epsilon\Omega - k^2)$ in $L^1(\mathbb{R}'_0)$. Therefore, the linearized integral operator

$$\hat{L}\hat{\Phi}_0 := \left(E_n + \epsilon\Omega - k^2\right)\hat{\Phi}_0(k) + \epsilon\sum_{n'\in\mathbb{Z}\setminus\{0\}}V_{n'}\chi_{\mathbb{R}'_0}(k)\hat{\Phi}_0(k - n')$$

is continuously invertible near $\epsilon = 0$ and

$$\forall\epsilon \in (0, \epsilon_0): \quad \exists C_n > 0: \quad \left\|\hat{L}^{-1}\chi_{\mathbb{R}'_0}\hat{f}\right\|_{L^1(\mathbb{R}'_0)} \leq C_n\epsilon^{-2/3}\|\hat{f}\|_{L^1(\mathbb{R}'_0)}. \qquad (2.2.32)$$

Recall that $\epsilon^{2/3} \gg \epsilon$ for sufficiently small $\epsilon > 0$. Inverting $\hat{L}$, we rewrite the integral equation (2.2.31) as a near-identity equation, where the right-hand side has the order of $\mathcal{O}(\epsilon^{1/3})$. By Lemma 2.6, the vector field of this integral equation consists of convolution integrals, which map elements of $L^1(\mathbb{R})$ to elements of $L^1(\mathbb{R})$.

Fix $\delta_0 > 0$ such that

$$\|\hat{\Phi}_+\|_{L^1(\mathbb{R}'_+)} + \|\hat{\Phi}_-\|_{L^1(\mathbb{R}'_-)} < \delta_0.$$

By the Implicit Function Theorem (Appendix B.7), there exists a unique map (2.2.29) for $\hat{\Phi}_\pm \in B_{\delta_0}(L^1(\mathbb{R}'_\pm))$ and $\epsilon \in (0, \epsilon_0)$. Since a unique trivial solution $\hat{\Phi}_0 \equiv 0$ exists if $\hat{\Phi}_\pm \equiv 0$, then the map satisfies $\Psi(0, 0, \epsilon) = 0$. The desired bound (2.2.30) follows from analyticity of the integral equation (2.2.31) in $\hat{\Phi}$ and ϵ and the bound (2.2.32). $\qquad\square$

Lemma 2.10 *Let* $V(-x) = V(x)$ *be given by (2.2.4) with* $\mathbf{V} \in l^1(\mathbb{Z})$. *Fix* $n \in \mathbb{N}$ *such that* $V_n \neq 0$. *Let* $A(X) = \bar{A}(-X) = \bar{B}(X)$ *be the localized mode (2.2.22) for* $\Omega \in (-|V_n|, |V_n|)$. *There exists* $\epsilon_0 > 0$ *and* $C_0 > 0$ *such that the integral equation (2.2.27) for any* $\epsilon \in (0, \epsilon_0)$ *admits a solution in the form (2.2.28), where* $\hat{\Phi}_0(k) = \Psi(\hat{\Phi}_+, \hat{\Phi}_-, \epsilon)(k)$ *is given by Lemma 2.9 and* $\hat{\Phi}_\pm(k)$ *satisfy the bound*

$$\left\|\hat{\Phi}_+ - \frac{1}{\epsilon}\hat{A}\left(\frac{\cdot - k_n}{\epsilon}\right)\right\|_{L^1(\mathbb{R}'_+)} + \left\|\hat{\Phi}_- - \frac{1}{\epsilon}\hat{B}\left(\frac{\cdot + k_n}{\epsilon}\right)\right\|_{L^1(\mathbb{R}'_-)} \leq C_0\epsilon^{1/3}.$$

Proof Let us first map the intervals $\mathbb{R}'_\pm$ for $\hat{\Phi}_\pm(k)$ to the normalized interval $\mathbb{R}_0 = \left[-\epsilon^{-1/3}, \epsilon^{-1/3}\right]$ for $\hat{\Psi}_\pm(p)$ given by

$$\hat{\Phi}_\pm(k) = \frac{1}{\epsilon}\hat{\Psi}_\pm\left(\frac{k \mp k_n}{\epsilon}\right). \qquad (2.2.33)$$

The new functions $\hat{\Psi}_\pm(p)$ have a compact support on $\mathbb{R}_0$ and satisfy

$$\|\hat{\Phi}_\pm\|_{L^1(\mathbb{R}'_\pm)} = \|\hat{\Psi}_\pm\|_{L^1(\mathbb{R}_0)},$$

thanks to the scaling invariance of the L^1 norm. The integral equation (2.2.27) is projected to the system of two integral equations for $p \in \mathbb{R}_0$:

$$(\Omega \mp np)\,\hat{\Psi}_\pm(p) + V_{\pm n}\hat{\Psi}_\mp(p)$$

$$- \int_{\mathbb{R}_0}\int_{\mathbb{R}_0} \left[\hat{\bar{\Psi}}_+(p_1)\hat{\bar{\Psi}}_+(p_2) + \hat{\bar{\Psi}}_-(p_1)\hat{\bar{\Psi}}_-(p_2)\right]\hat{\Psi}_\pm(p - p_1 + p_2)dp_1 dp_2$$

$$- \int_{\mathbb{R}_0}\int_{\mathbb{R}_0} \hat{\bar{\Psi}}_+(p_1)\hat{\bar{\Psi}}_-(p_2)\hat{\Psi}_\mp(p - p_1 + p_2)dp_1 dp_2$$

$$= \epsilon p^2 \hat{\Psi}_\pm(p) + \epsilon^{1/3}\hat{R}_\pm(\hat{\Psi}_+, \hat{\Psi}_-, \Psi_0(\hat{\Psi}_+, \hat{\Psi}_-, \epsilon)),$$

where the linear terms $p^2\hat{\Psi}_\pm(p)$ are controlled by the bounds

$$\epsilon\|p^2\hat{\Psi}_\pm(p)\|_{L^1(\mathbb{R}_0)} \le \epsilon^{1/3}\|\hat{\Psi}_\pm\|_{L^1(\mathbb{R}_0)}$$

and the remainder terms $\hat{R}_\pm$ are controlled by the bound (2.2.30) in Lemma 2.9,

$$\forall \epsilon \in (0, \epsilon_0): \quad \exists C_\pm > 0: \quad \|\hat{R}_\pm\|_{L^1(\mathbb{R}_0)} \le C_\pm\left(\|\hat{\Psi}_+\|_{L^1(\mathbb{R}_0)} + \|\hat{\Psi}_-\|_{L^1(\mathbb{R}_0)}\right).$$

Therefore, the system on $\hat{\Psi}_\pm(p)$ is a perturbation of the nonlinear Dirac equations (2.2.20) in Fourier space after it is truncated on $\mathbb{R}_0$ and perturbed by the remainder terms of the order of $\mathcal{O}(\epsilon^{1/3})$.

Let us now consider solutions of the system of integral equations for $\hat{\Psi}_\pm(p)$ for all $p \in \mathbb{R}$. We do this by extending the residual terms from $L^1(\mathbb{R}_0)$ to $L^1(\mathbb{R})$ with a compact support on $\mathbb{R}_0$. Let $\hat{\boldsymbol{\Psi}} = (\hat{\Psi}_+, \hat{\Psi}_-, \hat{\bar{\Psi}}_+, \hat{\bar{\Psi}}_-)^{\mathrm{T}}$ and denote the system of integral equations by an abstract notation

$$\hat{\mathbf{N}}(\hat{\boldsymbol{\Psi}}) = \hat{\mathbf{R}}(\hat{\boldsymbol{\Psi}}).$$

If $\hat{\boldsymbol{\Psi}} = \hat{\mathbf{A}} + \hat{\boldsymbol{\Theta}}$, where $\hat{\mathbf{A}}$ is a decaying solution of $\hat{\mathbf{N}}(\hat{\mathbf{A}}) = \mathbf{0}$, then $\hat{\boldsymbol{\Theta}}$ solves the nonlinear system in the form

$$\hat{L}\hat{\boldsymbol{\Theta}} = \hat{\mathbf{F}}(\hat{\boldsymbol{\Theta}}) := \hat{\mathbf{R}}(\hat{\mathbf{A}} + \hat{\boldsymbol{\Theta}}) - \left[\hat{\mathbf{N}}(\hat{\mathbf{A}} + \hat{\boldsymbol{\Theta}}) - \hat{L}\hat{\boldsymbol{\Theta}}\right], \tag{2.2.34}$$

where $\hat{L} = D_{\hat{\mathbf{A}}}\hat{\mathbf{N}}(\hat{\mathbf{A}})$ is a linearized operator and the nonlinear terms satisfy the bounds

$$\exists C > 0: \quad \|\hat{\mathbf{N}}(\hat{\mathbf{A}} + \hat{\boldsymbol{\Theta}}) - \hat{L}\hat{\boldsymbol{\Theta}}\|_{L^1(\mathbb{R})} \le C\|\hat{\boldsymbol{\Theta}}\|^2_{L^1(\mathbb{R})}$$

and

$$\forall \epsilon \in (0, \epsilon_0): \quad \exists C > 0: \quad \|\hat{\mathbf{R}}(\hat{\mathbf{A}} + \hat{\boldsymbol{\Theta}})\|_{L^1(\mathbb{R})} \le C\epsilon^{1/3}.$$

The linearized differential operator associated to the nonlinear Dirac equations (2.2.20) in physical space is given by a self-adjoint system of 4×4 component Dirac operators,

$$L = \begin{bmatrix} in\partial_X + D_0 & -A^2 & V_n - 2A\bar{B} & -2AB \\ -\bar{A}^2 & -in\partial_X + D_0 & -2\bar{A}\bar{B} & \bar{V}_n - 2AB \\ \bar{V}_n - 2\bar{A}B & -2AB & -in\partial_X + D_0 & -B^2 \\ -2\bar{A}\bar{B} & V_n - 2A\bar{B} & -\bar{B}^2 & in\partial_X + D_0 \end{bmatrix}, \tag{2.2.35}$$

where $D_0 = \Omega - 2(|A|^2 + |B|^2)$. The linearized operator (2.2.35) is block-diagonalized into two uncoupled 2×2 Dirac operators, each having a one-dimensional kernel (Theorem 4.20 in Section 4.3.5). The two-dimensional kernel of the linearized operator (2.2.35) is spanned by the eigenvectors

$$\left(A'(X), B'(X), \bar{A}'(X), \bar{B}'(X) \right), \quad \left(iA, iB, -i\bar{A}, -i\bar{B} \right), \tag{2.2.36}$$

which are related to the symmetries of the nonlinear Dirac equations (2.2.20) with respect to spatial translation and gauge transformation. The zero eigenvalue of the linearized operator (2.2.35) is bounded away from the rest of the spectrum of L if $\Omega \in (-|V_n|, |V_n|)$.

By the Fredholm Alternative Theorem (Appendix B.4), there exists a solution of system (2.2.34) if and only if the right-hand side $\hat{\mathbf{F}}(\hat{\boldsymbol{\Theta}})$ lies in the range of the linearized operator $\hat{L}$. We recall that the original stationary Gross–Pitaevskii equation (2.2.21) inherits only gauge invariance, so that we can uniquely fix the phase factor of $\Phi(x)$ by requiring that

$$\text{Im}\,\Phi(0) = 0.$$

On the other hand, if $V(-x) = V(x)$, then both $\Phi(x)$ and $\Phi(-x)$ are solutions of (2.2.21) and one can look for even solutions under the constraint

$$\text{Re}\,\Phi'(0) = 0.$$

These two constraints are compatible with the localized mode satisfying

$$A(X) = \bar{A}(-X) = \bar{B}(X) \quad \text{for all} \quad X \in \mathbb{R}.$$

Since eigenvectors (2.2.36) violate the aforementioned constraints, we infer that the operator L is invertible in the corresponding constrained space.

By the Banach Fixed-Point Theorem (Appendix B.2), there exists a unique solution

$$\hat{\boldsymbol{\Theta}} = \hat{L}^{-1}\hat{\mathbf{F}}(\hat{\boldsymbol{\Theta}}) \in X := L^1(\mathbb{R}) \times L^1(\mathbb{R}) \times L^1(\mathbb{R}) \times L^1(\mathbb{R})$$

such that

$$\forall \epsilon \in (0, \epsilon_0): \quad \exists C, C' > 0: \quad \|\hat{\boldsymbol{\Theta}}\|_X \leq C \|\hat{\mathbf{R}}(\hat{\mathbf{A}})\|_X \leq C' \epsilon^{1/3}.$$

As a result, for all $\epsilon \in (0, \epsilon_0)$, there exists $C > 0$ such that

$$\left\| \hat{\Psi}_+ - \hat{A} \right\|_{L^1(\mathbb{R})} + \left\| \hat{\Psi}_- - \hat{B} \right\|_{L^1(\mathbb{R})} \leq C \epsilon^{1/3}. \tag{2.2.37}$$

Lastly, we consider the system of integral equations for $\hat{\Psi}_\pm(p)$ for all $p \in \mathbb{R}_0$. This system is different from system (2.2.34) by the terms bounded by $\|\hat{\boldsymbol{\Theta}}\|_{L^1_1(\mathbb{R} \backslash \mathbb{R}_0, \mathbb{C}^4)}$. We need to show that these terms have the order of at least $\mathcal{O}(\epsilon^{1/3})$ to agree with the order of the residual term $\hat{\mathbf{R}}(\hat{\mathbf{A}})$. To show this, we recall that the inverse operator $\hat{L}^{-1}$ is a map from $L^1(\mathbb{R}, \mathbb{C}^4)$ to $L^1_1(\mathbb{R}, \mathbb{C}^4)$ thanks to the first derivatives in the differential operator (2.2.35). Therefore, for any $\epsilon \in (0, \epsilon_0)$ and any solution $\hat{\boldsymbol{\Theta}} = \hat{L}^{-1}\hat{\mathbf{F}}(\hat{\boldsymbol{\Theta}}) \in L^1_1(\mathbb{R}, \mathbb{C}^4)$, there exist $C, C' > 0$ such that

$$\|\hat{\boldsymbol{\Theta}}\|_{L^1_1(\mathbb{R} \backslash \mathbb{R}_0, \mathbb{C}^4)} \leq \|\hat{\boldsymbol{\Theta}}\|_{L^1_1(\mathbb{R}, \mathbb{C}^4)} \leq C \|\hat{\mathbf{R}}(\hat{\mathbf{A}})\|_{L^1(\mathbb{R}, \mathbb{C}^4)} \leq C' \epsilon^{1/3}.$$

The desired bound of Lemma 2.10 follows from the bound (2.2.37). $\qquad\square$

We can now prove Theorem 2.2 using Lemmas 2.9 and 2.10.

Proof of Theorem 2.2 When the solution $\Phi(x)$ is represented by the Fourier transform $\hat{\Phi}(k)$ and both scaling transformations of Lemma 2.9 and 2.10 are incorporated into the solution, we obtain the bound

$$\forall \epsilon \in (0, \epsilon_0): \quad \exists C > 0: \quad \left\| \hat{\Phi} - \frac{1}{\epsilon}\hat{A}\left(\frac{\cdot - k_n}{\epsilon}\right) - \frac{1}{\epsilon}\hat{B}\left(\frac{\cdot + k_n}{\epsilon}\right) \right\|_{L^1(\mathbb{R})} \le C\epsilon^{1/3}.$$

This bound in Fourier space implies the desired bound (2.2.25) in the original physical space.

It remains to prove that the solution $\Phi(x)$ constructed in this algorithm is real-valued and even in x. The real-valued property follows from the symmetry of the map $\Phi_0(k) = \Psi_0(\hat{\Phi}_+, \hat{\Phi}_-, \epsilon)$ constructed in Lemma 2.9 with respect to the interchange of $\hat{\Phi}_+(k - k_n)$ and $\hat{\Phi}_-(k + k_n)$ and complex conjugation. As a result, the system of integral equations for $\hat{\Psi}_\pm(k)$ has the symmetry reduction $\hat{\Psi}_+(p) = \bar{\hat{\Psi}}_-(p)$, which is satisfied by the localized mode (2.2.22). When the partition (2.2.28) is substituted into the Fourier transform with the symmetry $\hat{\Phi}_+(k - k_n) = \bar{\hat{\Phi}}_-(k + k_n)$, the resulting solution $\Phi(x)$ becomes real-valued. Similar arguments apply to prove that the solution $\Phi(x)$ is even in x, which corresponds to the symmetry reduction $\hat{\Phi}_+(k - k_n) = \hat{\Phi}_-(k + k_n)$ and the invariance of the map $\Phi_0(k) = \Psi_0(\hat{\Phi}_+, \hat{\Phi}_-, \epsilon)$ with respect to this reduction. □

Exercise 2.17 Thin intervals $\mathbb{R}'_\pm$ used in Theorem 2.2 have small length $c\epsilon^r$, where $c = 2$ and $r = \frac{2}{3}$. Generalize all proofs for any constant $c > 0$ and any scaling factor $\frac{1}{2} < r < 1$.

Persistence of non-symmetric decaying solutions in the stationary Gross–Pitaevskii equation (2.2.21) can be resolved at the algebraic order of the asymptotic expansion in ϵ. It does not involve beyond-all-orders expansions at the exponentially small order, unlike the case of the continuous NLS equation (Section 2.3).

Exercise 2.18 Assume the general case of non-symmetric $V(x)$ and find a constraint on parameter X_0 in (2.2.22) from persistence analysis of Lemma 2.10.

Exercise 2.19 Consider the stationary two-dimensional Gross–Pitaevskii equation with a small 2π-periodic potential in each coordinate

$$-(\partial_{x_1}^2 + \partial_{x_2}^2)\phi(x_1, x_2) + \epsilon V(x_1, x_2)\phi(x_1, x_2) + |\phi|^2\phi = \omega\phi(x_1, x_2). \qquad (2.2.38)$$

Characterize the set of resonant Fourier modes of the double Fourier series and derive the following system of algebraic equations,

$$\Omega a_1 + V_{2,2}a_2 + V_{0,2}a_3 + V_{2,0}a_4 = (|a_1|^2 + 2|a_2|^2 + 2|a_3|^2 + 2|a_4|^2)a_1 + 2\bar{a}_2 a_3 a_4,$$
$$\Omega a_2 + V_{-2,-2}a_1 + V_{-2,0}a_3 + V_{0,-2}a_4 = (2|a_1|^2 + |a_2|^2 + 2|a_3|^2 + 2|a_4|^2)a_2 + 2\bar{a}_1 a_3 a_4,$$
$$\Omega a_3 + V_{2,-2}a_4 + V_{0,-2}a_1 + V_{2,0}a_2 = (2|a_1|^2 + 2|a_2|^2 + |a_3|^2 + 2|a_4|^2)a_3 + 2\bar{a}_4 a_1 a_2,$$
$$\Omega a_4 + V_{-2,2}a_3 + V_{-2,0}a_1 + V_{0,2}a_2 = (2|a_1|^2 + 2|a_2|^2 + 2|a_3|^2 + |a_4|^2)a_4 + 2\bar{a}_3 a_1 a_2,$$

for four antiperiodic resonant modes

$$(k_1, k_2) \in \left\{ \left(\frac{1}{2}, \frac{1}{2} \right); \left(-\frac{1}{2}, -\frac{1}{2} \right); \left(\frac{1}{2}, -\frac{1}{2} \right); \left(-\frac{1}{2}, \frac{1}{2} \right) \right\}$$

corresponding to the same value $\omega - \frac{1}{2}$.

The system of four coupled equations in two-dimensional periodic equations was derived by Agueev & Pelinovsky [6]. It is important to realize that the two-dimensional Schrödinger operator

$$L = -\partial_{x_1}^2 - \partial_{x_2}^2 + \epsilon V(x_1, x_2)$$

has no spectral gaps for small $\epsilon > 0$. As a result, no localized modes exist in the stationary two-dimensional Gross–Pitaevskii equation (2.2.38) in the limit of small periodic potential. Hence the justification analysis of the differential system of four coupled equations must break down. Indeed, the operator $|k|^2 - \omega$ with $\omega = \frac{1}{4}$ and $k \in \mathbb{R}^2$ is not invertible in the neighborhood of a circle of radius $|k| = \frac{1}{2}$. Since only finitely many parts of the circle are excluded from the compact support of $\hat{\Phi}_0(k)$, Lemma 2.9 fails and no map from finitely many large parts of the solution $\hat{\Phi}(k)$ for the resonant Fourier modes to the small remainder part $\hat{\Phi}_0(k)$ exists.

2.3 Justification of the nonlinear Schrödinger equation

The second example of the justification analysis deals with a regular case of bounded potentials. The asymptotic multi-scale expansion method (Section 1.1.2) and the spectral theory of Bloch decomposition (Section 2.1.2) give everything we need to reduce the Gross–Pitaevskii equation with a periodic potential,

$$iu_t = -u_{xx} + V(x)u + |u|^2 u, \tag{2.3.1}$$

to the nonlinear Schrödinger equation,

$$ia_T = \alpha a_{XX} + \beta |a|^2 a, \tag{2.3.2}$$

where $a = a(X, T) : \mathbb{R} \times \mathbb{R} \to \mathbb{C}$ is an amplitude function in new (slow) variables and (α, β) are nonzero numerical coefficients.

Roughly speaking, the NLS equation (2.3.2) is valid for small-amplitude slowly modulated packets of the Bloch wave $u_{n_0}(x; k_0) e^{-i\omega_0 t}$, where $n_0 \in \mathbb{N}$ and $k_0 \in \mathbb{T}$ are fixed numbers for the spectral band and the quasi-momentum of the spectral problem $Lu_{n_0} = \omega_0 u_{n_0}$ corresponding to the eigenvalue $\omega_0 = E_{n_0}(k_0)$. The new variable X is related to the moving coordinate $x - E_{n_0}'(k_0)t$. Stationary solutions of the NLS equation (2.3.2) approximate moving pulse solutions of the Gross–Pitaevskii equation (2.3.1) if $E_{n_0}'(k_0) \neq 0$. However, moving pulse solutions do not generally exist in the periodic potentials V (Section 5.6).

This fact is not a contradiction, since the time-dependent NLS equation (2.3.2) is only justified for finite time intervals and the moving pulse solutions of the NLS equation (2.3.2) decay in amplitude during their time evolution in the Gross–Pitaevskii equation (2.3.1). Nevertheless, we shall avoid dealing with moving pulse solutions and shall choose $k_0 \in \mathbb{T}$ such that $E'_{n_0}(k_0) = 0$. In this case, the new variables X and T are rescaled versions of x and t.

By properties of $E_n(k)$ (Section 2.1.2), the only possibility of $E'_{n_0}(k_0) = 0$ is that $\omega_0 = E_{n_0}(k_0)$ is chosen at the band edge (either at $E^+_{n_0}$ or at $E^-_{n_0}$) of the n_0th spectral band. Moreover, if the n_0th spectral band is disjoint from another spectral band by a nonzero band gap, then $E'_{n_0}(k_0) = 0$. Hence we assume here that $\omega_0 = E_{n_0}(k_0)$ is chosen at the band edge between a nonzero band gap and the n_0th spectral band.

Exercise 2.20   Let $\omega_0 = E_{n_0}(k_0)$ be the band edge and assume that there exists only one linearly independent 2π-periodic solution u_0 of

$$Lu = \omega_0 u,$$

normalized by $\|u_0\|_{L^2_{\mathrm{per}}} = 1$. Use the Fredholm Alternative Theorem (Appendix B.4) and show that there exists a 2π-periodic solution of the inhomogeneous equation

$$Lu_1 = \omega_0 u_1 + 2u'_0(x),$$

which is uniquely defined by $\langle u_0, u_1 \rangle_{L^2_{\mathrm{per}}} = 0$. Expand solutions of

$$L_k w_{n_0}(x; k) = E_{n_0}(k) w_{n_0}(x; k)$$

in power series of k near $k_0 = 0$ using analyticity of $E_{n_0}(k)$ and $w_{n_0}(x; k)$ and prove that $E'_{n_0}(k_0) = 0$ and $E''_{n_0}(k_0) = 2 + 4\langle u_0, \partial_x u_1 \rangle_{L^2_{\mathrm{per}}}$.

A simple method exists to determine the numerical coefficient α of the NLS equation (2.3.2) from the spectral theory of operator $L = -\partial^2_x + V(x)$ without developing the nonlinear analysis of the problem. If the nonlinear term $|u|^2 u$ is dropped, the Gross–Pitaevskii equation (2.3.1) becomes the linear time-dependent Schrödinger equation which is solved with the Bloch wave function

$$u(x,t) = e^{-iE_{n_0}(k)t} u_{n_0}(x; k) = e^{ikx - iE_{n_0}(k)t} w_{n_0}(x; k),$$

where $w_{n_0}(x; k)$ is a solution of

$$L w_{n_0} - 2ik\partial_x w_{n_0} + k^2 w_{n_0} = E_{n_0} w_{n_0}.$$

For simplicity, let $k_0 = 0$ and consider 2π-periodic Bloch functions. Since the n_0th spectral band is disjoint from other spectral bands, there exists a unique 2π-periodic eigenfunction u_0 of operator L for $\omega_0 = E_{n_0}(0)$ and the functions $E_{n_0}(k)$ and $w_{n_0}(x; k)$ are analytic near $k = 0$. Using the Taylor series expansions (Exercise 2.20), we obtain

$$E_{n_0}(k) = \omega_0 + \omega_2 k^2 + \mathcal{O}(k^4), \quad w_{n_0}(x; k) = u_0(x) + iku_1(x) + \mathcal{O}(k^2),$$

where $\omega_2 = \frac{1}{2} E''_{n_0}(0) = 1 + 2\langle u_0, \partial_x u_1 \rangle_{L^2_{\mathrm{per}}}$.

Let us express the smallness of k using a formal small parameter $\epsilon > 0$ and the new ϵ-independent parameter p with the correspondence $k = \epsilon^{1/2} p$. If $k \in \mathbb{T} := \left[-\frac{1}{2}, \frac{1}{2}\right]$, then $p \in \mathbb{T}_\epsilon$, where

$$\mathbb{T}_\epsilon := \left[-\frac{1}{2\epsilon^{1/2}}, \frac{1}{2\epsilon^{1/2}}\right].$$

The Bloch wave function is now represented by the expansion in powers of $\epsilon^{1/2}$ as follows:

$$u(x,t) = e^{i\epsilon^{1/2}px - it(\omega_0 + \epsilon\omega_2 p^2 + \mathcal{O}(\epsilon^2))} \left(u_0(x) + i\epsilon^{1/2} p u_1(x) + \mathcal{O}(\epsilon) \right)$$
$$= e^{-i\omega_0 t} \left(a(X,T) u_0(x) + \epsilon^{1/2} a_X(X,T) u_1(x) + \mathcal{O}(\epsilon) \right),$$

where $X = \epsilon^{1/2}x$, $T = \epsilon t$, and $a(X,T) = e^{ipX - ip^2\omega_2 T}$ satisfies $ia_T = \alpha a_{XX}$ with

$$\alpha = -\omega_2 = -1 - 2\langle u_0, \partial_x u_1 \rangle_{L^2_{\text{per}}}.$$

To compute the numerical coefficient β of the NLS equation (2.3.2), we consider 2π-periodic solutions of the Gross–Pitaevskii equation (2.3.1) in the form

$$u(x,t) = \epsilon^{1/2} \left(a(T) u_0(x) + \mathcal{O}(\epsilon) \right) e^{-i\omega_0 t},$$

where $T = \epsilon t$. Substitution of the asymptotic expansion into the Gross–Pitaevskii equation (2.3.1) and projection to u_0 using the Fredholm Alternative Theorem (Appendix B.4) shows that $a(T)$ satisfies $ia_T = \beta |a|^2 a$ with $\beta = \|u_0\|_{L^4_{\text{per}}}^4$.

We still need to justify the NLS equation (2.3.2) for solutions $\phi(x,t)$ decaying to zero as $|x| \to \infty$. Our method will follow the work of Busch *et al.* [24] but the L^1 space for the Bloch transform will be used instead of the L^2 space. This approach is similar to the recent work of Dohnal *et al.* [50]. We will also look at the relevance of the stationary NLS equation for bifurcation of localized modes of the stationary Gross–Pitaevskii equation. Further works on the justification of the stationary NLS equation can be found in Dohnal & Uecker [51] for the cubic nonlinearity and in Ilan & Weinstein [90] for a more general power nonlinearity.

Other works, where the time-dependent NLS equation was justified in the context of spatially homogeneous hyperbolic equations, include the works of Colin [38], Kirrmann *et al.* [113], Lannes [129], and Schneider [184]. A difficulty that arises in problems with space-periodic coefficients compared to problems with space-constant coefficients is that the approximation equation (2.3.2) lives in a spatially homogeneous domain, where Fourier analysis is used, whereas the original system (2.3.1) lives in a spatially periodic domain, where Bloch analysis is applied. We will get around this difficulty by cutting the support of the Bloch transform around the resonance point $k = k_0$ for $n = n_0$ and by applying a scaling transformation suggested by the asymptotic multi-scale expansion method.

2.3.1 Justification of the time-dependent equation

Let us fix $n_0 \in \mathbb{N}$ and $\omega_0 = E_{n_0}(k_0)$ for either $k_0 = 0$ or $k_0 = \frac{1}{2}$. We define functions $u_0(x)$ and $u_1(x)$ as in Exercise 2.20. For simplicity, we will assume $k_0 = 0$ and work with 2π-periodic functions throughout this section. Our main result is the following justification theorem.

Theorem 2.3 *Let $V \in L^\infty_{\text{per}}([0, 2\pi])$. Fix $n_0 \in \mathbb{N}$ and assume that the n_0th spectral band of operator $L = -\partial_x^2 + V(x)$ is disjoint from other spectral bands of L. Fix*

$T_0 > 0$ *such that* $a(T) \in C([0,T_0], H^6(\mathbb{R}))$ *is a local solution of the NLS equation (2.3.2) with coefficients*

$$\alpha = -1 - 2\langle u_0, \partial_x u_1 \rangle_{L^2_{\mathrm{per}}}, \quad \beta = \|u_0\|^4_{L^4_{\mathrm{per}}}$$

and the initial data $a(0) \in H^6(\mathbb{R})$. *Fix* $C_0 > 0$ *and* $s \in \left(\frac{1}{2}, 1\right)$ *such that* $\hat{u}(0) \in L^1(\mathbb{T}, l^1_s(\mathbb{N}))$ *is the Bloch transform of* $u(0) \in C^0_b(\mathbb{R})$ *satisfying the bound*

$$\|u(0) - \epsilon^{1/2} a(\epsilon^{1/2} \cdot, 0) u_0\|_{L^\infty} \le C_0 \epsilon.$$

There exist $\epsilon_0 > 0$ *and* $C > 0$ *such that for all* $\epsilon \in (0, \epsilon_0)$, *the Gross–Pitaevskii equation (2.3.1) admits a solution* $u(t) \in C([0, \epsilon^{-1} T_0], C^0_b(\mathbb{R}))$ *satisfying the bound*

$$\|u(\cdot, t) - \epsilon^{1/2} a(\epsilon^{1/2} \cdot, \epsilon t) u_0 e^{-i\omega_0 t}\|_{L^\infty} \le C\epsilon, \quad t \in [0, \epsilon^{-1} T_0]. \tag{2.3.3}$$

Moreover, $u(x, t) \to 0$ *as* $|x| \to \infty$ *for all* $t \in [0, \epsilon^{-1} T_0]$.

Proof The proof of Theorem 2.3 consists of four steps.

Step 1: Decomposition. Let us represent a solution of the Gross–Pitaevskii equation (2.3.1) by the asymptotic multi-scale expansion,

$$u(x, t) = \epsilon^{1/2} \left(a(X, T) u_0(x) + \epsilon^{1/2} a_X(X, T) u_1(x) + \epsilon \varphi_\epsilon(x, t) \right) e^{-i\omega_0 t}, \tag{2.3.4}$$

where $X = \epsilon^{1/2} x$ and $T = \epsilon t$. Assuming that $a(X, T)$ solves the NLS equation (2.3.2), we find the time evolution problem for $\varphi_\epsilon(x, t)$ in the form

$$i\partial_t \varphi_\epsilon = (L - \omega_0)\varphi_\epsilon + F_\epsilon(a, \varphi_\epsilon), \tag{2.3.5}$$

where

$$F_\epsilon(a, \varphi_\epsilon) = -2a_{XX} \left(\partial_x u_1 - \langle u_0, \partial_x u_1 \rangle_{L^2_{\mathrm{per}}} u_0 \right) - \epsilon^{1/2} \left(i a_{XT} u_1 + a_{XXX} u_1 \right)$$
$$+ |au_0 + \epsilon^{1/2} a_X u_1 + \epsilon \varphi_\epsilon|^2 (au_0 + \epsilon^{1/2} a_X u_1 + \epsilon \varphi_\epsilon) - \beta |a|^2 a u_0.$$

The term $\epsilon \varphi_\epsilon$ of the asymptotic expansion (2.3.4) is referred to as the *remainder term* and the vector field $F_\epsilon(a, \varphi_\epsilon)$ of the time evolution problem (2.3.5) is referred to as the *residual term*.

Step 2: Normal form transformation. Justification of the NLS equation (2.3.2) is based on the application of the Gronwall inequality (Appendix B.6) to the time evolution problem (2.3.5). However, a simple inspection of this equation shows that the residual term is of the order of $\mathcal{O}(1)$ as $\epsilon \to 0$ and it will result in the remainder term $\epsilon \varphi_\epsilon$ of the same order $\mathcal{O}(1)$ as the leading-order term au_0 at the scale $t = \mathcal{O}(\epsilon^{-1})$ or $T = \mathcal{O}(1)$ as $\epsilon \to 0$. Therefore, before proceeding with the nonlinear analysis, let us remove the leading-order part of the residual term with a simple transformation

$$\epsilon \varphi_\epsilon = \epsilon \varphi_1 + \epsilon^{1/2} \psi_\epsilon,$$

where φ_1 is a solution of the equation

$$(L - \omega_0)\varphi_1 = 2a_{XX} \left(\partial_x u_1 - \langle u_0, \partial_x u_1 \rangle_{L^2_{\mathrm{per}}} u_0 \right) - |a|^2 a(u_0^3 - \langle u_0, u_0^3 \rangle_{L^2_{\mathrm{per}}} u_0) \tag{2.3.6}$$

and $\epsilon^{1/2}\psi_\epsilon$ is a new remainder term, which solves a new time-dependent problem

$$i\partial_t\psi_\epsilon = (L - \omega_0)\psi_\epsilon + \epsilon G_\epsilon(a, \psi_\epsilon) \qquad (2.3.7)$$

with a new residual term in the form

$$G_\epsilon(a, \psi_\epsilon) = -i\epsilon^{1/2}\partial_T\varphi_1 - 2\partial^2_{xX}\varphi_1 - \epsilon^{1/2}\partial^2_{XX}\varphi_1 - ia_{TX}u_1 - a_{XXX}u_1 + \epsilon^{-1/2}$$
$$\times\left(|au_0 + \epsilon^{1/2}a_Xu_1 + \epsilon\varphi_1 + \epsilon^{1/2}\psi_\epsilon|^2(au_0 + \epsilon^{1/2}a_Xu_1 + \epsilon\varphi_1 + \epsilon^{1/2}\psi_\epsilon) - |a|^2au_0^3\right).$$

By the Fredholm Alternative Theorem (Appendix B.4), a linear inhomogeneous equation

$$(L - \omega_0)\varphi = h$$

admits a unique solution $\varphi \in H^2_{\text{per}}(\mathbb{R})$ for a given $h \in L^2_{\text{per}}([0, 2\pi])$ if and only if $\langle u_0, h\rangle_{L^2_{\text{per}}} = 0$. (Recall that $u_0 \in L^2_{\text{per}}([0, 2\pi])$ is the only 2π-periodic solution of the homogeneous equation $(L - \omega_0)u_0 = 0$.) To eliminate the homogeneous solution and to determine $\varphi \in H^2_{\text{per}}([0, 2\pi])$ uniquely, one needs to add the orthogonality condition $\langle u_0, \varphi\rangle_{L^2_{\text{per}}} = 0$. As a result, the linear inhomogeneous equation (2.3.6) admits a unique solution in the form

$$\varphi_1 = a_{XX}f_1(x) + |a|^2af_2(x),$$

where unique $f_{1,2} \in H^2_{\text{per}}(\mathbb{R})$ satisfy $\langle u_0, f_{1,2}\rangle_{L^2_{\text{per}}} = 0$. Note that the choice of coefficients (α, β) in the NLS equation (2.3.2) gives the orthogonality of the right-hand side of the inhomogeneous equation (2.3.6) to u_0 in $L^2_{\text{per}}([0, 2\pi])$.

Remark 2.4 Non-resonance conditions such as

$$\exists C > 0 : \quad \inf_{n\in\mathbb{N}\setminus\{n_0\}} |E_n(3k_0) - 3E_{n_0}(k)| \geq C$$

are required typically to ensure the existence of the normal form transformation [24]. We do not need any non-resonance conditions thanks to the gauge invariance of the Gross–Pitaevskii equation (2.3.1), which allows us to detach the factor $e^{-i\omega_0 t}$ from the solution $u(x, t)$.

The theorem will be proved if we can show that the remainder term $\epsilon^{1/2}\psi_\epsilon$ has the order of $\mathcal{O}(\epsilon^{1/2})$ at the scale $t = \mathcal{O}(\epsilon^{-1})$ as $\epsilon \to 0$.

Step 3: Fixed-point iterations. It is now time to make all formal computations of the asymptotic multi-scale expansion method rigorous. We shall work in the space $L^1(\mathbb{T}, l^1_s(\mathbb{N}))$ for the Bloch transform of the solution of the Gross–Pitaevskii equation (2.3.1) for a fixed $t \geq 0$. By Lemma 2.2, if $\hat{\phi} \in L^1(\mathbb{T}, l^1_s(\mathbb{N}))$ with $s > \frac{1}{2}$, then $\phi \in C^0_b(\mathbb{R})$ and $\phi(x) \to 0$ as $|x| \to \infty$.

Let us apply the Bloch transform to the time evolution equation (2.3.7) and write it in the abstract form

$$i\partial_t\hat{\psi}_\epsilon = (\hat{L} - \omega_0)\hat{\psi}_\epsilon + \epsilon\hat{G}_\epsilon(\hat{a}, \hat{\psi}_\epsilon), \qquad (2.3.8)$$

where $\hat{L}$ is the image of L after the Bloch transform, $\hat{\psi}_\epsilon$ is the Bloch transform of ψ_ϵ, and $\hat{a}$ is the Fourier transform of a. We note that the differential operator L

becomes a pseudo-differential operator $\hat{L}$ after the Bloch transform, whose symbol is a multiple-valued dispersion relation expressed by $\{E_n(k)\}_{n \in \mathbb{N}}$ for $k \in \mathbb{T}$.

We need to show that the residual term $\hat{G}_\epsilon(\hat{a}, \hat{\psi}_\epsilon)$ for a fixed $\hat{a}$ maps an element $\hat{\psi}_\epsilon$ in the ball $B_\delta(L^1(\mathbb{T}, l_s^1(\mathbb{N})))$ of radius $\delta > 0$ to an element of the same ball and it is Lipschitz continuous in $B_\delta(L^1(\mathbb{T}, l_s^1(\mathbb{N})))$. Generally, it is only possible under some conditions on $\hat{a}$ and we will identify these conditions in our analysis.

If we have a product of two functions $\phi(x)$ and $\varphi(x)$ of $x \in \mathbb{R}$, then the Bloch transform of the product is given by the convolution operator

$$\left(\hat{\phi} \star \hat{\varphi}\right)_n (k) = \int_{\mathbb{T}} \int_{\mathbb{T}} \sum_{m_1 \in \mathbb{N}} \sum_{m_2 \in \mathbb{N}} K_{n,n_1,n_2}(k, k_1, k_2) \hat{\phi}_{n_1}(k_1) \hat{\varphi}_{n_2}(k_2) dk_1 dk_2 \quad (2.3.9)$$

and

$$K_{n,n_1,n_2}(k, k_1, k_2) = \int_{\mathbb{R}} \bar{u}_n(x; k) u_{n_1}(x; k_1) u_{n_2}(x; k_2) dx.$$

Since the convolution operator (2.3.9) is not a standard discrete or continuous convolution, it would have been a difficult task to study how nonlinear terms are mapped by the residual term if not for the remarkable approximate convolution formula found in Appendix A of Busch *et al.* [24]. They proved that, if $V \in L^2_{\text{per}}([0, 2\pi])$, then, for any fixed $p \in (0, 2)$, there exists a positive constant C_p such that

$$\left| \int_{\mathbb{R}} \bar{u}_n(x; k), u_{n_1}(x; k_1) u_{n_2}(x; k_2) dx \right| \leq \frac{C_p}{(1 + |n - n_1 - n_2|)^p}, \quad (2.3.10)$$

for all $(n, n_1, n_2) \in \mathbb{N}^3$ and $(k, k_1, k_2) \in \mathbb{T}^3$. Using this result, we prove that $L^1(\mathbb{T}, l_s^1(\mathbb{N}))$ is a Banach algebra with respect to the convolution operator (2.3.9).

Lemma 2.11 *Let $V \in L^\infty_{\text{per}}([0, 2\pi])$ and fix $s \in (0, 1)$. There exists a constant $C > 0$ such that*

$$\forall \hat{\phi}, \hat{\varphi} \in L^1(\mathbb{T}, l_s^1(\mathbb{N})): \quad \|\hat{\phi} \star \hat{\varphi}\|_{L^1(\mathbb{T}, l_s^1(\mathbb{N}))} \leq C \|\hat{\phi}\|_{L^1(\mathbb{T}, l_s^1(\mathbb{N}))} \|\hat{\varphi}\|_{L^1(\mathbb{T}, l_s^1(\mathbb{N}))}. \quad (2.3.11)$$

Proof Using the bound (2.3.10), we develop explicit computations and obtain

$$\|\hat{\phi} \star \hat{\varphi}\|_{L^1(\mathbb{T}, l_s^1(\mathbb{N}))}$$

$$\leq \int_{\mathbb{T}} \int_{\mathbb{T}} \int_{\mathbb{T}} \sum_{n \in \mathbb{N}} (1 + n)^s \sum_{n_1 \in \mathbb{N}} \sum_{n_2 \in \mathbb{N}} |K_{n,n_1,n_2}(k, k_1, k_2)| \|\hat{\phi}_{n_1}(k_1)\| \|\hat{\varphi}_{n_2}(k_2)\| dk dk_1 dk_2$$

$$\leq C \sum_{n_1 \in \mathbb{N}} \sum_{n_2 \in \mathbb{N}} (1 + n_1)^s (1 + n_2)^s \sum_{n \in \mathbb{N}} \left(\frac{1 + n}{(1 + n_1)(1 + n_2)}\right)^s \frac{1}{(1 + |n - n_1 - n_2|)^p}$$

$$\times \int_{\mathbb{T}} \int_{\mathbb{T}} |\hat{\phi}_{n_1}(k_1)| |\hat{\varphi}_{n_2}(k_2)| dk_1 dk_2$$

$$\leq C' \sum_{n_1 \in \mathbb{N}} \sum_{n_2 \in \mathbb{N}} (1 + n_1)^s (1 + n_2)^s \sum_{n \in \mathbb{N}} \left(1 + \frac{n^s}{(1 + n_1)^s (1 + n_2)^s}\right) \frac{1}{(1 + n)^p}$$

$$\times \int_{\mathbb{T}} \int_{\mathbb{T}} |\hat{\phi}_{n_1}(k_1)| |\hat{\varphi}_{n_2}(k_2)| dk_1 dk_2$$

for some $C, C' > 0$. If $p > 1$ and $p - s > 1$ (that is $0 < s < 1$), the bound is completed as follows

$$\|\hat{\phi} \star \hat{\varphi}\|_{L^1(\mathbb{T}, l^1_s(\mathbb{N}))} \leq C_1 \|\hat{\phi}\|_{L^1(\mathbb{T}, l^1_s(\mathbb{N}))} \|\hat{\varphi}\|_{L^1(\mathbb{T}, l^1_s(\mathbb{N}))} + C_2 \|\hat{\phi}\|_{L^1(\mathbb{T}, l^1(\mathbb{N}))} \|\hat{\varphi}\|_{L^1(\mathbb{T}, l^1(\mathbb{N}))}$$

$$\leq (C_1 + C_2) \|\hat{\phi}\|_{L^1(\mathbb{T}, l^1_s(\mathbb{N}))} \|\hat{\varphi}\|_{L^1(\mathbb{T}, l^1_s(\mathbb{N}))}$$

for some $C_1, C_2 > 0$.   □

In view of Lemmas 2.2 and 2.11, we shall work in the space $L^1(\mathbb{T}, l^1_s(\mathbb{N}))$ for any fixed $s \in \left(\frac{1}{2}, 1\right)$. By comparison with the Wiener space $W^s(\mathbb{R})$ which forms a Banach algebra with respect to pointwise multiplication for any $s \geq 0$ (Appendix B.16), there are reasons to believe that the upper bound on s in Lemma 2.11 is artificial, but no improvement of it has been made so far.

We shall now consider how the Bloch transform of product terms in the residual term $\hat{G}_\epsilon(\hat{a}, \hat{\psi}_\epsilon)$ are bounded in space $L^1(\mathbb{T}, l^1_s(\mathbb{N}))$. The residual term has several product terms, which involve powers and derivatives of $a(X, T)$ and bounded 2π-periodic functions $u(x)$ for a fixed $T \in [0, T_0]$. Since $X = \epsilon^{1/2}x$, we shall represent $a(X)$ by the Fourier transform

$$\hat{a}(p) = \frac{1}{(2\pi)^{1/2}} \int_\mathbb{R} a(X) e^{-ipX} dX.$$

If $A(x) = a(\epsilon^{1/2}x)$, then

$$\hat{A}(k) = \frac{1}{\epsilon^{1/2}} \hat{a}\left(\frac{k}{\epsilon^{1/2}}\right),$$

which shows that the Brillouin zone $\mathbb{T} \ni k$ corresponds to the new interval $\mathbb{T}_\epsilon \ni p$ and that $\mathbb{T}_\epsilon \to \mathbb{R}$ as $\epsilon \to 0$. If the Bloch transform in k is supported on $\mathbb{T}$, it makes sense therefore to consider the compact support of $\hat{a}(p)$ in $\mathbb{T}_\epsilon$. Note that the L^1 norm for the Fourier transform of $A(x) = a(X)$ is invariant with respect to the small parameter ϵ in the sense

$$\|\hat{A}\|_{L^1} = \int_\mathbb{T} |\hat{A}(k)| dk = \int_{\mathbb{T}_\epsilon} |\hat{a}(p)| dp = \|\hat{a}\|_{L^1}.$$

The NLS equation (2.3.2) can be written in Fourier space using the convolution integrals,

$$i\partial_T \hat{a}(p) + \alpha p^2 \hat{a}(p) = \beta \int_\mathbb{R} \int_\mathbb{R} \hat{a}(p_1) \bar{\hat{a}}(p_2) a(p + p_2 - p_1) dp_1 dp_2, \quad p \in \mathbb{R}, \quad (2.3.12)$$

where the T dependence of $\hat{a}(p, T)$ is not written. The previous discussion shows, however, that we are going to consider approximations of solutions of this equation on the compact set $\mathbb{T}_\epsilon$, that is, we will deal with solutions of the ϵ-dependent equation

$$i\partial_T \hat{a}_{app}(p) + \alpha p^2 \hat{a}_{app}(p) = \beta \int_{\mathbb{T}_\epsilon} \int_{\mathbb{T}_\epsilon} \hat{a}_{app}(p_1) \bar{\hat{a}}_{app}(p_2) a_{app}(p + p_2 - p_1) dp_1 dp_2,$$

$$(2.3.13)$$

for any $p \in \mathbb{T}_\epsilon$. The following lemma describes the difference in weighted L^1 norms between the solution $\hat{a}(p)$ of the NLS equation (2.3.12) and the approximation $\hat{a}_{app}(p)$ of the truncated NLS equation (2.3.13).

Lemma 2.12 *Fix $s \geq 0$. Assume that there exist $T_0 > 0$ and a local solution $\hat{a}(T) \in C([0, T_0], L^1_{s+1}(\mathbb{R}))$ of the NLS equation in the Fourier form (2.3.12). There exists a solution $\hat{a}_{\mathrm{app}}(T) \in C([0, T_0], L^1_s(\mathbb{T}_\epsilon))$ of the truncated NLS equation (2.3.13) such that $\hat{a}_{\mathrm{app}}(0) = \hat{a}(0)$ for any $p \in \mathbb{T}_\epsilon$ and for any small $\epsilon > 0$, there is $C > 0$ such that*

$$\|\hat{a}_{\mathrm{app}}(T) - \hat{a}(T)\|_{L^1_s(\mathbb{T}_\epsilon)} \leq C\epsilon^{1/2}\|\hat{a}(T)\|_{L^1_{s+1}(\mathbb{R})}, \quad T \in [0, T_0]. \tag{2.3.14}$$

Proof Let us give the proof for $s = 0$. The proof for $s > 0$ is similar. To estimate the difference in $\|\hat{a}_{\mathrm{app}}(T) - \hat{a}(T)\|_{L^1(\mathbb{T}_\epsilon)}$, we first estimate $\|\hat{a}\|_{L^1(\mathbb{R}\backslash\mathbb{T}_\epsilon)}$ and use the fact that $L^1(\mathbb{R})$ is a Banach algebra with respect to the convolution integrals (Appendix B.16). For any small $\epsilon > 0$, there is $C > 0$ such that

$$\int_{\mathbb{R}\backslash D_\epsilon} |\hat{a}(p)|dp \leq \int_{|p| \geq \frac{1}{2\epsilon^{1/2}}} \frac{1}{(1+p^2)^{1/2}}(1+p^2)^{1/2}|\hat{a}(p)|dp \leq C\epsilon^{1/2}\|\hat{a}\|_{L^1_1(\mathbb{R})}.$$

By variation of the constant, differential equation (2.3.12) can be written in the integral form,

$$\hat{a}(p, T) = e^{i\alpha p^2 T}\hat{a}(p, 0) - i\beta \int_0^T e^{i\alpha p^2(T-T')}$$
$$\times \left(\int_\mathbb{R} \int_\mathbb{R} \hat{a}(p_1, T')\bar{\hat{a}}(p_2, T')a(p + p_2 - p_1, T')dp_1 dp_2 \right) dT'.$$

It follows from the integral equation that there are $C_1, C_2 > 0$ such that

$$\|\hat{a}(T) - \hat{a}_{\mathrm{app}}(T)\|_{L^1(\mathbb{T}_\epsilon)} \leq \|\hat{a}(0) - \hat{a}_{\mathrm{app}}(0)\|_{L^1(\mathbb{T}_\epsilon)}$$
$$+ C_1 \int_0^T \|\hat{a}(T') - \hat{a}_{\mathrm{app}}(T')\|^3_{L^1(\mathbb{T}_\epsilon)}dT' + C_2\epsilon^{1/2} \sup_{T' \in [0,T]} \|\hat{a}(T')\|_{L^1_1(\mathbb{R})}.$$

We assume that $\hat{a}_{\mathrm{app}}(0) = \hat{a}(0)$ for any $p \in \mathbb{T}_\epsilon$. Then, bound (2.3.14) holds by the Gronwall inequality (Appendix B.6) and fixed-point arguments. $\qquad\square$

It remains to bound the Bloch transform of the product terms in the residual $G_\epsilon(a, \psi_\epsilon)$ in terms of weighted L^1 norms of $\hat{a}(T)$. According to bound (2.3.14), we can replace $\hat{a}(T)$ by compactly supported $\hat{a}_{\mathrm{app}}(T)$. We shall also project all product terms to the n_0th spectral band so that we only need the n_0th component of the Bloch transform.

Lemma 2.13 *Let $\phi(x) = a_{\mathrm{app}}(X)u_0(x)$, where $X = \epsilon^{1/2}x$, $u_0 \in C^0_{\mathrm{per}}([0, 2\pi])$, and $\hat{a}_{\mathrm{app}}(p)$ is compactly supported on $\mathbb{T}_\epsilon$. For sufficiently small $\epsilon > 0$, there is $C > 0$ such that*

$$\|\hat{\phi}_{n_0}\|_{L^1(\mathbb{T})} \leq C\|\hat{a}_{\mathrm{app}}\|_{L^1(\mathbb{T}_\epsilon)},$$

where $\hat{\phi}_{n_0}$ is the Bloch transform of ϕ at the n_0th spectral band.

Proof Using definition (2.1.21) for the Bloch transform of $\phi(x)$, we compute

$$\hat{\phi}_{n_0}(k) = \langle \phi, u_{n_0}(\cdot; k) \rangle_{L^2} = \int_{\mathbb{R}} a_{\mathrm{app}}(X) u_0(x) \bar{u}_{n_0}(x; k) dx$$

$$= \int_{\mathbb{R}} a_{\mathrm{app}}(X) u_0(x) \bar{w}_{n_0}(x; k) e^{-ikx} dx$$

$$= \int_{\mathbb{R}} a_{\mathrm{app}}(X) \left(\sum_{n \in \mathbb{Z}} c_n(k) e^{-i(k-n)x} \right) dx$$

$$= \sum_{n \in \mathbb{Z}} c_n(k) \frac{1}{\epsilon^{1/2}} \hat{a}_{\mathrm{app}} \left(\frac{k-n}{\epsilon^{1/2}} \right),$$

where

$$c_n(k) = \frac{1}{2\pi} \int_0^{2\pi} u_0(x) \bar{w}_{n_0}(x; k) e^{-inx} dx, \quad n \in \mathbb{Z}$$

is the nth coefficient of the Fourier series for 2π-periodic function $u_0(x) \bar{w}_{n_0}(x; k)$. Because of the compact support of $\hat{a}_{\mathrm{app}}(p)$ on $\mathbb{T}_\epsilon$ and the normalization

$$\|w_{n_0}(\cdot; k)\|_{L^2_{\mathrm{per}}} = 1, \quad k \in \mathbb{T},$$

we obtain

$$\hat{\phi}_{n_0}(k) = c_0(k) \frac{1}{\epsilon^{1/2}} \hat{a}_{\mathrm{app}} \left(\frac{k}{\epsilon^{1/2}} \right) \quad \Rightarrow \quad \|\hat{\phi}_{n_0}\|_{L^1(\mathbb{T})} \leq \|c_0\|_{L^\infty(\mathbb{T})} \|\hat{a}_{\mathrm{app}}\|_{L^1(\mathbb{T}_\epsilon)},$$

where $\|c_0\|_{L^\infty(\mathbb{T})} \leq \frac{1}{2\pi} \|u_0\|_{L^2_{\mathrm{per}}}$. $\qquad\square$

Exercise 2.21 Under the conditions of Lemma 2.13, show that if $\phi(x) = a'_{\mathrm{app}}(X) u_1(x)$, then for small $\epsilon > 0$, there is $C > 0$ such that

$$\|\hat{\phi}_{n_0}\|_{L^1(\mathbb{T})} \leq C \|\hat{a}_{\mathrm{app}}\|_{L^1_1(\mathbb{T}_\epsilon)}.$$

The above estimates are sufficient for the main result on the residual term $\hat{G}_\epsilon(\hat{a}, \hat{\psi}_\epsilon)$ of the time evolution problem (2.3.8).

Lemma 2.14 *Let $\hat{a}_{\mathrm{app}}$ be compactly supported on $\mathbb{T}_\epsilon$ and all product terms be projected to the n_0th spectral band. Then, for any fixed $s \in (\frac{1}{2}, 1)$, we have*

$$\hat{G}_\epsilon(\hat{a}_{\mathrm{app}}, \hat{\psi}_\epsilon) : L^1_4(\mathbb{R}) \times L^1(\mathbb{T}, l^1_s(\mathbb{N})) \to L^1(\mathbb{T}, l^1_s(\mathbb{N})).$$

Fix $\delta_0, \delta > 0$ such that $\hat{a} \in B_{\delta_0}(L^1_4(\mathbb{R}))$ and $\hat{\psi}_\epsilon \in B_\delta(L^1(\mathbb{T}, l^1_s(\mathbb{N})))$. There exist positive constants $C_1(\delta_0, \delta)$ and $C_2(\delta_0, \delta)$ such that

$$\|\hat{G}_\epsilon(\hat{a}_{\mathrm{app}}, \hat{\psi}_\epsilon)\|_{L^1(\mathbb{T}, l^1_s(\mathbb{N}))} \leq C_1(\delta_0, \delta) \left(\|\hat{a}_{\mathrm{app}}\|_{L^1_4(\mathbb{R})} + \|\hat{\psi}_\epsilon\|_{L^1(\mathbb{T}, l^1_s(\mathbb{N}))} \right), \tag{2.3.15}$$

$$\|\hat{G}_\epsilon(\hat{a}_{\mathrm{app}}, \hat{\psi}_\epsilon) - \hat{G}_\epsilon(\hat{a}_{\mathrm{app}}, \tilde{\hat{\psi}}_\epsilon)\|_{L^1(\mathbb{T}, l^1_s(\mathbb{N}))} \leq C_2(\delta_0, \delta) \|\hat{\psi}_\epsilon - \tilde{\hat{\psi}}_\epsilon\|_{L^1(\mathbb{T}, l^1_s(\mathbb{N}))}. \tag{2.3.16}$$

Proof According to the explicit form, the residual term $G_\epsilon(a, \psi_\epsilon)$ contains product terms involving powers of a and up to the fourth derivative of a in $X = \epsilon^{1/2} x$ and bounded 2π-periodic functions of x. By Lemma 2.13 and properties of Wiener spaces (Appendix B.16), all product terms are controlled by the L_4^1 norm on $\hat{a}_{\mathrm{app}}$. By Lemma 2.11, every term of the residual $\hat{G}_\epsilon(\hat{a}, \hat{\psi}_\epsilon)$ belongs to $L^1(\mathbb{T}, l_s^1(\mathbb{N}))$ for a fixed $s \in \left(\frac{1}{2}, 1\right)$. Bounds (2.3.15) and (2.3.16) are found from analyticity of $G_\epsilon(a, \psi_\epsilon)$ in variables a and ψ_ϵ. $\qquad\square$

Step 4: Control on the error bound. By Theorem 1.1 (Section 1.3.1), the initial-value problem for the NLS equation (2.3.2) is locally well-posed in space $H^s(\mathbb{R})$ for any $s > \frac{1}{2}$. By Lemma 2.14, we need to require that $\hat{a}_{\mathrm{app}}(T) \in L_4^1(\mathbb{R})$ remains for all $T \in [0, T_0]$ for some fixed $T_0 > 0$. By Lemma 2.12 with $s = 4$, we need $\hat{a}(T)$ to belong to $L_5^1(\mathbb{R})$ for all $T \in [0, T_0]$.

Recall that there exists a positive constant $C_{s,q}$ such that

$$\|\hat{a}\|_{L_s^1} \leq C_{s,q} \|\hat{a}\|_{L_{s+q}^2}, \quad s \geq 0, \quad q > \frac{1}{2}.$$

Therefore, we shall use the local well-posedness of the NLS equation (2.3.2) in $H^s(\mathbb{R})$ for any $s > 5 + \frac{1}{2}$, for instance for the nearest integer $s = 6$.

Using the Duhamel principle, we rewrite the time evolution problem (2.3.8) after $\hat{a}$ is replaced by the approximation $\hat{a}_{\mathrm{app}}$ in the integral form

$$\hat{\psi}_\epsilon(t) = e^{-\mathrm{i}t(\hat{L}-\omega_0)}\hat{\psi}_\epsilon(0) - \mathrm{i}\epsilon \int_0^t e^{-\mathrm{i}(t-t')(\hat{L}-\omega_0)} \hat{G}_\epsilon(\hat{a}_{\mathrm{app}}(\epsilon t'), \hat{\psi}_\epsilon(t')) dt'.$$

Since $\hat{L} \in \mathbb{R}$ is a multiplication operator, we have the norm preservation of the semi-group $e^{-\mathrm{i}t(\hat{L}-\omega_0)}$ as follows:

$$\forall \hat{\psi} \in L^1(\mathbb{T}, l_s^1(\mathbb{N})): \quad \|e^{-\mathrm{i}t(\hat{L}-\omega_0)}\hat{\psi}\|_{L^1(\mathbb{T},l_s^1(\mathbb{N}))} = \|\hat{\psi}\|_{L^1(\mathbb{T},l_s^1(\mathbb{N}))}, \quad t \in \mathbb{R}.$$

We shall choose $\delta_0 > 0$ and $\delta > 0$ large enough for $\|\hat{a}_{\mathrm{app}}(T)\|_{L_4^1} \leq \delta_0$ and $\|\hat{\psi}_\epsilon(t)\|_{L^1(\mathbb{T},l_s^1(\mathbb{N}))} \leq \delta$ so that bounds (2.3.15) and (2.3.16) of Lemma 2.14 can be used for any fixed $T \geq 0$ and $t \geq 0$. By the Banach Fixed-Point Theorem (Appendix B.2), existence and uniqueness of solutions in space $C([0, t_0], L^1(\mathbb{T}, l_s^1(\mathbb{N})))$ are proved from the integral equation if

$$\epsilon C_2(\delta_0, \delta) t_0 < 1 \quad \Rightarrow \quad C_2(\delta_0, \delta) T_0 < 1, \tag{2.3.17}$$

where we denote $T_0 = \epsilon t_0$. We also find that

$$\|\hat{\psi}_\epsilon(t)\|_{L^1(\mathbb{T},l_s^1(\mathbb{N}))} \leq \|\hat{\psi}_\epsilon(0)\|_{L^1(\mathbb{T},l_s^1(\mathbb{N}))}$$
$$+ C_1(\delta_0, \delta)\epsilon \int_0^t \left(\|\hat{a}_{\mathrm{app}}(\epsilon t')\|_{L_4^1} + \|\hat{\psi}_\epsilon(t')\|_{L^1(\mathbb{T},l_s^1(\mathbb{N}))} \right) dt'.$$

By Gronwall's inequality (Appendix B.6), we have

$$\sup_{t\in[0,t_0]} \|\hat{\psi}_\epsilon(t)\|_{L^1(\mathbb{T},l_s^1(\mathbb{N}))} \leq \delta$$

if

$$\left(\|\hat{\psi}_\epsilon(0)\|_{L^1(\mathbb{T},l_s^1(\mathbb{N}))} + C_1(\delta_0, \delta)T_0 \sup_{T\subset[0,T_0]} \|\hat{a}_{\mathrm{app}}(T)\|_{L_4^1} \right) e^{C_1(\delta_0,\delta)T_0} \leq \delta. \tag{2.3.18}$$

Since none of the parameters T_0, δ_0, and δ depends on ϵ, there is always a choice of $\delta_0 > 0$ and $\delta > 0$ to satisfy inequalities (2.3.17) and (2.3.18).

The proof of Theorem 2.3 is now complete.   □

Exercise 2.22 Consider the Gross–Pitaevskii equation with a potential $V(x)$ that depends on parameter η such that $E_{n_0}(k_0) = E_{n_0+1}(k_0)$ for $\eta = \eta_0 > 0$ but $E_{n_0}(k_0) < E_{n_0+1}(k_0)$ for $\eta > \eta_0$. Let $\epsilon = \eta - \eta_0 > 0$ be a small parameter and justify the nonlinear Dirac equations with first-order derivative terms.

Exercise 2.23 Let $\phi(x) = a(X)u_0(x)$, where $a \in H^1(\mathbb{R})$, $X = \epsilon^{1/2}x$, and $u_0 \in L^\infty_{\mathrm{per}}([0, 2\pi])$. Prove that

$$\|\phi\|_{H^1} \le C\epsilon^{-1/4}(\|u_0\|_{L^\infty_{\mathrm{per}}} + \|u_0'\|_{L^\infty_{\mathrm{per}}})\|a\|_{H^1}. \qquad (2.3.19)$$

Remark 2.5 Because of the bound (2.3.19), Theorem 2.3 can also be proved in Sobolev space $H^1(\mathbb{R})$ with the error bound

$$\|u(\cdot, t) - \epsilon^{1/2}a(\epsilon^{1/2}\cdot, \epsilon t)u_0 e^{-i\omega_0 t}\|_{H^1} \le C_0\epsilon^{3/4}.$$

Note that the error bound is larger compared to the one in Theorem 2.3.

2.3.2 Justification of the stationary equation

Consider the stationary reduction of the NLS equation (2.3.2),

$$a(X, T) = A(X)e^{-i\Omega T},$$

where Ω is a real-valued parameter and $A(X)$ is a real-valued solution of the stationary NLS equation,

$$\alpha A''(X) + \beta A^3(X) = \Omega A(X), \quad X \in \mathbb{R}. \qquad (2.3.20)$$

Decaying solutions of this equation, called *NLS solitons*, exist for

$$\mathrm{sign}(\alpha) = \mathrm{sign}(\beta) = \mathrm{sign}(\Omega) = 1,$$

where we recall that $\beta = \|u_0\|^4_{L^4_{\mathrm{per}}} > 0$. Since their analytical form is remarkably simple, we are tempted to write it explicitly as

$$A(X) = A_0 \,\mathrm{sech}(KX), \quad A_0 = \left(\frac{2\Omega}{\beta}\right)^{1/2}, \quad K = \left(\frac{\Omega}{\alpha}\right)^{1/2}, \qquad (2.3.21)$$

although the exact analytical form is not used in our analysis. We consider persistence of the localized mode of the stationary NLS equation (2.3.20) as a localized mode of the stationary Gross–Pitaevskii equation,

$$-\Phi''(x) + V(x)\Phi(x) + \Phi^3(x) = \omega\Phi(x), \quad x \in \mathbb{R}. \qquad (2.3.22)$$

Equation (2.3.22) follows from the time-dependent Gross–Pitaevskii equation (2.3.1) after the substitution

$$u(x, t) = \Phi(x)e^{-i\omega t},$$

where ω is close to the band edge $\omega_0 = E_{n_0}(k_0)$ of an isolated n_0th spectral band of $L = -\partial_x^2 + V(x)$ and $\Phi(x)$ is a real-valued function.

To simplify the analysis, we assume that $V(-x) = V(x)$ and look for even solutions

$$\Phi(-x) = \Phi(x), \quad x \in \mathbb{R}.$$

To understand the role of the constraint $V(-x) = V(x)$, we should realize that the stationary NLS equation (2.3.20) is invariant with respect to spatial translation (if $A(X)$ is a solution, so is $A(X - X_0)$ for any $X_0 \in \mathbb{R}$), whereas the stationary Gross–Pitaevskii equation (2.3.22) is not. Therefore, the NLS soliton (2.3.21) can be extended as a one-parameter family of localized modes of the space-homogeneous equation (2.3.20) but only some solutions of this family persist in the space-periodic equation (2.3.22). Without the constraint $V(x) = V(-x)$, the persistence analysis becomes more complicated and involves the beyond-all-orders asymptotic expansion (Section 3.2.2). If we impose the above constraint on $V(x)$, the proof of the main result becomes simpler. The following theorem represents the main result of the justification analysis.

Theorem 2.4 *Let $V \in L^\infty_{\mathrm{per}}([0, 2\pi])$ satisfy $V(-x) = V(x)$. Fix $n_0 \in \mathbb{N}$ such that $E'_{n_0}(k_0) = 0$, $E''_{n_0}(k_0) < 0$, and the n_0th spectral band is disjoint from other spectral bands. Let $A(X) = A(-X)$ be a localized mode of the stationary NLS equation (2.3.20) for $\Omega > 0$. There exists $\epsilon_0 > 0$ and $C_0 > 0$ such that for any $\epsilon \in (0, \epsilon_0)$, the stationary Gross–Pitaevskii equation (2.3.22) with $\omega = \omega_0 + \epsilon\Omega$ admits a real-valued solution $\Phi \in C^0_b(\mathbb{R})$, even in x, decaying to zero as $|x| \to \infty$, satisfying the bound*

$$\|\Phi - \epsilon^{1/2}A(\epsilon^{1/2}\cdot)u_{n_0}(\cdot; k_0)\|_{L^\infty} \le C_0\epsilon^{5/6}. \tag{2.3.23}$$

Remark 2.6 Since $\alpha = -\frac{1}{2}E''_n(k_0) > 0$ and $\Omega > 0$, the localized mode of the Gross–Pitaevskii equation (2.3.22) is generated by the localized mode of the stationary NLS equation (2.3.20) near the maximum values of the band curves inside the corresponding band gap with $\omega > \omega_0$.

Proof The proof of Theorem 2.4 consists of three steps.

Step 1: Decomposition. The only relevant spectral band of operator $L = -\partial^2_x + V(x)$ is the n_0th spectral band with the resonant mode at $k = k_0$. To single out this mode, we expand the solution $\Phi(x)$ using the Bloch decomposition,

$$\Phi(x) = \epsilon^{1/2} \int_{\mathbb{T}} \sum_{n \in \mathbb{N}} \hat{\Phi}_n(k)u_n(x; k)dk, \tag{2.3.24}$$

where the scaling $\epsilon^{1/2}$ is introduced for convenience of handling the cubic nonlinear term. By Lemma 2.2, if $\hat{\phi} \in L^1(\mathbb{T}, l^1_s(\mathbb{N}))$ with $s > \frac{1}{2}$, then $\phi \in C^0_b(\mathbb{R})$ and $\phi(x) \to 0$ as $|x| \to \infty$. Substituting (2.3.24) into (2.3.22) with $\omega = \omega_0 + \epsilon\Omega$, we find a system of integral equations, which is diagonal with respect to the linear terms

$$[E_n(k) - \omega_0 - \epsilon\Omega]\,\hat{\Phi}_n(k) =$$

$$-\epsilon \int_{\mathbb{T}^3} \sum_{(n_1, n_2, n_3) \in \mathbb{N}^3} M_{n, n_1, n_2, n_3}(k, k_1, k_2, k_3)\hat{\Phi}_{n_1}(k_1)\bar{\hat{\Phi}}_{n_2}(k_2)\hat{\Phi}_{n_3}(k_3)dk_1dk_2dk_3$$

$$\tag{2.3.25}$$

for all $n \in \mathbb{N}$ and $k \in \mathbb{T}$, where

$$M_{n,n_1,n_2,n_3}(k,k_1,k_2,k_3) = \int_{\mathbb{R}} \bar{u}_n(x;k)\bar{u}_{n_2}(x;k_2)u_{n_1}(x;k_1)u_{n_3}(x;k_3)dk_1dk_2dk_3,$$

for all $(n,n_1,n_2,n_3) \in \mathbb{N}^4$ and $(k,k_1,k_2,k_3) \in \mathbb{T}^4$. By Lemma 2.11, the cubic nonlinear terms of the integral equations (2.3.25) map an element of $L^1(\mathbb{T}, l_s^1(\mathbb{N}))$ to an element of $L^1(\mathbb{T}, l_s^1(\mathbb{N}))$ for any fixed $s \in (0,1)$. Therefore, we shall consider solutions of the integral equations (2.3.25) in the space $L^1(\mathbb{T}, l_s^1(\mathbb{N}))$ for a fixed $s \in \left(\frac{1}{2}, 1\right)$.

The multiplication operator $E_{n_0}(k) - \omega_0$ vanishes at the point $k = k_0$, which results in a singularity of $\hat{\Phi}_{n_0}(k)$ near $k = k_0$ when $\epsilon = 0$. To single out this singularity, we decompose $\hat{\Phi}(k)$ into three parts

$$\hat{\Phi}_n(k) = \left(\hat{\Phi}_0(k)\chi_{D_0}(k) + \hat{\Phi}_{n_0}(k)(1 - \chi_{D_0}(k))\right)\delta_{n,n_0} + \hat{\Phi}_n(k)(1 - \delta_{n,n_0}), \quad (2.3.26)$$

where δ_{n,n_0} is the Kronecker symbol, $\chi_{D_0}(k)$ is the characteristic function, and

$$D_0 = \{k \in \mathbb{R} : |k - k_0| < \epsilon^r\} \subset \mathbb{T},$$

for a fixed $r > 0$. Note that the 1-periodic continuation must be used in k-space if $k_0 = \frac{1}{2}$ to ensure that $D_0 \subset \mathbb{T}$.

Step 2: Non-singular part of the solution. We can now characterize $\hat{\Phi}_n(k)$ for all $n \in \mathbb{N}$ and $k \in \mathbb{T}$ in terms of $\hat{\Phi}_0(k)$ defined for $n = n_0$ near $k = k_0$.

Lemma 2.15   *Let $V \in L^\infty_{\mathrm{per}}([0, 2\pi])$ and fix $r \in \left(0, \frac{1}{2}\right)$ and $s \in \left(\frac{1}{2}, 1\right)$. There exist $\epsilon_0 > 0$, $\delta_0 > 0$, and $C_0 > 0$ such that for any $\hat{\Phi}_0 \in B_{\delta_0}(L^1(D_0))$ and $\epsilon \in (0, \epsilon_0)$, there exists a unique continuous map*

$$\Psi(\hat{\Phi}_0, \epsilon) : L^1(D_0) \times \mathbb{R} \to L^1(\mathbb{T}, l_s^1(\mathbb{N})) \qquad (2.3.27)$$

such that $\hat{\Phi}(k) = \Psi(\hat{\Phi}_0, \epsilon)$ solves (2.3.25) and satisfies the bound

$$\begin{cases} \|\hat{\Phi}_{n_0}\|_{L^1(\mathbb{T})} \leq C_0\epsilon^{1-2r}\|\hat{\Phi}_0\|_{L^1(D_0)}, \\ \|\hat{\Phi}_n\|_{L^1(\mathbb{T}, l_s^1(\mathbb{N}))} \leq C_0\epsilon\|\hat{\Phi}_0\|_{L^1(D_0)}, \quad n \neq n_0. \end{cases} \qquad (2.3.28)$$

Proof   Since $E'_{n_0}(k_0) = 0$ and $E''_{n_0}(k_0) \neq 0$, for any $\epsilon \in (0, \epsilon_0)$, there exists constant $C_0 > 0$ such that

$$\min_{k \in \mathrm{supp}(\hat{\Phi}_{n_0})} |E_{n_0}(k) - \omega_0| \geq C_0\epsilon^{2r}, \qquad \inf_{n \in \mathbb{N}\setminus\{n_0\}} \min_{k \in \mathbb{T}} |E_n(k) - \omega_0| \geq C_0. \qquad (2.3.29)$$

If $r < \frac{1}{2}$, then the lower bound (2.3.29) is larger than the cubic nonlinear terms of system (2.3.25). By the Implicit Function Theorem (Appendix B.7), there exists a unique continuous map (2.3.27) for $\hat{\Phi}_0 \in B_{\delta_0}(L^1(D_0))$ and $\epsilon \in (0, \epsilon_0)$. The right-hand side of the integral equation (2.3.25) is a homogeneous cubic polynomial of $\hat{\Phi}_n(k)$ multiplied by ϵ, so that bound (2.3.29) gives bound (2.3.28). $\square$

Using the map of Lemma 2.15, we eliminate all components but $\hat{\Phi}_0$ from the integral equations (2.3.25). To recover the stationary NLS equation (2.3.20) in Fourier space from the integral equation (2.3.25) at $n = n_0$ and $k \in D_0$, we use the stretched variable p in the scaling transformation,

$$\hat{\Phi}_0(k) = \frac{1}{\epsilon^{1/2}}\hat{a}(p), \quad p = \frac{k - k_0}{\epsilon^{1/2}}, \tag{2.3.30}$$

where the scaling exponents are inspired by the scaling of variables in the asymptotic expansion (2.3.4). If we apply now the scaling transformation (2.3.30) to the decomposition (2.3.24) and (2.3.26), we obtain at the leading order

$$\epsilon^{1/2}\int_{D_0}\hat{\Phi}_0(k)u_{n_0}(x;k)dk = \epsilon^{1/2}\int_{|p|\leq\epsilon^{r-1/2}}\hat{a}(p)w_{n_0}(x;k_0+\epsilon^{1/2}p)e^{ik_0x}e^{ipX}dp$$

$$= \epsilon^{1/2}a(X)u_0(x) + r(x),$$

where

$$u_0(x) := u_{n_0}(x;k_0) = w_{n_0}(x;k_0)e^{ik_0x},$$

$$a(X) = \int_{|p|\leq\epsilon^{r-1/2}}\hat{a}(p)e^{ipX}dp,$$

and the remainder term $r(x)$ enjoys the estimate

$$\|r\|_{L^\infty} = \epsilon^{1/2}\sup_{x\in\mathbb{R}}\left|\int_{|p|\leq\epsilon^{r-1/2}}\hat{a}(p)\left[w_{n_0}(x;k_0+\epsilon^{1/2}p) - w_{n_0}(x;k_0)\right]e^{ik_0x}e^{ipX}dp\right|$$

$$\leq C\epsilon^{1/2+r}\|\hat{a}\|_{L^1(D_\epsilon)},$$

thanks to analyticity of $w_{n_0}(x;k)$ in k near $k = k_0$. The interval $D_0 \subset \mathbb{T}$ is now mapped to the interval $D_\epsilon \subset \mathbb{R}$, where

$$D_\epsilon = \{p \in \mathbb{R} : |p| \leq \epsilon^{r-1/2}\} \subset \mathbb{R}.$$

The new domain covers the entire line $\mathbb{R}$ as $\epsilon \to 0$ if $r < \frac{1}{2}$. The scaling transformation (2.3.30) preserves the L^1 norm $\|\hat{\Phi}_0\|_{L^1(D_0)} = \|\hat{a}\|_{L^1(D_\epsilon)}$, which means that the bounds in Lemma 2.15 do not lose any power of ϵ. Applying the scaling transformation to the integral equation (2.3.25) at $n = n_0$ and $k \in D_0$, we obtain for $p \in D_\epsilon$

$$(-\alpha p^2 - \Omega)\hat{a}(p) + \epsilon^{1/2}\int_{D_0^3}M_{n_0,n_0,n_0,n_0}(k,k_1,k_2,k_3)\hat{\Phi}_0(k_1)\bar{\hat{\Phi}}_0(k_2)\hat{\Phi}_0(k_3)dk_1dk_2dk_3$$

$$= \frac{1}{\epsilon}\left[E_{n_0}(k_0+\epsilon^{1/2}p) - E_{n_0}(k_0) - \frac{1}{2}\epsilon p^2 E_{n_0}''(k_0)\right]\hat{a}(p) + \epsilon^{1-2r}\hat{P}_\epsilon(\hat{a}), \tag{2.3.31}$$

where, thanks to Lemma 2.15, for any $\hat{a} \in B_\delta(L^1(D_\epsilon))$, there is a constant $C(\delta) > 0$ such that

$$\|\hat{P}_\epsilon(\hat{a})\|_{L^1(D_\epsilon)} \leq C(\delta)\|\hat{a}\|_{L^1(D_\epsilon)}.$$

On the other hand, thanks to the analyticity of $E_{n_0}(k)$ near $k = k_0$, there is $C > 0$ such that

$$\frac{1}{\epsilon} \int_{D_\epsilon} \left| E_{n_0}(k_0 + \epsilon^{1/2}p) - E_{n_0}(k_0) - \frac{1}{2}\epsilon p^2 E_{n_0}''(k_0) \right| |\hat{a}(p)| dp \leq C\epsilon \int_{D_\epsilon} p^4 |\hat{a}(p)| \, dp$$

$$\leq C\epsilon^{4r-1} \|\hat{a}\|_{L^1(D_\epsilon)}.$$

If $r > \frac{1}{4}$, the linear residual term is small. Thus, we need $r \in (\frac{1}{4}, \frac{1}{2})$ to ensure that both remainder terms are small. They have the same order $\mathcal{O}(\epsilon^{1/3})$ if $r = \frac{1}{3}$.

We shall now simplify the nonlinear term in the left-hand side of the integral equation (2.3.31). We will find that the triple integration in the Bloch space results in the double convolution integral in Fourier space at the leading order according to the following computation

$$\epsilon^{1/2} \int_{D_0^3} M_{n_0,n_0,n_0,n_0}(k, k_1, k_2, k_3) \hat{\Phi}_0(k_1) \bar{\hat{\Phi}}_0(k_2) \hat{\Phi}_0(k_3) dk_1 dk_2 dk_3$$

$$= \epsilon^{1/2} \int_{D_\epsilon^3} \int_{\mathbb{R}} \bar{w}_{n_0}(x; k_0 + \epsilon^{1/2}p) \bar{w}_{n_0}(x; k_0 + \epsilon^{1/2}p_2) w_{n_0}(x; k_0 + \epsilon^{1/2}p_1)$$

$$\times w_{n_0}(x; k_0 + \epsilon^{1/2}p_2) \hat{a}(p_1) \bar{\hat{a}}(p_2) \hat{a}(p_3) e^{i(p_1+p_3-p_2-p)X} dx dp_1 dp_2 dp_3$$

$$= \int_{D_\epsilon^3} \int_{\mathbb{R}} |u_0(x)|^4 \hat{a}(p_1) \bar{\hat{a}}(p_2) \hat{a}(p_3) e^{i(p_1+p_3-p_2-p)X} dX dp_1 dp_2 dp_3 + \hat{Q}_\epsilon(\hat{a}),$$

where $u_0 = u_{n_0}(x; k_0)$ and, for any $\hat{a} \in B_\delta(L^1(D_\epsilon))$, there is a constant $C(\delta) > 0$ such that

$$\|\hat{Q}_\epsilon(\hat{a})\|_{L^1(D_\epsilon)} \leq C(\delta)\epsilon^{r+1/2}\|\hat{a}\|_{L^1(D_\epsilon)}.$$

The bound for $\hat{Q}_\epsilon(\hat{a})$ is obtained from the analyticity of $w_n(x;k)$ in k near $k = k_0$ thanks to the fact that $L^1(D_\epsilon)$ forms a Banach algebra with respect to convolution integrals (Appendix B.1). Let us now represent $|u_0|^4 \in C_{\mathrm{per}}^0([0, 2\pi])$ in the Fourier series

$$|u_0(x)|^4 = \frac{\beta}{2\pi} + \sum_{n \in \mathbb{Z}\backslash\{0\}} c_n e^{inx},$$

where $\beta = \|u_0\|_{L_{\mathrm{per}}^4}^4$ and $\{c_n\}_{n \in \mathbb{Z}}$ are some Fourier coefficients. The zero term of the Fourier series results in the leading-order term

$$\frac{\beta}{2\pi} \int_{D_\epsilon^3} \int_{\mathbb{R}} \hat{a}(p_1) \bar{\hat{a}}(p_2) \hat{a}(p_3) e^{i(p_1+p_3-p_2-p)X} dX dp_1 dp_2 dp_3$$

$$= \beta \int_{D_\epsilon^3} \hat{a}(p_1) \bar{\hat{a}}(p_2) \hat{a}(p_3) \delta(p_1 + p_3 - p_2 - p) dp_1 dp_2 dp_3$$

$$= \beta \int_{D_\epsilon^2} \hat{a}(p_1) \bar{\hat{a}}(p_2) \hat{a}(p - p_1 + p_2) dp_1 dp_2,$$

where $\delta(p_1 + p_3 - p_2 - p)$ is the Dirac delta function in the sense of distributions. All nonzero terms of the Fourier series for $|u_0(x)|^4$ with $n \in \mathbb{Z}\backslash\{0\}$ give zero contribution to the triple integral since the singularity of the delta function at

$$p_1 + p_3 - p_2 - p = \frac{n}{\epsilon^{1/2}}$$

is beyond the compact support of $\hat{a}(p_1)\bar{\hat{a}}(p_2)\hat{a}(p_3)$. Combining the previous results, we write the reduced equation for $\hat{a}(p)$ in the form

$$(-\alpha p^2 - \Omega)\hat{a}(p) + \beta \int_{D_\epsilon} \int_{D_\epsilon} \hat{a}(p_1)\hat{a}(p_2)\hat{a}(p_1 + p_2 - p)dp_1 dp_2 = R_\epsilon(\hat{a}), \quad (2.3.32)$$

where

$$\forall \epsilon \in (0, \epsilon_0) : \quad \exists C > 0 \quad \|\hat{R}_\epsilon(\hat{a})\|_{L^1(D_\epsilon)} \le C\epsilon^{\min(4r-1,1-2r)}\|\hat{a}\|_{L^1(D_\epsilon)}.$$

Step 3: Singular part of the solution. The integral equation (2.3.32) is different from the stationary NLS equation (2.3.20) rewritten in the Fourier domain in two aspects. First, the convolution integrals are truncated on $D_\epsilon \subset \mathbb{R}$. Second, the residual terms of order $\epsilon^{\min(4r-1,1-2r)}$ are present. The first source of error is small if solutions of the stationary NLS equation (2.3.20) decay fast in Fourier space as $|p| \to \infty$, which is the case for the NLS solitons (2.3.21) (they decay exponentially in Fourier space). The second source of error can be handled with the Implicit Function Theorem. For notational simplicity, we write the stationary NLS equation (2.3.20) as

$$F(A) := \alpha A'' + \beta A^3 - \Omega A = 0$$

and the linearized Schrödinger operator at a solution $A = A(X)$ as

$$L_A = D_A F(A) := \alpha \partial_X^2 + 3\beta A^2(X) - \Omega.$$

These quantities in the Fourier domain are rewritten as $\hat{F}(\hat{A}) = 0$ and $\hat{L}_A = D_{\hat{A}}\hat{F}(\hat{A})$. The integral equation (2.3.32) can be written as $\hat{F}(\hat{a}) = \hat{R}_\epsilon(\hat{a})$, where $\hat{a}$ is compactly supported in D_ϵ.

Lemma 2.16 *Let $V \in L_{per}^\infty([0, 2\pi])$ satisfy $V(-x) = V(x)$. Let $A(X) = A(-X)$ be a localized mode of the stationary NLS equation (2.3.20) for $\Omega > 0$. Fix $r = \frac{1}{3}$ (for simplicity). For any $\epsilon \in (0, \epsilon_0)$, there exist $C > 0$ and a solution $\hat{a} \in L^1(D_\epsilon)$ of the integral equation (2.3.32) such that*

$$\|\hat{a} - \hat{A}\|_{L^1(D_\epsilon)} \le C\epsilon^{1/3}. \quad (2.3.33)$$

Proof First, we consider the integral equation (2.3.32) with $r = \frac{1}{3}$ in $L^1(\mathbb{R})$ by extending the residual terms to $\mathbb{R}$ with a compact support on $D_\epsilon \subset \mathbb{R}$. Decompose the solution of the extended system $\hat{F}(\hat{a}) = \hat{R}_\epsilon(\hat{a})$ into $\hat{a} = \hat{A} + \hat{b}$ and write the nonlinear problem for $\hat{b}$ in the form

$$\hat{L}_A \hat{b} = \hat{N}(\hat{b}) := \hat{R}_\epsilon(\hat{A} + \hat{b}) - \left[\hat{F}(\hat{A} + \hat{b}) - \hat{L}\hat{b}\right], \quad (2.3.34)$$

where $\hat{L}_A = D_{\hat{A}}\hat{F}(\hat{A})$ is a linearized operator, $\hat{F}(\hat{A} + \hat{b}) - \hat{L}\hat{b}$ is quadratic in $\hat{b}$, and $\hat{R}_\epsilon(\hat{A} + \hat{b})$ maps an element $L^1(\mathbb{R})$ to itself.

The linearized Schrödinger operator L_A has a one-dimensional kernel spanned by the eigenvector $A'(X)$ and it is bounded away from the rest of the spectrum of L_A. Imposing the constraint $A(-X) = A(X)$ on solutions of the stationary NLS equation (2.3.20), we obtain a constrained space of $L^1(\mathbb{R})$, where the linearized operator L_A is continuously invertible. Since the constraint $\Phi(-x) = \Phi(x)$ is preserved by solutions of the stationary Gross–Pitaevskii equation (2.3.22) if $V(-x) = V(x)$, the fixed-point problem $\hat{b} = \hat{L}_A^{-1}\hat{N}(\hat{b})$ is closed in the corresponding constrained space. By the Implicit Function Theorem (Appendix B.7), for any $\epsilon \in (0, \epsilon_0)$, there are $C, C' > 0$ and a unique even solution b such that

$$\|\hat{b}\|_{L^1} \le C\|\hat{R}_\epsilon(\hat{A})\|_{L^1} \le C'\epsilon^{1/3}.$$

This result implies the desired bound (2.3.33) if the error of the residual terms on $\mathbb{R}\backslash D_\epsilon$ does not exceed the error of the residual terms on D_ϵ, which is of the order of $\mathcal{O}(\epsilon^{1/3})$. The error comes from the terms $\|\hat{b}\|_{L_2^1(\mathbb{R}\backslash D_\epsilon)}$ generated by the second-order differential operators of the stationary NLS equation and from the terms $\|\hat{b}\|_{L^1(\mathbb{R}\backslash D_\epsilon)}$ generated by the convolution operators of the stationary NLS equation. The error coming from $\|\hat{A}\|_{L^1(\mathbb{R}\backslash D_\epsilon)}$ is negligible thanks to the fast (exponential) decay of $\hat{A}(p)$ as $|p| \to \infty$. Since $\hat{b} = \hat{L}_A^{-1}\hat{N}(\hat{b})$ and $\hat{L}_A^{-1}$ is a map from $L^1(\mathbb{R})$ to $L_2^1(\mathbb{R})$, for any $\epsilon \in (0, \epsilon_0)$, there are $C, C', C'' > 0$ such that

$$\|\hat{b}\|_{L_2^1(\mathbb{R}\backslash D_\epsilon)} \le \|\hat{b}\|_{L_2^1(\mathbb{R})} \le C\|\hat{N}(\hat{b})\|_{L^1(\mathbb{R})} \le C'\|\hat{R}_\epsilon(\hat{A})\|_{L^1(\mathbb{R})} \le C''\epsilon^{1/3}.$$

Therefore, the truncation does not modify the order of the error of the residual terms of the integral equation (2.3.32). □

We now conclude the proof of Theorem 2.4.

Fix $r = \frac{1}{3}$ and $s \in (\frac{1}{2}, 1)$. Using the triangle inequality and the bounds of Lemma 2.15, we obtain

$$\|\Phi - \epsilon^{1/2}A(\epsilon^{1/2}\cdot)u_0\|_{L^\infty} \le \epsilon^{1/2}\left(C_1\epsilon^{1/3} + C_2\|a - A\|_{L^\infty}\right),$$

for some $C_1, C_2 > 0$. By the Hölder inequality, we have

$$\|a - A\|_{L^\infty} \le \|\hat{a} - \hat{A}\|_{L^1} = \|\hat{a} - \hat{A}\|_{L^1(D_\epsilon)} + \|\hat{A}\|_{L^1(\mathbb{R}\backslash D_\epsilon)},$$

with the last equality due to a compact support of $\hat{a}$ on D_ϵ. The first term is bounded by Lemma 2.16 and the second term is smaller than any power of ϵ if $\hat{A}$ decays exponentially as $|p| \to \infty$. Therefore, the solution Φ constructed in Lemmas 2.15 and 2.16 satisfies the desired bound.

The proof of Theorem 2.4 is now complete. □

The justification analysis described above is limited in the space $C_b^0(\mathbb{R})$ (since $s \in (\frac{1}{2}, 1)$) and does not specify any information on the regularity of the constructed solutions in $C_b^r(\mathbb{R})$ for $r \ge 1$. Dohnal & Uecker [51] constructed solutions with better regularity in $H^s(\mathbb{R})$ for $s \ge 1$ at the price of losing too many powers of ϵ in the bound on the error term. Ilan & Weinstein [90] incorporated more terms of the asymptotic multi-scale expansion in order to get a better power of ϵ in the bound on the error term.

Exercise 2.24 Prove that there exists a non-trivial solution $\Phi(x)$ of the stationary Gross–Pitaevskii equation (2.3.22) in $C_{\mathrm{per}}^0([0, 2\pi])$ such that

$$\exists C > 0: \quad \left\|\Phi - \epsilon^{1/2}Au_0\right\|_{L^\infty} \le C\epsilon^{3/2},$$

if and only if there exists a non-trivial solution for A of the bifurcation equation

$$\beta A^3 = \Omega A + \epsilon R_\epsilon(A),$$

where $R_\epsilon(A)$ is an analytic function of ϵ and A such that $R_\epsilon(0) = 0$.

Exercise 2.25 Justify the system of three two-dimensional coupled NLS equations,

$$
\begin{aligned}
(\Omega - \beta_1)A_1 &+ \left(\alpha_1 \partial_{X_1}^2 + \alpha_2 \partial_{X_2}^2\right) A_1 \\
&= \gamma_1 |A_1|^2 A_1 + \gamma_2 (2|A_2|^2 A_1 + A_2^2 \bar{A}_1) + \gamma_3 (2|A_3|^2 A_1 + A_3^2 \bar{A}_1), \\
(\Omega - \beta_1)A_2 &+ \left(\alpha_2 \partial_{X_1}^2 + \alpha_1 \partial_{X_2}^2\right) A_2 \\
&= \gamma_1 |A_2|^2 A_2 + \gamma_2 (2|A_1|^2 A_2 + A_1^2 \bar{A}_2) + \gamma_3 (2|A_3|^2 A_2 + A_3^2 \bar{A}_2), \\
(\Omega - \beta_2)A_3 &+ \alpha_3 \left(\partial_{X_1}^2 + \partial_{X_2}^2\right) A_3 \\
&= \gamma_4 |A_3|^2 A_3 + 2\gamma_3 (|A_1|^2 + |A_2|^2)A_3 + \gamma_3 (A_1^2 + A_2^2)\bar{A}_3,
\end{aligned}
$$

with some numerical coefficients $(\beta_1, \beta_2, \alpha_1, \alpha_2, \alpha_3, \gamma_1, \gamma_2, \gamma_3, \gamma_4)$ and parameter Ω starting with the stationary two-dimensional Gross–Pitaevskii equation with the separable potential,

$$
-(\partial_{x_1}^2 + \partial_{x_2}^2)\phi(x_1, x_2) + \eta\left(V(x_1) + V(x_2)\right)\phi(x_1, x_2) + |\phi|^2\phi = \omega\phi(x_1, x_2)
$$

where η and ω are parameters. Assume that there exists η_0 such that at $\eta = \eta_0$,

$$
\omega_0 := E_1(0) + E_3\left(\frac{1}{2}\right) = E_3\left(\frac{1}{2}\right) + E_1(0) = 2E_2\left(\frac{1}{2}\right),
$$

whereas a narrow band gap between the three adjacent spectral bands bifurcates for $\eta > \eta_0$.

2.4 Justification of the DNLS equation

The third and last example of the justification analysis deals with the discrete nonlinear Schrödinger equation,

$$
i\dot{a}_m = \alpha\left(a_{m+1} + a_{m-1}\right) + \beta|a_m|^2 a_m, \tag{2.4.1}
$$

where $\{a_m(T)\}_{m\in\mathbb{Z}}$ is a set of complex-valued functions in a new time variable T, whereas α and β are some nonzero numerical coefficients. The DNLS equation (2.4.1) is relevant for the Gross–Pitaevskii equation,

$$
iu_t = -u_{xx} + \epsilon^{-2}V(x)u + |u|^2 u \tag{2.4.2}
$$

in the limit of *large* periodic potentials measured by a small parameter $\epsilon > 0$. In this context, $a_m(T)$ represents $u(x,t)$ in the mth potential well located for $x \in [2\pi m, 2\pi(m + 1)]$ and it evolves in a slow time T. The limit of large potentials is generally singular for the spectral theory of Schrödinger operators (Section 2.1). For instance, if V is the piecewise-constant periodic potential (2.1.15), then V reduces to a periodic sequence of infinite walls of nonzero width in the limit $b \to \infty$. If V is the squared-sine function (2.1.16), then V becomes unbounded in the limit $V_0 \to \infty$.

Since the limit $\epsilon \to 0$ is singular, we have to ensure that the properties of the Wannier decomposition (Section 2.1.3) remain valid uniformly in $(0, \epsilon_0)$ for small

$\epsilon_0 > 0$. To be precise and to extend the discussion beyond the particular examples (2.1.15) and (2.1.16), we begin by postulating the main assumptions when the DNLS equation (2.4.1) can be justified in the context of the Gross–Pitaevskii equation (2.4.2).

Recall that Sobolev space $H^1(\mathbb{R})$ is the energy space for the Gross–Pitaevskii equation (2.4.2) (Section 1.3.1). However, because $\epsilon^{-2}V$ becomes singular as $\epsilon \to 0$, the H^1 norm is not equivalent to the norm induced by the quadratic form generated by operator $L = -\partial_x^2 + \epsilon^{-2}V(x)$. Since the transformation

$$u(x,t) \mapsto u(x,t)e^{-i\omega_0 t}, \quad \epsilon^{-2}V(x) \mapsto \epsilon^{-2}V(x) + \omega_0$$

does not change solutions of the Gross–Pitaevskii equation (2.4.2), we may assume without loss of generality that V is bounded from below. For convenience, let us assume that $V(x) \geq 0$ for all $x \in \mathbb{R}$. Then, $\sigma(L) \subset \mathbb{R}_+$ and it is more convenient to work in the function space $\mathcal{H}^1(\mathbb{R})$ equipped with the operator norm

$$\|\phi\|_{\mathcal{H}^1} := \|(I+L)^{1/2}\phi\|_{L^2} = \left(\int_{\mathbb{R}} \left(|\phi'|^2 + \epsilon^{-2}V|\phi|^2 + |\phi|^2 \right) dx \right)^{1/2}, \qquad (2.4.3)$$

so that $\|\phi\|_{H^1} \leq \|\phi\|_{\mathcal{H}^1}$.

Let us assume that there is a small parameter $\mu = \mu(\epsilon) > 0$ such that $\mu \to 0$ if $\epsilon \to 0$. Pick the nth spectral band for a fixed $n \in \mathbb{N}$ and assume that the functions $E_n(k)$ and $u_n(x;k)$ and their Fourier series (2.1.29) and (2.1.30) satisfy the following properties for all $0 < \mu \ll 1$.

(1) A center of the nth spectral band remains bounded as $\mu \to 0$:

$$\exists E_0 < \infty : \quad |\hat{E}_{n,0}| \leq E_0.$$

(2) The width of the nth spectral band is small:

$$\exists C_1^{\pm}, C_2 > 0 : \quad C_1^- \mu \leq |\hat{E}_{n,1}| \leq C_1^+ \mu, \quad |\hat{E}_{n,m}| \leq C_2 \mu^2, \quad m \geq 2.$$

(3) The nth spectral band is bounded away from other spectral bands:

$$\exists C_0 > 0 : \quad \inf_{n' \in \mathbb{N} \setminus \{n\}} \inf_{k \in \mathbb{T}} |E_{n'}(k) - \hat{E}_{n,0}| \geq C_0.$$

(4) There exists a nonzero $\hat{u}_0 \in L^2(\mathbb{R})$ such that:

$$\exists C_0 > 0 : \quad \|\hat{u}_{n,0} - \hat{u}_0\|_{L^\infty} \leq C_0 \mu.$$

(5) The Wannier function $\hat{u}_{n,0}(x)$ satisfies the exponential decay:

$$\exists C > 0 : \quad \sup_{x \in [-2\pi m, -2\pi(m-1)] \cup [2\pi m, 2\pi(m+1))} |\hat{u}_{n,0}(x)| \leq C \mu^m, \quad m \geq 1.$$

The limit $\mu \to 0$, which is characterized by properties (2) and (3), is usually referred to as the *tight-binding approximation* of the nth spectral band. It is proved in Appendices B and C of [165] that the above properties are satisfied by the piecewise-constant potential (2.1.15) with $b = \epsilon^{-2}$ and $\mu = \epsilon e^{-\pi/\epsilon}$. The small parameter μ is exponentially small in terms of the small parameter $\epsilon = b^{-1/2}$.

Exercise 2.26 Consider an L-periodic potential $V(x + L) = V(x)$ and prove that, if V on $[0, L]$ consists of two constant non-equal steps, then the limit $L \to \infty$ corresponds to the limit $b \to \infty$ for the potential (2.1.15) after rescaling of t, x, V, and ϕ.

Exercise 2.27 Consider a 2π-periodic potential supported at a single point on $[0, 2\pi]$, e.g. $V(x) = \delta(x - \pi)$ for all $x \in [0, 2\pi]$, and show that property (2) is not satisfied for this potential.

Exercise 2.28 Consider the 2π-periodic squared-sine potential (2.1.16) with $V_0 = \epsilon^{-2}$ and show that properties (2) and (3) are satisfied with $\mu = \epsilon^{-3/2} e^{-8/\epsilon}$.

Recent justification of the DNLS equation for a finite lattice was performed in the context of the Gross–Pitaevskii equation with an N-well trapping potential by Bambusi & Sacchetti [13]. Infinite lattices for C^∞ potentials in the semi-classical limit were studied by Aftalion & Helffer [4], where the DNLS equation was justified among other models. Analysis of the coupled N-wave equations for a sequence of modulated pulses in the semi-classical limit of the Gross–Pitaevskii equation with a periodic potential was performed by Giannoulis *et al.* [67]. In a similar context, a nonlinear heat equation with a periodic diffusive term was reduced to a discrete heat equation by Scheel & Van Vleck [180] using the invariant manifold reductions. Lattice differential equations were also justified for an infinite sequence of interacting pulses with a large separation in a general system of reaction–diffusion equations by Zelik & Mielke [221] using projection methods.

Our treatment of the problem is based on the work of Pelinovsky *et al.* [165] and Pelinovsky & Schneider [164]. As in Sections 2.2 and 2.3, the justifications of the time-dependent DNLS equation and the stationary DNLS equation are separated into two different subsections.

2.4.1 Justification of the time-dependent equation

Let us fix $n \in \mathbb{N}$ and use assumptions (1)–(5) listed at the beginning of this section. The main goal of this section is to prove the following theorem.

Theorem 2.5 *Fix $s > \frac{1}{2}$ and let $\mathbf{a}(T) \in C^1(\mathbb{R}, l_s^2(\mathbb{Z}))$ be a global solution of the DNLS equation (2.4.1) with initial data $\mathbf{a}(0) = \mathbf{a}_0 \in l_s^2(\mathbb{Z})$. Let $u_0 \in \mathcal{H}^1(\mathbb{R})$ and assume that there is a $C_0 > 0$ such that*

$$\left\| u_0 - \mu^{1/2} \sum_{m \in \mathbb{Z}} a_m(0) \hat{u}_{n,m} \right\|_{\mathcal{H}^1} \leq C_0 \mu^{3/2}.$$

For sufficiently small $\mu > 0$, there are $T_0 > 0$ and $C > 0$ such that the Gross–Pitaevskii equation (2.4.2) admits a unique solution $u(t) \in C([0, \mu^{-1}T_0], \mathcal{H}^1(\mathbb{R}))$ such that $u(0) = u_0$ and

$$\left\| u(\cdot, t) - \mu^{1/2} e^{-i\hat{E}_{n,0}t} \sum_{m \in \mathbb{Z}} a_m(T) \hat{u}_{n,m} \right\|_{\mathcal{H}^1} \leq C\mu^{3/2}, \quad t \in \left[0, \mu^{-1}T_0\right]. \quad (2.4.4)$$

Proof The proof of Theorem 2.5 consists of four steps.

Step 1: Decomposition. Let us represent a solution of the Gross–Pitaevskii equation (2.4.2) by the asymptotic multi-scale expansion,

$$u(x,t) = \mu^{1/2} \left(\varphi_0(x,T) + \mu \varphi_\mu(x,t) \right) e^{-\mathrm{i}\hat{E}_{n,0}t},$$

where $T = \mu t$, φ_μ is a remainder term to be controlled, and φ_0 is the leading-order term given by

$$\varphi_0(x,T) = \sum_{m \in \mathbb{Z}} a_m(T) \hat{u}_{n,m}(x).$$

According to the decomposition (2.1.39), $\varphi_0(\cdot, T) \in \mathcal{E}_n \subset L^2(\mathbb{R})$ if the sequence $\{a_m(T)\}_{m \in \mathbb{Z}}$ is squared summable for this value of $T \in \mathbb{R}$. It does not imply, however, that $\varphi_\mu(\cdot, t)$ lies in the complement of $\mathcal{E}_n$ in $L^2(\mathbb{R})$. Therefore, we do not restrict φ_μ but impose a restriction on the sequence $\{a_m(T)\}_{m \in \mathbb{Z}}$ to satisfy the DNLS equation (2.4.1) with

$$\alpha = \frac{\hat{E}_{n,1}}{\mu}, \qquad \beta = \|\hat{u}_{n,0}\|_{L^4}^4. \tag{2.4.5}$$

The value of α is uniformly bounded and nonzero in $0 < \mu \ll 1$ thanks to property (2). Since

$$\|\hat{u}_{n,0}\|_{\mathcal{H}^1}^2 = 1 + \hat{E}_{n,0},$$

property (1) ensures that $\|\hat{u}_{n,0}\|_{H^1} \leq \|\hat{u}_{n,0}\|_{\mathcal{H}^1} \leq (1 + E_0)^{1/2}$ uniformly in $0 < \mu \ll 1$. By the Sobolev Embedding Theorem (Appendix B.10), the value of β is bounded. Property (4) ensures that $\beta \neq 0$ in $0 < \mu \ll 1$. As a result, the DNLS equation (2.4.1) has nonzero coefficients α and β as $\mu \to 0$.

Substitution of $\phi(x,t)$ into the Gross–Pitaevskii equation (2.4.2) generates an equation for the remainder term $\varphi_\mu(x,t)$, which depends on the behavior of $\varphi_0(x,T)$. The time evolution problem takes the form

$$\mathrm{i}\partial_t \varphi_\mu = (L - \hat{E}_{n,0})\varphi_\mu + \frac{1}{\mu} \sum_{m \in \mathbb{Z}} \sum_{m' \in \mathbb{Z} \setminus \{m-1,m,m+1\}} \hat{E}_{n,m'-m} a_{m'} \hat{u}_{n,m}$$

$$+ \left(|\varphi_0 + \mu \varphi_\mu|^2 \left(\varphi_0 + \mu \varphi_\mu \right) - \beta \sum_{m \in \mathbb{Z}} |a_m|^2 a_m \hat{u}_{n,m} \right), \tag{2.4.6}$$

where $L = -\partial_x^2 + \epsilon^{-2} V(x)$. The term in large parentheses gives non-vanishing projections to the complement of the selected nth spectral band in $L^2(\mathbb{R})$ as $\mu \to 0$. Therefore, before analyzing the time evolution problem (2.4.6), we need to remove this non-vanishing term by means of a normal form transformation.

Step 2: Normal form transformation. Let Π_n be an orthogonal projection from $L^2(\mathbb{R})$ to $\mathcal{E}_n \subset L^2(\mathbb{R})$. We shall project $|\varphi_0|^2 \varphi_0$ onto $\mathcal{E}_n$ and its complement in $L^2(\mathbb{R})$, i.e.

$$|\varphi_0|^2 \varphi_0 = \Pi_n |\varphi_0|^2 \varphi_0 + (I - \Pi_n)|\varphi_0|^2 \varphi_0.$$

The following lemma is useful to control the projection $(I - \Pi_n)|\varphi_0|^2 \varphi_0$.

Lemma 2.17 *There exists a unique solution $\varphi \in \mathcal{H}^1(\mathbb{R})$ of the inhomogeneous equation*

$$\left(L - \hat{E}_{n,0}\right)\varphi = (I - \Pi_n)f, \qquad (2.4.7)$$

for any $f \in L^2(\mathbb{R})$ such that

$$\langle \varphi, \psi \rangle_{L^2} = 0, \quad \psi \in \mathcal{E}_n,$$

and for all $0 < \mu \ll 1$, there is $C > 0$ such that

$$\|\varphi\|_{\mathcal{H}^1} \leq C\|f\|_{L^2}. \qquad (2.4.8)$$

Proof Denote $\mathcal{E}_n^{\perp} = L^2(\mathbb{R})\backslash\mathcal{E}_n$. By property (3), $\hat{E}_{n,0} \notin \sigma(L|_{\mathcal{E}_n^{\perp}})$, so that if φ is a solution of (2.4.7), then $\varphi \in L^2(\mathbb{R})$ for any $f \in L^2(\mathbb{R})$. Existence of solutions of (2.4.7) can be found by means of the Bloch decomposition (2.1.22) in the form

$$\varphi(x) = \int_{\mathbb{T}} \sum_{n' \in \mathbb{N}\backslash\{n\}} \frac{\hat{f}_{n'}(k)}{E_{n'}(k) - \hat{E}_{n,0}} u_{n'}(x; k)dk,$$

which implies that there is $C > 0$ such that

$$\|\varphi\|_{L^2}^2 = \int_{\mathbb{T}} \sum_{n' \in \mathbb{N}\backslash\{n\}} \frac{|\hat{f}_{n'}(k)|^2}{(E_{n'}(k) - \hat{E}_{n,0})^2}dk \leq C \int_{\mathbb{T}} \sum_{n' \in \mathbb{N}\backslash\{n\}} |\hat{f}_{n'}(k)|^2 dk \leq C\|f\|_{L^2}^2.$$

By property (1), we also obtain

$$\|\varphi\|_{\mathcal{H}^1}^2 \leq (1 + \hat{E}_{n,0})\|\varphi\|_{L^2}^2 + \|f\|_{L^2}^2\|\varphi\|_{L^2}^2 \leq C\|f\|_{L^2}^2,$$

for some $C > 0$. Therefore, if $f \in L^2(\mathbb{R})$, then $\varphi \in \mathcal{H}^1(\mathbb{R})$ and the bound (2.4.8) holds. Uniqueness of φ follows from the fact that the operator $(I-\Pi_n)(L-\hat{E}_{n,0})(I-\Pi_n)$ is invertible. $\qquad\square$

Using Lemma 2.17, we decompose $\varphi_\mu(x,t)$ into two parts,

$$\varphi_\mu(x,t) = \varphi_1(x,T) + \psi_\mu(x,t), \qquad (2.4.9)$$

where φ_1 is a unique solution of

$$(L - \hat{E}_{n,0})\varphi_1 = -(I - \Pi_n)|\varphi_0|^2\varphi_0.$$

Now the leading-order term in large parentheses of system (2.4.6), which gives a non-vanishing projection to $\mathcal{E}_n^{\perp}$ as $\mu \to 0$, is removed with the normal form transformation (2.4.9). The same leading-order term also gives a nonzero projection to $\mathcal{E}_n$. However, thanks to property (5), this projection is as small as $\mathcal{O}(\mu)$ in $\mathcal{H}^1$ norm. Therefore, we substitute the decomposition (2.4.9) into system (2.4.6) and write the time evolution problem for $\psi_\mu(x,t)$ in the abstract form

$$i\partial_t \psi_\mu = \left(L - \hat{E}_{n,0}\right)\psi_\mu + \mu R(\mathbf{a}) + \mu N(\mathbf{a}, \psi_\mu). \qquad (2.4.10)$$

Thanks to the choice of α and β in (2.4.5), we have $R(\mathbf{a}) = \sum_{m\in\mathbb{Z}} r_m(\mathbf{a})\hat{u}_{n,m}(x)$ with

$$r_m(\mathbf{a}) = \frac{1}{\mu^2} \sum_{m'\in\mathbb{Z}\setminus\{m-1,m,m+1\}} \hat{E}_{n,m'-m} a_{m'}$$

$$+ \frac{1}{\mu} \sum_{(m_1,m_2,m_3)\in\mathbb{Z}^3\setminus\{(m,m,m)\}} K_{m,m_1,m_2,m_3} a_{m_1}\bar{a}_{m_2} a_{m_3},$$

where $K_{m,m_1,m_2,m_3} = \langle \hat{u}_{n,m}, \hat{u}_{n,m_1}\hat{u}_{n,m_2}\hat{u}_{n,m_3}\rangle_{L^2}$. On the other hand, the nonlinear vector field is given by

$$N(\mathbf{a},\psi_\mu) = -\mathrm{i}\partial_T\varphi_1 + 2|\varphi_0|^2(\varphi_1 + \psi_\mu) + \varphi_0^2(\bar\varphi_1 + \bar\psi_\mu)$$

$$+\mu\left(2|\varphi_1 + \psi_\mu|^2\varphi_0 + (\varphi_1 + \psi_\mu)^2\bar\varphi_0\right) + \mu^2|\varphi_1 + \psi_\mu|^2(\varphi_1 + \psi_\mu),$$

where φ_0 and φ_1 denote

$$\varphi_0 = \sum_{m\in\mathbb{N}} a_m \hat{u}_{n,m}, \qquad \varphi_1 = -(I - \Pi_n)(L - \hat{E}_{n,0})^{-1}(I - \Pi_n)|\varphi_0|^2\varphi_0.$$

Step 3: Fixed-point iterations. The abstract representation (2.4.10) suggests that $R(\mathbf{a})$ and $N(\mathbf{a},\psi_\mu)$ remain uniformly bounded in $\mathcal{H}^1$ norm for $0 < \mu \ll 1$. To control these terms, we consider a local ball of radius $\delta_0 > 0$ in $l^1(\mathbb{Z})$ denoted by $B_{\delta_0}(l^1(\mathbb{Z}))$ and a local ball of radius $\delta > 0$ in $\mathcal{H}^1$ denoted by $B_\delta(\mathcal{H}^1(\mathbb{R}))$. The following lemma gives a bound on the vector field of the evolution problem (2.4.10).

Lemma 2.18   *Fix $\delta_0, \delta > 0$ such that $\mathbf{a}, \dot{\mathbf{a}} \in B_{\delta_0}(l^1(\mathbb{Z}))$ and $\psi_\mu, \tilde\psi_\mu \in B_\delta(\mathcal{H}^1(\mathbb{R}))$. For any $0 < \mu \ll 1$, there exist positive constants $C_R(\delta_0)$, $C_N(\delta_0,\delta)$, and $K_N(\delta_0,\delta)$ such that*

$$\|R(\mathbf{a})\|_{\mathcal{H}^1} \le C_R(\delta_0)\|\mathbf{a}\|_{l^1}, \tag{2.4.11}$$

$$\|N(\mathbf{a},\psi_\mu)\|_{\mathcal{H}^1} \le C_N(\delta_0,\delta)\left(\|\mathbf{a}\|_{l^1} + \|\psi_\mu\|_{\mathcal{H}^1}\right), \tag{2.4.12}$$

$$\|N(\mathbf{a},\psi_\mu) - N(\mathbf{a},\tilde\psi_\mu)\|_{\mathcal{H}^1} \le K_N(\delta_0,\delta)\|\psi_\mu - \tilde\psi_\mu\|_{\mathcal{H}^1}. \tag{2.4.13}$$

Proof   By Lemma 2.4 (extended to $\mathcal{H}^1(\mathbb{R})$ since $\hat{u}_{n,0} \in \mathcal{H}^1(\mathbb{R})$), if $\mathbf{a} \in l^1(\mathbb{Z})$, then $\varphi_0 \in \mathcal{H}^1(\mathbb{R})$ and there is $C_0 > 0$ such that $\|\varphi_0\|_{\mathcal{H}^1} \le C_0\|\mathbf{a}\|_{l^1}$. Furthermore, since

$$\|u\|_{\mathcal{H}^1}^2 = \|u\|_{H^1}^2 + \epsilon^{-2}\|V^{1/2}u\|_{L^2}^2,$$

the Sobolev Embedding Theorem (Appendix B.10) implies that $\mathcal{H}^1(\mathbb{R})$ forms a Banach algebra with respect to pointwise multiplication, so that

$$\exists C > 0: \quad \forall u,v \in \mathcal{H}^1(\mathbb{R}): \quad \|uv\|_{\mathcal{H}^1} \le C\|u\|_{\mathcal{H}^1}\|v\|_{\mathcal{H}^1}.$$

Therefore, $\||\varphi_0|^2\varphi_0\|_{\mathcal{H}^1} \le C^3\|\varphi_0\|_{\mathcal{H}^1}^3 \le C^3 C_0^3\|\mathbf{a}\|_{l^1}^3$ and, by Lemma 2.17, there is $C_1 > 0$ such that $\|\varphi_1\|_{\mathcal{H}^1} \le C_1\|\mathbf{a}\|_{l^1}^3$.

Bound (2.4.11) on the vector field $R(\mathbf{a})$ is proved if for any $\phi \in B_{\delta_0}(l^1(\mathbb{Z}))$ there is $C_R(\delta_0) > 0$ such that $\|\mathbf{r}(\mathbf{a})\|_{l^1} \le C_R(\delta_0)\|\mathbf{a}\|_{l^1}$. The first term in $\mathbf{r}(\mathbf{a})$ is estimated as follows:

$$\left\| \sum_{m' \in \mathbb{Z} \setminus \{m-1, m, m+1\}} \hat{E}_{n, m'-m} a_{m'} \right\|_{l^1} \leq \sum_{m \in \mathbb{Z}} \sum_{m' \in \mathbb{Z} \setminus \{m-1, m, m+1\}} |\hat{E}_{n, m'-m}| |a_{m'}|$$

$$\leq K_1 \|\mathbf{a}\|_{l^1},$$

where

$$K_1 = \sup_{m \in \mathbb{Z}} \sum_{m' \in \mathbb{Z} \setminus \{m-1, m, m+1\}} |\hat{E}_{n, m'-m}| = \sum_{l \in \mathbb{Z} \setminus \{-1, 0, 1\}} |\hat{E}_{n, l}|.$$

Because $E_n(k)$ is analytically continued along the Riemann surface on $k \in \mathbb{T}$, we have $E_n \in H^s(\mathbb{T})$ for any $s \geq 0$ and hence $K_1 < \infty$ for $\mu > 0$. By property (2), $\hat{E}_{n, l} = \mathcal{O}(\mu^2)$ for all $l \geq 2$, so that K_1/μ^2 is uniformly bounded as $\mu \to 0$.

The second term in $\mathbf{r}(\mathbf{a})$ is estimated as follows:

$$\left\| \sum_{(m_1, m_2, m_3) \in \mathbb{Z}^3 \setminus \{(m, m, m)\}} K_{m, m_1, m_2, m_3} a_{m_1} \bar{a}_{m_2} a_{m_3} \right\|_{l^1}$$

$$\leq \sum_{m \in \mathbb{Z}} \sum_{(m_1, m_2, m_3) \in \mathbb{Z}^3 \setminus \{(m, m, m)\}} |K_{m, m_1, m_2, m_3}| |a_{m_1}| |a_{m_2}| |a_{m_3}| \leq K_2 \|\mathbf{a}\|_{l^1}^3,$$

where

$$K_2 = \sup_{(m_1, m_2, m_3) \in \mathbb{Z}^3 \setminus \{(m, m, m)\}} \sum_{m \in \mathbb{Z}} |K_{m, m_1, m_2, m_3}|.$$

Using the exponential decay (2.1.38), we obtain

$$\exists A_n > 0 : \quad \forall x \in \mathbb{R} : \quad \sum_{m \in \mathbb{Z}} |\hat{u}_{n, m}(x)| \leq C_n \sum_{m \in \mathbb{Z}} e^{-\eta_n |x - 2\pi m|} \leq A_n.$$

For all $(m_1, m_2, m_3) \in \mathbb{Z}^3$, we have

$$\sum_{m \in \mathbb{Z}} |K_{m, m_1, m_2, m_3}| \leq A_n \int_{\mathbb{R}} |\hat{u}_{n, m_1}(x)| |\hat{u}_{n, m_2}(x)| |\hat{u}_{n, m_3}(x)| dx \leq A_n \|\hat{u}_{n, 0}\|_{\mathcal{H}^1}^3,$$

and hence $K_2 < \infty$ for $\mu > 0$. By property (5),

$$K_{m, m_1, m_2, m_3} = \mathcal{O}\left(\mu^{|m_1 - m| + |m_2 - m| + |m_3 - m| + |m_2 - m_1| + |m_3 - m_1| + |m_3 - m_2|} \right),$$

so that K_2/μ is uniformly bounded as $\mu \to 0$ and bound (2.4.11) is proved.

Bound (2.4.12) follows from the fact that both $\mathcal{H}^1(\mathbb{R})$ and $l^1(\mathbb{Z})$ form Banach algebras with respect to pointwise multiplication (Appendix B.1). If $\mathbf{a}, \dot{\mathbf{a}} \in B_{\delta_0}(l^1(\mathbb{Z}))$, then $\varphi_0, \varphi_1 \in B_\delta(\mathcal{H}^1(\mathbb{R}))$ and $N(\mathbf{a}, \psi_\mu)$ maps $\psi_\mu \in \mathcal{H}^1(\mathbb{R})$ to an element of $\mathcal{H}^1(\mathbb{R})$. Moreover, $N(\mathbf{a}, \psi_\mu)$ is uniformly bounded as $\mu \to 0$. The proof of bound (2.4.13) also follows from the explicit expression for $N(\mathbf{a}, \psi_\mu)$. $\qquad \square$

Step 4: Control on the error bound. To work with solutions of the time evolution equation (2.4.10), we need to prove that the initial-value problem for this equation is locally well-posed if $\mathbf{a}(T)$ is a C^1 function on $[0, T_0]$ in a ball $B_{\delta_0}(l^1(\mathbb{Z}))$, where T_0 may depend on δ_0.

Lemma 2.19   *Fix $T_0 > 0$ and $\delta_0 > 0$ such that $\mathbf{a}(T) \in C^1([0, T_0], B_{\delta_0}(l^1(\mathbb{Z})))$ for some $T_0 > 0$ and $\delta_0 > 0$. Fix $\delta > 0$ such that $\psi_\mu(0) \in B_\delta(\mathcal{H}^1(\mathbb{R}))$. There exist $t_0 > 0$ and $\mu_0 > 0$ such that, for any $\mu \in (0, \mu_0)$, the time evolution problem (2.4.10) admits a unique solution $\psi_\mu(t) \in C([0, t_0], \mathcal{H}^1(\mathbb{R}))$.*

Proof   Since L is a self-adjoint operator in $L^2(\mathbb{R})$, the operator $e^{-it(L - \hat{E}_{n,0})}$ forms a unitary evolution group satisfying

$$\forall \psi \in \mathcal{H}^1(\mathbb{R}): \quad \|e^{-it(L - \hat{E}_{n,0})}\psi\|_{\mathcal{H}^1} = \|\psi\|_{\mathcal{H}^1}, \quad t \in \mathbb{R}.$$

By Duhamel's principle, the time evolution problem (2.4.10) is rewritten in the integral form

$$\psi_\mu(t) = e^{-it(L - \hat{E}_{n,0})}\psi_\mu(0)$$
$$+ \mu \int_0^t e^{-i(t-s)(L - \hat{E}_{n,0})} \left(R(\mathbf{a}(\mu s)) + N(\mathbf{a}(\mu s), \psi_\mu(s)) \right) ds. \quad (2.4.14)$$

The vector field of the integral equation (2.4.14) maps an element of $\mathcal{H}^1(\mathbb{R})$ to an element of $\mathcal{H}^1(\mathbb{R})$ thanks to bounds (2.4.11) and (2.4.12). By the Banach Fixed-Point Theorem (Appendix B.2), there exists a unique fixed point of the integral equation (2.4.14) in space $\psi_\mu(t) \in C([0, t_0], \mathcal{H}^1(\mathbb{R}))$ for sufficiently small $t_0 > 0$ such that

$$\mu K_N(\delta_0, \delta) t_0 < 1,$$

thanks to bound (2.4.13).   □

By Theorem 1.3 (Section 1.3.2), the DNLS equation (2.4.1) is globally well-posed in $l_s^2(\mathbb{Z})$ for any $s \geq 0$ and the Cauchy problem admits a unique solution $\mathbf{a}(T) \in C^1(\mathbb{R}, l_s^2(\mathbb{Z}))$. If $s > \frac{1}{2}$, there there is $C_s > 0$ such that

$$\|\mathbf{a}(T)\|_{l^1} \leq C_s \|\mathbf{a}(T)\|_{l_s^2}.$$

Because of the relation $T = \mu t$ and the constraint in the proof of Lemma 2.19, we need only local well-posedness of the DNLS equation (2.4.1) on the time interval $[0, T_0]$, where $T_0 = \mu t_0 < 1/K_N$. We need to select δ_0 and δ large enough so that the solutions $\mathbf{a}(T)$ and $\psi_\mu(t)$ remain in $B_{\delta_0}(l^1(\mathbb{Z}))$ and $B_\delta(\mathcal{H}^1(\mathbb{R}))$ on $[0, T_0]$ and $[0, t_0]$ respectively.

Existence of a unique solution $\psi_\mu(t)$ of the time evolution equation (2.4.10) in $C([0, T_0/\mu], \mathcal{H}^1(\mathbb{R}))$ follows from Lemma 2.19 since $\varphi_0, \varphi_1 \in \mathcal{H}^1(\mathbb{R})$ if $\mathbf{a}(T) \in C^1([0, T_0], B_{\delta_0}(l^1(\mathbb{Z})))$. We need to bound the component $\psi_\mu(t)$ in $B_\delta(\mathcal{H}^1(\mathbb{R}))$ on $[0, t_0]$ assuming that $\psi_\mu(0) \in B_\delta(\mathcal{H}^1(\mathbb{R}))$. Using bounds (2.4.11) and (2.4.12) of Lemma 2.18 and the integral equation (2.4.14), we obtain

$$\|\psi_\mu(t)\|_{\mathcal{H}^1} \leq \|\psi_\mu(0)\|_{\mathcal{H}^1} + \mu(C_R(\delta_0) + C_N(\delta_0, \delta)) \int_0^t \|\mathbf{a}(\mu s)\|_{l^1} ds$$
$$+ C_N(\delta_0, \delta) \int_0^t \|\psi_\mu(s)\|_{\mathcal{H}^1} ds.$$

By the Gronwall inequality (Appendix B.6), a local solution $\psi_\mu(t)$ of the integral equation (2.4.14) satisfies the bound

$$\sup_{t\in[0,T_0/\mu]} \|\psi_\mu(t)\|_{\mathcal{H}^1}$$

$$\leq \left(\|\psi_\mu(0)\|_{\mathcal{H}^1} + (C_R(\delta_0) + C_N(\delta_0,\delta))T_0 \sup_{T\in[0,T_0]} \|\mathbf{a}(T)\|_{l^1} \right) e^{C_N(\delta_0,\delta)T_0}$$

$$\leq (C_0 + (C_R(\delta_0) + C_N(\delta_0,\delta))T_0\delta_0)\, e^{C_N(\delta_0,\delta)T_0} \leq \delta.$$

Therefore, for given $\delta_0 > 0$ and $\delta > 0$, one can find a $T_0 \in \left(0, K_N^{-1}(\delta_0,\delta)\right)$ such that the bound (2.4.4) is justified for some constant $C > 0$ uniformly in $0 < \mu \ll 1$.

The proof of Theorem 2.5 is now complete. $\qquad\square$

Exercise 2.29 Using the scaling transformation of $\{a_m(T)\}_{m\in\mathbb{Z}}$, reduce the DNLS equation (2.4.1) to the normalized form,

$$i\dot{u}_m = \sigma_0(u_{m+1} + u_{m-1}) + |u_m|^2 u_m,$$

where $\sigma_0 = \text{sign}(\alpha\beta)$.

Remark 2.7 The DNLS equation (2.4.1) has numerical coefficients α and β which may depend on μ. Exercise 2.29 shows that this dependence is irrelevant for the justification analysis as long as parameters α and β remain nonzero and bounded as $\mu \to 0$.

Exercise 2.30 Show that the DNLS equation (2.4.1) can be justified for the potential (2.1.16) with $V_0 = \epsilon^{-2}$ after the modification of the asymptotic multiscale expansion with the algebraic factor

$$u(x,t) = \epsilon^{1/2}\mu^{1/2} \left(\varphi_0(x,T) + \mu\varphi_\mu(x,t)\right) e^{-i\hat{E}_{n,0}t},$$

where $\mu = \epsilon^{-3/2}e^{-8/\epsilon}$ is defined in Exercise 2.28.

Recently, Belmonte-Beitia and Pelinovsky [18] gave details of the semi-classical analysis needed for the justification of the DNLS equations for the squared-sine potential (2.1.16) in the context of a Gross–Pitaevskii equation with periodic sign-varying nonlinearity coefficient.

2.4.2 Justification of the stationary equation

Consider the stationary reduction of the DNLS equation (2.4.1),

$$\mathbf{a}(T) = \mathbf{A}e^{-i\Omega T},$$

where Ω is a real parameter and $\mathbf{A} \in l^1(\mathbb{Z})$. Theorem 2.5 implies that the Gross–Pitaevskii equation (2.4.2) admits a time-dependent solution, which remains close to the stationary solution

$$u(x,t) = \Phi(x)e^{-i\omega t},$$

where $\omega = \hat{E}_{n,0} + \mu\Omega$ and $\Phi \in \mathcal{H}^1(\mathbb{R})$. This correspondence is valid for a small $\mu > 0$ and on a finite time interval $[0, \mu^{-1}T_0]$. The error bounds cannot be controlled beyond the finite time interval, so it is not clear if the Gross–Pitaevskii equation (2.4.2) admits a true stationary solution in the above form. To answer definitely and positively on this question, we consider the stationary DNLS equation,

$$\alpha \left(A_{m+1} + A_{m-1}\right) + \beta |A_m|^2 A_m = \Omega A_m, \quad m \in \mathbb{Z}, \qquad (2.4.15)$$

and the stationary Gross–Pitaevskii equation,

$$-\Phi''(x) + V(x)\Phi(x) + |\Phi(x)|^2 \Phi(x) = \omega \Phi(x), \quad x \in \mathbb{R}. \qquad (2.4.16)$$

Our main result is given by the following theorem.

Theorem 2.6 *Let* $\mathbf{A} \in l^1(\mathbb{Z})$ *be a solution of the stationary DNLS equation (2.4.15) and assume that the linearized equation at* $\mathbf{A}$ *admits a one-dimensional kernel in* $l^1(\mathbb{Z})$ *spanned by* $\{i\mathbf{A}\}$ *and isolated from the rest of the spectrum. There exist* $\mu_0 > 0$ *and* $C > 0$ *such that, for any* $\mu \in (0, \mu_0)$*, the stationary Gross–Pitaevskii equation (2.4.16) with* $\omega = \hat{E}_{n,0} + \mu\Omega$ *admits a solution* $\Phi \in \mathcal{H}^1(\mathbb{R})$ *satisfying the bound*

$$\left\| \Phi - \mu^{1/2} \sum_{m \in \mathbb{Z}} A_m \hat{u}_{n,m} \right\|_{\mathcal{H}^1} \leq C\mu^{3/2}. \qquad (2.4.17)$$

Moreover, $\Phi(x)$ *decays to zero exponentially fast as* $|x| \to \infty$ *if* $\{A_n\}_{n \in \mathbb{Z}}$ *decays to zero exponentially fast as* $|n| \to \infty$.

Proof The proof of Theorem 2.6 consists of three steps.

Step 1: Decomposition. We decompose solutions of (2.4.16) in the form

$$\Phi(x) = \mu^{1/2} \left(\varphi_\mu(x) + \mu\psi_\mu(x)\right), \qquad (2.4.18)$$

where $\varphi_\mu \in \mathcal{E}_n$ and $\psi_\mu \in \mathcal{E}_n^\perp$. Unlike the standard method of Lyapunov–Schmidt reductions (Section 3.2), we have an infinite-dimensional kernel since $\text{Dim}(\mathcal{E}_n) = \infty$. Projecting equation (2.4.16) to $\mathcal{E}_n$ and $\mathcal{E}_n^\perp$ using projector operators Π_n and $(I - \Pi_n)$ from Lemma 2.17, we obtain a system of two equations

$$\left(L - \hat{E}_{n,0} - \mu\Omega\right) \varphi_\mu = -\mu\Pi_n |\varphi_\mu + \mu\psi_\mu|^2 (\varphi_\mu + \mu\psi_\mu), \qquad (2.4.19)$$

$$\left(L - \hat{E}_{n,0} - \mu\Omega\right) \psi_\mu = -\mu(I - \Pi_n)|\varphi_\mu + \mu\psi_\mu|^2 (\varphi_\mu + \mu\psi_\mu), \qquad (2.4.20)$$

where $L = -\partial_x^2 + \epsilon^{-2} V(x)$ and $\omega = \hat{E}_{n,0} + \mu\Omega$. Using the Wannier decomposition,

$$\varphi_\mu(x) = \sum_{m \in \mathbb{Z}} a_m \hat{u}_{n,m}(x),$$

we can rewrite equation (2.4.19) in the lattice form

$$\frac{1}{\mu} \sum_{l \geq 1} \hat{E}_{n,l} \left(a_{m+l} + a_{m-l}\right) + \sum_{(m_1, m_2, m_3) \in \mathbb{Z}^3} K_{m,m_1,m_2,m_3} a_{m_1} \bar{a}_{m_2} a_{m_3}$$

$$= \Omega a_m + R_m(\mathbf{a}, \psi_\mu), \quad m \in \mathbb{Z}, \qquad (2.4.21)$$

where $K_{m,m_1,m_2,m_3} = \langle \hat{u}_{n,m}, \hat{u}_{n,m_1} \hat{u}_{n,m_2} \hat{u}_{n,m_3} \rangle_{L^2}$ and $R_m(\mathbf{a}, \psi_\mu)$ is given by

$$R_m(\mathbf{a}, \psi_\mu) = -\int_{\mathbb{R}} \hat{u}_{n,m} \left(|\varphi_\mu + \mu\psi_\mu|^2 (\varphi_\mu + \mu\psi_\mu) - |\varphi_\mu|^2 \varphi_\mu\right) dx.$$

The idea of the decomposition (2.4.18) is to invert the non-singular operator in equation (2.4.20) and to eliminate the component ψ_μ from equation (2.4.19). In this algorithm, the problem reduces to the lattice equation (2.4.21) in coordinates $\{a_m\}_{m \in \mathbb{Z}}$, which is regarded as a perturbed stationary DNLS equation (2.4.15). The Implicit Function Theorem (Appendix B.7) is used both to single out the component ψ_μ and to analyze persistence of localized solutions in the perturbed lattice equation (2.4.21).

Step 2: Non-singular part of the solution. Let us consider again $\mathbf{a}$ and ψ_μ in balls $B_{\delta_0}(l^1(\mathbb{Z}))$ and $B_\delta(\mathcal{H}^1(\mathbb{R}))$, similarly to the time-dependent case.

Lemma 2.20 *Fix $\delta_0 > 0$. For all $\mathbf{a} \in B_{\delta_0}(l^1(\mathbb{Z}))$, there exists a unique smooth map $\hat\psi : l^1(\mathbb{Z}) \times \mathbb{R} \to \mathcal{H}^1(\mathbb{R})$ such that $\psi_\mu = \hat\psi(\mathbf{a}, \mu)$ solves (2.4.20), and for any small $\mu > 0$ and any $\mathbf{a} \in B_{\delta_0}(l^1(\mathbb{Z}))$, there is $C_0 > 0$ such that*

$$\|\psi_\mu\|_{\mathcal{H}^1} \le \mu C_0 \|\mathbf{a}\|_{l^1}^3. \tag{2.4.22}$$

Moreover, $\psi_\mu(x)$ decays exponentially as $|x| \to \infty$.

Proof By Lemma 2.17, it follows that operator $(I - \Pi_n)(L - \hat E_{n,0} - \mu\Omega)(I - \Pi_n)$ is continuously invertible for a sufficiently small $\mu > 0$ and a μ-independent Ω. Therefore, there exist $\mu_0 > 0$ and $C_0 > 0$ such that for any $\mu \in (0, \mu_0)$

$$\|(I - \Pi_n)(L - \hat E_{n,0} - \mu\Omega)^{-1}(I - \Pi_n)\|_{L^2 \to \mathcal{H}^1} \le C_0.$$

Equation (2.4.20) can be rewritten in the form

$$\psi_\mu = -\mu(I - \Pi_n)\left(L - \hat E_{n,0} - \mu\Omega\right)^{-1}(I - \Pi_n)|\varphi_\mu + \mu\psi_\mu|^2(\varphi_\mu + \mu\psi_\mu).$$

Because $\mathcal{H}^1(\mathbb{R})$ is a Banach algebra with respect to pointwise multiplication (Appendix B.1), the right-hand side of this equation maps an element of $\mathcal{H}^1(\mathbb{R})$ to an element of $\mathcal{H}^1(\mathbb{R})$ if $\mathbf{a} \in B_{\delta_0}(l^1(\mathbb{Z}))$. Existence of a unique smooth map $\psi_\mu = \hat\psi(\mathbf{a}, \mu)$ satisfying (2.4.22) follows by the Implicit Function Theorem (Appendix B.7). By the property of elliptic equations, if $f \in L^2(\mathbb{R})$ decays exponentially as $|x| \to \infty$, then

$$(I - \Pi_n)\left(L - \hat E_{n,0} - \mu\Omega\right)^{-1}(I - \Pi_n)f$$

also decays exponentially as $|x| \to \infty$. Therefore, $\psi_\mu(x)$ decays exponentially as $|x| \to \infty$. $\qquad\square$

Using the map in Lemma 2.20, we close the lattice equation (2.4.21) in variable $\mathbf{a}$. The residual term of this equation is estimated by the following lemma.

Lemma 2.21 *There are $\mu_0 > 0$ and $C_0 > 0$ such that for any $\mu \in (0, \mu_0)$ and any $\mathbf{a} \in B_{\delta_0}(l^1(\mathbb{Z}))$,*

$$\|\mathbf{R}(\mathbf{a}, \hat\psi(\mathbf{a}, \mu))\|_{l^1} \le \mu C_0 \|\mathbf{a}\|_{l^1}^5. \tag{2.4.23}$$

Proof Using the uniform bound $\sum_{m\in\mathbb{Z}} |\hat{u}_{n,m}(x)| \leq A_n$ for all $x \in \mathbb{R}$ from Lemma 2.18, there are $C, C' > 0$ such that

$$\|\mathbf{R}(\mathbf{a}, \hat{\psi}_\mu(\phi, \mu))\|_{l^1} \leq A_n \int_{\mathbb{R}} \left| |\varphi_\mu + \mu\psi_\mu|^2(\varphi_\mu + \mu\psi_\mu) - |\varphi_\mu|^2\varphi_\mu \right| dx$$

$$\leq \mu C \|\varphi_\mu\|_{\mathcal{H}^1}^2 \|\psi_\mu\|_{\mathcal{H}^1}$$

$$\leq \mu C' \|\mathbf{a}\|_{l^1}^5,$$

where we have used bound (2.4.22) of Lemma 2.20. $\qquad\square$

Step 3: Singular part of the solution. By the estimate in the proof of Lemma 2.18, the linear term on the left-hand side of the lattice equation (2.4.21) can be written in the form

$$\frac{1}{\mu} \sum_{l\geq 1} \hat{E}_{n,l} (a_{m+l} + a_{m-l}) = \alpha (a_{m+1} + a_{m-1}) + \mu L_n(\mathbf{a}, \mu),$$

where $\alpha = \hat{E}_{n,1}/\mu$ and there exists a μ-independent constant $C > 0$ such that

$$\|\mathbf{L}(\mathbf{a}, \mu)\|_{l^1} \leq C \|\mathbf{a}\|_{l^1}. \qquad (2.4.24)$$

Similarly, the cubic nonlinear term on the left-hand side of the lattice equation (2.4.21) can be written in the form

$$\sum_{(m_1, m_2, m_3)\in\mathbb{Z}^3} K_{m,m_1,m_2,m_3} a_{m_1} \bar{a}_{m_2} a_{m_3} = \beta |a_m|^2 \phi_m + \mu Q_n(\mathbf{a}, \mu),$$

where $\beta = \|\hat{u}_{n,0}\|_{L^4}^4$ and there exists a μ-independent constant $C > 0$ such that

$$\|\mathbf{Q}(\mathbf{a}, \mu)\|_{l^1} \leq C \|\mathbf{a}\|_{l^1}^3. \qquad (2.4.25)$$

By bounds (2.4.23), (2.4.24) and (2.4.25), we write the lattice equation (2.4.21) in the form

$$\alpha (a_{m+1} + a_{m-1}) + \beta |a_m|^2 a_m - \Omega a_m = \mu N_m(\mathbf{a}, \mu), \quad m \in \mathbb{Z}, \qquad (2.4.26)$$

where the left-hand side is the stationary DNLS equation (2.4.15) and the right-hand side is a perturbation term. For any $\mu \in (0, \mu_0)$ and $\mathbf{a} \in B_{\delta_0}(l^1(\mathbb{Z}))$, there is $C_0 > 0$ such that

$$\|\mathbf{N}(\mathbf{a}, \mu)\|_{l^1} \leq C_0 \|\mathbf{a}\|_{l^1}. \qquad (2.4.27)$$

We now conclude the proof of Theorem 2.6.

Let $\mathbf{A} \in l^1(\mathbb{Z})$ be a solution of the stationary DNLS equation (2.4.15) and the linearized equation at $\mathbf{A}$ admits a one-dimensional kernel in $l^1(\mathbb{Z})$ spanned by $\{i\mathbf{A}\}$ and isolated from the rest of the spectrum. Eigenvector $\{i\mathbf{A}\}$ is always present due to the invariance of the stationary DNLS equation (2.4.15) with respect to the gauge transformation,

$$\mathbf{A} \mapsto \mathbf{A}e^{i\theta}, \quad \theta \in \mathbb{R}.$$

The perturbed lattice equation (2.4.26) is also invariant with respect to this transformation, since it is inherited from the properties of the stationary Gross–Pitaevskii equation (2.4.16). Let us fix the gauge factor θ by picking up an $n_0 \in \mathbb{Z}$ such that $|A_{n_0}| \neq 0$ and fixing

$$\text{Im}(a_{n_0}) = \text{Im}(A_{n_0}) = 0. \tag{2.4.28}$$

The linearized operator is invertible subject to the constraint (2.4.28). The vector field of the perturbed lattice equation (2.4.26) is a smooth map from $l^1(\mathbb{Z})$ to $l^1(\mathbb{Z})$ which preserves the constraint (2.4.28). By the Implicit Function Theorem (Appendix B.7), there exists a smooth continuation of the solution $\mathbf{A}$ of the stationary DNLS equation (2.4.15) into the solution $\mathbf{a}$ of the perturbed lattice equation (2.4.26) with respect to μ and there are $\mu_0 > 0$ and $C_0 > 0$ such that for any $\mu \in (0, \mu_0)$,

$$\|\mathbf{a} - \mathbf{A}\|_{l^1} \leq C\mu. \tag{2.4.29}$$

By Lemmas 2.4 and 2.20, if $\mathbf{a} \in l^1(\mathbb{Z})$, then $\varphi_\mu, \psi_\mu \in \mathcal{H}^1(\mathbb{R})$, so that both φ_μ and ψ_μ are bounded continuous functions of $x \in \mathbb{R}$ that decay to zero as $|x| \to \infty$. By Lemma 2.5, $\varphi_\mu(x)$ decays to zero exponentially fast as $|x| \to \infty$ if $\{a_m\}_{m \in \mathbb{Z}}$ decays to zero exponentially fast as $|m| \to \infty$. The same properties hold for $\psi_\mu = \hat{\psi}(\varphi_\mu, \mu)$ thanks to Lemma 2.20, and thus to the full solution Φ.

The proof of Theorem 2.6 is now complete. $\qquad\square$

Note that the assumption of Theorem 2.6 on non-degeneracy of the linearized equation at a solution $\mathbf{A}$ of the stationary DNLS equation (2.4.15) is common in bifurcation theory. If the linearized equation has a higher-dimensional kernel, it signals that the branch of localized solutions $\mathbf{A}$ undertakes a bifurcation and may not persist with respect to parameter continuation for a small nonzero μ. Note that the assumption on the non-degeneracy of the linearized equation is satisfied for *any* solution $\mathbf{A} \in l^1(\mathbb{Z})$ of the stationary one-dimensional DNLS equation (2.4.15) in the anticontinuum limit $\alpha \to 0$ but it need not be satisfied for higher-dimensional DNLS equations (Section 3.2.5).

Exercise 2.31 Justify the stationary two-dimensional DNLS equation,

$$\alpha_1 \left(a_{m_1+1,m_2} + a_{m_1-1,m_2}\right) + \alpha_2 \left(a_{m_1,m_2+1} + a_{m_1,m_2-1}\right)$$
$$+ \beta|a_{m_1,m_2}|^2 a_{m_1,m_2} = \Omega a_{m_1,m_2}, \quad (m_1, m_2) \in \mathbb{Z}^2,$$

starting with the Gross–Pitaevskii equation with the separable potential,

$$-(\partial_{x_1}^2 + \partial_{x_2}^2)\phi + (V_1(x_1) + V_2(x_2))\phi + |\phi|^2\phi = \omega\phi, \quad (x_1, x_2) \in \mathbb{R}^2,$$

in space $H^2(\mathbb{R}^2)$, where V_1 and V_2 are bounded 2π-periodic potentials.

Existence of localized modes in periodic potentials

We can't solve problems by using the same kind of thinking we used when we created them.
– Albert Einstein.

The irrationality of a thing is no argument against its existence, rather a condition of it.
– Friedrich Nietzsche.

Nonlinear evolution equations were less attractive if they would not admit special solutions observed in many laboratory and computer experiments. These solutions are stationary localized modes, which are periodic in time and decaying to zero at infinity in space coordinates. Depending on whom we are speaking with, such stationary localized modes are known as *gap solitons* or *nonlinear bound states* or *discrete breathers* or, more abstractly, *solitary waves*.

This chapter covers the main aspects of the existence theory of stationary localized modes of the Gross–Pitaevskii equation,

$$iu_t = -\nabla^2 u + V(x)u + \sigma|u|^2 u, \quad x \in \mathbb{R}^d, \quad t \in \mathbb{R},$$

where $\sigma \in \{1, -1\}$, $V(x) : \mathbb{R}^d \to \mathbb{R}$ is bounded and 2π-periodic in each coordinate, and $u(x, t) : \mathbb{R}^d \times \mathbb{R} \to \mathbb{C}$ is the wave function.

Thanks to the gauge invariance of the Gross–Pitaevskii equation and the time invariance of the potential function $V(x)$, the time and space coordinates can be separated for the stationary solutions in the form

$$u(x, t) = \phi(x)e^{-i\omega t}, \quad x \in \mathbb{R}^d, \quad t \in \mathbb{R},$$

where $\omega \in \mathbb{R}$ is a free parameter and $\phi(x) : \mathbb{R}^d \to \mathbb{C}$ is a function decaying to zero as $|x| \to \infty$. Clearly, stationary solutions are time-periodic with period $\tau = 2\pi/\omega$, but this time dependence is trivial in the sense that it can be detached by the gauge transformation $u(x, t) = \tilde{u}(x, t)e^{-i\omega t}$, where $\tilde{u}(x, t)$ solves the same Gross–Pitaevskii equation with new potential $\tilde{V}(x) = V(x) - \omega$. The stationary solution for $\tilde{u}(x, t)$ is now time-independent but the potential $\tilde{V}(x)$ has a parameter ω.

Stationary solutions ϕ for a fixed value of $\omega \in \mathbb{R}$ can be considered in two different senses. Recall that $L = -\Delta + V(x)$ is a self-adjoint operator from $H^2(\mathbb{R}^d) \subset \text{Dom}(L)$ to $L^2(\mathbb{R}^d)$ (Section 2.1) and that $H^1(\mathbb{R}^d)$ is the energy space of the Gross–Pitaevskii equation, where its conserved quantities are defined (Section 1.3).

Definition 3.1 We say that $\phi \in H^2(\mathbb{R}^d)$ is a *strong solution* of the stationary Gross–Pitaevskii equation,

$$-\nabla^2\phi(x) + V(x)\phi(x) + \sigma|\phi(x)|^2\phi(x) = \omega\phi(x), \quad x \in \mathbb{R}^d,$$

if

$$\| -\nabla^2\phi + V\phi + \sigma|\phi|^2\phi - \omega\phi \|_{L^2} = 0.$$

We say that $\phi \in H^1(\mathbb{R})$ is a *weak solution* of the stationary Gross–Pitaevskii equation if it is a critical point of the energy functional $E_\omega(u) = H(u) - \omega Q(u)$, where the Hamiltonian $H(u)$ and the power $Q(u)$ are given by

$$H(u) = \int_{\mathbb{R}^d} \left[|\nabla u|^2 + V|u|^2 + \frac{\sigma}{2}|u|^4 \right] dx, \quad Q(u) = \int_{\mathbb{R}^d} |u|^2 dx.$$

The stationary Gross–Pitaevskii equation has the same role as the stationary Schrödinger equation in quantum mechanics. It coincides with the Euler–Lagrange equation generated by the Gateaux derivative of the energy functional $E_\omega(u) = H(u) - \omega Q(u)$,

$$\nabla_{\bar{u}}E_\omega(\phi) = -\nabla^2\phi + V\phi + \sigma|\phi|^2\phi - \omega\phi = 0.$$

Thanks to the Sobolev Embedding Theorem (Appendix B.10), if $d = 1$, then either $\phi \in H^2(\mathbb{R})$ or $\phi \in H^1(\mathbb{R})$ implies that $\phi \in C_b^0(\mathbb{R})$ and $\lim_{x\to\pm\infty}\phi(x) = 0$. The continuity and decay properties come from *bootstrapping* arguments for $d \geq 2$.

Note that if ϕ is a solution of the stationary Gross–Pitaevskii equation in any sense, then $e^{-i\theta}\phi$ is also a solution of the same equation for any $\theta \in \mathbb{R}$. This fact is due to the gauge invariance and it shows that the localized mode in a periodic potential has at least one arbitrary parameter θ in addition to parameter ω.

Recall that the stationary Gross–Pitaevskii equation with a periodic potential can be reduced to the nonlinear Dirac equations, the nonlinear Schrödinger equation, and the discrete nonlinear Schrödinger equation (Chapter 2). Instead of dealing with the space-periodic stationary equations, one can work with the reduced space-homogeneous stationary equations. Solutions of the reduced stationary equations can be constructed explicitly and they give useful analytical approximations to solutions of the original stationary Gross–Pitaevskii equation. Therefore, in this chapter, we shall look at the construction of localized modes not only for the stationary Gross–Pitaevskii equation with a periodic potential but also for the reduced stationary equations.

3.1 Variational methods

Let us consider the stationary Gross–Pitaevskii equation,

$$-\nabla^2\phi(x) + V(x)\phi(x) + \sigma|\phi(x)|^2\phi(x) = \omega\phi(x), \quad x \in \mathbb{R}^d, \tag{3.1.1}$$

where $\sigma \in \{1, -1\}$, $\omega \in \mathbb{R}$, and $V(x)$ is a bounded, continuous, and 2π-periodic function. Because the spectrum of the linear Schrödinger operator $L = -\nabla^2 + V(x)$

in $L^2(\mathbb{R}^d)$ is bounded from below, we say that the localized mode $\phi(x)$ exists in the semi-infinite gap if $\omega < \inf \sigma(L)$ and in a finite band gap if $\omega > \inf \sigma(L)$ and $\omega \notin \sigma(L)$.

We shall be looking for weak solutions $\phi \in H^1(\mathbb{R}^d)$ of the stationary Gross–Pitaevskii equation (3.1.1) in the sense of Definition 3.1. Therefore, we add a trial function $w \in H^1(\mathbb{R}^d)$ to a critical point $\phi \in H^1(\mathbb{R}^d)$ of the energy functional $E_\omega(u)$ and require that

$$\forall w \in H^1(\mathbb{R}^d): \quad \left. \frac{d}{d\epsilon} E_\omega(\phi + \epsilon w) \right|_{\epsilon=0} = 0.$$

Direct computations show that $\phi \in H^1(\mathbb{R})$ is a critical point of $E_\omega(u)$ if

$$\forall w \in H^1(\mathbb{R}^d): \quad \int_{\mathbb{R}^d} \left(\nabla \phi \cdot \nabla \bar{w} + V \phi \bar{w} + \sigma |\phi|^2 \phi \bar{w} - \omega \phi \bar{w} \right) dx = 0. \quad (3.1.2)$$

If ϕ is also a strong solution, that is, if $\phi \in H^2(\mathbb{R}^d)$, then integration by parts shows that

$$\forall w \in L^2(\mathbb{R}^d): \quad \langle -\nabla^2 \phi + V \phi + \sigma |\phi|^2 \phi - \omega \phi, w \rangle_{L^2} = 0. \quad (3.1.3)$$

By the Gagliardo–Nirenberg inequality (Appendix B.5), there is $C_{p,d} > 0$ such that

$$\|u\|_{L^{2p}}^{2p} \leq C_{p,d} \|u\|_{L^2}^{d+p(2-d)} \|\nabla u\|_{L^2}^{d(p-1)}. \quad (3.1.4)$$

Since $E_\omega(u)$ contains $\|u\|_{L^4}^4$ ($p = 2$), for any $1 \leq d \leq 3$ and $V \in L^\infty(\mathbb{R}^d)$, Sobolev space $H^1(\mathbb{R}^d)$ is the space of admissible functions for the functional $E_\omega(u)$ in the sense that $|E_\omega(u)| < \infty$ if $\|u\|_{H^1} < \infty$. The following lemma shows that all solutions in $H^1(\mathbb{R}^d)$ belong to $H^2(\mathbb{R}^d)$ if $1 \leq d \leq 3$ and $V \in L^\infty(\mathbb{R}^d)$.

Lemma 3.1 *Assume $1 \leq d \leq 3$ and $V \in L^\infty(\mathbb{R}^d)$. The weak and strong solutions of the stationary Gross–Pitaevskii equation (3.1.1) are equivalent.*

Proof Since $H^2(\mathbb{R}^d)$ is embedded into $H^1(\mathbb{R}^d)$, the statement is trivial in one direction. In the opposite direction, we let $\phi \in H^1(\mathbb{R}^d)$ and consider

$$F := V\phi + \sigma |\phi|^2 \phi - \omega \phi.$$

By the Gagliardo–Nirenberg inequality (3.1.4) for $p = 3$, we have $F \in L^2(\mathbb{R}^d)$ if $1 \leq d \leq 3$ and $V \in L^\infty(\mathbb{R}^d)$. Therefore, if $\phi \in H^1(\mathbb{R}^d)$ satisfies the weak formulation (3.1.2), then $\nabla^2 \phi \equiv F \in L^2(\mathbb{R}^d)$ and $\phi \in H^2(\mathbb{R}^d)$ satisfies the strong formulation (3.1.3). $\qquad \square$

Existence of localized modes in the semi-infinite gap is similar to the existence of nonlinear bound states in the nonlinear Schrödinger equation with $V(x) \equiv 0$. We will prove that no nonzero weak solution of the stationary equation (3.1.2) in $H^1(\mathbb{R}^d)$ exists if $\omega < \inf \sigma(L)$ and $\sigma = +1$ (defocusing or repulsive case), whereas there exists always a nonzero weak solution $\phi(x)$ if $\omega < \inf \sigma(L)$ and $\sigma = -1$ (focusing or attractive case). In what follows, we will always assume $1 \leq d \leq 3$ and $V \in L^\infty(\mathbb{R}^d)$.

Theorem 3.1 *Let $\omega < \inf \sigma(L)$ and $\sigma = +1$. No nonzero critical points of $E_\omega(u)$ exist.*

Proof Let $\phi \in H^1(\mathbb{R}^d)$ be a critical point of $E_\omega(u)$ for $\sigma = +1$. Substitution of $w = \phi$ into the stationary equation (3.1.2) gives

$$\|L^{1/2}\phi\|_{L^2}^2 - \omega\|\phi\|_{L^2}^2 + \|\phi\|_{L^4}^4 = 0,$$

where

$$\|L^{1/2}\phi\|_{L^2}^2 := \int_{\mathbb{R}^d} \left(|\nabla\phi|^2 + V|\phi|^2\right) dx.$$

By Rayleigh–Ritz's variational principle for the self-adjoint operator L, we have

$$\inf \sigma(L) = \inf_{u \in H^1} \frac{\|L^{1/2}u\|_{L^2}^2}{\|u\|_{L^2}^2}.$$

Therefore,

$$(\inf \sigma(L) - \omega)\|\phi\|_{L^2}^2 + \|\phi\|_{L^4}^4 \leq 0,$$

but since $\inf \sigma(L) > \omega$, we have $\|\phi\|_{L^2} = \|\phi\|_{L^4} = 0$. As a result, we have

$$0 \leq \|\nabla\phi\|_{L^2}^2 \leq \int_{\mathbb{R}^d} \left(|\nabla\phi|^2 + V|\phi|^2 + \|V\|_{L^\infty}|\phi|^2\right) dx = 0,$$

which proves that $\|\nabla\phi\|_{L^2} = 0$, so that $\|\phi\|_{H^1} = 0$. Therefore, the only critical point $\phi \in H^1(\mathbb{R}^d)$ is the zero point $\phi = 0$. $\qquad\square$

Theorem 3.2 *Let $\omega < \inf \sigma(L)$ and $\sigma = -1$. There exists a nonzero critical point ϕ of $E_\omega(u)$.*

Remark 3.1 If $V(x) \equiv 0$, a nonzero critical point ϕ of $E_\omega(u)$ exists for any $\omega < 0$.

Two different variational methods can be applied to the proof of Theorem 3.2 that rely on the existence of a minimizer of a constrained quadratic functional (Section 3.1.1) and the existence of a critical point of a functional with mountain pass geometry (Section 3.1.2). Equivalence between the least energy solutions of the first method and the critical points of the second method is shown by JeanJean & Tanaka [98].

Another variational technique based on the concentration compactness principle was developed by Lions [134]. Using the concentration compactness principle, the weak solutions of the stationary equation (3.1.2) is obtained from a minimizer of the energy

$$H(u) = \int_{\mathbb{R}^d} \left(|\nabla u|^2 + V|u|^2 - \frac{1}{2}|u|^4\right) dx \qquad (3.1.5)$$

under the fixed power $Q(u) = \|u\|_{L^2}^2$. Parameter ω becomes the Lagrange multiplier of the constrained variational problem. This technique not only gives the existence of a localized mode but also predicts the orbital stability of the minimizer (Section 4.4.2). Details of the concentration compactness principle and the orbital stability of the nonlinear waves in nonlinear dispersive equations are reviewed in the text of Angulo Pava [11].

It is more delicate to prove existence of localized modes in a finite band gap associated with the periodic potential $V(x)$. Earlier results on this topic date back to the 1990s, collected together by Stuart [198], who proved existence of critical points using the Mountain Pass Theorem. Similar variational results are also obtained by Alama & Liu [9]. Bifurcations of bound states were analyzed by Heinz, Küpper and Stuart [87, 124] who proved that, depending on the sign of the nonlinear term, lower or upper band edges of the spectral bands of $L = -\nabla^2 + V(x)$ become points of local bifurcations of the localized modes for small values of Q.

The same problem in a more general stationary Gross–Pitaevskii equation has recently received another wave of interest, after Pankov [148] applied the technique of the Linking Theorem to the proof of existence of localized modes in periodic potentials. The following theorem gives the existence of localized modes in every finite band gap of $\sigma(L)$.

Theorem 3.3 *Let ω be in a finite gap of $\sigma(L)$. There exists a nonzero critical point ϕ of $E_\omega(u)$. Moreover $\phi(x)$ is a bounded continuous function for all $x \in \mathbb{R}^d$ that decays to zero exponentially fast as $|x| \to \infty$.*

All the above methods turn out to be useful in the discrete setting, in the context of localized modes of the stationary DNLS equation,

$$-(\Delta\phi)_n + \sigma|\phi_n|^2\phi_n = \omega\phi_n, \quad n \in \mathbb{Z}^d, \tag{3.1.6}$$

where $d \geq 1$ is any dimension of the cubic lattice, Δ is the discrete Laplacian on $l^2(\mathbb{Z}^d)$, $\sigma \in \{1, -1\}$, and $\omega \in \mathbb{R}$. The concentration compactness principle was applied to the proof of existence of localized modes in the stationary DNLS equation (3.1.6) by Weinstein [212]. Localized modes of a more general lattice equation were also constructed by Pankov using the Linking Theorem in [149] and the Nehari manifolds in [150]. Recent works in the development of these results include Shi & Zhang [190], Zhang [222], and Zhou *et al.* [223].

3.1.1 Minimizers of a constrained quadratic functional

We shall prove Theorem 3.2, which states the existence of a critical point ϕ of the energy functional $E_\omega(u)$ for the stationary Gross–Pitaevskii equation (3.1.1) if $\omega < \inf \sigma(L)$ and $\sigma = -1$. The proof relies on the decomposition

$$E_\omega(u) = I(u) - \frac{1}{2}J(u),$$

where $I(u)$ is a quadratic functional

$$I(u) := \|L^{1/2}u\|_{L^2}^2 - \omega\|u\|_{L^2}^2 = \int_{\mathbb{R}^d} \left(|\nabla u|^2 + V|u|^2 - \omega|u|^2\right) dx, \tag{3.1.7}$$

and $J(u) := \|u\|_{L^4}^4$. We shall consider a constrained minimization problem

$$\text{find } \min_{u \in H^1} I(u) \quad \text{such that} \quad J(u) = J_0, \tag{3.1.8}$$

where $J_0 > 0$ is a fixed number.

Unfortunately, the lack of compactness of the embedding of $H^1(\mathbb{R}^d)$ into $L^4(\mathbb{R}^d)$ is an obstacle in the proof of Theorem 3.2 in an unbounded domain. To bypass this obstacle, we shall split the proof into two steps.

At the first step, we shall prove the existence of a minimizer of the constrained minimization problem (3.1.8) in $H^1_{\text{per}}(\mathbb{T}^d_n)$, where $\mathbb{T}^d_n$ is the cube in $\mathbb{R}^d$ of the edge length $2\pi n$, centered at the origin, where $n \in \mathbb{N}$. Note that the edge length $2\pi n$ is an n-multiple of the 2π-period of the potential $V(x)$. The embedding of $H^1_{\text{per}}(\mathbb{T}^d_n)$ into $L^4_{\text{per}}(\mathbb{T}^d_n)$ is compact and the calculus of variations is hence applied.

At the second step, we shall prove that the minimizer of the constrained minimization problem (3.1.8) in $H^1_{\text{per}}(\mathbb{T}^d_n)$ persists in the singular limit $n \to \infty$ as a minimizer of the same variational problem in $H^1(\mathbb{R}^d)$.

The first step relies on the following lemma.

Lemma 3.2 *Fix $\omega < \inf \sigma(L)$ and $\sigma = -1$. Let $\mathbb{T}^d_n$ be the cube in $\mathbb{R}^d$ of the edge length $2\pi n$ for a fixed $n \in \mathbb{N}$, centered at the origin. There exists a minimizer $\psi_n \in H^1_{\text{per}}(\mathbb{T}^d_n)$ of the constrained minimization problem (3.1.8) for $J_0 > 0$.*

Proof It is clear from the variational characterization of eigenvalues that

$$I(u) = \langle (L - \omega I)u, u \rangle_{L^2} \geq (\inf \sigma(L) - \omega)\|u\|^2_{L^2}.$$

Therefore, $I(u)$ is bounded from below by zero if $\omega < \inf \sigma(L)$. Therefore, there is an infimum of $I(u)$ in $H^1_{\text{per}}(\mathbb{T}^d_n)$, which we denote by

$$m := \inf_{u \in H^1} I(u).$$

Furthermore, if $\omega < \inf \sigma(L)$, there exists $C > 0$ that depends on the distance $|\inf \sigma(L) - \omega|$ such that

$$I(u) \geq C\|u\|^2_{H^1}. \tag{3.1.9}$$

The proof of the *coercivity* condition (3.1.9) is trivial if $\omega < \min_{x \in \mathbb{R}^d} V(x)$ but it also holds for any $\omega < \inf \sigma(L)$. Thanks to the coercivity of the functional $I(u)$, any minimizing sequence $\{u_k\}_{k \geq 1}$ in $H^1_{\text{per}}(\mathbb{T}^d_n)$ such that $\lim_{k \to \infty} I(u_k) = m$ belongs to a bounded subset of $H^1_{\text{per}}(\mathbb{T}^d_n)$ and satisfies

$$\sup_{k \geq 1} \|u_k\|_{H^1} < \infty.$$

Because $H^1_{\text{per}}(\mathbb{T}^d_n)$ is not compact, boundedness of the sequence $\{u_k\}_{k \geq 1}$ in $H^1_{\text{per}}(\mathbb{T}^d_n)$ does *not* imply that there is an element $u \in H^1_{\text{per}}(\mathbb{T}^d_n)$ to which the sequence *converges strongly* in the sense

$$u_k \to u \text{ in } H^1 \quad \Longleftrightarrow \quad \lim_{k \to \infty} \|u_k - u\|_{H^1} = 0.$$

Instead, by the Weak Convergence Theorem (Appendix B.14), the boundedness of the sequence $\{u_k\}_{k \geq 1}$ in $H^1_{\text{per}}(\mathbb{T}^d_n)$ only implies that there exists a subsequence $\{u_{k_j}\}_{j \geq 1} \subset \{u_k\}_{k \geq 1}$ and a function $u \in H^1_{\text{per}}(\mathbb{T}^d_n)$ such that $\{u_{k_j}\}_{j \geq 1}$ *converges weakly* to u in the sense

$$u_{k_j} \rightharpoonup u \text{ in } H^1 \quad \Longleftrightarrow \quad \forall v \in H^1 : \lim_{j \to \infty} \langle v, u_{k_j} - u \rangle_{H^1} = 0,$$

where

$$\langle v, u \rangle_{H^1} := \langle \nabla v, \nabla u \rangle_{L^2} + \langle v, u \rangle_{L^2}.$$

Thanks to the boundedness of $\|V\|_{L^\infty}$ and the coercivity (3.1.9), the functional $I(u)$ is equivalent to the squared norm $\|u\|_{H^1}^2$ in the sense

$$\exists C \geq 1 : \quad C^{-1}\|u\|_{H^1}^2 \leq I(u) \leq C\|u\|_{H^1}^2. \tag{3.1.10}$$

Inherited from the norm $\|u\|_{H^1}$, the functional $I(u)$ is hence *weakly lower semi-continuous* in the sense

$$I(u) \leq \liminf_{j\to\infty} I(u_{k_j}) \tag{3.1.11}$$

whenever $\{u_{k_j}\}_{j\geq 1}$ converges weakly to u in $H^1_{\mathrm{per}}(\mathbb{T}_n^d)$. Since $\{u_k\}_{k\geq 1}$ is a minimizing sequence for $I(u)$, the weak semi-continuity (3.1.11) implies that $I(u) \leq m$ and since $u \in H^1_{\mathrm{per}}(\mathbb{T}_n^d)$, we have $I(u) \geq m$. Therefore, $I(u)$ attains its infimum at $u \equiv \psi$ in $H^1_{\mathrm{per}}(\mathbb{T}_n^d)$.

Unfortunately, the infimum of $I(u)$ in the unconstrained minimization problem is simply the zero solution $\psi \equiv 0$ since $I(0) = 0 = m$. To construct a nonzero ψ for the constrained minimization problem (3.1.8) with $J_0 > 0$, we shall define the space of admissible functions

$$H_0 := \left\{ u \in H^1_{\mathrm{per}}(\mathbb{T}_n^d) : \quad \|u\|_{L^4}^4 = J_0 \right\}.$$

It is obvious that this space is non-empty for any $J_0 > 0$.

Thanks to the Gagliardo–Nirenberg inequality (3.1.4), $H^1_{\mathrm{per}}(\mathbb{T}_n^d)$ is embedded into $L^4_{\mathrm{per}}(\mathbb{T}_n^d)$ and this embedding is compact in the periodic (bounded) domain. As a result, weak convergence in $H^1_{\mathrm{per}}(\mathbb{T}_n^d)$ implies strong convergence in $L^4_{\mathrm{per}}(\mathbb{T}_n^d)$ (Theorem 8.6 in [130]), that is, for any minimizing sequence $\{u_k\}_{k\geq 1}$ in $H^1_{\mathrm{per}}(\mathbb{T}_n^d)$, there exist a subsequence $\{u_{k_j}\}_{j\geq 1} \subset \{u_k\}_{k\geq 1}$ and a function $u \in H^1_{\mathrm{per}}(\mathbb{T}_n^d)$ such that

$$u_{k_j} \rightharpoonup u \text{ in } H^1_{\mathrm{per}}(\mathbb{T}_n^d) \quad \Rightarrow \quad u_{k_j} \to u \text{ in } L^4_{\mathrm{per}}(\mathbb{T}_n^d).$$

Let $\{u_k\}_{k\geq 1}$ be a minimizing sequence in H_0 so that $J(u_k) = J_0$ for any $k \geq 1$. By the strong convergence in $L^4_{\mathrm{per}}(\mathbb{T}_n^d)$, we have $J(u) = J_0$, so that the limit of $\{u_{k_j}\}_{j\geq 1}$, which is $u \in H^1_{\mathrm{per}}(\mathbb{T}_n^d)$, is actually in H_0. Therefore, $u \equiv \psi$ is a minimizer of the constrained minimization problem (3.1.8) in $H^1_{\mathrm{per}}(\mathbb{T}_n^d)$. $\square$

We shall now consider the convergence of the sequence of minimizers $\{\psi_n\}_{n\in\mathbb{N}}$ as $n \to \infty$. We note the following result.

Lemma 3.3 *Let $\psi_n \in H^1_{\mathrm{per}}(\mathbb{T}_n^d)$ be a minimizer in Lemma 3.2. Then,*

$$\phi_n = \frac{\sqrt{I(\psi_n)}}{\sqrt{J(\psi_n)}}\psi_n \in H^1_{\mathrm{per}}(\mathbb{T}_n^d)$$

is a critical point of $E_\omega(u) = I(u) - \frac{1}{2}J(u)$ and

$$E_\omega(\phi_n) = \frac{1}{2}I(\phi_n) = \frac{1}{2}J(\phi_n) = \frac{1}{2}J_0.$$

Proof Using the technique of Lagrange multipliers, a minimizer of the constrained variational problem (3.1.8) is a critical point of the functional

$$\hat{I}(u) := I(u) - \lambda \left(J(u) - J_0 \right), \tag{3.1.12}$$

where λ is the Lagrange multiplier. If there exists a critical point (λ, ψ) for $\hat{I}(u)$, then

$$\partial_\lambda \hat{I}(u)|_{u=\psi} = J_0 - J(\psi) = 0$$

and

$$\forall w \in H^1: \quad \int \left(\nabla \psi \cdot \nabla \bar{w} + V \psi \bar{w} - \omega \psi \bar{w} - 2\lambda |\psi|^2 \psi \bar{w} \right) dx = 0.$$

For $w = \psi$, this gives

$$2\lambda = \frac{I(\psi)}{J(\psi)} > 0,$$

which allows us to substitute $\psi(x) = (1/\sqrt{2\lambda})\phi(x)$ into the variational equations. On comparison with the weak formulation (3.1.2), we infer that ϕ is a critical point of the energy functional $E_\omega(u)$ with $I(\phi) = J(\phi) = J_0$. Observing that $E_\omega(u) = I(u) - \frac{1}{2}J(u)$, we arrive at the statement of the lemma. $\square$

When we pass to the limit of $n \to \infty$, the principal difficulty is to show that the limit of a sequence of critical points $\{\phi_n\}_{n \in \mathbb{N}}$ is a nonzero function $\phi \in H^1(\mathbb{R}^d)$. For this purpose, we need to use the concentration compactness arguments, which ensure that the critical point $\phi_n(x)$ is confined in a bounded region of $\mathbb{R}^d$ as $n \to \infty$.

Lemma 3.4 *Let $\{\phi_n\}_{n \in \mathbb{N}}$ with $\phi_n \in H^1_{\mathrm{per}}(\mathbb{T}^d_n)$ be a sequence of critical points of $E_\omega(u)$ in Lemma 3.3. Then, there is a sequence $\{b_n\}_{n \in \mathbb{N}}$ with $b_n \in \mathbb{Z}^d$ such that $\{\phi_n(x - 2\pi b_n)\}_{n \in \mathbb{N}}$ converges weakly in $H^1_{\mathrm{loc}}(\mathbb{R}^d)$ to a nonzero weak solution $\phi \in H^1(\mathbb{R}^d)$ of the stationary Gross–Pitaevskii equation (3.1.2).*

Proof By Lemma 3.3 and the norm equivalence (3.1.10), the norms $\|\phi_n\|_{H^1_{\mathrm{per}}}$ are bounded and nonzero as $n \to \infty$. Let $Q_r(x_0)$ be a cube in $\mathbb{R}^d$ with the fixed edge length $2\pi r$, centered at the point $x_0 \in \mathbb{R}^d$. We will show that there exist a sequence of points $x_n \in \mathbb{R}^d$ and positive numbers r and η such that

$$\|u_n\|_{L^2(Q_r(x_n))} \geq \eta, \quad n \in \mathbb{N}. \tag{3.1.13}$$

We show this by contradiction. If the condition (3.1.13) does not hold, then there is $r > 0$ such that

$$\lim_{n \to \infty} \sup_{x \in \mathbb{R}^d} \|u_n\|_{L^2(Q_r(x))} = 0.$$

By Lemma I.1 in Lions [134], $\lim_{n \to \infty} \|u_n\|_{L^4_{\mathrm{per}}(\mathbb{T}^d_n)} = 0$. On the other hand, this contradicts the fact that $\phi_n \in H^1_{\mathrm{per}}(\mathbb{T}^d_n)$ is a critical point of $E_\omega(u)$ with $J(\phi_n) = \|u_n\|^4_{L^4_{\mathrm{per}}(\mathbb{T}^d_n)} = J_0$ for a fixed $J_0 > 0$.

Thanks to the condition (3.1.13), there exists a sequence of integer vectors $b_n \in \mathbb{Z}^d$ such that the sequence of cubes $Q_r(x_n - 2\pi b_n)$ is confined in a bounded region and $\tilde{\phi}_n(x) := \phi_n(x - 2\pi b_n)$ is a critical point of $E_\omega(u)$ so that the norms $\|\tilde{\phi}_n\|_{H^1_{\mathrm{per}}}$ are bounded and nonzero. Passing to a subsequence, it follows that $\{\tilde{\phi}_n\}_{n \in \mathbb{N}}$ converges weakly in $H^1_{\text{loc}}(\mathbb{R}^d)$ to a function $\phi \in H^1(\mathbb{R}^d)$. By the compactness of the Sobolev embedding, we have $\tilde{\phi}_n \to \phi$ strongly in $L^4_{\text{loc}}(\mathbb{R}^d)$. As a result, we have

$$\forall w \in H^1(\mathbb{R}^d) : \quad \int_{\mathbb{R}^d} \left(\nabla\phi \cdot \nabla\bar{w} + V\phi\bar{w} - \omega\phi\bar{w} - |\phi|^2\phi\bar{w} \right) dx$$

$$= \lim_{n\to\infty} \int_{\mathbb{R}^d} \left(\nabla\tilde{\phi}_n \cdot \nabla\bar{w} + V\tilde{\phi}_n\bar{w} - \omega\tilde{\phi}_n\bar{w} - |\tilde{\phi}_n|^2\tilde{\phi}_n\bar{w} \right) dx = 0.$$

Therefore, $\phi \in H^1(\mathbb{R}^d)$ is a weak nonzero solution of equation (3.1.2). □

Lemma 3.4 concludes the proof of Theorem 3.2.

Exercise 3.1   Use a constrained minimization of a discrete quadratic functional and prove existence of a localized mode $\phi \in l^2(\mathbb{Z}^d)$ of the stationary DNLS equation (3.1.6) for any $\omega < 0$, $\sigma = -1$, and $d \geq 1$.

Exercise 3.2   Use a constrained maximization of a discrete quadratic functional and prove existence of a localized mode $\phi \in l^2(\mathbb{Z}^d)$ of the stationary DNLS equation (3.1.6) for any $\omega > 4d$, $\sigma = +1$, and $d \geq 1$.

3.1.2 Critical points of a functional with mountain pass geometry

Lemma 3.3 shows that a minimizer ψ of the constrained variational problem (3.1.8) gives uniquely a critical point ϕ of the energy functional $E_\omega(u)$, which is nonzero if $J_0 > 0$. This observation is important in the alternative proof of Theorem 3.2 from the existence of a critical point ϕ of the energy functional $E_\omega(u)$ in $H^1(\mathbb{R}^d)$ for $\omega < \inf \sigma(L)$. Let us write the corresponding functional again:

$$E_\omega(u) = \int_{\mathbb{R}^d} \left(|\nabla u|^2 + V|u|^2 - \omega|u|^2 - \frac{1}{2}|u|^4 \right) dx. \qquad (3.1.14)$$

We recall that if $1 \leq d \leq 3$, $V \in L^\infty(\mathbb{R}^d)$, and $\|u\|_{H^1} < \infty$, then $|E_\omega(u)| < \infty$ thanks to the Gagliardo–Nirenberg inequality (3.1.4). Since all terms are powers, $E_\omega(u)$ is a C^∞ map from $H^1(\mathbb{R}^d)$ to $\mathbb{R}$.

Because of the negative sign in front of the quartic term, the functional $E_\omega(u)$ is unbounded from below. Therefore, no global minimizer of $E_\omega(u)$ exists in $H^1(\mathbb{R}^d)$.

Nevertheless, $\phi = 0$ is a local minimizer of $E_\omega(u)$ because $E_\omega(0) = 0$, $\nabla_{\bar{u}} E_\omega(0) = 0$, and the quadratic part of $E_\omega(u)$ is positive definite for any $\omega < \inf \sigma(L)$. This fact suggests that we can look for a nonzero critical point $\phi \in H^1(\mathbb{R}^d)$ of $E_\omega(u)$. We shall prove that the functional $E_\omega(u)$ has the mountain pass geometry in $H^1(\mathbb{R}^d)$ and satisfies the Palais–Smale compactness condition. These conditions enable us to use the Mountain Pass Theorem (Appendix B.8).

Again, the lack of compactness is an obstacle on the Palais–Smale condition in the unbounded domain. Similarly to Section 3.1.1, we shall first consider an approximation of the critical point of $E_\omega(u)$ in the periodic domain with the period $2\pi n$, where $n \in \mathbb{N}$. Then, we shall prove that the non-trivial critical point of $E_\omega(u)$ persists in the limit $n \to \infty$.

Lemma 3.5   *Fix $\omega < \inf \sigma(L)$. The functional $E_\omega(u)$ given by (3.1.14) satisfies conditions of the mountain pass geometry in $H^1(\mathbb{R}^d)$.*

Proof Let us start with the three conditions of the mountain pass geometry in Definition B.4 (Appendix B.8).

(i) From (3.1.14), it follows that $E_\omega(0) = 0$.

(ii) Using bounds (3.1.9) and (3.1.4), we find that for any $\omega < \inf \sigma(L)$, there exist positive constants C_1 and C_2 such that

$$E_\omega(u) \geq C_1 \|u\|_{H^1}^2 - C_2 \|u\|_{H^1}^4.$$

For any $u \in H^1(\mathbb{R}^d)$ with $\|u\|_{H^1} = r < (C_1/C_2)^{1/2}$, there is $a > 0$ such that $E_\omega(u) \geq a$.

(iii) Let $v = tu_0$ with $t > 0$ and $u_0 \in H^1(\mathbb{R}^d)$ such that $\|u_0\|_{H^1} = 1$. Then, we have

$$E_\omega(v) = t^2 \left(\|L^{1/2} u_0\|_{L^2}^2 - \omega \|u_0\|_{L^2}^2 \right) - \frac{1}{2} t^4 \|u_0\|_{L^4}^4 \leq C_3 t^2 - \frac{1}{2} t^4 \|u_0\|_{L^4}^4.$$

If $t \geq \left(2C_3 / \|u_0\|_{L^4}^4 \right)^{1/2}$, then $E_\omega(v) \leq 0$. It is always possible to choose $u_0 \in H^1(\mathbb{R}^d)$ such that $\|u_0\|_{H^1} = 1$ and

$$t^2 \geq \frac{2C_3}{\|u_0\|_{L^4}^4} \geq \frac{C_1}{C_2} > r^2,$$

so that condition (iii) is satisfied. $\square$

Lemma 3.6 *Fix $\omega < \inf \sigma(L)$. Let $\mathbb{T}_n^d$ be the cube in $\mathbb{R}^d$ of the edge length $2\pi n$ for a fixed $n \in \mathbb{N}$, centered at the origin. The functional $E_\omega(u)$ given by (3.1.14) satisfies the Palais–Smale compactness condition in $H_{\mathrm{per}}^1(\mathbb{T}_n^d)$.*

Proof We shall now verify that $E_\omega(u)$ satisfies the Palais–Smale condition in Definition B.5 (Appendix B.8).

We consider a sequence $\{u_k\}_{k \geq 1} \in H_{\mathrm{per}}^1(\mathbb{T}_n^d)$ and assume properties (a) and (b) of the Palais–Smale condition. Therefore, we assume that there are positive constants M and ϵ such that

(a) $\quad |E_\omega(u_k)| \leq M$

and for any $w \in H_{\mathrm{per}}^1(\mathbb{T}_n^d)$,

(b) $\quad \left| \int_{\mathbb{T}_n^d} \left(\nabla u_k \cdot \nabla \bar{w} + V(x) u_k \bar{w} - \omega u_k \bar{w} - |u_k|^2 u_k \bar{w} \right) dx \right| \leq \epsilon \|w\|_{H^1}.$

Moreover, $\epsilon > 0$ is small and

$$\left| I(u_k) - \|u_k\|_{L^4}^4 \right| \to 0 \quad \text{as} \quad k \to \infty, \tag{3.1.15}$$

where $I(u) := \|L^{1/2} u\|_{L^2}^2 - \omega \|u\|_{L^2}^2$.

We need to show that the sequence is bounded in $H_{\mathrm{per}}^1(\mathbb{T}_n^d)$. Choosing $w = u_k$ in condition (b), we have

$$\left| I(u_k) - \|u_k\|_{L^4}^4 \right| \leq \epsilon \|u_k\|_{H^1}.$$

Using now condition (a), we have

$$M + \frac{1}{2} \epsilon \|u_k\|_{H^1} \geq E_\omega(u_k) - \frac{1}{2} I(u_k) + \frac{1}{2} \|u_k\|_{L^4}^4 = \frac{1}{2} I(u_k) \geq \frac{1}{2} C_1 \|u_k\|_{H^1}^2.$$

As a result, there is a positive constant $C(M, \epsilon)$ such that $\|u_k\|_{H^1} \leq C(M, \epsilon)$ for any $k \geq 1$. By the Weak Convergence Theorem (Appendix B.14), there is a subsequence $\{u_{k_j}\}_{j \geq 1} \subset \{u_k\}_{k \geq 1}$ and an element $u \in H^1_{\mathrm{per}}(\mathbb{T}^d_n)$ such that $u_{k_j} \rightharpoonup u$ as $j \to \infty$ in $H^1_{\mathrm{per}}(\mathbb{T}^d_n)$.

Since $H^1_{\mathrm{per}}(\mathbb{T}^d_n)$ is compactly embedded into $L^4_{\mathrm{per}}(\mathbb{T}^d_n)$, we have $u_{k_j} \to u$ as $j \to \infty$ in $L^4_{\mathrm{per}}(\mathbb{T}^d_n)$. From strong convergence in $L^4_{\mathrm{per}}(\mathbb{T}^d_n)$, we have $\|u_{k_j}\|^4_{L^4} \to \|u\|^4_{L^4}$ as $j \to \infty$. Using condition (3.1.15), we infer that

$$I(u_{k_j}) \to I(u) \quad \text{as} \quad j \to \infty.$$

Thanks to the equivalence of $I(u)$ to $\|u\|^2_{H^1}$, the subsequence $\{u_{k_j}\}_{j \geq 1}$ converges strongly to u in $H^1_{\mathrm{per}}(\mathbb{T}^d_n)$. The sequence $\{u_k\}_{k \geq 1}$ satisfying conditions (a) and (b) has hence a convergent subsequence in $H^1_{\mathrm{per}}(\mathbb{T}^d_n)$. $\square$

By the Mountain Pass Theorem (Appendix B.8), there is a critical point of the functional $E_\omega(u)$ in $H^1_{\mathrm{per}}(\mathbb{T}^d_n)$. Repeating the same arguments as in Section 3.1.1, we obtain that the critical point persists in the limit $n \to \infty$ and it recovers the critical point of $E_\omega(u)$ in $H^1(\mathbb{R}^d)$. These facts conclude the alternative proof of Theorem 3.2.

Exercise 3.3   Prove existence of a localized mode $\phi \in l^2(\mathbb{Z}^d)$ of the stationary DNLS equation (3.1.6) for any $\omega < 0$, $\sigma = -1$, and $d \geq 1$ using the critical point theory for a discrete energy functional.

Exercise 3.4   Similarly to Exercise 3.3, prove existence of a localized mode $\phi \in l^2(\mathbb{Z}^d)$ of the stationary DNLS equation (3.1.6) for any $\omega > 4d$, $\sigma = +1$, and $d \geq 1$.

Let ϕ be a weak solution of the stationary Gross–Pitaevskii equation (3.1.2). If we take $w = \bar{\phi}$, we obtain

$$I(\phi) = \|L^{1/2}\phi\|^2_{L^2} - \omega\|\phi\|^2_{L^2} = \|\phi\|^4_{L^4} = J(\phi).$$

This recovers the result of Lemma 3.3,

$$E_\omega(\phi) = H(\phi) - \omega Q(\phi) = I(\phi) - \frac{1}{2}J(\phi) = \frac{1}{2}\|\phi\|^4_{L^4} > 0, \qquad (3.1.16)$$

and shows that if $E_\omega(u)$ has the mountain pass geometry and 0 is a local minimizer of $E_\omega(u)$ with $E_\omega(0) = 0$, then the nonzero critical point ϕ of $E_\omega(u)$ can be either another local minimum or a saddle point or a local maximum of $E_\omega(\phi)$.

3.1.3 Approximations for localized modes

A different method of finding nonzero critical points of $E_\omega(u)$, called the *concentration compactness principle*, is based on another constrained variational problem:

$$\text{find} \quad \min_{u \in H^1} E(u) \quad \text{such that} \quad Q(u) = Q_0 \qquad (3.1.17)$$

for a fixed $Q_0 > 0$. Since $E(0) = 0$, a nonzero minimizer ϕ of $E(u)$ for any $Q_0 > 0$ is supposed to have $E(\phi) < 0$. It is proved by Lions [134] under some technical conditions that if $E(u)$ is bounded from below under the constraint $Q(u) = Q_0$ and there is at least one $v \in H^1(\mathbb{R}^d)$ such that $E(v) < 0$, then the minimum of the constrained variational problem is achieved at ϕ with $E(\phi) < 0$. The existence of a minimizer is related to stability of ϕ with respect to the gradient flow (Section 3.4.2), and with respect to the time evolution in the Gross–Pitaevskii equation (Section 4.4.2).

Exercise 3.5 Let $V(x) \equiv 0$. Using the Gagliardo–Nirenberg inequality (3.1.4), prove that $E(u)$ is bounded from below under the constraint $Q(u) = Q_0$ for $d = 1$ and for $d = 2$ and sufficiently small Q_0.

Remark 3.2 If $V(x) \equiv 0$, no bounds from below exist on $E(u)$ under the constraint $Q(u) = Q_0$ for $d = 2$ and sufficiently large Q_0 and for $d = 3$. As a result, no solutions $\phi \in H^1(\mathbb{R}^d)$ of the constrained minimization problem (3.1.17) exist in these cases. At the same time, the critical points of $E_\omega(u)$ always exist for any $\omega < \inf \sigma(L) = 0$ using the approach based on the mountain pass geometry.

In the discrete setting, Weinstein [212] showed that the minimizers of the discrete energy $E(\mathbf{u})$ under fixed discrete power $Q(\mathbf{u})$ always exist for the stationary DNLS equation (3.1.6) with $\sigma = -1$ but Q_0 must exceed a nonzero *excitation threshold* for $d = 2$ or $d = 3$ (no excitation threshold exists for $d = 1$).

Similar results hold for the stationary Gross–Pitaevskii equation (3.1.1) with a nonzero periodic $V(x)$. Numerical computations of Ilan & Weinstein [90] for $d = 2$ show that there exists a minimizer of $E(u)$ under fixed power $Q(u)$ in the semi-infinite gap of $\sigma(L)$ for large negative values of ω. No problems with existence of global minimizers in the constrained variational problem (3.1.17) arise for $d = 1$.

Figure 3.1 shows numerical approximations of localized modes in the semi-infinite and finite band gaps for $d = 1$, $\sigma = -1$, and $V(x) = \sin^2(x/2)$. Figure 3.2 gives similar results for $\sigma = 1$. These numerical approximations are obtained by Pelinovsky *et al.* [161] using a shooting method.

Figures 3.1 and 3.2 illustrate that localized modes exist in each finite band gap, according to Theorem 3.3, for both $\sigma = -1$ and $\sigma = 1$. The difference between these two cases is the bifurcation of localized modes with small values of power $Q(\phi)$. This bifurcation occurs from the lower band edges of each spectral band for $\sigma = -1$ and from the upper band edges for $\sigma = 1$. (Note that $\omega = \mu$ and the horizontal axes of Figures 3.1 and 3.2 are drawn for $-\mu$.) Similarly, localized modes in the semi-infinite gap exist only for $\sigma = -1$, according to Theorems 3.1 and 3.2. Note also that two distinct branches of localized modes exist in each gap. The multiplicity of branches is not captured by the results of the variational theory.

We shall now discuss the correspondence between the branches of localized modes in Figures 3.1 and 3.2 and the results of the asymptotic approximations from Chapter 2.

In the defocusing case $\sigma = 1$, Theorem 2.4 (Section 2.3.2) establishes the correspondence between the localized mode of the stationary Gross–Pitaevskii equation (2.3.22) near *each* band edge with $E_n''(k_0) < 0$ (the band edge of the nth spectral band at the point $k = k_0$ is a local maximum) and the localized mode of the stationary NLS equation (2.3.20). This corresponds to bifurcation of localized modes above the nth spectral band. For the focusing case $\sigma = -1$, the conclusion is

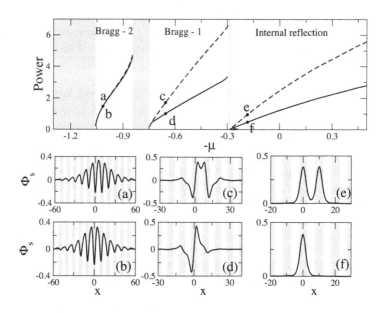

Figure 3.1 Numerical approximations of localized modes for $\sigma = -1$. Reproduced from [161].

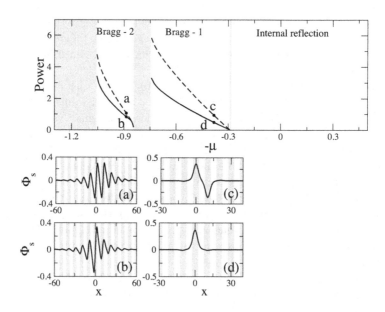

Figure 3.2 Numerical approximations of localized modes for $\sigma = 1$. Reproduced from [161].

the opposite: the localized mode exists near *each* band edge with $E_n''(k_0) > 0$, that is, below the nth spectral band. In particular, it exists in the semi-infinite gap.

Theorem 2.2 (Section 2.2.2) gives more precise information about stationary solutions of the Gross–Pitaevskii equation (2.2.21) compared to Theorem 3.3. In particular, analytical approximation of the localized mode is given by the exponentially decaying functions (2.2.22), which shows that the localized modes reside *everywhere* in the finite spectral gap. One band edge becomes the small-amplitude limit of the localized mode and the other band edge is the finite-amplitude limit of the algebraically decaying soliton. The asymptotic correspondence of Theorem 2.2 is only justified in the limit $\|V\|_{L^\infty} \to 0$ but the same pattern holds in Figures 3.1 and 3.2.

Similarly, Theorem 2.6 (Section 2.4.2) recovers the existence of the localized mode in the stationary Gross–Pitaevskii equation (2.4.16) from the localized mode of the stationary DNLS equation (2.4.15). Additionally, localized modes in the stationary DNLS equation (2.4.15) can be supported at different lattice nodes, which correspond to multi-pulse localized modes of the stationary Gross–Pitaevskii equation (2.4.16) supported in different wells of the periodic potential V. The asymptotic correspondence of Theorem 2.6 is justified in the limit $\|V\|_{L^\infty} \to \infty$.

3.2 Lyapunov–Schmidt reductions

Variational methods (Section 3.1) allow us to predict existence of localized modes ϕ of the stationary Gross–Pitaevskii equation with a periodic potential in spectral gaps of the Schrödinger operator $L = -\nabla^2 + V(x)$. However, these methods lack information about multiplicity of solution branches, dependence of their Hamiltonian $H(\phi)$ and power $Q(\phi)$ versus parameter ω, and analytical approximations. To study localized modes in more detail, we shall now apply local bifurcation methods and consider strong solutions of the stationary Gross–Pitaevskii equation in the sense of Definition 3.1. In comparison with topological arguments of the variational theory, the local bifurcation methods rely on the analytical technique, which is generally referred to as the *Lyapunov–Schmidt reduction method*.

The algorithm of Lyapunov–Schmidt reductions has been applied to many bifurcation problems and is described in several texts [56, 71, 93]. In an abstract setting, it is used to trace roots of a nonlinear operator function $F(x, \epsilon) : X \times \mathbb{R} \to Y$, where X and Y are Banach spaces and $\epsilon \in \mathbb{R}$ is a small parameter. The value $\epsilon = 0$ is considered to be a bifurcation point.

Let us assume that:

- $F(x, \epsilon)$ is C^2 in x and ϵ;
- there is a root $x_0 \in X$ of $F(x, 0) = 0$;
- the Jacobian operator $J = D_x F(x_0, 0) : X \to Y$ has a finite-dimensional kernel and the rest of its spectrum is bounded away from zero.

To simplify the formalism, let us assume that $\text{Ker}(J) = \text{span}\{\psi_0\} \subset X$ is one-dimensional, $X \subset Y = L^2$, and J is a self-adjoint operator in L^2. For strong solutions of the stationary Gross–Pitaevskii equation (3.1.1), we have $X = H^2(\mathbb{R}^d)$

and $Y = L^2(\mathbb{R}^d)$, whereas ϵ is related to the distance of parameter ω from a particular bifurcation point ω_0, for which J is not invertible in $L^2(\mathbb{R}^d)$. The method of Lyapunov–Schmidt reductions consists of three steps.

Step 1: Decomposition. Let $P : L^2 \to \mathrm{Ker}(J) \subset L^2$ be the orthogonal projection operator and $Q = \mathrm{Id} - P$ be the projection operator to the orthogonal complement of $\mathrm{Ker}(J)$. If we normalize $\|\psi_0\|_{L^2} = 1$, then

$$\forall f \in X : \quad P(f) = \langle \psi_0, f \rangle_{L^2} \psi_0, \quad Q(f) = f - \langle \psi_0, f \rangle_{L^2} \psi_0.$$

Let us decompose the solution of the root finding problem $F(x, \epsilon) = 0$ to the sum of three terms

$$\forall x \in X : \quad x = x_0 + a\psi_0 + y,$$

where $y \in [\mathrm{Ker}(J)]^\perp$, that is, $\langle \psi_0, y \rangle_{L^2} = 0$. With the decomposition of the solution, we can also expand the C^2 function $F(x, \epsilon)$ near $x = x_0$ and $\epsilon = 0$:

$$
\begin{aligned}
F(x, \epsilon) &= F(x_0, 0) + J(a\psi_0 + y) + \epsilon \partial_\epsilon F(x_0, 0) + N(a\psi_0 + y, \epsilon) \\
&= Jy + \epsilon \partial_\epsilon F(x_0, 0) + N(a\psi_0 + y, \epsilon),
\end{aligned}
$$

where $\|N(a\psi_0 + y, \epsilon)\|_{L^2} = \mathcal{O}(\|a\psi_0 + y\|_X^2 + \epsilon^2)$ as $a, \epsilon, \|y\|_X \to 0$. Using the same projection operators, the root finding problem is now decomposed into two equations for $a \in \mathbb{R}$ and $y \in X$ as follows:

$$QJQy + \epsilon Q\partial_\epsilon F(x_0, 0) + QN(a\psi_0 + y, \epsilon) = 0, \tag{3.2.1}$$

$$\langle \psi_0, Jy + \epsilon \partial_\epsilon F(x_0, 0) + N(a\psi_0 + y, \epsilon) \rangle_{L^2} = 0. \tag{3.2.2}$$

Step 2: Solution of the non-singular equation. When we deal with the first equation (3.2.1), we quickly realize that the linear operator QJQ no longer has a non-trivial kernel in X, whereas the other terms of the equation are linear in ϵ and quadratic in $(a, y) \in \mathbb{R} \times X$. If $a = 0$ and $\epsilon = 0$, there exists a trivial solution $y = 0$ of equation (3.2.1) since $x = x_0$ satisfies $F(x, 0) = 0$. By the Implicit Function Theorem (Appendix B.7), the zero solution is uniquely continued for nonzero but small values of $(a, \epsilon) \in \mathbb{R}^2$, so that there exists a unique solution $y = Y(a, \epsilon) : \mathbb{R}^2 \to X$ of equation (3.2.1) for small $(a, \epsilon) \in \mathbb{R}^2$. The map $\mathbb{R}^2 \ni (a, \epsilon) \mapsto y \in X$ is C^2 and

$$Y(a, \epsilon) = \epsilon y_1 + \tilde{Y}(a, \epsilon), \quad y_1 = -QJ^{-1}Q\partial_\epsilon F(x_0, 0) \in X,$$

where $\|\tilde{Y}(a, \epsilon)\|_X = \mathcal{O}(a^2 + \epsilon^2)$ as $a, \epsilon \to 0$.

Step 3: Solution of the bifurcation equation. We can now eliminate the component $y \in X$ in the second equation (3.2.2) using the map $y = Y(a, \epsilon)$ constructed in Step 2. Since J is self-adjoint and $J\psi_0 = 0$, we realize that the second equation (3.2.2) can be written as the root finding problem for a scalar function of two variables

$$f(a, \epsilon) := \epsilon \langle \psi_0, \partial_\epsilon F(x_0, 0) \rangle_{L^2} + \langle \psi_0, N(a\psi_0 + Y(a, \epsilon), \epsilon) \rangle_{L^2} = 0,$$

where $f(a, \epsilon) : \mathbb{R}^2 \to \mathbb{R}$ is a C^2 function in variables (a, ϵ). In a generic case, it is expanded to the first nonzero powers as follows:

$$f(a, \epsilon) = \epsilon \alpha_1 + a^2 \alpha_2 + \tilde{f}(a, \epsilon),$$

where

$$\alpha_1 = \langle \psi_0, \partial_\epsilon F(x_0, 0) \rangle_{L^2}, \quad \alpha_2 = \frac{1}{2} \langle \psi_0, \partial_{y^2}^2 N(0,0) \psi_0^2 \rangle_{L^2},$$

and $|\tilde{f}(a, \epsilon)| = \mathcal{O}(|a|^3 + \epsilon^2)$ as $a, \epsilon \to 0$. If $(\alpha_1, \alpha_2) \neq (0, 0)$, roots of $f(a, \epsilon) = 0$ near $(a, \epsilon) = (0, 0)$ are found immediately from the leading-order terms of $f(a, \epsilon)$,

$$\epsilon = -\frac{\alpha_2}{\alpha_1} a^2 + \mathcal{O}(a^3) \quad \text{as} \quad a \to 0.$$

There exists no root a of $f(a, \epsilon) = 0$ for one sign of ϵ, exactly one zero root $a = 0$ for $\epsilon = 0$, and exactly two nonzero roots $a = a_\pm(\epsilon)$ for the opposite sign of ϵ, so that $|a_\pm(\epsilon)| = \mathcal{O}(\epsilon^{1/2})$. This scenario corresponds to the *fold* bifurcation (also known as *tangent* or *saddle-node* bifurcations) when two branches of solutions merge together and disappear in parameter continuation.

We summarize that the orthogonal projections and the Implicit Function Theorem allow us to reduce the root finding problem on a Banach space X to a bifurcation problem on a finite-dimensional subspace of the kernel of J, which is studied with the power expansions of nonlinear functions.

Fold bifurcations occur less often when we deal with the stationary Gross–Pitaevskii equation which has gauge invariance: given a solution $\phi \in H^2(\mathbb{R}^d)$, there is a continuous family of solutions $e^{i\theta}\phi \in H^2(\mathbb{R}^d)$ for any $\theta \in \mathbb{R}$. For such problems, pitchfork (symmetry-breaking) bifurcations occur more often.

We will apply the general algorithm of the Lyapunov–Schmidt reduction method to five particular examples of bifurcations of localized modes in the context of the stationary Gross–Pitaevskii equation: small-amplitude localized modes (Section 3.2.1), broad localized modes (Section 3.2.2), narrow localized modes (Section 3.2.3), multi-pulse localized modes (Section 3.2.4), and localized modes in the anticontinuum limit (Section 3.2.5).

These five examples have broader applicability than just the existence of localized modes in periodic potentials. The readers may develop similar algorithms for stationary localized and periodic solutions in other nonlinear evolution equations.

For all examples except for the last one, we will be dealing with the one-dimensional version of the stationary Gross–Pitaevskii equation (3.1.1). This gives a good simplification of the nonlinear operator function $F(x, \epsilon)$ since it is sufficient to consider strong real-valued solutions of the second-order ordinary differential equation,

$$-\phi''(x) + V(x)\phi(x) + \sigma\phi^3(x) = \omega\phi(x), \quad x \in \mathbb{R}. \tag{3.2.3}$$

The following lemma justifies this simplification.

Lemma 3.7 *Let $d = 1$ and $\phi \in H^2(\mathbb{R})$ be a strong solution of the stationary Gross–Pitaevskii equation (3.1.1). There is $\theta \in \mathbb{R}$ such that $e^{i\theta}\phi : \mathbb{R} \to \mathbb{R}$.*

Proof Multiplying equation (3.1.1) by $\bar{\phi}$ and subtracting a complex conjugate equation, we obtain for $d = 1$,

$$-\bar{\phi}(x)\phi''(x) + \bar{\phi}''(x)\phi(x) = \frac{d}{dx}\left(\bar{\phi}(x)\phi'(x) - \bar{\phi}'(x)\phi(x)\right) = 0, \quad x \in \mathbb{R}.$$

By Sobolev's embedding of $H^2(\mathbb{R})$ to $C^1(\mathbb{R})$ (Appendix B.10), both $\phi(x)$ and $\phi'(x)$ are continuous and decay to zero at infinity, so that

$$\bar{\phi}(x)\phi'(x) - \bar{\phi}'(x)\phi(x) = 0, \quad x \in \mathbb{R}.$$

Since both $\phi(x)$ and $\phi'(x)$ may not vanish at the same point $x \in \mathbb{R}$, there are $x_1, x_2 \in \mathbb{R}$ such that

$$\frac{d}{dx} \log \frac{\phi(x)}{\bar{\phi}(x)} = 0 \quad \Rightarrow \quad \arg(\phi(x)) = \theta, \quad x \in (x_1, x_2),$$

where θ is constant in x. Using the gauge transformation $\phi(x) \mapsto \phi(x)e^{i\theta}$, $\theta \in \mathbb{R}$ for solutions of the stationary equation (3.1.1), we can set $\theta = 0$ without loss of generality. Then, continuity of $\phi'(x)$ implies that

$$\mathrm{Im}\,\phi(x) = \mathrm{Im}\,\phi'(x) = 0, \quad x \in (x_1, x_2).$$

In this case, it follows from equation (3.1.1) that $\mathrm{Im}\,\phi''(x) = 0$ for all $x \in (x_1, x_2)$ and the trivial solution $\mathrm{Im}\,\phi(x) = 0$ is extended globally for all $x \in \mathbb{R}$. $\qquad \square$

Note that if $d = 2$ or $d = 3$, complex-valued localized modes of the stationary Gross–Pitaevskii equation (3.1.1) also exist. Such solutions have non-trivial phase dependence across a contour enclosing their centers although they still decay to zero at infinity. These localized modes are often referred to as *vortices*. Vortices are only included in the last example of the stationary DNLS equation (Section 3.2.5).

3.2.1 Small-amplitude localized modes

Let us start with the simplest Lyapunov–Schmidt reduction for small-amplitude solutions of the stationary Gross–Pitaevskii equation (3.2.3). A 2π-periodic potential $V(x)$ can be represented in many cases by a periodic sequence of identical potentials $V_0(x)$ in the form

$$V(x) = \sum_{n\in\mathbb{Z}} V_0(x - 2\pi n), \quad x \in \mathbb{R},$$

where $V_0(x)$ decay to zero as $|x| \to \infty$ exponentially fast.

The infinite sequence of identical potentials lies actually beyond the applicability of the method of Lyapunov–Schmidt reductions because the kernel of the linear operator $L = -\partial_x^2 + V(x)$ becomes infinite-dimensional. Therefore, we shall only deal with the bifurcation of small-amplitude localized modes supported by the potential $V_0(x)$.

Using similar techniques but much lengthy computation, one can work with the sum of two identical potentials $V_0(x) + V_0(x - 2\pi)$, three identical potentials $V_0(x - 2\pi) + V_0(x) + V_0(x + 2\pi)$, and then with N identical potentials, ultimately approaching to the limit of large N. This program was undertaken in the works of Kapitula *et al.* for a potential with two wells [103], three wells [104], and N wells [106]. Nevertheless, the limit to infinitely many wells is singular as it appears beyond the Lyapunov–Schmidt reduction method.

We hence consider strong solutions of the stationary Gross–Pitaevskii equation

$$-\phi''(x) + V_0(x)\phi(x) + \sigma\phi^3(x) = \omega\phi(x), \quad x \in \mathbb{R}. \tag{3.2.4}$$

Thanks to the exponentially fast decay of $V_0(x)$ to zero as $|x| \to \infty$, we shall assume that there is a single eigenvalue $\omega_0 < 0$ of the Schrödinger operator $L_0 = -\partial_x^2 + V_0(x)$ with the corresponding L^2-normalized eigenfunction ϕ_0 (Section 4.1).

Let us rewrite the stationary problem (3.2.4) as the root finding problem for the operator function $F(\phi, \omega) : H^2(\mathbb{R}) \times \mathbb{R} \to L^2(\mathbb{R})$ in the form

$$F(\phi, \omega) = (-\partial_x^2 + V_0 - \omega)\phi + \sigma\phi^3. \tag{3.2.5}$$

We will show that a nonzero localized mode ϕ of the stationary equation (3.2.4) bifurcates from the zero solution $\phi = 0$ in the direction of ϕ_0 for ω near ω_0.

Theorem 3.4 *Let $\omega_0 < 0$ be the smallest eigenvalue of $L_0 = -\partial_x^2 + V_0(x)$. There exist $\epsilon > 0$ and $C > 0$ such that the stationary equation (3.2.4) for each $\omega \in \mathcal{I}$ has a unique nonzero solution $\phi \in H^2(\mathbb{R})$ satisfying*

$$\|\phi\|_{H^2} \le C|\omega - \omega_0|^{1/2},$$

where $\mathcal{I} = (\omega_0 - \epsilon, \omega_0)$ for $\sigma = -1$ and $\mathcal{I} = (\omega_0, \omega_0 + \epsilon)$ for $\sigma = +1$. Moreover, the map $\mathcal{I} \ni \omega \mapsto \phi \in H^2(\mathbb{R})$ is C^1.

Proof The operator function $F(\phi, \omega) : H^2(\mathbb{R}) \times \mathbb{R} \to L^2(\mathbb{R})$ is analytic in both ϕ and ω. The Fréchet derivative of $F(\phi, \omega)$ with respect to $\phi \in H^2(\mathbb{R})$ at $\phi = 0 \in H^2(\mathbb{R})$ is

$$D_\phi F(0, \omega) = L_0 - \omega.$$

Let ϕ_0 be the L^2-normalized eigenfunction of L_0 corresponding to the eigenvalue $\omega_0 < 0$. Then, $D_\phi F(0, \omega_0)$ is a Fredholm operator of index zero with $\mathrm{Ker}(L_0 - \omega_0) = \mathrm{span}\{\phi_0\}$ and $\mathrm{Ran}(L_0 - \omega_0) = [\mathrm{Ker}(L_0 - \omega_0)]^\perp$ (Appendix B.4). Let P be the orthogonal projection operator from $L^2(\mathbb{R})$ to $\mathrm{Ker}(L_0 - \omega_0) \subset L^2(\mathbb{R})$ and $Q = \mathrm{Id} - P$. Then, we represent $\phi = a\phi_0 + \psi$, where

$$\forall \phi \in H^2(\mathbb{R}) : \quad a = \langle \phi_0, \phi \rangle_{L^2} \quad \text{and} \quad \psi = Q\phi = \phi - \langle \phi_0, \phi \rangle_{L^2}\phi_0.$$

The root finding equation $F(\phi, \omega) = 0$ is equivalent to the following system:

$$QF(a\phi_0 + \psi, \omega) = 0, \quad PF(a\phi_0 + \psi, \omega) = 0.$$

The first equation of the system can be written explicitly as

$$Q(L - \omega)Q\psi + \sigma Q(a\phi_0 + \psi)^3 = 0. \tag{3.2.6}$$

By the Implicit Function Theorem (Appendix B.7), there is a unique solution $\psi = \psi_0(a, \omega) \in H^2(\mathbb{R})$ of equation (3.2.6) for any small $a \in \mathbb{R}$ and $\omega - \omega_0 \in \mathbb{R}$. From the same Implicit Function Theorem, the map

$$\mathbb{R} \times \mathbb{R} \ni (a, \omega) \mapsto \psi \in H^2(\mathbb{R})$$

is at least C^3 and there is $C > 0$ such that $\|\psi\|_{H^2} \le C|a|^3$. Substituting

$$\psi = a^3 \tilde{\psi}(a, \omega), \quad \|\tilde{\psi}(a, \omega)\|_{H^2} \le C, \tag{3.2.7}$$

into the equation $PF(a\phi_0 + \psi, \omega) = 0$, we obtain a scalar equation in the real variables a and $\omega - \omega_0$, which can be written in the form

$$(\omega_0 - \omega) + \sigma a^2 \langle \phi_0, (\phi_0 + a^2 \tilde{\psi}(a, \omega))^3 \rangle_{L^2} = 0. \tag{3.2.8}$$

Thanks to bound (3.2.7), there is a unique root $\omega = \omega_0$ of equation (3.2.8) for $a = 0$. By the scalar version of the Implicit Function Theorem, the root is uniquely continued to (at least) the C^4 function $\omega = \tilde{\omega}(a)$, which satisfies the expansion

$$\tilde{\omega}(a) = \omega_0 + \sigma a^2 \|\phi_0\|_{L^4}^4 + \mathcal{O}(a^4) \quad \text{as} \quad a \to 0.$$

As a result, $\omega > \omega_0$ for $\sigma = +1$ and $\omega < \omega_0$ for $\sigma = -1$ with $|a| = \mathcal{O}(|\omega - \omega_0|^{1/2})$ as $\omega \to \omega_0$. $\qquad\square$

Remark 3.3 Since $F(\phi, \omega)$ is analytic with respect to ϕ and ω, the map $\mathcal{I} \ni \omega \mapsto \phi \in H^2(\mathbb{R})$ is actually analytic.

Exercise 3.6 Consider the Jacobian operator along the solution (ω, ϕ) of Theorem 3.4,

$$L_+ = D_\phi F(\phi, \omega) = -\partial_x^2 + V_0(x) - \omega + 3\sigma\phi^2(x)$$

and prove that L_+ is invertible for all $\omega \in \mathcal{I}$ and has exactly one negative eigenvalue for $\sigma = -1$ and no negative eigenvalues for $\sigma = 1$.

Remark 3.4 The result of Exercise 3.6 implies that the localized mode ϕ is stable in the time evolution of the Gross–Pitaevskii equation for $\omega \in \mathcal{I}$ for both $\sigma = +1$ and $\sigma = -1$ (Section 4.3).

Exercise 3.7 Assume that the operator $L = -\partial_x^2 + V_0(x)$ has two negative eigenvalues $\omega_0 < \omega_1 < 0$ and prove that there exists a local bifurcation of a localized mode of the stationary equation (3.2.4) for ω near ω_1. Furthermore, prove that L_+ has two negative eigenvalues for $\sigma = -1$ and one negative eigenvalue for $\sigma = +1$ for ω near ω_1.

If a localized potential V_0 is replaced by a 2π-periodic potential $V(x)$ and ω_0 is the band edge of a spectral band of $L = -\partial_x^2 + V(x)$, then Theorem 3.4 is not applicable because $\phi_0 \notin H^2(\mathbb{R})$ (Section 2.1.2). However, the interval $\mathcal{I}$ suggests heuristically that bifurcation of small-amplitude localized modes can be expected in the domain $\omega < \omega_0$ if $\sigma = -1$ and in the domain $\omega > \omega_0$ if $\sigma = +1$ provided that the domain is not occupied by the spectral band of operator L. This is exactly the conclusion that we can draw from the reduction of the stationary Gross–Pitaevskii equation with a periodic potential to the nonlinear Schrödinger equation (Section 2.3.2). Therefore, we can use the NLS equation to describe bifurcation of localized modes from the band edge of a spectral band of $L = -\partial_x^2 + V(x)$.

3.2.2 Broad localized modes

Fix $n_0 \in \mathbb{N}$ and a band edge $\omega_0 = E_{n_0}(k_0)$ of the particular n_0th spectral band of $L = -\partial_x^2 + V(x)$ with either $k_0 = 0$ or $k_0 = \frac{1}{2}$. Recall that the Gross–Pitaevskii equation,

$$iu_t = -u_{xx} + V(x)u + \sigma|u|^2 u,$$

can be reduced with the asymptotic multi-scale expansion,

$$u(x,t) = \epsilon^{1/2}\left(a(X,T)u_0(x) + \mathcal{O}_{L^\infty}(\epsilon^{1/2})\right)e^{-i\omega_0 t}, \quad X = \epsilon^{1/2}x, \quad T = \epsilon t,$$

to the NLS equation,

$$ia_T = \alpha a_{XX} + \sigma\beta|a|^2 a, \qquad (3.2.9)$$

where $u_0 = u_{n_0}(\cdot; k_0) \in L^2_{\mathrm{per}}([0, 2\pi])$, $\alpha = -\frac{1}{2}E''_{n_0}(k_0)$, and $\beta = \|u_0\|^4_{L^4_{\mathrm{per}}} > 0$ (Section 2.3.1).

There exists an exact localized mode of the NLS equation (3.2.9) if $\sigma = \mathrm{sign}(\alpha)$ in the form

$$a(X,T) = A(X - X_0)e^{-i\Omega T},$$

where $X_0 \in \mathbb{R}$ is arbitrary, $\Omega \in \mathbb{R}$ satisfies $\mathrm{sign}(\Omega) = \sigma$, and $A(X)$ is given by

$$A(X) = A_0\,\mathrm{sech}(KX), \quad A_0 = \left(\frac{2\Omega}{\sigma\beta}\right)^{1/2}, \quad K = \left(\frac{\Omega}{\alpha}\right)^{1/2}. \qquad (3.2.10)$$

If $\sigma = -1$, the localized mode (3.2.10) exists for $E''_{n_0}(k_0) > 0$ and $\Omega < 0$. In other words, the n_0th spectral band is located to the right of point ω_0 whereas the localized mode with $\omega = \omega_0 + \epsilon\Omega < \omega_0$ exists to the left of point ω_0. If $\sigma = 1$, the conclusion is the opposite. This property of bifurcations of localized modes is supported by numerical results of Figures 3.1 and 3.2.

Theorem 2.4 (Section 2.3.2) gives the persistence of the localized mode (3.2.10) as the asymptotic solution,

$$\phi(x) = \epsilon^{1/2}\left(A(\epsilon^{1/2}(x - x_0))u_0(x) + \mathcal{O}_{L^\infty}(\epsilon^{1/3})\right), \qquad (3.2.11)$$

of the stationary Gross–Pitaevskii equation with $\omega = \omega_0 + \epsilon\Omega$,

$$-\phi''(x) + V(x)\phi(x) + \sigma\phi^3(x) = \omega\phi(x), \quad x \in \mathbb{R}, \qquad (3.2.12)$$

under the condition that $V(x)$ is symmetric about $x = 0$ and $x_0 = 0$. If $V(x)$ has two points of symmetry such as $x = 0$ and $x = \pi$, the same theorem guarantees existence of two different branches of localized modes with centers of symmetries at either $x = 0$ or $x = \pi$. Here we show that the existence of exactly two localized modes on the fundamental period of $V(x)$ is a general situation when we are dealing with periodic potentials.

Because of the translational invariance of the continuous NLS equation (3.2.9), the localized mode (3.2.10) can be centered at any $X_0 = \epsilon^{1/2}x_0 \in \mathbb{R}$. However, this parameter determines the center of the localized mode $\phi(x)$ relative to the center of the Bloch function $u_0(x)$ and, therefore, with respect to the potential $V(x)$. As a result, parameter X_0 cannot be arbitrary but must be determined from a bifurcation equation in the third step of the Lyapunov–Schmidt reduction method. Unfortunately, this happens beyond all powers of the asymptotic expansion (3.2.11). At this point, we apply formal computations developed by Pelinovsky et al. [161].

First, we note the following lemma.

Lemma 3.8   *Let $\phi \in H^2(\mathbb{R})$ be a strong solution of the stationary Gross–Pitaevskii equation (3.2.12) and $V \in C^1(\mathbb{R})$. Then,*

$$\int_{\mathbb{R}} V'(x)\phi^2(x)dx = 0. \tag{3.2.13}$$

Proof   Equation (3.2.12) is a spatial Hamiltonian system, which is related to the Hamiltonian function

$$H(\phi, \phi', x) = -\left(\frac{d\phi}{dx}\right)^2 + (V - \omega)\phi^2 + \frac{1}{2}\sigma\phi^4.$$

The spatial Hamiltonian system is not autonomous in $x \in \mathbb{R}$. The change of $H(\phi, \phi', x)$ at the trajectory (ϕ, ϕ') is computed from

$$\lim_{x \to \infty} H(\phi, \phi', x) - \lim_{x \to -\infty} H(\phi, \phi', x) = \int_{\mathbb{R}} \frac{dH}{dx} dx = \int_{\mathbb{R}} V'(x)\phi^2(x)dx.$$

For any strong solution $\phi \in H^2(\mathbb{R})$, both $\phi(x)$ and $\phi'(x)$ decay to zero as $|x| \to \infty$, which gives $\lim_{x \to \pm\infty} H(\phi, \phi', x) = 0$ and the constraint (3.2.13). $\qquad\square$

Exercise 3.8   Let $u_0 \in L^2_{\mathrm{per}}([0, 2\pi])$ be a solution of the equation

$$-u_0''(x) + V(x)u_0(x) = \omega_0 u_0(x), \quad x \in \mathbb{R}.$$

Prove that

$$C_0 := \frac{1}{2\pi} \int_0^{2\pi} V'(x)u_0^2(x)dx = 0.$$

Substitution of the leading order of the asymptotic solution (3.2.11) into the constraint (3.2.13) gives the scalar function

$$f(x_0, \epsilon) = \epsilon \int_{\mathbb{R}} V'(x)A^2(\epsilon^{1/2}(x - x_0))u_0^2(x)dx. \tag{3.2.14}$$

The presence of the small parameter ϵ for the broad localized mode ϕ allows us to simplify computations of $f(x_0, \epsilon)$ at the leading order, which becomes exponentially small with respect to ϵ. This computation is summarized in the following lemma.

Lemma 3.9   *Let*

$$C_1 := \frac{1}{2\pi} \int_0^{2\pi} V'(x)u_0^2(x)e^{-ix}dx \neq 0.$$

There exist exactly two simple roots of $f(x_0, \epsilon) = 0$ for $x_0 \in [0, 2\pi)$ and sufficiently small $\epsilon > 0$.

Proof   We shall prove that $f(x_0, \epsilon)$ is exponentially small with respect to ϵ and compute the leading order of $f(x_0, \epsilon)$. Let us use the Fourier series for $V'u_0^2 \in L^2_{\mathrm{per}}([0, 2\pi])$ given by

$$V'(x)u_0^2(x) = \sum_{n \in \mathbb{Z}} C_n e^{inx}$$

and the Fourier transform for $A^2 \in L^2(\mathbb{R})$ given by

$$\hat{A}^2(p) = \int_{\mathbb{R}} A^2(X) e^{ipX} dX.$$

Function $f(x_0, \epsilon)$ is rewritten in the form

$$f(x_0, \epsilon) = \epsilon \sum_{n \in \mathbb{Z}} C_n \int_{\mathbb{R}} A^2(\epsilon^{1/2}(x - x_0)) e^{inx} dx$$

$$= \epsilon^{1/2} \sum_{n \in \mathbb{Z}} C_n \hat{A}^2(n\epsilon^{-1/2}) e^{inx_0}. \qquad (3.2.15)$$

It follows from the analyticity of $A^2(X)$ in X that $\hat{A}^2(p)$ decays to zero exponentially fast as $|p| \to \infty$. Therefore, series (3.2.15) is hierarchic in the sense that the higher-order terms with larger values of $|n|$ are exponentially smaller compared to the lower-order terms with smaller values of $|n|$ as $\epsilon \to 0$. From Exercise 3.8, the zero-order term is absent since

$$C_0 = \frac{1}{2\pi} \int_0^{2\pi} V'(x) u_0^2(x) dx = 0.$$

Therefore, the leading order of series (3.2.15) becomes

$$f(x_0, \epsilon) = 2\epsilon^{1/2} |C_1| \hat{A}^2(\epsilon^{-1/2}) \cos(x_0 + \arg(C_1)) + \dots.$$

Assuming that $C_1 \neq 0$, the leading order of $f(x_0, \epsilon)$ has precisely two simple roots for $x_0 \in [0, 2\pi)$. Persistence of simple roots of $f(x_0, \epsilon) = 0$ for sufficiently small $\epsilon > 0$ is proved using the Implicit Function Theorem. $\qquad \square$

If $V(x)$ is symmetric about the points $x = 0$ and $x = \pi$, then $\psi_0^2(x)$ is also symmetric about these points and $C_1 \in i\mathbb{R}$. In this case, the two roots of $f(x_0, \epsilon)$ occur at the extremal points of $V(x)$, that is, for $x_0 = 0$ and $x_0 = \pi$, in full correspondence to the result of Theorem 2.4.

The limit of broad localized modes is simultaneously the limit of small powers $Q(\phi) = \|\phi\|_{L^2}^2$. If we substitute the leading order of the asymptotic expansion (3.2.11) with $x_0 = 0$ and perform computations similar to the proof of Lemma 3.9, we obtain the leading order of $Q(\phi)$,

$$\epsilon \int_{\mathbb{R}} A^2(\epsilon^{1/2} x) u_0^2(x) dx = \epsilon^{1/2} \sum_{n \in \mathbb{Z}} D_n \hat{A}^2(n\epsilon^{-1/2}) = \epsilon^{1/2} \left(D_0 \hat{A}^2(0) + \dots \right),$$

where $\{D_n\}_{n \in \mathbb{Z}}$ are coefficients of the Fourier series for $u_0^2 \in L^2_{\text{per}}([0, 2\pi])$ and the remainder terms are exponentially small with respect to ϵ. Since

$$D_0 = \frac{1}{2\pi} \int_0^{2\pi} u_0^2(x) dx > 0,$$

we can see that

$$Q(\phi) = \mathcal{O}(\epsilon^{1/2}) = \mathcal{O}(|\omega - \omega_0|^{1/2}).$$

There are several distinctive differences between bifurcations of broad localized modes considered here and bifurcations of small-amplitude localized modes considered in Section 3.2.1. First, the bifurcation point ω_0 is not isolated from the spectrum of the Schrödinger operator $L = -\partial_x^2 + V(x)$. Second, two distinctive branches may bifurcate from the point ω_0 here, compared to one branch in Section 3.2.1. Third, $Q(\phi) = \mathcal{O}(|\omega-\omega_0|^{1/2})$ as $\omega \to \omega_0$ here, compared to $Q(\phi) = \mathcal{O}(|\omega-\omega_0|)$ in Section 3.2.1.

Exercise 3.9 Consider the Schrödinger operator

$$L_+ = -\partial_x^2 + V(x) - \omega + 3\sigma\phi^2(x), \quad \omega < \inf \sigma(-\partial_x^2 + V),$$

and compute $\langle L_+ A'u_0, A'u_0\rangle_{L^2}$ in terms of $\partial_{x_0} f(x_0, \epsilon)$. Prove, for $\sigma = -1$, that the operator L_+ has one negative eigenvalue if $\partial_{x_0} f(x_0, \epsilon) > 0$ and two negative eigenvalues if $\partial_{x_0} f(x_0, \epsilon) < 0$.

Remark 3.5 If $\arg(C_1) = -\frac{\pi}{2}$, the root x_0 with $\partial_{x_0} f(x_0, \epsilon) > 0$ corresponds to $x_0 = 0$ (that is, the minimum of $V(x)$) and the root x_0 with $\partial_{x_0} f(x_0, \epsilon) < 0$ corresponds to $x_0 = \pi$ (that is, the maximum of $V(x)$). The result of Exercise 3.9 implies that the localized mode with $x_0 = \pi$ is unstable in the time evolution of the Gross–Pitaevskii equation, whereas the localized mode with $x_0 = 0$ is stable in the time evolution (Section 4.3).

If the potential $V(x)$ is special such that $C_1 = 0$ but $C_2 \neq 0$, the leading-order terms in series (3.2.15) give

$$f(x_0, \epsilon) = 2\epsilon^{1/2}|C_2|\hat{A}^2(2\epsilon^{-1/2}) \cos(2x_0 + \arg(C_2)) + \dots.$$

Four branches of localized modes may bifurcate from four roots of the function $\cos(2x_0)$ on $[0, 2\pi)$. Examples of such bifurcations were considered by Kominis & Hizanidis [121] for the Gross–Pitaevskii equation with a *small* periodic potential.

3.2.3 Narrow localized modes

There is another limit we can explore to clarify the question of multiplicity of branches in the stationary Gross–Pitaevskii equation (3.2.3). If $V(x) \equiv 0$, the stationary equation (3.2.3) with $\sigma = -1$ has a localized mode for any $\omega < 0$,

$$\phi(x) = (2|\omega|)^{1/2} \mathrm{sech}(|\omega|^{1/2}(x - x_0)), \quad x_0 \in \mathbb{R}. \tag{3.2.16}$$

This motivates the use of a perturbation theory for small $V(x)$ to identify persistence of this localized mode for nonzero potentials, as in Kapitula [101].

Even if $V(x)$ is not small, there is a simple way to transform the stationary equation (3.2.3) with $\sigma = -1$ to the same form with a small parameter in front of the potential $V(x)$. The technique is based on the scaling transformation, which captures the limit of a narrow localized mode ϕ with a large power $Q(\phi)$. The approach has been known since the work of Floer & Weinstein [58] but it has received much attention in recent years, for instance, in the work of Sivan *et al.* [191]. Using this approach, we will show that a narrow localized mode exists in a neighborhood of any non-degenerate extremum of $V(x)$.

Let us fix the center $x_0 \in \mathbb{R}$ of the localized mode (3.2.16) and apply the scaling transformation,

$$\omega = V(x_0) - E, \quad \xi = E^{1/2}(x - x_0), \quad \psi(\xi) = (2E)^{-1/2}\phi(x). \tag{3.2.17}$$

The limit $\omega \to -\infty$ corresponds to the limit $E \to \infty$. Therefore, we define a small positive parameter $\epsilon = E^{-1}$. If $\phi(x)$ satisfies the stationary equation (3.2.3) with $\sigma = -1$, then $\psi(\xi)$ satisfies a new stationary equation

$$-\psi''(\xi) + \tilde{V}_\epsilon(\xi)\psi(\xi) - 2\psi^3(\xi) + \psi(\xi) = 0, \qquad (3.2.18)$$

where

$$\tilde{V}_\epsilon(\xi) := \epsilon \left(V(x_0 + \epsilon^{1/2}\xi) - V(x_0) \right).$$

Since $\|\tilde{V}_\epsilon\|_{L^\infty} = 2\epsilon\|V\|_{L^\infty}$, the new potential $\tilde{V}_\epsilon(\xi)$ is small as $\epsilon \to 0$ in L^∞ norm. Recall that there exists a unique (up to translation in $\xi \in \mathbb{R}$) exponentially decaying solution $\psi_0(\xi) = \text{sech}(\xi)$ of the stationary equation (3.2.18) for $\epsilon = 0$. The following result specifies the conditions on $V(x)$ and x_0, under which the solution ψ_0 persists for $\epsilon \neq 0$.

Theorem 3.5 *Let $V(x) \in L^\infty(\mathbb{R}) \cap C^2(\mathbb{R})$. For each $x_0 \in \mathbb{R}$ such that $V'(x_0) \neq 0$, no solutions $\psi \in H^2(\mathbb{R})$ of the stationary equation (3.2.18) exist for sufficiently small $\epsilon > 0$. For each $x_0 \in \mathbb{R}$ such that*

$$V'(x_0) = 0, \quad V''(x_0) \neq 0$$

there exists a $\epsilon_0 > 0$ such that for any $\epsilon \in (0, \epsilon_0)$, there exists a unique solution $\psi \in H^2(\mathbb{R})$ of the stationary equation (3.2.18) and a constant $C > 0$ such that

$$\|\psi - \psi_0\|_{H^2} \leq C\epsilon^2,$$

where $\psi_0(\xi) = \text{sech}(\xi)$.

Proof Let $\psi(\xi) = \psi_0(\xi - s) + \varphi(\xi)$, where $s \in \mathbb{R}$ is arbitrary and $\psi_0(\xi) = \text{sech}(\xi)$. We rewrite (3.2.18) in the form

$$L_0\varphi = -\tilde{V}_\epsilon(\psi_0(\cdot - s) + \varphi) + N(\varphi), \qquad (3.2.19)$$

where

$$L_0 := -\partial_\xi^2 + 1 - 6\,\text{sech}^2(\xi - s)$$

and

$$\begin{aligned}\frac{1}{2}N(\varphi) &:= (\psi_0(\cdot - s) + \varphi)^3 - \psi_0^3(\cdot - s) - 3\psi_0^2(\cdot - s)\varphi \\ &= 3\psi_0(\cdot - s)\varphi^2 + \varphi^3.\end{aligned}$$

Because $H^2(\mathbb{R})$ is a Banach algebra with respect to pointwise multiplication (Appendix B.1), the nonlinear vector field enjoys the bound

$$\|N(\varphi)\|_{L^2} \leq 6\|\varphi\|_{H^2}^2 + 2\|\varphi\|_{H^2}^3.$$

Since $V \in L^\infty(\mathbb{R})$, there is also $C > 0$ such that

$$\|\tilde{V}_\epsilon(\psi_0(\cdot - s) + \varphi)\|_{L^2} \leq C\epsilon\|\psi_0(\cdot - s) + \varphi\|_{H^2}.$$

Because of the exact equation $L_0\psi_0'(\xi - s) = 0$, operator L_0 has zero eigenvalue, which is isolated from the rest of the spectrum of L_0. Therefore, the operator L_0 is not invertible in $L^2(\mathbb{R})$ but the method of Lyapunov–Schmidt reductions can be applied. Although we are supposed to decompose φ into the direction of $\mathrm{Ker}(L_0)$ and into the orthogonal direction, we note that parameter s is not defined at this point. If $\psi_0(x - s)$ is expanded in power series of s, then the linear term of this expansion is parallel to $\mathrm{Ker}(L_0)$. Therefore, we can keep s as a free parameter and add the constraint,

$$\langle \psi_0'(\cdot - s), \varphi \rangle_{L^2} = 0.$$

The values of $s \in \mathbb{R}$ are defined by projecting of equation (3.2.19) to the direction of $\mathrm{Ker}(L_0) \subset L^2(\mathbb{R})$,

$$-\langle \psi_0'(\cdot - s), \tilde{V}_\epsilon(\psi_0(\cdot - s) + \varphi) \rangle_{L^2} + \langle \psi_0'(\cdot - s), N(\varphi) \rangle_{L^2} = 0. \qquad (3.2.20)$$

Let $Q : L^2(\mathbb{R}) \to [\mathrm{Ker}(L_0)]^\perp \subset L^2(\mathbb{R})$ be the orthogonal projection operator. The non-singular equation for φ is given by

$$F(\varphi, \epsilon, s) := (QL_0Q)\varphi + Q\tilde{V}_\epsilon(\psi_0(\cdot - s) + \varphi) - QN(\varphi) = 0. \qquad (3.2.21)$$

The vector field $F(\varphi, \epsilon, s) : H^2(\mathbb{R}) \times \mathbb{R} \times \mathbb{R} \to L^2(\mathbb{R})$ is analytic in φ and s and C^1 in ϵ. We know that $F(0, 0, s) = 0$ and $J = D_\varphi F(0, 0, s)$ is invertible for any $s \in \mathbb{R}$. By the Implicit Function Theorem (Appendix B.7), there exists a unique solution $\varphi = \varphi_0(\epsilon, s) : \mathbb{R} \times \mathbb{R} \to H^2(\mathbb{R})$ of equation (3.2.21) for small $\epsilon \in \mathbb{R}$ and any $s \in \mathbb{R}$. Moreover, the map $\mathbb{R}^2 \ni (\epsilon, s) \mapsto \varphi_0 \in H^2(\mathbb{R})$ is C^1 in ϵ, smooth in s, and for sufficiently small $\epsilon > 0$, there is $C > 0$ such that

$$\|\varphi_0(\epsilon, s)\|_{H^2} \le C\epsilon. \qquad (3.2.22)$$

We substitute $\varphi = \varphi_0(\epsilon, s)$ into the bifurcation equation (3.2.20) and obtain

$$f(\epsilon, s) = f_0(\epsilon, s) + f_1(\epsilon, s) = 0,$$

where

$$f_0(\epsilon, s) = -\langle \psi_0'(\cdot - s), \tilde{V}_\epsilon \psi_0(\cdot - s) \rangle_{L^2},$$
$$f_1(\epsilon, s) = -\langle \psi_0'(\cdot - s), \tilde{V}_\epsilon \varphi_0(\epsilon, s) \rangle_{L^2} + \langle \psi_0'(\cdot - s), N(\varphi_0(\epsilon, s)) \rangle_{L^2}.$$

For sufficiently small $\epsilon > 0$, we have

$$f_0(\epsilon, s) = \frac{1}{2}\epsilon^{3/2} \int_\mathbb{R} V'(x_0 + \epsilon^{1/2}\xi)\psi_0^2(\xi - s)ds = \frac{1}{2}\epsilon^{3/2}\left(V'(x_0)\|\psi_0\|_{L^2}^2 + \mathcal{O}(\epsilon^{1/2})\right)$$

and $|f_1(\epsilon, s)| = \mathcal{O}(\epsilon^2)$ for any $s \in \mathbb{R}$.

If $V'(x_0) \ne 0$, no root of $f(\epsilon, s) = 0$ exists near $\epsilon = 0$ for any $s \in \mathbb{R}$. Assuming $V'(x_0) = 0$ and $V''(x_0) \ne 0$, we represent $\tilde{V}_\epsilon \in C^2(\mathbb{R})$ locally near $\xi = 0$ by

$$\tilde{V}_\epsilon(\xi) = \frac{1}{2}\epsilon^2 V''(x_0)\xi^2 + \tilde{R}_\epsilon(\xi),$$

where $\|\tilde{R}_\epsilon\|_{L^\infty_{\mathrm{loc}}} = o(\epsilon^2)$ near $\xi = 0$. To improve bound (3.2.22), we define $\varphi_0 = \varphi_1 + \theta$, where $\varphi_1 \in H^2(\mathbb{R})$ solves the inhomogeneous equation

$$(QL_0Q)\varphi_1 = -Q\tilde{V}_\epsilon\psi_0(\cdot - s) \qquad (3.2.23)$$

and the remainder term $\theta \in H^2(\mathbb{R})$ solves

$$(QL_0Q)\theta + \tilde{V}_\epsilon(\varphi_1 + \theta) + N(\varphi_1 + \theta) = 0. \qquad (3.2.24)$$

Because of the exponential decay of $\psi_0(\xi)$ to zero as $\xi \to \pm\infty$, we have

$$\xi^2 \psi_0(\xi - s) \in L^2(\mathbb{R}) \quad \text{and} \quad \|\tilde{R}_\epsilon \psi_0\|_{L^2} = o(\epsilon^2).$$

Solving (3.2.23) and (3.2.24), we infer that there is a $C > 0$ such that

$$\|\varphi_1\|_{H^2} \leq C\epsilon^2, \quad \|\theta\|_{H^2} \leq C\epsilon^3,$$

which improves bound (3.2.22). Using this estimate for $f_0(\epsilon, s)$ and $f_1(\epsilon, s)$, we obtain

$$f_0(\epsilon, s) = \frac{1}{2}\epsilon^{3/2} \int_{\mathbb{R}} V'(x_0 + \epsilon^{1/2}\xi)\psi_0^2(\xi - s)ds = \frac{1}{2}\epsilon^2 \left(V''(x_0)s\|\psi_0\|_{L^2}^2 + o(1) \right)$$

and $|f_1(\epsilon, s)| = \mathcal{O}(\epsilon^4)$ for any $s \in \mathbb{R}$. If $V''(x_0) \neq 0$, a unique root of $f(\epsilon, s) = 0$ exists near $\epsilon = 0$ and $s = \mathcal{O}(\epsilon^2)$ as $\epsilon \to 0$. $\qquad\square$

Exercise 3.10 Under the conditions of Theorem 3.5, prove that there exists $\epsilon_0 > 0$ such that for any $\epsilon \in (0, \epsilon_0)$, the second eigenvalue of the Jacobian operator along the solution branch

$$L_+ = -\partial_x^2 + V(x) - \omega - 3\phi^2(x)$$

is negative if $V''(x_0) < 0$ and positive if $V''(x_0) > 0$.

Remark 3.6 The result of Exercise 3.10 suggests that the localized mode at the minimum of $V(x)$ is stable in the time evolution of the Gross–Pitaevskii equation, whereas the localized mode at the maximum of $V(x)$ is unstable in the time evolution (Section 4.3).

If $V(x)$ is a smooth 2π-periodic potential with only one minimum and maximum on the fundamental period $[0, 2\pi)$, there are only two narrow localized modes in the semi-infinite gap, which correspond to the non-degenerate extremal points of $V(x)$. Because of the scaling transformation (3.2.17), we have

$$Q(\phi) = \|\phi\|_{L^2}^2 = 2E^{1/2}\|\psi\|_{L^2}^2 = \mathcal{O}(E^{1/2}) \quad \text{as} \quad E \to \infty.$$

Therefore, the limit of narrow localized modes corresponds to the limit of large powers $Q(\phi)$ far away from the lowest band edge of the spectrum of the Schrödinger operator $L = -\partial_x^2 + V(x)$.

The scaling transformation (3.2.17) ensures that the new potential $\tilde{V}_\epsilon(x)$ is both small in amplitude and slowly varying with respect to x. If we only require $V(x)$ to be small in amplitude, then an *effective* potential determines existence of localized modes of the stationary equation (3.2.3).

Recall from Lemma 3.8 that if $\phi(x)$ is a localized mode of the stationary equation (3.2.3), then

$$\int_{\mathbb{R}} V'(x)\phi^2(x)dx = 0.$$

If $V(x)$ is small in the sense that it is multiplied by a small parameter ϵ, then $\phi(x)$ is close to the localized mode (3.2.16) for any $\omega < 0$. This implies that the persistence of the solution ϕ for $\epsilon \neq 0$ can be related to the derivative of the effective potential,

$$M(x_0) = \int_{\mathbb{R}} V(x + x_0)\phi_0^2(x)dx, \quad a \in \mathbb{R}.$$

This is indeed the case, as the following exercise shows.

Exercise 3.11   Assume that $V(x)$ is multiplied by a small parameter ϵ and prove that for each $x_0 \in \mathbb{R}$ such that

$$M'(x_0) = 0, \quad M''(x_0) \neq 0,$$

there exists a $\epsilon_0 > 0$ such that for any $\epsilon \in (0, \epsilon_0)$, there exists a unique solution $\phi \in H^2(\mathbb{R})$ of the stationary equation (3.2.3) with $\sigma = -1$ satisfying bounds

$$\exists C > 0 : \quad \|\phi - \phi_0(\cdot - x_0)\|_{H^2} \leq C\epsilon.$$

Note that if $V(x)$ is multiplied by a small parameter, then narrow band gaps bifurcate between two adjacent spectral bands near $\omega = \frac{n^2}{4} > 0$ for $n \in \mathbb{N}$. Therefore, besides localized modes in the semi-infinite gap, there exist localized modes in narrow band gaps according to the nonlinear Dirac equations (Section 2.2.2).

3.2.4 Multi-pulse localized modes

Single-pulse localized modes ϕ are also called the *fundamental* localized modes. Multi-pulse localized modes can be thought to be compositions of fundamental modes with sufficient separation between the individual pulses. If $\phi(x) \to 0$ as $|x| \to \infty$ exponentially fast, overlapping between the tails of individual pulses is exponentially small in terms of the large distance between the pulses. This brings us to the theory of tail-to-tail interaction between individual pulses, which was pioneered by Gorshkov & Ostrovsky [73] with the use of formal asymptotic computations. The rigorous justification of this theory using geometric ideas was developed by Sandstede [179] in the context of pulses in reaction–diffusion equations. Infinite sequences of pulses in reaction–diffusion equations were recently considered by Zelik & Mielke [221].

Let us consider again the stationary Gross–Pitaevskii equation (3.2.3) with a 2π-periodic potential $V(x)$ in the focusing case $\sigma = -1$ for $\omega < \omega_0 := \inf \sigma(L)$, where $L = -\partial_x^2 + V(x)$. As we have seen in Sections 3.2.2 and 3.2.3, there exist at least two branches of fundamental localized modes ϕ. If $V(x)$ has a spatial symmetry

$$V(x) = V(-x) = V(2\pi - x), \quad x \in \mathbb{R},$$

the two localized modes $\phi(x)$ are symmetric with respect to points $x = 0$ and $x = \pi$. We have also seen that the spectrum of the Jacobian operator L_+ for the stationary Gross–Pitaevskii equation evaluated at the localized mode ϕ is bounded away from zero and includes a finite number of negative eigenvalues and no zero eigenvalue.

We now construct a multi-pulse localized mode Φ of the same stationary Gross–Pitaevskii equation,

$$-\Phi''(x) + V(x)\Phi(x) - \Phi^3(x) = \omega\Phi(x), \quad x \in \mathbb{R}, \qquad (3.2.25)$$

which consists of two individual pulses ϕ centered at two particular points s_1 and s_2 with a large separation distance $|s_2 - s_1|$. Since $V(x)$ is 2π-periodic, we can set $s_1 = 0$ and $s_2 = 2\pi n$ for an integer $n \in \mathbb{N}$. The following theorem shows that this superposition always exists in the asymptotic limit of large $n \in \mathbb{N}$.

Theorem 3.6 *Let $\omega < \omega_0$ and assume that there exists a fundamental localized mode ϕ of the stationary equation (3.2.3) with an exponential decay to zero as $|x| \to \infty$. There exists an infinite countable set $\{\Phi_n\}_{n\in\mathbb{N}}$ of two-pulse localized modes of the stationary equation (3.2.25) in the form*

$$\Phi_n(x) = \phi(x) + \phi(x - 2\pi n) + \varphi_n(x), \quad n \in \mathbb{N}, \qquad (3.2.26)$$

where for any small $\epsilon > 0$ there is $N \geq 1$ such that

$$\forall n \geq N : \quad \exists C > 0 : \quad \|\varphi_n\|_{H^2} \leq C\epsilon. \qquad (3.2.27)$$

Remark 3.7 Because two distinct branches of fundamental localized modes ϕ_1 and ϕ_2 of the stationary equation (3.2.3) generally exist, one can use the same technique to prove existence of a two-pulse localized mode $\Phi(x)$, which is represented by the sum of $\phi_1(x)$, $\phi_2(x - 2\pi n)$, and a small remainder term.

Proof Substitution of the decomposition (3.2.26) into the stationary equation (3.2.25) gives an equation for $\varphi_n(x)$ written in the abstract form

$$L\varphi_n = f_n + N(\varphi_n), \qquad (3.2.28)$$

where

$$L := -\partial_x^2 + V(x) - \omega - 3(\phi + \phi_n)^2,$$
$$f_n := 3\phi\phi_n(\phi + \phi_n),$$
$$N(\varphi) := 3(\phi + \phi_n)\varphi^2 + \varphi^3,$$

and $\phi_n(x) = \phi(x - 2\pi n)$. Thanks to the exponential decay of $\phi(x)$ to zero as $|x| \to \infty$, we know that

$$\|\phi\phi_n\|_{L^\infty} \to 0 \quad \text{as} \quad n \to \infty.$$

Therefore, f_n is small in $L^2(\mathbb{R})$ for large $n \in \mathbb{N}$ and for any small $\epsilon > 0$ there is $N \geq 1$ such that

$$\forall n \geq N : \quad \exists C > 0 : \quad \|f_n\|_{L^2} \leq C\epsilon.$$

On the other hand, because $\sigma(L_+)$ associated with a fundamental localized mode ϕ is bounded away from zero, $\sigma(L)$ is also bounded away from zero for large $n \in \mathbb{N}$. Therefore, we can rewrite (3.2.28) as the fixed-point equation

$$\varphi \in H^2(\mathbb{R}): \quad \varphi = F(\varphi) := L^{-1}f + L^{-1}N(\varphi). \qquad (3.2.29)$$

Since $H^2(\mathbb{R})$ is a Banach algebra with respect to pointwise multiplication (Appendix B.1), there is $C > 0$ such that

$$\|N(\varphi)\|_{L^2} \leq C\|\varphi\|_{H^2}^2 + \|\varphi\|_{H^2}^3.$$

The vector field $F(\varphi)$ is analytic in $\varphi \in H^2(\mathbb{R})$, maps a ball of radius $\delta > 0$ in $H^2(\mathbb{R})$ to itself for large $n \in \mathbb{N}$ and small $\delta > 0$, and it is a contraction in the ball. By the Banach Fixed-Point Theorem (Appendix B.2), there exists a unique solution $\varphi_n \in H^2(\mathbb{R})$ for sufficiently large $n \in \mathbb{N}$ such that φ_n solves the fixed-point equation (3.2.29) and satisfies bound (3.2.27). $\qquad \square$

Exercise 3.12   Consider the Jacobian operator

$$L_+ = -\partial_x^2 + V(x) - \omega - 3\Phi^2(x),$$

evaluated at the two-pulse localized mode $\Phi(x)$ and prove that L_+ has twice as many negative eigenvalues as the Jacobian operator L_+ evaluated at the fundamental localized mode $\phi(x)$.

Although Theorem 3.6 only covers a two-pulse localized mode, the construction of a multi-pulse localized mode is a straightforward exercise.

Exercise 3.13   Prove existence of a localized mode $\Phi(x)$ in the stationary Gross–Pitaevskii equation (3.2.25), which consists of the sum of N individual pulses $\phi(x - 2\pi n_j)$ for distinct integers $n_1 < n_2 < ... < n_N$ in the asymptotic limit $|n_{j+1} - n_j| \to \infty$ for all $j \in \{1, 2, ..., N-1\}$.

The two-pulse localized mode $\Phi(x)$ exists at the balance between two opposite forces: the fundamental localized modes $\phi(x)$ repel each other by their monotonically decaying exponential tails but each localized mode is trapped by the potential $V(x)$ of the stationary Gross–Pitaevskii equation (3.2.25). If $V(x) \equiv 0$, no balance between two individual pulses exists unless the repulsive and attractive forces arise due to oscillatory tails of the fundamental localized modes.

We note that the technique of the Lyapunov–Schmidt reduction method was avoided in the previous discussion because the operator L_+ for a fundamental localized mode ϕ had a trivial kernel thanks to the broken translational invariance of the stationary Gross–Pitaevskii equation (3.2.25) with a nonzero potential $V(x)$. Let us now consider a homogeneous stationary equation, where the translational invariance induces a non-trivial kernel of L_+.

If we simply take $V(x) \equiv 0$, then the second-order differential equation (3.2.25) has the fundamental localized mode (3.2.16) and it is the only solution decaying to zero. This follows from the phase-plane analysis of trajectories on the phase plane $(\phi, \phi') \in \mathbb{R}^2$. Since the trajectories do not intersect at any (ϕ, ϕ') unless it is an equilibrium point, homoclinic trajectories approaching the zero equilibrium may have at most one loop in the phase plane. If we now consider a stationary Gross–Pitaevskii equation modified by the fourth-order dispersion term,

$$-\phi''(x) + \epsilon^2 \phi''''(x) - \phi^3(x) = \omega \phi(x), \quad x \in \mathbb{R}, \tag{3.2.30}$$

then trajectories live in the phase space $(\phi, \phi', \phi'', \phi''') \in \mathbb{R}^4$ and may approach the zero equilibrium after several loops in the phase space.

Typically ϵ is a small parameter in equation (3.2.30) because the fourth-order dispersion term is considered to be smaller compared to the second-order dispersion term in the asymptotic multi-scale expansion method (Section 1.1). However, we can use the scaling transformation,

$$\phi(x) = \epsilon^{-1}\psi(\xi), \quad \xi = \epsilon^{-1}x, \quad \omega = -\epsilon^{-2}c,$$

which brings equation (3.2.30) to the form

$$\psi''''(\xi) - \psi''(\xi) + c\psi(\xi) = \psi^3(x), \quad \xi \in \mathbb{R}, \tag{3.2.31}$$

where $c \in \mathbb{R}$ is an arbitrary parameter. Note that equation (3.2.31) coincides with the reduction of the fifth-order modified Korteweg–de Vries equation for a traveling wave [23, 79].

Linearization of equation (3.2.31) at the zero equilibrium with the substitution $\psi(\xi) = \psi_0 e^{\kappa z}$ leads to four admissible values of κ found from the roots of the characteristic equation

$$\kappa^4 - \kappa^2 + c = 0. \tag{3.2.32}$$

When $c < 0$, one pair of roots κ is purely imaginary and the other pair is purely real. When $0 \leq c \leq \frac{1}{4}$, two pairs of roots κ are real-valued. When $c > \frac{1}{4}$, the four complex-valued roots κ are located symmetrically about the axes, that is, at $\kappa = \pm\kappa_0 \pm ik_0$, where κ_0 and k_0 are real positive numbers.

It is proved with the variational method by Buffoni & Sere [23] and Groves [79] that the fourth-order equation (3.2.31) has a fundamental localized mode $\psi \in C^\infty(\mathbb{R})$ for $c > 0$, which is even on $\mathbb{R}$ and decays to zero as $|\xi| \to \infty$ exponentially fast.

For $c > \frac{1}{4}$, the fourth-order equation (3.2.31) has also an infinite countable set of two-pulse localized modes $\Psi \in C^\infty(\mathbb{R})$, which look like two copies of the fundamental localized modes ψ separated by finitely many oscillations close to the zero equilibrium [23]. The members of the set are distinguished by the distance between the two individual pulses.

The following theorem gives an asymptotic construction of the two-pulse solution Ψ, which confirms this variational result using the method of Lyapunov–Schmidt reductions. This application of the method was developed by Chugunova & Pelinovsky [34] in the context of the fourth-order equation (3.2.31) with a quadratic nonlinear term.

Theorem 3.7 *Let $c > \frac{1}{4}$ and denote the quartet of roots of the characteristic equation (3.2.32) by $\kappa = \pm\kappa_0 \pm ik_0$. Assume that there exists a fundamental localized mode $\psi \in H^4(\mathbb{R})$ of the stationary equation (3.2.31) such that the Jacobian operator*

$$L_+ = c - \partial_\xi^2 + \partial_\xi^4 - 3\psi^2(\xi) \tag{3.2.33}$$

has a one-dimensional isolated kernel in $L^2(\mathbb{R})$ spanned by $\psi'(\xi)$. Let $2s$ be the distance between two copies of the single-pulse solution $\psi(\xi)$ in the decomposition

$$\Psi(\xi) = \psi(\xi - s) + \psi(\xi + s) + \theta(\xi), \qquad (3.2.34)$$

where $\theta(\xi)$ is a remainder term. Assume that the effective potential function $W(s)$: $\mathbb{R}_+ \to \mathbb{R}$ is given by

$$W(s) := \int_{\mathbb{R}} \psi^3(\xi)\psi(\xi + 2s)dx. \qquad (3.2.35)$$

There exists an infinite countable set of extrema $\{s_n\}_{n\in\mathbb{N}}$ of $W(s)$ such that

$$\lim_{n\to\infty} |s_{n+1} - s_n| = \frac{\pi}{2k_0}.$$

Fix $n \in \mathbb{N}$ and assume that $W''(s_n) \neq 0$. Then, there exists $C_n > 0$ and a unique two-pulse solution $\Psi \in H^4(\mathbb{R})$ of the stationary equation (3.2.31) such that $\Psi(\xi)$ is even on $\mathbb{R}$ and

$$|s - s_n| \leq C_n e^{-2\kappa_0 s}, \quad \|\theta\|_{H^4} \leq C_n e^{-2\kappa_0 s}. \qquad (3.2.36)$$

Proof When the tails of the fundamental localized mode $\psi(\xi)$ are decaying and oscillatory (i.e. when $c > \frac{1}{4}$), the function $W(s)$ defined by (3.2.35) is decaying and oscillatory in s. As a result, there exists an infinite set of extrema $\{s_n\}_{n\in\mathbb{N}}$. Let us pick s_n for a fixed value of $n \in \mathbb{N}$ such that $W'(s_n) = 0$ and $W''(s_n) \neq 0$. When the decomposition (3.2.34) is substituted into the stationary equation (3.2.31), we obtain

$$\left(c - \partial_\xi^2 + \partial_\xi^4 - 3(\psi_+ + \psi_-)^2\right)\theta - 3(\psi_+ + \psi_-)\theta^2 - \theta^3 = 3\psi_+\psi_-(\psi_+ + \psi_-), \qquad (3.2.37)$$

where $\psi_\pm = \psi(\xi \pm s)$.

Let $\epsilon = e^{-2\kappa_0 s}$ be a small parameter that measures the L^∞ norm of the overlapping term in the sense that for each $\epsilon > 0$ there exist constants $C_0 > 0$ and $s_0 > 0$ such that

$$\|\psi(\cdot - s)\psi(\cdot + s)\|_{L^\infty} \leq C_0\epsilon, \quad s \geq s_0. \qquad (3.2.38)$$

We rewrite equation (3.2.37) as a fixed-point equation

$$L\theta = \epsilon f + N(\theta), \qquad (3.2.39)$$

where

$$L := c - \partial_\xi^2 + \partial_\xi^4 - 3(\psi_+ + \psi_-)^2,$$
$$\epsilon f := 3\psi_+\psi_-(\psi_+ + \psi_-),$$
$$N(\theta) := 3(\psi_+ + \psi_-)\theta^2 + \theta^3.$$

Since $H^4(\mathbb{R})$ is a Banach algebra with respect to pointwise multiplication (Appendix B.1), there is $C > 0$ such that

$$\|N(\theta)\|_{L^2} \leq C\|\theta\|_{H^4}^2 + \|\theta\|_{H^4}^3. \qquad (3.2.40)$$

On the other hand, because L_+ for a fundamental localized mode $\psi(\xi)$ has an isolated kernel spanned by $\psi'(\xi)$ and s is large, operator L has two eigenvalues in the neighborhood of 0 which correspond to the even and odd eigenvectors

$$\theta_\pm(\xi) = \psi'(\xi + s) \pm \psi'(\xi - s) + \mathcal{O}_{L^\infty}(\epsilon) \quad \text{as} \quad \epsilon \to 0.$$

We are looking for an even solution of the fixed-point equation (3.2.39) in the constrained H^4 space

$$H_c^4 = \left\{ \theta \in H^4(\mathbb{R}) : \quad \langle \theta_+, \theta \rangle_{L^2} = 0 \right\}.$$

Therefore, we impose the Lyapunov–Schmidt bifurcation equation

$$F(\epsilon, s, \theta) = \epsilon \langle \theta_+, f \rangle_{L^2} + \langle \theta_+, N(\theta) \rangle_{L^2} = 0. \tag{3.2.41}$$

Under condition (3.2.41), there exists a unique C^1 map

$$\mathbb{R}_+ \times \mathbb{R}_+ \ni (\epsilon, s) \mapsto \theta \in H_c^4$$

such that θ solves the fixed-point equation (3.2.39) and satisfies

$$\exists C > 0 : \quad \|\theta\|_{H^2} \le C\epsilon. \tag{3.2.42}$$

The first term of the bifurcation equation (3.2.41) can be simplified as follows:

$$
\begin{aligned}
\epsilon \langle \theta_+, f \rangle_{L^2} &= 3 \langle \theta_+, 3\psi_+ \psi_- (\psi_+ + \psi_-) \rangle_{L^2} \\
&= 3 \langle \psi'_+, \psi_+^2 \psi_- \rangle_{L^2} + 3 \langle \psi'_-, \psi_-^2 \psi_+ \rangle_{L^2} + \mathcal{O}(\epsilon^2) \\
&= -\langle \psi_+^3, \psi'_- \rangle_{L^2} - \langle \psi_-^3, \psi'_+ \rangle_{L^2} + \mathcal{O}(\epsilon^2) \\
&= -W'(s) + \mathcal{O}(\epsilon^2).
\end{aligned}
$$

Combining with bounds (3.2.40) and (3.2.42), we conclude that

$$F(\epsilon, s, \theta) = -W'(s) + \mathcal{O}(\epsilon^2) = 0,$$

where $W'(s) = \mathcal{O}(\epsilon)$ as $\epsilon \to 0$. Using the Implicit Function Theorem for the equation $\epsilon^{-1} F(\epsilon, s, \theta) = 0$, we conclude that there is a unique root s near each s_n, for which $W'(s_n) = 0$ and $W''(s_n) \ne 0$. This construction concludes the proof of bound (3.2.36). $\qquad\square$

Exercise 3.14 Under the conditions of Theorem 3.7, prove that for any small $\epsilon > 0$, the Jacobian operator along the solution branch

$$L_+ = c - \partial_\xi^2 + \partial_\xi^4 - 3\Psi^2(\xi)$$

has a simple zero eigenvalue with the odd eigenvector $\Psi'(\xi)$ and a small eigenvalue in the neighborhood of 0 with an even eigenvector. Moreover, prove that the small eigenvalue is negative if $W''(s_n) > 0$ and positive if $W''(s_n) < 0$.

An infinite countable sequence of two-pulse solutions $\Psi(\xi)$ exists if the interaction potential $W(s)$ between two individual pulses has an alternating sequence of non-degenerate maxima and minima (which corresponds to the case when the fundamental localized mode $\psi(\xi)$ has an oscillatory decaying tail at infinity). The distance between the two pulses occurs near the extremal points of the interaction potential $W(s)$. Three-pulse solutions can be constructed as a bi-infinite countable sequence of three individual pulses where each pair of two adjacent pulses is located approximately at a distance defined by the two-pulse solution. Similarly, N-pulse solutions can be formed by an $(N-1)$-infinite countable sequence of N individual pulses.

3.2.5 Localized modes in the anticontinuum limit

Our final example of the Lyapunov–Schmidt reduction method deals with the stationary DNLS equation,

$$-\Delta u_n + \sigma |u_n|^2 u_n = \omega u_n, \quad n \in \mathbb{Z}^d, \tag{3.2.43}$$

where Δ is the discrete Laplacian on a cubic d-dimensional lattice,

$$\Delta u_n = \sum_{j=1}^{d} \left(u_{n+e_j} + u_{n-e_j} - 2u_n \right), \quad n \in \mathbb{Z}^d,$$

and $\{e_1, ..., e_d\}$ are standard unit vectors on $\mathbb{Z}^d$.

For the stationary DNLS equation (3.2.43), it is sufficient to consider one sign of σ, because the *staggering* transformation,

$$\omega = 4d - \tilde{\omega}, \quad u_n = (-1)^{n_1 + \cdots + n_d} \tilde{u}_n, \quad n = (n_1, \ldots, n_d) \in \mathbb{Z}^d,$$

transforms a solution $(\omega, \{u_n\}_{n \in \mathbb{Z}^d})$ of the stationary DNLS equation (3.2.43) for a fixed sign of σ to another solution $(\tilde{\omega}, \{\tilde{u}_n\}_{n \in \mathbb{Z}^d})$ of the same equation (3.2.43) for the opposite sign of $-\sigma$. Therefore, we need only consider the focusing case $\sigma = -1$.

Recall that $\sigma(-\Delta) \in [0, 4d]$. Localized modes outside the spectral band must have either $\omega < 0$ or $\omega > 4d$. We will show that the stationary DNLS equation (3.2.43) with $\sigma = -1$ admits no localized modes for $\omega > 4d$ and a variety of localized modes for $\omega < 0$. This picture resembles the existence of localized modes in the semi-infinite gap and in each finite gap of the stationary Gross–Pitaevskii equation (3.1.1) with $\sigma = -1$ (Section 3.1), for which the stationary DNLS equation (3.2.43) is an asymptotic reduction in the tight-binding limit (Section 2.4.2).

Lemma 3.10 *No nonzero localized modes of the stationary DNLS equation (3.2.43) with $\sigma = -1$ and $\omega > 4d$ exist.*

Proof Let us multiply (3.2.43) by $\bar{u}_n$ and sum over $\mathbb{Z}^d$ to obtain

$$\langle \mathbf{u}, (\omega + \Delta)\mathbf{u} \rangle_{l^2} + \|\mathbf{u}\|_{l^4}^4 = 0.$$

If $\omega > 4d$, then $(\omega + \Delta)$ is a positive operator and the left-hand side may vanish only for $\|\mathbf{u}\|_{l^2} = \|\mathbf{u}\|_{l^4} = 0$. In view of the discrete embedding of $l^2(\mathbb{Z}^d)$ to $l^\infty(\mathbb{Z}^d)$ (Appendix B.1), we have $\mathbf{u} = \mathbf{0}$, that is, no nonzero localized modes exist for $\omega > 4d$. $\qquad\square$

Let us then fix $\sigma = -1$, $\omega < 0$ in the stationary DNLS equation (3.2.43) and consider the scaling transformation,

$$\omega = -\epsilon^{-1}, \quad \mathbf{u} = \epsilon^{-1/2} \boldsymbol{\phi},$$

where $\epsilon > 0$ is a new parameter and vector ϕ is a solution of the difference equation

$$(1 - |\phi_n|^2)\phi_n = \epsilon \sum_{j=1}^{d} \left(\phi_{n+e_j} + \phi_{n-e_j} - 2\phi_n \right), \quad n \in \mathbb{Z}^d. \tag{3.2.44}$$

The limit $\epsilon \to 0$ is referred to as the *anticontinuum* limit of weakly coupled oscillators. Analysis of time-periodic localized modes in this limit was developed by MacKay & Aubry [136]. Classification of localized modes is proposed by Alfimov *et al.* [10]. Our study follows the work of Pelinovsky *et al.* in one [155], two [156] and three [135] dimensions.

The difference equation (3.2.44) for $\epsilon = 0$ has a general set of localized modes

$$\phi_n^{(0)} = \begin{cases} e^{i\theta_n}, & n \in S, \\ 0, & n \in S^{\perp}, \end{cases} \tag{3.2.45}$$

where $S \subset \mathbb{Z}^d$ is a bounded set of nodes on the lattice, $S^{\perp} = \mathbb{Z}^d \backslash S$, and $\{\theta_n\}_{n \in S}$ is a set of phase configurations. The set $\{\theta_n\}_{n \in S}$ is arbitrary for $\epsilon = 0$.

Definition 3.2 The limiting solution (3.2.45) is called a *soliton configuration* if $\theta_n \in \{0, \pi\}$ for all $n \in S$.

The main question is: For what soliton and non-soliton configurations can the localized mode (3.2.45) be continued in the difference equation (3.2.44) for small nonzero ϵ? The main idea of the reduction method is to track how the extra symmetries for $\epsilon = 0$ are destroyed for $\epsilon \neq 0$. Indeed, the complex phases $\{\theta_n\}_{n \in S}$ are arbitrary if $\epsilon = 0$ but they are related if $\epsilon \neq 0$ by the one-parameter gauge transformation $\phi \to e^{i\alpha}\phi$ for any $\alpha \in \mathbb{R}$.

Similarly to the case of the stationary Gross–Pitaevskii equation (3.2.3) in one dimension, all localized modes of the stationary DNLS equation (3.2.43) are real-valued modulo the gauge transformation. Therefore, only soliton configurations have to be considered if $d = 1$.

Lemma 3.11 *Let $d = 1$ and $\phi \in l^2(\mathbb{Z})$ be a nonzero solution of the stationary DNLS equation (3.2.44). There is $\theta \in \mathbb{R}$ such that $e^{i\theta}\phi : \mathbb{R} \to \mathbb{R}$.*

Proof Because of the gauge transformation, it is sufficient to choose $\theta = 0$. Since ϕ is nonzero, there is $n_0 \in \mathbb{Z}$ such that $\phi_{n_0} \in \mathbb{R} \backslash \{0\}$.

Let $\{\phi_n\}_{n \in \mathbb{Z}}$ solves the stationary DNLS equation (3.2.44). Multiplying the equation by $\bar{\phi}_n$, summing over $\mathbb{Z}$, and subtracting the complex conjugate equation, we can see that

$$J_n = \bar{\phi}_n \phi_{n+1} - \phi_n \bar{\phi}_{n+1} = J_{n-1}, \quad n \in \mathbb{Z},$$

is constant for all $n \in \mathbb{Z}$. If $\lim_{|n| \to \infty} \phi_n = 0$ or if there exists $n_1 \in \mathbb{Z}$ such that $\phi_{n_1} = 0$, then $J_n = 0$ for all $n \in \mathbb{Z}$. If $\phi_{n_1}, \phi_{n_1+1} \neq 0$ for some $n_1 \in \mathbb{Z}$, then $J_{n_1} = 0$ implies that

$$2 \arg(\phi_{n_1+1}) = 2 \arg(\phi_{n_1}) \mod(2\pi).$$

Therefore, if $\phi_{n_1} \in \mathbb{R}$, then $\phi_{n_1+1} \in \mathbb{R}$.

On the other hand, because (3.2.44) with $d = 1$ is a second-order difference map, if $\{\phi_n\}_{n \in \mathbb{Z}}$ is nonzero identically, there exists at most one consequent node, say $n_1 \in \mathbb{Z}$, with $\phi_{n_1} = 0$. In this case, $\phi_{n_1+1} = -\phi_{n_1-1} \neq 0$, hence, if $\phi_{n_1-1} \in \mathbb{R}$, then $\phi_{n_1+1} \in \mathbb{R}$.

Combining both cases together, we summarize that if $\phi_{n_0} \in \mathbb{R} \backslash \{0\}$ for at least one $n_0 \in \mathbb{Z}$, then $\phi_n \in \mathbb{R}$ for any $n \in \mathbb{Z}$. $\square$

The following result shows that the persistence question has a simple solution for $d = 1$.

Theorem 3.8 *Let $d = 1$, $\phi^{(0)}$ be given by (3.2.45), and $\theta_n \in \{0, \pi\}$ for all $n \in S$. There exists a unique real-valued solution $\phi \in l^2(\mathbb{Z})$ of the stationary DNLS equation (3.2.44) for any small $\epsilon \neq 0$. Moreover, the mapping $\mathbb{R} \ni \epsilon \mapsto \phi \in l^2(\mathbb{Z})$ is analytic near $\epsilon = 0$ and there is $C > 0$ such that*

$$\|\phi - \phi^{(0)}\|_{l^2} \leq C\epsilon. \qquad (3.2.46)$$

Proof Let $\mathbf{F}(\phi, \epsilon) : l^2(\mathbb{Z}) \times \mathbb{R} \to l^2(\mathbb{Z})$ be given by

$$F_n(\phi, \epsilon) = (1 - \phi_n^2)\phi_n - \epsilon(\phi_{n+1} + \phi_{n-1} - 2\phi_n), \quad n \in \mathbb{Z}.$$

The map $\mathbf{F}$ is analytic in ϕ and ϵ. It is clear that $\mathbf{F}(\phi^{(0)}, 0) = \mathbf{0}$ and $J = D_\phi \mathbf{F}(\phi^{(0)}, 0)$ is represented by a diagonal matrix with entries

$$1 - 3(\phi_n^{(0)})^2 \in \{1, -2\}.$$

By the Implicit Function Theorem (Appendix B.7), there exists a unique analytic continuation ϕ near $\phi^{(0)}$ in $l^2(\mathbb{Z})$ for small $\epsilon \neq 0$ satisfying the bound (3.2.46). $\square$

We shall now consider persistence of soliton and non-soliton configurations for $d \geq 2$. Let $N = \dim(S)$ and $\mathbb{T}^N$ be the torus on $[0, 2\pi]^N$ for the vector $\boldsymbol{\theta}$ of phase components $\{\theta_n\}_{n \in S}$. We define the nonlinear vector field $\mathbf{F}(\phi, \epsilon) : l^2(\mathbb{Z}^d) \times \mathbb{R} \to l^2(\mathbb{Z}^d)$ by

$$\mathbf{F}_n(\phi, \epsilon) = (1 - |\phi_n|^2)\phi_n - \epsilon \sum_{j=1}^{d} (\phi_{n+e_j} + \phi_{n-e_j} - 2\phi_n), \quad n \in \mathbb{Z}^d. \qquad (3.2.47)$$

The Jacobian operator is given by

$$\mathcal{H} := \begin{bmatrix} D_\phi \mathbf{F}(\phi, \epsilon) & D_{\bar{\phi}} \mathbf{F}(\phi, \epsilon) \\ D_\phi \overline{\mathbf{F}(\phi, \epsilon)} & D_{\bar{\phi}} \overline{\mathbf{F}(\phi, \epsilon)} \end{bmatrix} : l^2(\mathbb{Z}^d) \times l^2(\mathbb{Z}^d) \to l^2(\mathbb{Z}^d) \times l^2(\mathbb{Z}^d).$$

This operator is not block-diagonal due to the presence of the shift operators but we can still use a formal notation $\mathcal{H}_n$ for the "2-block" of $\mathcal{H}$ at the node $n \in \mathbb{Z}^d$,

$$\mathcal{H}_n = \begin{bmatrix} 1 - 2|\phi_n|^2 & -\phi_n^2 \\ -\bar{\phi}_n^2 & 1 - 2|\phi_n|^2 \end{bmatrix} - \epsilon \sum_{j=1}^{d} (\delta_{+e_j} + \delta_{-e_j} - 2) \begin{bmatrix} 1 & 0 \\ 0 & 1 \end{bmatrix}, \qquad (3.2.48)$$

where $\delta_{e_j} u_n = u_{n+e_j}$. This notation allows us to write the matrix–vector form $\mathcal{H}\psi$ in the component form $\mathcal{H}_n \psi_n$ for each ψ_n, $n \in \mathbb{Z}^d$.

Exercise 3.15 Recall that the DNLS equation

$$i\dot{u}_n = -\epsilon \sum_{j=1}^{d}(u_{n+e_j} + u_{n-e_j} - 2u_n) - |u_n|^2 u_n, \quad n \in \mathbb{Z}^d,$$

has the Hamiltonian function

$$H(\mathbf{u}) = \epsilon \sum_{j=1}^{d} \|u_{n+e_j} - u_n\|_{l^2}^2 - \frac{1}{2}\|u_n\|_{l^4}^4$$

and enjoys gauge invariance with the conserved power

$$Q(\mathbf{u}) = \|\mathbf{u}\|_{l^2}^2.$$

Define the energy functional $E_\omega(\mathbf{u}) = H(\mathbf{u}) + Q(\mathbf{u})$ and show that

$$\forall \boldsymbol{\varphi} \in l^2(\mathbb{Z}^d): \quad E_\omega(\boldsymbol{\phi} + \boldsymbol{\varphi}) = E_\omega(\boldsymbol{\phi}) + \frac{1}{2}\langle \mathcal{H}\boldsymbol{\varphi}, \boldsymbol{\varphi}\rangle_{l^2} + \mathcal{O}(\|\boldsymbol{\varphi}\|_{l^2}^3).$$

Let $\boldsymbol{\phi}^{(0)} = \boldsymbol{\phi}^{(0)}(\boldsymbol{\theta})$ be the limiting configuration (3.2.45). By explicit computations, $\mathbf{F}(\boldsymbol{\phi}^{(0)}, 0) = \mathbf{0}$ and $\mathcal{H}^{(0)}$ is block-diagonal with the 2-block at the node $n \in \mathbb{Z}^d$,

$$(\mathcal{H}^{(0)})_n = \begin{bmatrix} 1 & 0 \\ 0 & 1 \end{bmatrix}, \; n \in S^\perp, \quad (\mathcal{H}^{(0)})_n = \begin{bmatrix} -1 & -e^{2i\theta_n} \\ -e^{-2i\theta_n} & -1 \end{bmatrix}, \; n \in S.$$

We note that $\mathcal{H}^{(0)}\mathbf{e}_n = \mathbf{0}$ and $\mathcal{H}^{(0)}\hat{\mathbf{e}}_n = -2\hat{\mathbf{e}}_n$ for all $n \in S \subset \mathbb{Z}^d$, where the 2-blocks of eigenvectors $\mathbf{e}_n$ and $\hat{\mathbf{e}}_n$ at the node $k \in \mathbb{Z}^d$ are given by

$$(\mathbf{e}_n)_k = i\begin{bmatrix} e^{i\theta_n} \\ -e^{-i\theta_n} \end{bmatrix}\delta_{k,n}, \quad (\hat{\mathbf{e}}_n)_k = \begin{bmatrix} e^{i\theta_n} \\ e^{-i\theta_n} \end{bmatrix}\delta_{k,n}.$$

Let $\mathcal{P} : l^2(\mathbb{Z}^d) \times l^2(\mathbb{Z}^d) \to \mathrm{Ker}(\mathcal{H}^{(0)}) \subset l^2(\mathbb{Z}^d) \times l^2(\mathbb{Z}^d)$ be an orthogonal projection operator to the N-dimensional kernel of $\mathcal{H}^{(0)}$. In explicit form, the projection operator $\mathcal{P}$ is expressed by

$$\forall \mathbf{f} \in l^2(\mathbb{Z}^d): \quad (\mathcal{P}\mathbf{f})_n = \frac{1}{2i}\begin{cases} \left(e^{-i\theta_n}(\mathbf{f})_n - e^{i\theta_n}(\bar{\mathbf{f}})_n\right), & \\ -\left(e^{i\theta_n}(\bar{\mathbf{f}})_n - e^{-i\theta_n}(\mathbf{f})_n\right), & \end{cases} \quad n \in S, \qquad (3.2.49)$$

and $(\mathcal{P}\mathbf{f})_n = 0$ for all $n \in S^\perp$.

Since the operator $\mathcal{H}^{(0)}$ is a self-adjoint Fredholm operator of zero index, the decomposition $l^2(\mathbb{Z}^d) \times l^2(\mathbb{Z}^d) = \mathrm{Ker}(\mathcal{H}^{(0)}) \oplus \mathrm{Ran}(\mathcal{H}^{(0)})$ is well-defined. By using the Lyapunov–Schmidt reduction algorithm, we consider the decomposition

$$\boldsymbol{\phi} = \boldsymbol{\phi}^{(0)}(\boldsymbol{\theta}) + \sum_{n \in S}\alpha_n \mathbf{e}_n + \boldsymbol{\varphi},$$

where $(\boldsymbol{\varphi}, \bar{\boldsymbol{\varphi}}) \in \mathrm{Ran}(\mathcal{H}^{(0)})$ and $\alpha_n \in \mathbb{R}$ for each $n \in S$. We note that

$$\forall \boldsymbol{\theta}_0 \in \mathbb{T}^N: \quad \boldsymbol{\phi}^{(0)}(\boldsymbol{\theta}_0) + \sum_{n \in S}\alpha_n \mathbf{e}_n = \boldsymbol{\phi}^{(0)}(\boldsymbol{\theta}_0 + \boldsymbol{\alpha}) + \mathcal{O}(\|\boldsymbol{\alpha}\|^2). \qquad (3.2.50)$$

Since the values of $\boldsymbol{\theta}$ in $\boldsymbol{\phi}^{(0)}(\boldsymbol{\theta})$ have not been defined yet, we can set $\alpha_n = 0$ for all $n \in S$ without loss of generality. The projection equations in the Lyapunov–Schmidt reduction algorithm are

$$\mathcal{P}\mathbf{F}(\boldsymbol{\phi}^{(0)}(\boldsymbol{\theta}) + \boldsymbol{\varphi}, \epsilon) = 0, \qquad (\mathcal{I} - \mathcal{P})\mathbf{F}(\boldsymbol{\phi}^{(0)}(\boldsymbol{\theta}) + \boldsymbol{\varphi}, \epsilon) = 0. \qquad (3.2.51)$$

We note that $\mathbf{F}(\boldsymbol{\phi}, \epsilon)$ is analytic in $\epsilon \in \mathbb{R}$ and $(\mathcal{I} - \mathcal{P})\mathcal{H}(\mathcal{I} - \mathcal{P}) : \mathrm{Ran}(\mathcal{H}^{(0)}) \to \mathrm{Ran}(\mathcal{H}^{(0)})$ is invertible at $\epsilon = 0$. By the Implicit Function Theorem (Appendix B.7), there exists a unique solution $\boldsymbol{\varphi} \in l^2(\mathbb{Z}^d)$ of the second equation of system (3.2.51), which is analytic in ϵ near $\epsilon = 0$ and depends on $\boldsymbol{\theta} \in \mathbb{T}^N$. Let us write this map as

$$\mathbb{T}^N \times \mathbb{R} \ni (\boldsymbol{\theta}, \epsilon) \mapsto \boldsymbol{\varphi} \in l^2(\mathbb{Z}^d).$$

From the Implicit Function Theorem, we know that, for sufficiently small $\epsilon \in \mathbb{R}$, there is $C > 0$ such that

$$\|\boldsymbol{\varphi}^{(0)}(\boldsymbol{\theta}, \epsilon)\|_{l^2} \le C\epsilon, \quad \boldsymbol{\theta} \in \mathbb{T}^N.$$

As a result, there exists the nonlinear vector field $\mathbf{g}(\boldsymbol{\theta}, \epsilon) : \mathbb{T}^N \times \mathbb{R} \to \mathbb{R}^N$, which generates the reduced bifurcation equations

$$\mathbf{g}(\boldsymbol{\theta}, \epsilon) = \mathcal{P}\mathbf{F}(\boldsymbol{\phi}^{(0)}(\boldsymbol{\theta}) + \boldsymbol{\varphi}^{(0)}(\boldsymbol{\theta}, \epsilon), \epsilon) = 0. \qquad (3.2.52)$$

By construction, the function $\mathbf{g}(\boldsymbol{\theta}, \epsilon)$ is analytic in ϵ near $\epsilon = 0$ and $\mathbf{g}(\boldsymbol{\theta}, 0) = \mathbf{0}$ for any $\boldsymbol{\theta} \in \mathbb{T}^N$. Therefore, the Taylor series for $\mathbf{g}(\boldsymbol{\theta}, \epsilon)$ can be written in the form,

$$\mathbf{g}(\boldsymbol{\theta}, \epsilon) = \sum_{k=1}^{\infty} \epsilon^k \mathbf{g}^{(k)}(\boldsymbol{\theta}). \qquad (3.2.53)$$

Exercise 3.16 Show that

$$\mathbf{g}_n^{(1)}(\boldsymbol{\theta}) = e^{-i\theta_n} \sum_{j=1}^{d} \left(\phi_{n+e_j} + \phi_{n-e_j} \right) - e^{i\theta_n} \sum_{j=1}^{d} \left(\bar{\phi}_{n+e_j} + \bar{\phi}_{n-e_j} \right), \quad n \in S$$

and $\mathbf{g}_n^{(1)}(\boldsymbol{\theta}) = 0$ for all $n \in S^\perp$.

The persistence question is now answered in the following theorem.

Theorem 3.9 *The limiting configuration (3.2.45) for $\boldsymbol{\phi}^{(0)}(\boldsymbol{\theta})$ can be continued for any small nonzero ϵ if and only if there exists a root $\boldsymbol{\theta}_* \in \mathbb{T}^N$ of the vector field $\mathbf{g}(\boldsymbol{\theta}, \epsilon)$ for this ϵ. Moreover, if the root $\boldsymbol{\theta}_*$ is analytic in ϵ with $\boldsymbol{\theta}_* = \boldsymbol{\theta}_0 + \mathcal{O}(\epsilon)$, the solution $\boldsymbol{\phi}$ of the difference equation (3.2.44) is analytic in ϵ and*

$$\boldsymbol{\phi} = \boldsymbol{\phi}^{(0)}(\boldsymbol{\theta}_*) + \boldsymbol{\varphi}^{(0)}(\boldsymbol{\theta}_*, \epsilon) = \boldsymbol{\phi}^{(0)}(\boldsymbol{\theta}_0) + \sum_{k=1}^{\infty} \epsilon^k \boldsymbol{\phi}^{(k)}, \qquad (3.2.54)$$

where $\{\boldsymbol{\phi}^{(k)}\}_{k \in \mathbb{N}}$ are independent of ϵ.

Exercise 3.17 Fix $\boldsymbol{\theta}_*$ at the soliton configuration in the sense of Definition 3.2 and prove that there exists a continuation of the soliton configuration in $\epsilon \neq 0$.

For the non-soliton configurations, computations of roots of the function $\mathbf{g}(\boldsymbol{\theta}, \epsilon)$ are based on the analysis of the convergent Taylor series expansions (3.2.53) and (3.2.54). Let $\mathcal{M} = D_{\boldsymbol{\theta}}\mathbf{g}(\boldsymbol{\theta}, \epsilon) : \mathbb{R}^N \to \mathbb{R}^N$ be the Jacobian matrix evaluated at the vector $\boldsymbol{\theta} \in \mathbb{T}^N$. By the symmetry of the shift operators, the matrix $\mathcal{M}$ is symmetric. By the gauge symmetry, the function $\mathbf{g}(\boldsymbol{\theta}, \epsilon)$ satisfies the following relation:

$$\forall \alpha_0 \in \mathbb{R}, \ \forall \boldsymbol{\theta} \in \mathbb{T}^N : \qquad \mathbf{g}(\boldsymbol{\theta} + \alpha_0 \mathbf{p}_0, \epsilon) = \mathbf{g}(\boldsymbol{\theta} + \alpha_0 \mathbf{p}_0, \epsilon), \qquad (3.2.55)$$

where $\mathbf{p}_0 = (1, 1, ..., 1) \in \mathbb{R}^N$. As a result, $\mathcal{M}\mathbf{p}_0 = \mathbf{0}$ and the spectrum of $\mathcal{M}$ always includes a zero eigenvalue. If the zero eigenvalue of $\mathcal{M}$ is simple, we can use again the Implicit Function Theorem and conclude that zeros $\boldsymbol{\theta}_*$ of the function $\mathbf{g}(\boldsymbol{\theta}, \epsilon)$ are uniquely continued in ϵ modulo the gauge transformation (3.2.55). However, if the zero eigenvalue is not simple for the first non-trivial matrix $\mathcal{M}^{(k_0)} = D_{\boldsymbol{\theta}}\mathbf{g}^{(k_0)}(\boldsymbol{\theta}_0)$ for some $k_0 \in \mathbb{N}$, we need to compute $\mathbf{g}^{(k)}(\boldsymbol{\theta}_0)$ for $k > k_0$ and ensure that the Fredholm condition

$$\mathbf{g}^{(k)}(\boldsymbol{\theta}_0) \perp \mathrm{Ker}(\mathcal{M}^{(k_0)})$$

is satisfied as, otherwise, the solution cannot be continued for small nonzero ϵ. Combining all these facts together, we formulate the criterion for persistence of non-soliton configurations.

Theorem 3.10 *Suppose that $\mathbf{g}^{(k)}(\boldsymbol{\theta}) \equiv \mathbf{0}$ for $k = 1, 2, ..., k_0 - 1$ and $\mathbf{g}^{(k_0)}(\boldsymbol{\theta}) \neq \mathbf{0}$ for some $k_0 \in \mathbb{N}$. Let $\boldsymbol{\theta}_0$ be the root of $\mathbf{g}^{(k_0)}(\boldsymbol{\theta})$.*

(1) *If $\mathrm{Ker}(\mathcal{M}^{(k_0)}) = \mathrm{Span}\{\mathbf{p}_0\} \subset \mathbb{R}^N$, then the configuration (3.2.45) is uniquely continued in $\epsilon \neq 0$ modulo the gauge transformation (3.2.55).*
(2) *If $\mathrm{Ker}(\mathcal{M}^{(k_0)}) = \mathrm{Span}\{\mathbf{p}_0, \mathbf{p}_1, ..., \mathbf{p}_{d_{k_0}}\} \subset \mathbb{R}^N$ with $1 \leq d_{k_0} \leq N - 1$, then:*

   (i) *if $\mathbf{g}^{(k_0+1)}(\boldsymbol{\theta}_0) \notin \mathrm{Ran}(\mathcal{M}^{(k_0)})$, the configuration (3.2.45) does not persist for any $\epsilon \neq 0$;*
   (ii) *if $\mathbf{g}^{(k_0+1)}(\boldsymbol{\theta}_0) \in \mathrm{Ran}(\mathcal{M}^{(k_0)})$, the configuration (3.2.45) is continued beyond $\mathcal{O}(\epsilon^{k_0})$.*

The criterion of Theorem 3.10 can be used in the computational algorithm, which provides a binary answer to whether the configuration persists beyond $\epsilon \neq 0$ or terminates at $\epsilon = 0$. Let us consider a square lattice $\mathbb{Z}^2$ and a discrete contour S for a *vortex cell*,

$$S = \{(1, 1), (2, 1), (2, 2), (1, 2)\}. \qquad (3.2.56)$$

Let $\theta_1 = \theta_{1,1}$, $\theta_2 = \theta_{2,1}$, $\theta_3 = \theta_{2,2}$, and $\theta_4 = \theta_{1,2}$ along the discrete contour S. Periodic boundary conditions can be used for $\theta_0 = \theta_4$, $\theta_5 = \theta_1$ if necessary. By the gauge transformation, we can always set $\theta_1 = 0$ for convenience. We also denote $\theta_2 = \theta$ and choose $\theta \in [0, \pi]$ for convenience.

Definition 3.3 *Let S be a closed contour on $\mathbb{Z}^2$ and $\Delta\theta_j = \theta_{j+1} - \theta_j \in (-\pi, \pi]$ be the phase difference between two successive nodes on S. The total number of 2π phase shifts across the closed contour S is called the* vortex charge L.

From Exercise 3.16, we compute for the contour (3.2.56),

$$\mathbf{g}_j^{(1)}(\boldsymbol{\theta}) = \sin(\theta_j - \theta_{j+1}) + \sin(\theta_j - \theta_{j-1}), \quad 1 \le j \le 4. \qquad (3.2.57)$$

Therefore, $\mathbf{g}^{(1)}(\boldsymbol{\theta}) = \mathbf{0}$ if and only if

$$\sin(\theta_2 - \theta_1) = \sin(\theta_3 - \theta_2) = \sin(\theta_4 - \theta_3) = \sin(\theta_1 - \theta_4). \qquad (3.2.58)$$

Upon our convention on $\theta_1 = 0$ and $\theta_2 = \theta \in [0, \pi]$, there are eight *soliton* configurations with $\theta_j \in \{0, \pi\}$, one *symmetric vortex* configuration of charge $L = 1$ with

$$\theta_j = \frac{\pi(j-1)}{2}, \quad j \in \{1, 2, 3, 4\}, \qquad (3.2.59)$$

and three *asymmetric vortex* configurations of charge $L = 1$ with

(a) $\qquad\qquad\qquad \theta_1 = 0,\ \theta_2 = \theta,\ \theta_3 = \pi,\ \theta_4 = \pi + \theta;$

(b) $\qquad\qquad\qquad \theta_1 = 0,\ \theta_2 = \theta,\ \theta_3 = 2\theta,\ \theta_4 = \pi + \theta;$

(c) $\qquad\qquad\qquad \theta_1 = 0,\ \theta_2 = \theta,\ \theta_3 = \pi,\ \theta_4 = 2\pi - \theta.$

We can now compute the Jacobian matrix $\mathcal{M}^{(1)} = D_{\boldsymbol{\theta}}\mathbf{g}^{(1)}(\boldsymbol{\theta})$ from explicit expression (3.2.57):

$$(\mathcal{M}^{(1)})_{i,j} = \begin{cases} \cos(\theta_{j+1} - \theta_j) + \cos(\theta_{j-1} - \theta_j), & i = j, \\ -\cos(\theta_j - \theta_i), & i = j \pm 1, \\ 0, & |i - j| \ge 2, \end{cases} \qquad (3.2.60)$$

subject to the periodic boundary conditions.

Remark 3.8 Although $\mathcal{M}^{(1)}$ is just a 4×4 matrix, it has a generic structure, which is defined by the coefficients $a_j = \cos(\theta_{j+1} - \theta_j)$ for $1 \le j \le N$, where $N = 4$ in our case.

Let n_0, z_0, and p_0 be the numbers of negative, zero and positive terms of $a_j = \cos(\theta_{j+1} - \theta_j)$, $1 \le j \le N$. Let $n(\mathcal{M}^{(1)})$, $z(\mathcal{M}^{(1)})$, and $p(\mathcal{M}^{(1)})$ be the numbers of negative, zero and positive eigenvalues of the matrix $\mathcal{M}^{(1)}$. The following lemma proved by Sandstede [179] tells us how to count eigenvalues of $\mathcal{M}^{(1)}$.

Lemma 3.12 *Assume that* $z_0 = 0$ *and*

$$A_1 = \sum_{i=1}^{N} \prod_{j \ne i} a_j = \left(\prod_{i=1}^{N} a_i\right)\left(\sum_{i=1}^{N} \frac{1}{a_i}\right) \ne 0.$$

Then, $z(\mathcal{M}^{(1)}) = 1$, *and either* $n(\mathcal{M}^{(1)}) = n_0 - 1$, $p(\mathcal{M}^{(1)}) = p_0$ *or* $n(\mathcal{M}^{(1)}) = n_0$, $p(\mathcal{M}^{(1)}) = p_0 - 1$. *Moreover,* $n(\mathcal{M}^{(1)})$ *is even if* $A_1 > 0$ *and is odd if* $A_1 < 0$.

Exercise 3.18 Let $a_j = a$ for all $1 \le j \le N$. Use the discrete Fourier transform and prove that eigenvalues of $\mathcal{M}^{(1)}$ are given by

$$\lambda_n = 4a \sin^2 \frac{\pi n}{N}, \quad 1 \le n \le N.$$

If $a > 0$, then $z(\mathcal{M}^{(1)}) = 1$, $p(\mathcal{M}^{(1)}) = N - 1$, and $n(\mathcal{M}^{(1)}) = 0$.

Exercise 3.19 Let N be even and $a_j = (-1)^j a$ for all $1 \leq j \leq N$. Use the discrete Fourier transform and prove that eigenvalues of $\mathcal{M}^{(1)}$ are given by

$$\lambda_n = -\lambda_{n+N/2} = 2a \sin \frac{2\pi n}{N}, \quad 1 \leq n \leq \frac{N}{2}.$$

Hence, $n(\mathcal{M}^{(1)}) = \frac{1}{2}N - 1$, $z(\mathcal{M}^{(1)}) = 2$, and $p(\mathcal{M}^{(1)}) = \frac{1}{2}N - 1$.

Let us apply these results to various solutions of system (3.2.58).

For soliton configurations with non-equal numbers of 1 and -1 in $a_j \in \{-1, 1\}$, for $j \in \{1, 2, 3, 4\}$, we have $z_0 = 0$ and $A_1 \neq 0$. By Lemma 3.12, we have $z(\mathcal{M}^{(1)}) = 1$. Theorem 3.10 states the existence of a unique continuation of the soliton configuration in $\epsilon \neq 0$. The resulting solution is real-valued thanks to the result in Exercise 3.17. Soliton configurations with equal numbers of 1 and -1 have $A_1 = 0$ and $z(\mathcal{M}^{(1)}) = 2$. Therefore, these configurations must be considered beyond the first-order reduction $\mathcal{O}(\epsilon)$.

For the symmetric vortex configuration, all coefficients a_j are the same and zero. Therefore, $\mathcal{M}^{(1)} = 0$ and we need to compute the next-order corrections $\mathbf{g}^{(2)}(\boldsymbol{\theta})$ and the Jacobian $\mathcal{M}^{(2)} = D_{\boldsymbol{\theta}} \mathbf{g}^{(2)}(\boldsymbol{\theta})$.

For the asymmetric vortex configurations, there are two coefficients $a_j = \cos \theta$ and two coefficients $a_j = -\cos \theta$, which are nonzero if $\theta \neq \frac{\pi}{2}$. The result of Exercise 3.19 shows that $n(\mathcal{M}^{(1)}) = 1$, $z(\mathcal{M}^{(1)}) = 2$, and $p(\mathcal{M}^{(1)}) = 1$. The additional zero eigenvalue is related to the derivative of the family of asymmetric vortex configurations with respect to the parameter θ. Therefore, continuations of these configurations are also considered beyond the first-order reduction $\mathcal{O}(\epsilon)$.

To compute $\mathbf{g}^{(2)}(\boldsymbol{\theta})$, we use Taylor series (3.2.54) and obtain the inhomogeneous problem

$$(1 - 2|\phi_n^{(0)}|^2)\phi_n^{(1)} - \phi_n^{(0)2}\bar{\phi}_n^{(1)} = \phi_{n+e_1}^{(0)} + \phi_{n-e_1}^{(0)} + \phi_{n+e_2}^{(0)} + \phi_{n-e_2}^{(0)} - 4\phi_n^{(0)}, \quad n \in \mathbb{Z}^2.$$

To solve the inhomogeneous problem, we remove the homogeneous solutions, which would only redefine parameters $\boldsymbol{\theta}$. This is equivalent to the constraint $\phi_n^{(1)} = u_n^{(1)} e^{i\theta_n}$ with $u_n^{(1)} \in \mathbb{R}$ for all $n \in S$. In this way, we obtain the unique solution

$$\phi_n^{(1)} = \begin{cases} -\frac{1}{2} \left(\cos(\theta_{j-1} - \theta_j) + \cos(\theta_{j+1} - \theta_j) - 4 \right) e^{i\theta_j}, & n \in S, \\ e^{i\theta_j}, & n \in S_1, \\ 0, & n \notin S \cup S_1, \end{cases}$$

where the index j enumerates the node n on the contour S and the nodes $n \in S_1 \subset S^\perp$ are adjacent to the jth node on the contour S.

Substituting the first-order correction term $\phi_n^{(1)}$ into the bifurcation function (3.2.52), we find the correction term $\mathbf{g}^{(2)}(\boldsymbol{\theta})$ in the Taylor series (3.2.53)

$$\mathbf{g}_j^{(2)}(\boldsymbol{\theta}) = \frac{1}{2} \sin(\theta_{j+1} - \theta_j) \left[\cos(\theta_j - \theta_{j+1}) + \cos(\theta_{j+2} - \theta_{j+1}) \right]$$

$$+ \frac{1}{2} \sin(\theta_{j-1} - \theta_j) \left[\cos(\theta_j - \theta_{j-1}) + \cos(\theta_{j-2} - \theta_{j-1}) \right], \quad 1 \leq j \leq N.$$

For the three asymmetric vortex configurations, we have

$$
\text{(a) } \mathbf{g}^{(2)} = \begin{bmatrix} 0 \\ 0 \\ 0 \\ 0 \end{bmatrix}, \quad \text{(b) } \mathbf{g}^{(2)} = \begin{bmatrix} 2 \\ 0 \\ -2 \\ 0 \end{bmatrix} \sin\theta\cos\theta, \quad \text{(c) } \mathbf{g}^{(2)} = \begin{bmatrix} 0 \\ -2 \\ 0 \\ 2 \end{bmatrix} \sin\theta\cos\theta.
$$

The kernel of $\mathcal{M}_1$ for each configuration is two-dimensional with the eigenvectors $\mathbf{p}_0 = (1,1,1,1)$ and $\mathbf{p}_1$ obtained from the derivative of the vortex configurations in θ,

$$
\text{(a) } \mathbf{p}_1 = \begin{bmatrix} 0 \\ 1 \\ 0 \\ 1 \end{bmatrix}, \quad \text{(b) } \mathbf{p}_1 = \begin{bmatrix} 0 \\ 1 \\ 2 \\ 1 \end{bmatrix}, \quad \text{(c) } \mathbf{p}_1 = \begin{bmatrix} 0 \\ 1 \\ 0 \\ -1 \end{bmatrix}.
$$

The Fredholm alternative $\langle \mathbf{p}_1, \mathbf{g}^{(2)} \rangle_{\mathbb{R}^4} = 0$ is satisfied for the solution (a) but fails for the solutions (b) and (c), unless $\theta = \{0, \frac{\pi}{2}, \pi\}$. The latter cases belong to the cases of soliton and symmetric vortex configurations. By Theorem 3.10, asymmetric vortex configurations (b) and (c) cannot be continued in $\epsilon \neq 0$, while (a) must be considered beyond the second-order reductions $\mathcal{O}(\epsilon^2)$.

For the symmetric vortex configuration (3.2.59), the Jacobian matrix $\mathcal{M}^{(1)} = D_{\boldsymbol\theta}\mathbf{g}^{(1)}(\boldsymbol\theta)$ is identically zero and we need to compute $\mathcal{M}^{(2)} = D_{\boldsymbol\theta}\mathbf{g}^{(2)}(\boldsymbol\theta)$. This matrix is computed explicitly by

$$
\mathcal{M}^{(2)} = \begin{bmatrix} 1 & 0 & -1 & 0 \\ 0 & 1 & 0 & -1 \\ -1 & 0 & 1 & 0 \\ 0 & -1 & 0 & 1 \end{bmatrix}.
$$

The matrix $\mathcal{M}^{(2)}$ has four eigenvalues: $\lambda_1 = \lambda_2 = 0$ and $\lambda_3 = \lambda_4 = 2$. The two eigenvectors for the zero eigenvalue are $\mathbf{p}_1 = (0,1,0,1)$ and $\mathbf{p}_2 = (1,0,1,0)$. The eigenvector $\mathbf{p}_1$ corresponds to the derivative of the asymmetric vortex configuration (a) with respect to parameter θ, while the eigenvector $\mathbf{p}_0 = \mathbf{p}_1 + \mathbf{p}_2$ corresponds to the shift due to gauge transformation. The presence of double zero eigenvalue for $\mathcal{M}^{(2)}$ confirms the result of the continuation of the asymmetric vortex configuration (a) beyond the second-order reductions.

The presence of the arbitrary parameter θ in the family (a) of asymmetric vortex configurations is not supported by the symmetries of the discrete contour S in the stationary DNLS equation (3.2.43). Therefore, we would expect that this family terminates in higher-order reductions. In order to confirm this conjecture, a symbolic algorithm was developed [156, 135]. It was found that the vector $\mathbf{g}^{(k)}(\boldsymbol\theta)$ computed at the asymmetric vortex configuration (a) is zero for $k = 1, 2, 3, 4, 5$ and nonzero for $k = 6$. Moreover, $\langle \mathbf{p}_1, \mathbf{g}^{(6)} \rangle_{\mathbb{R}^4} \neq 0$ for any $\theta \neq \{0, \frac{\pi}{2}, \pi\}$, where $\mathbf{p}_1 = (0,1,0,1)$ is an eigenvector in $\mathrm{Ker}(\mathcal{M}^{(1)})$. Therefore, the family (a) of asymmetric vortex configurations terminates at the sixth-order reduction $\mathcal{O}(\epsilon^6)$.

Nevertheless, all soliton configurations and the symmetric vortex configuration in the discrete contour S persist beyond all orders of the reductive algorithm.

Exercise 3.20 Consider the discrete contour for the vortex cross,

$$S = \{(-1,0), (0,-1), (1,0), (0,1)\} \subset \mathbb{Z}^2.$$

Compute reduction functions $\mathbf{g}^{(2)}(\boldsymbol{\theta})$ and $\mathbf{g}^{(4)}(\boldsymbol{\theta})$ for the family (a) of asymmetric vortex configurations and show that the family (a) terminates for any $\theta \notin \{0, \frac{\pi}{2}, \pi\}$.

Persistence of time-periodic localized modes of the Klein–Gordon lattices is considered using slow manifolds for Hamiltonian systems in action–angle variables by MacKay [137] and Koukouloyannis & Kevrekidis [120]. Essentially the same computations as in this section are performed in [137, 120] with the formalism of canonical transformations.

3.3 Other analytical methods

The best proof of existence of a stationary localized mode is achieved if an exact solution can be constructed in an analytical form. In some exceptional circumstances, exact solutions for localized modes can be expressed in terms of elementary (exponential, hyperbolic and algebraic) functions. Unfortunately, the analytical solutions in the stationary Gross–Pitaevskii equation with a periodic potential can only be constructed for isolated values of the frequency parameter ω, whereas those for the DNLS equation can only be found for the integrable case of the Ablowitz–Ladik lattice or for some special discretizations of the nonlinear functions (Section 5.2). For reduced evolution equations such as the nonlinear Dirac and Schrödinger equations, the exact analytical solutions for localized modes can be found for all parameter values, but the asymptotic validity of these equations is only justified in a subset of parameter values.

We shall review here several methods to construct analytical solutions for the stationary Gross–Pitaevskii equation and its reductions. Given a variety of recent computational techniques based on explicit substitutions (such as the tanh–sech method, the homogeneous balance method, the F-expansion method, and the like), we will focus on the methods that are inspired by the dynamical system theory and have a broader applicability to other problems involving nonlinear evolution equations and localized modes. In other words, we shall avoid techniques which rely solely on trials and substitutions.

Even if analytical expressions for localized modes cannot be found, the dynamical system methods allow us to construct qualitative solutions. These solutions clarify better the properties of the localized modes compared to the variational methods in Section 3.1 and have a wider applicability compared to the Lyapunov–Schmidt reduction methods in Section 3.2. In particular, the dynamical system methods are not constrained by the presence of a small parameter in the stationary equation.

We shall consider the construction of localized modes using symmetry reductions of the stationary Gross–Pitaevskii equation (Section 3.3.1), shooting methods for the differential equations (Section 3.3.2), shooting methods for the difference equations (Section 3.3.3), and exact solutions of the nonlinear Dirac equations (Section 3.3.4).

3.3.1 Symmetry reductions of the stationary Gross–Pitaevskii equation

Consider the stationary Gross–Pitaevskii equation with periodic coefficients,

$$-u''(x) + V(x)u(x) + G(x)u^3(x) = \omega u(x), \quad x \in \mathbb{R}, \tag{3.3.1}$$

where $V(x)$ and $G(x)$ are real-valued, 2π-periodic, and bounded functions.

We shall use the point transformations (Section 1.4.2) and find special coefficients $V(x)$ and $G(x)$, for which exact solutions of the stationary Gross–Pitaevskii equation (3.3.1) can be constructed. Our presentation follows Belmonte-Beitia *et al.* [17]. A general algorithm of construction of exact solutions from symmetry reductions of differential equations is described in the text of Bluman & Kumei [21].

We view the differential equation (3.3.1) in the abstract form

$$A(x, u, u_x, u_{xx}) := -u_{xx} + Vu + Gu^3 - \omega u = 0,$$

and look for the point transformations defined by the generator,

$$M = \xi(x, u)\frac{\partial}{\partial x} + \eta(x, u)\frac{\partial}{\partial u}. \tag{3.3.2}$$

Since the differential equation (3.3.1) is of the second order, we need to compute the second prolongation of operator M,

$$M^{(2)} = M + \eta^{(1)}(x, u)\frac{\partial}{\partial u_x} + \eta^{(2)}(x, u)\frac{\partial}{\partial u_{xx}},$$

where

$$\eta^{(1)} = \eta_x + u_x(\eta_u - \xi_x) - (u_x)^2\xi_u,$$

$$\eta^{(2)} = \eta_{xx} + u_x(2\eta_{ux} - \xi_{xx}) + (u_x)^2(\eta_{uu} - 2\xi_{xu}) - (u_x)^3\xi_{uu}$$
$$+ u_{xx}(\eta_u - 2\xi_x) - 3u_x u_{xx}\xi_u.$$

The differential equation (3.3.1) is invariant under the point transformation if $M^{(2)}A = 0$. Expressing $u_{xx} = Vu + Gu^3 - \omega u$ and removing terms $(u_x)^3$, $(u_x)^2$, u_x, and $(u_x)^0 \equiv 1$, we obtain the following system of equations on $\xi(x, u)$ and $\eta(x, u)$:

$$\begin{cases} \xi_{uu} = 0, \\ \eta_{uu} - 2\xi_{ux} = 0, \\ 2\eta_{xu} - \xi_{xx} - 3\xi_u(Vu + Gu^3 - \omega u) = 0, \\ \eta_{xx} - \xi(V'u + G'u^3) - \eta(V + 3Gu^2 - \omega) + (\eta_u - 2\xi_x)(Vu + Gu^3 - \omega u) = 0. \end{cases}$$

The first two equations of this system show that

$$\xi(x, u) = a(x)u + b(x), \quad \eta(x, u) = a'(x)u^2 + c(x)u + d(x),$$

where (a, b, c, d) are arbitrary functions. The third equation of the system shows that if $G \neq 0$, then $a(x) \equiv 0$ and $2c'(x) - b''(x) = 0$. Integrating this equation, we write

$$c(x) = \frac{1}{2}b'(x) + C,$$

where C is an arbitrary constant. The last equation of the system of point transformations shows that if $G \neq 0$, then $d(x) \equiv 0$. Equating terms with u^3 and u to zero, we obtain the following two constraints

$$\begin{cases} 2c(x)G(x) + b(x)G'(x) + 2b'(x)G(x) = 0, \\ c''(x) - b(x)V'(x) - 2b'(x)(V(x) - \omega) = 0. \end{cases} \quad (3.3.3)$$

The first equation of system (3.3.3) can be integrated explicitly,

$$G(x) = \frac{G_0}{b^3(x)} \exp\left(-2C \int_0^x \frac{dx'}{b(x')}\right), \quad (3.3.4)$$

where G_0 is an arbitrary constant. Thanks to the scaling transformation, this constant can always be normalized either to $G_0 = 1$ or to $G_0 = -1$. The second equation of system (3.3.3) relates the potential $V(x)$ to the nonlinearity function $G(x)$ via the function $b(x)$:

$$b'''(x) - 2b(x)V'(x) - 4b'(x)(V(x) - \omega) = 0. \quad (3.3.5)$$

This equation can be integrated in the form

$$\frac{1}{2}b(x)b''(x) - \frac{1}{4}(b'(x))^2 + (\omega - V(x))b^2(x) = \Omega, \quad (3.3.6)$$

where Ω is a new integration constant.

Exercise 3.21 Show that the change of variables

$$X = X(x) = \int_0^x \frac{dx'}{b(x')}, \quad u(x, U) = b^{1/2}(x)U,$$

transforms the stationary equation with variable coefficients (3.3.1) to the stationary equation with constant coefficients,

$$-U''(X) + G_0 U^3(X) = \Omega U(X), \quad X \in \mathbb{R}. \quad (3.3.7)$$

Show that $M = \partial/\partial X$ in new variables and $C = 0$ in (3.3.4).

If $b(x)$ is positive and bounded away from 0, then $X(x) \to \pm\infty$ as $x \to \pm\infty$. For instance, if

$$b(x) = 1 + \alpha \cos(x), \quad \alpha \in (-1, 1),$$

then the periodic potentials (3.3.4) and (3.3.6) become

$$V(x) = \omega - \frac{\Omega}{(1 + \alpha \cos(x))^2} - \frac{2\alpha \cos(x) + 3\alpha^2 \cos^2(x)}{4(1 + \alpha \cos(x))^2}, \quad G(x) = \frac{G_0}{(1 + \alpha \cos(x))^3}.$$

An exact localized mode of the stationary equation with constant coefficients (3.3.7) exists for $\text{sign}(\Omega) = \text{sign}(G_0) = -1$. If $G_0 = -1$, then

$$U(X) = \sqrt{2|\Omega|} \, \text{sech}(\sqrt{|\Omega|}X), \quad X \in \mathbb{R}. \quad (3.3.8)$$

This solution generates the exact localized mode

$$\phi(x) = (1 + \alpha \cos(x))^{1/2} U(X), \quad X = \int_0^x \frac{dx'}{1 + \alpha \cos(x')},$$

of the stationary equation with variable coefficients (3.3.1). The localized mode exists in the semi-infinite gap of the spectrum of $L = -\partial_x^2 + V(x)$. Unfortunately, if $V(x)$ and $G(x)$ are fixed, this localized mode only exists for a single value of ω in the semi-infinite gap. No exact solution exists for *all values* of ω in the semi-infinite gap, that is, for all $\omega < \omega_0 = \inf \sigma(L)$.

Exercise 3.22 Substitute $b(x) = \cosh(x)$ and find the potential functions $V(x)$ and $G(x)$ of the stationary equation (3.3.1) that decay to zero as $|x| \to \infty$ and admit the exact localized mode $\phi(x)$ generated from the exact solution (3.3.8).

Exercise 3.23 Consider the stationary cubic–quintic equation with periodic coefficients

$$-u''(x) + V(x)u(x) + G(x)u^3(x) + H(x)u^5(x) = \omega u(x), \quad x \in \mathbb{R}. \tag{3.3.9}$$

Show that if

$$V(x) = -\frac{2\alpha(1 + \alpha \cos(x))}{(1 + 2\alpha \cos(x))^2}, \quad G(x) = \frac{B}{(1 + 2\alpha \cos(x))^3}, \quad H(x) = \frac{A}{(1 + 2\alpha \cos(x))^4},$$

for a fixed $\alpha \in \left(-\frac{1}{2}, \frac{1}{2}\right)$ and arbitrary (A, B), the stationary cubic–quintic equation with variable coefficients reduces to the stationary equation with constant coefficients,

$$-U''(X) + BU^3(X) + AU^5(X) = -\alpha U(X), \quad X \in \mathbb{R},$$

in new variables X and $U(X)$.

Computations of symmetry reductions for the stationary cubic–quintic Gross–Pitaevskii equation (3.3.9) are developed by Belmonte-Beitia & Cuevas [19]. Symmetry reductions for the discrete versions of the NLS equation are studied by Hernández-Heredero & Levi [88].

3.3.2 Shooting methods for the stationary Gross–Pitaevskii equation

Consider the stationary Gross–Pitaevskii equation with the power nonlinearity,

$$-\phi''(x) + V(x)\phi(x) + \sigma \phi^{2p+1}(x) = \omega \phi(x), \quad x \in \mathbb{R}, \tag{3.3.10}$$

where $\sigma \in \{1, -1\}$, $p > 0$, and $\omega \in \mathbb{R}$. Recall from Lemma 3.8 that if ϕ is the localized mode, then

$$\int_{\mathbb{R}} V'(x)\phi^2(x)dx = 0.$$

This constraint is trivially satisfied if $V(-x) = V(x)$ and $\phi(-x) = \phi(x)$.

If $V(-x) = V(x)$ and $\phi(x)$ is a solution of the stationary Gross–Pitaevskii equation (3.3.10), then $\phi(-x)$ is another solution of the same equation. This suggests that we can look for a reversible localized mode $\phi(x)$ which is only defined for $x \leq 0$ with the conditions that $\lim_{x \to -\infty} \phi(x) = 0$ and $\phi'(0) = 0$. This mode is extended to $x \geq 0$ using the symmetric reflection of $\phi(x)$ about the center of symmetry $x = 0$. Because one initial condition $\phi(0)$ serves as a parameter to be determined from the condition $\lim_{x \to -\infty} \phi(x) = 0$, this method is referred to as the *shooting method*: we change the shooting parameter $\phi(0)$ to shoot along the solution with $\phi'(0) = 0$ and $\lim_{x \to -\infty} \phi(x) = 0$.

If the potential $V(x)$ has another point of symmetry on the interval $[0, 2\pi]$, e.g. $V(x)$ has a minimum at $x = 0$ and a maximum at $x = \pi$,

$$V(-x) = V(x) \quad \text{and} \quad V(2\pi - x) = V(x),$$

then the same method allows us to construct another reversible localized mode $\phi(x)$ such that $\phi(2\pi - x) = \phi(x)$ with the reversibility condition $\phi'(\pi) = 0$. These two distinct localized modes ϕ of the Gross–Pitaevskii equation (3.3.10) with a symmetric periodic potential V have been obtained in the method of Lyapunov–Schmidt reductions near the band edges (Section 3.2.2) and far from the band edges (Section 3.2.3).

We will now develop a global proof of the existence of these two distinct localized modes for any $\omega < \inf_{x \in \mathbb{R}} V(x)$ in the case $\sigma = -1$. Although these results do not cover the entire existence interval $\omega < \inf \sigma(L)$, where $L = -\partial_x^2 + V(x)$, they still show robustness of the two localized modes with different spatial symmetries.

Theorem 3.11 *Let* $\sigma = -1$, $V(-x) = V(x)$ *for all* $x \in \mathbb{R}$, *and* $\omega < \inf_{x \in \mathbb{R}} V(x)$. *There exists a strong solution* $\phi \in H^2(\mathbb{R})$ *of the stationary equation (3.3.10) such that* $\phi(-x) = \phi(x)$ *for all* $x \in \mathbb{R}$.

Proof If $\omega < \inf_{x \in \mathbb{R}} V(x) \leq \inf \sigma(L)$, the linear problem

$$-\phi''(x) + V(x)\phi(x) = \omega\phi(x)$$

has two real Floquet multipliers (Section 2.1.1), one of which is inside the unit circle and the other one is outside. By the Unstable Manifold Theorem (Appendix B.13), there exists a one-parameter family of solutions of the stationary equation (3.3.10) such that

$$\phi_C(x) \sim Cw_+(x)e^{\kappa x} \quad \text{as} \quad x \to -\infty,$$

where $\kappa > 0$ is the characteristic exponent, $w_+(x)$ is a 2π-periodic bounded eigenfunction, and $C > 0$ is parameter. Since $\omega < \inf_{x \in \mathbb{R}} V(x)$, one can normalize $w_+ \in C_{\text{per}}([0, 2\pi])$ such that there is at least one point $x_0 \in \mathbb{R}$ for large negative x_0 such that $\phi_C(x_0) > 0$ and $\phi_C'(x_0) > 0$.

For all $\phi \in (0, \phi_1)$ with $\phi_1 = (\inf_{x \in \mathbb{R}} V(x) - \omega)^{\frac{1}{2p}}$, we have

$$\phi''(x) = (V(x) - \omega)\phi(x) - \phi^{2p+1}(x) \geq (\phi_1^{2p} - \phi^{2p})\phi > 0.$$

Therefore, $\phi_C(x)$ remains an increasing function of x for large negative $x \in \mathbb{R}$ as long as $0 < \phi_C(x) < \phi_1$. Therefore, there is $x_1(C) \in \mathbb{R}$ such that $\phi(x_1(C)) = \phi_1$. The map $C \mapsto x_1$ is continuous for any $C > 0$ (thanks to the existence of a C^1

invariant unstable manifold by the Unstable Manifold Theorem) with the following two limits:

$$x_1(C) \to -\infty \quad \text{as} \quad C \to \infty, \quad x_1(C) \to +\infty \quad \text{as} \quad C \to 0.$$

On the other hand, for all $\phi > \phi_2$ with $\phi_2 = (\sup_{x \in \mathbb{R}} V(x) - \omega)^{\frac{1}{2p}} > \phi_1$, we have

$$\phi''(x) = (V(x) - \omega)\phi(x) - \phi^{2p+1}(x) \le (\phi_2^{2p} - \phi^{2p})\phi < 0.$$

Therefore, there is a turning point $x_2(C) > x_1(C)$ such that $\phi_C'(x_2(C)) = 0$ with the limits,

$$x_2(C) \to -\infty \quad \text{as} \quad C \to \infty, \quad x_2(C) \to +\infty \quad \text{as} \quad C \to 0.$$

Continuity of the map $C \mapsto x_2$ implies that there exists a value $0 < C_0 < \infty$ such that $x_2(C_0) = 0$. We have hence constructed a trajectory $\phi_{C_0}(x)$ such that $\lim_{x \to -\infty} \phi_{C_0}(x) = 0$ and $\phi_{C_0}'(0) = 0$. After reflection in x, this trajectory gives the symmetric localized mode $\phi \in H^2(\mathbb{R})$. □

Note that the localized mode $\phi_C(x)$ is monotonically increasing for all $x < 0$ and decreasing for all $x > 0$. If $\inf_{x \in \mathbb{R}} V(x) < \omega < \inf \sigma(L)$, then $\phi_C''(x)$ cannot be controlled for small positive ϕ and the family of solutions $\phi_C(x)$ may be non-monotonic. As a result, it is not possible to control the map $C \mapsto x_1$, at which the trajectory $\phi_C(x)$ leaves the neighborhood of $\phi = 0$. It is also impossible to uniquely define the turning point $x_2(C)$, where $\phi_C'(x_2(C)) = 0$, because of these oscillations.

Figure 3.3 shows two solutions of the stationary Gross–Pitaevskii equation (3.3.10) with $p = 1$, $\sigma = -1$, and $V(x) = 1 - \cos(x)$ for $\omega = -0.2$ (top) and $\omega = 0.2$ (bottom). In the first case, $\phi(x)$ is monotonic for $x < 0$ and $x > 0$ because $\omega = -0.2 < \inf_{x \in \mathbb{R}} V(x) = 0$. In the second case, $\phi(x)$ has no monotonic behavior for $x < 0$ or $x > 0$.

Results of Sections 3.2.2 and 3.2.3 suggest that a similar structure of two distinct localized modes of the stationary Gross–Pitaevskii equation may exist for non-symmetric potentials $V(x)$ but the reversibility symmetry cannot be used to guarantee the existence of the two localized modes. In particular, one cannot rule out situations when the two localized modes disappear in the saddle-node bifurcation or additional localized modes coexist with the two fundamental modes.

Exercise 3.24 Consider the stationary generalized NLS equation,

$$-\phi''(x) - f'(\phi^2)\phi(x) = \omega\phi(x), \quad x \in \mathbb{R},$$

where $f(\phi^2) : \mathbb{R}_+ \to \mathbb{R}$ is C^1 with $f'(0) = 0$ and $f'(x) > 0$ for $x > 0$, and prove the existence of a localized mode for any $\omega < 0$ from the phase-plane analysis.

3.3.3 Shooting methods for the stationary DNLS equation

Consider the stationary DNLS equation with the power nonlinearity,

$$-\Delta\phi_n + \sigma|\phi_n|^{2p}\phi_n = \omega\phi_n, \quad n \in \mathbb{Z}, \tag{3.3.11}$$

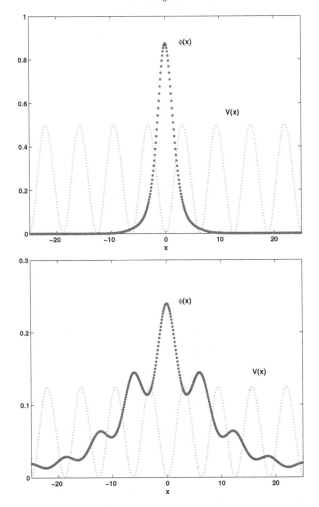

Figure 3.3 The numerical approximation of the localized mode $\phi(x)$ and the periodic potential $V(x)$ for $\omega = -0.2$ (top) and $\omega = 0.2$ (bottom).

where $\Delta\phi_n := \phi_{n+1} - 2\phi_n + \phi_{n-1}$, $\sigma \in \{1, -1\}$, $p > 0$, and $\omega \in \mathbb{R}$. Since $\sigma(-\Delta) \in [0, 4]$, the localized mode with an exponential decay rate may only exist either for $\omega < 0$ or for $\omega > 4$. Because of the staggering transformation (Section 3.2.5), it is sufficient to consider the case $\sigma = -1$, for which no localized modes exist for $\omega > 4$ (Lemma 3.10).

All localized modes $\phi \in l^2(\mathbb{Z})$ are real-valued in the space of one dimension modulo multiplication by $e^{i\theta}$ for any $\theta \in \mathbb{R}$ (Lemma 3.11). Therefore, the stationary DNLS equation (3.3.11) with $\sigma = -1$ can be written in the form

$$-\Delta\phi_n - \phi_n^{2p+1} = \omega\phi_n, \quad n \in \mathbb{Z}. \tag{3.3.12}$$

We shall prove that there exists at least two families of positive localized modes of the stationary DNLS equation (3.3.12) for any $\omega < 0$ and any $p > 0$. One localized mode, known as the *site-symmetric* soliton (or the *on-site* soliton) has the property

$$\phi_{-n} = \phi_n, \quad n \in \mathbb{Z}, \tag{3.3.13}$$

whereas the other mode, known as the *bond-symmetric* soliton (or the *inter-site* soliton) has the property

$$\phi_{-n} = \phi_{n+1}, \quad n \in \mathbb{Z}. \tag{3.3.14}$$

The first solution is symmetric about the central site $n = 0$, whereas the second solution is symmetric about the midpoint between two adjacent sites $n = 0$ and $n = 1$. The two distinct solutions (3.3.13) and (3.3.14) resemble the two branches of localized modes in the stationary Gross–Pitaevskii equation with the periodic potential (Section 3.3.2) thanks to the correspondence between the two models (Section 2.4.2).

Exercise 3.25   Consider a localized mode in the stationary cubic DNLS equation

$$\phi_{n+1} - 2\phi_n + \phi_{n-1} + \phi_n^3 + \omega\phi_n = 0, \quad n \in \mathbb{Z}.$$

Using a scaling transformation, show that the limit $|\omega| \to 0$ corresponds to the small-amplitude slowly varying localized mode with small $\|\phi\|_{l^2}$, whereas the limit $\omega \to -\infty$ corresponds to the compact localized mode with large $\|\phi\|_{l^2}$.

The stationary DNLS equation (3.3.12) can be simplified in the two limits $|\omega| \to 0$ and $|\omega| \to \infty$ (Exercise 3.25). In the first limit $|\omega| \to 0$, the stationary DNLS equation can be reduced to the continuous stationary NLS equation (Section 5.5),

$$\phi''(x) + \phi^3(x) + \omega\phi(x) = 0, \quad x \in \mathbb{R}. \tag{3.3.15}$$

Both localized modes (3.3.13) and (3.3.14) of the stationary DNLS equation (3.3.11) correspond to the same translationally invariant localized mode of the stationary NLS equation (3.3.15). In the second limit $|\omega| \to \infty$, the two localized modes can be constructed by the Lyapunov–Schmidt reduction methods from the limiting configurations $\phi_n^{(0)} = \delta_{n,0}$ and $\phi_n^{(0)} = \delta_{n,0} + \delta_{n,1}$ (Section 3.2.5).

We shall here exploit dynamical system methods and the symmetry properties of reversible discrete maps to construct localized modes of the stationary DNLS equation (3.3.12) for all values of $\omega < 0$. The technique is based on the work of Qin & Xiao [175].

Let $x_n = \phi_{n-1}$ and $y_n = \phi_n$. The stationary DNLS equation (3.3.12) can be formulated as the reversible second-order discrete map

$$\begin{bmatrix} x_{n+1} \\ y_{n+1} \end{bmatrix} = \begin{bmatrix} y_n \\ 2y_n + f(y_n) - x_n \end{bmatrix}, \tag{3.3.16}$$

where $f(y_n) := -\omega y_n - y_n^{2p+1}$. The discrete map (3.3.16) is reversible with respect to the *involution*

$$\mathbf{R} = \begin{bmatrix} 0 & 1 \\ 1 & 0 \end{bmatrix}$$

in the sense that if the map (3.3.16) is represented by

$$\mathbf{x}_{n+1} = \mathbf{T}(\mathbf{x}_n), \quad \mathbf{x}_n = \begin{bmatrix} x_n \\ y_n \end{bmatrix}, \quad \mathbf{T}(\mathbf{x}_n) = \begin{bmatrix} y_n \\ 2y_n + f(y_n) - x_n \end{bmatrix},$$

then

$$\mathbf{R}\mathbf{x}_{n+1} = \left(\mathbf{R}\mathbf{T}\mathbf{R}^{-1}\right)\mathbf{R}\mathbf{x}_n = \mathbf{T}^{-1}\mathbf{R}\mathbf{x}_n, \quad \mathbf{T}^{-1}(\mathbf{x}_n) = \begin{bmatrix} 2x_n + f(x_n) - y_n \\ x_n \end{bmatrix}.$$

Fixed points of the involution $\mathbf{R}\mathbf{x} = \mathbf{x}$ belong to the one-dimensional set $x = y$, which corresponds to the symmetry constraint $\phi_0 = \phi_1$ for the two consequent nodes $n = 0$ and $n = 1$. Therefore the involution $\mathbf{R}$ is useful to generate the bond-symmetric localized mode (3.3.14).

To generate the site-symmetric localized mode (3.3.13), we can see that the map $\mathbf{T}$ is also reversible with respect to the involution $\mathbf{R}\mathbf{T}$, where the fixed points of $\mathbf{R}\mathbf{T}$ belong to the one-dimensional set $2x = 2y + f(y)$, which corresponds to the symmetry constraint $\phi_{-1} = \phi_1$ for the two nodes adjacent to the central site $n = 0$.

Exercise 3.26 Consider the involution

$$\tilde{\mathbf{R}} = \begin{bmatrix} 0 & -1 \\ -1 & 0 \end{bmatrix}$$

and show that the fixed points of the involutions $\tilde{\mathbf{R}}$ and $\tilde{\mathbf{R}}\mathbf{T}$ may be useful to generate two kink solutions $\{\phi_n\}_{n \in \mathbb{Z}}$ such that $\lim_{n \to \pm\infty} \phi_n = \pm\phi_\infty$, where $f(\phi_\infty) = 0$ for some $\phi_\infty > 0$. The two kink solutions associated with the fixed points of $\tilde{\mathbf{R}}$ and $\tilde{\mathbf{R}}\mathbf{T}$ have the symmetries

$$\phi_{-n} = -\phi_{n+1} \quad \text{and} \quad \phi_{-n} = -\phi_n \quad \text{for all} \quad n \in \mathbb{Z}.$$

We shall construct the localized mode ϕ of the stationary equation (3.3.12) as a homoclinic orbit of the second-order discrete map (3.3.16), which is an intersection of discrete trajectories along stable and unstable manifolds from the hyperbolic equilibrium point $\mathbf{x} = \mathbf{0}$. If an unstable trajectory for all $n \leq 0$ intersects with the set of fixed points of the involution $\mathbf{R}$ (or those of the involution $\mathbf{R}\mathbf{T}$) at $n = 0$, then the stable trajectory can be constructed for all $n > 0$ by the symmetric reflection of the unstable trajectory, thanks to the reversibility of the map $\mathbf{T}$ with respect to $\mathbf{R}$ (or $\mathbf{R}\mathbf{T}$). This construction gives the following theorem.

Theorem 3.12 *Assume that $f(x)$ is an odd C^1 function with $f'(0) > 0$ that has only one positive zero at x_* with $f(x) < 0$ for some $x > x_*$. Then, the second-order discrete map (3.3.16) has a homoclinic orbit symmetric with respect to the involution $\mathbf{R}$ and a homoclinic orbit symmetric with respect to the involution $\mathbf{R}\mathbf{T}$.*

Proof It is sufficient to consider a homoclinic orbit symmetric with respect to the involution $\mathbf{R}$, that is, with respect to the diagonal $y = x$ on the phase plane (x, y). Let $W^u(0)$ denote the discrete trajectory along the unstable manifold from the hyperbolic fixed point $(0, 0)$, thanks to $f'(0) > 0$. The linearized trajectory $E^u(0)$ is located at the straight line $y = \lambda x$ with $\lambda > 1$ and $W^u(0)$ is tangent to E^u near the origin $(0, 0)$ by the Unstable Manifold Theorem (Appendix B.13).

The proof consists of two steps. In the first step, we need to show that trajectory $W^u(0)$ intersects the horizontal line segment $y = x_*$ for $x \in (0, x_*)$, where x_* is the only positive zero of $f(x)$. This is obvious from the property that $f(x) > 0$ for all $x \in (0, x_*)$, which gives $y_n > x_n$ as long as $x_n \in (0, x_*)$. In the second step, we need to show that $W^u(0)$ intersects the diagonal $y = x$ for some $x > x_*$. This follows from the condition that $f(x) < 0$ for all $x > x_*$. $\square$

Exercise 3.27 Show that if $f(x)$ is an odd C^1 function with $f'(0) < 0$ that has only one positive zero x_* with $f'(x_*) > 0$, then the second-order discrete map (3.3.16) has a heteroclinic orbit from $-x_*$ to x_*, which is symmetric with respect to the involution $\tilde{\mathbf{R}}$.

For the stationary DNLS equation (3.3.12), we have $f(x) = -\omega x - x^{2p+1}$. The zero solution $\boldsymbol{\phi} = \mathbf{0}$ is a hyperbolic point either for $\omega < 0$ or for $\omega > 4$ but we have seen that no localized modes exist for $\omega > 4$. For any $\omega < 0$ and $p > 0$, we have $f'(0) = -\omega > 0$ and there exists exactly one positive zero $x_* = |\omega|^{1/2p}$ of $f(x)$. By Theorem 3.12, the stationary DNLS equation (3.3.12) with $\omega < 0$ and $p > 0$ admits the localized modes (3.3.13) and (3.3.14).

Exercise 3.28 Consider the saturable DNLS equation

$$\phi_{n+1} - 2\phi_n + \phi_{n-1} + \gamma \frac{|\phi_n|^2 \phi_n}{1 + |\phi_n|^2} + \omega \phi_n = 0, \quad n \in \mathbb{Z}$$

for $\gamma > 0$ and show that the localized modes are real and exist for any $\omega \in (-\gamma, 0)$.

3.3.4 Exact solutions of the nonlinear Dirac equations

Consider the stationary nonlinear Dirac equations,

$$\begin{cases} iu'(x) + \omega u(x) + v(x) = \partial_{\bar{u}} W(u, \bar{u}, v, \bar{v}), \\ -iv'(x) + \omega v(x) + u(x) = \partial_{\bar{v}} W(u, \bar{u}, v, \bar{v}), \end{cases} \quad x \in \mathbb{R}, \qquad (3.3.17)$$

where $(u, v) \in \mathbb{C}^2$ and $W(u, \bar{u}, v, \bar{v}) : \mathbb{C}^4 \to \mathbb{R}$ can be written in the form (Section 1.2.3)

$$W = W(|u|^2 + |v|^2, |u|^2|v|^2, u\bar{v} + v\bar{u}). \qquad (3.3.18)$$

Exercise 3.29 Show that a localized mode of the stationary equations (3.3.17) is a critical point of the energy functional,

$$E_\omega(u, \bar{u}, v, \bar{v}) = H(u, \bar{u}, v, \bar{v}) - \omega Q(u, \bar{u}, v, \bar{v}),$$

where H and Q are the Hamiltonian and power invariants of the nonlinear Dirac equations (Section 1.2.3). Prove that $E_\omega(u, \bar{u}, v, \bar{v})$ is not bounded either from above or from below in a neighborhood of $u = v = 0$.

It is difficult to prove existence of critical points of the sign-indefinite energy functional $E_\omega(u, \bar{u}, v, \bar{v})$ using variational methods. Some results in this direction are reported by Esteban & Séré [54]. We can still find exact analytical solutions of the stationary nonlinear Dirac equations (3.3.17) using elementary integration techniques.

Lemma 3.13 *Assume that there exists a localized mode (u, v) of system (3.3.17) with (3.3.18) that is continuous for all $x \in \mathbb{R}$ and decays to zero as $|x| \to \infty$. Then*

$$u(x) = \bar{v}(x) \quad \text{for all} \quad x \in \mathbb{R} \tag{3.3.19}$$

up to the multiplication of (u, v) by $e^{i\alpha}$, $\alpha \in \mathbb{R}$.

Proof If W satisfies (3.3.18), then

$$(u\partial_u - \bar{u}\partial_{\bar{u}} + v\partial_v - \bar{v}\partial_{\bar{v}}) W(|u|^2 + |v|^2, |u|^2|v|^2, u\bar{v} + v\bar{u}) = 0. \tag{3.3.20}$$

System (3.3.17), symmetry (3.3.20), and the decay of (u, v) to zero as $|x| \to \infty$ imply that

$$\frac{d}{dx}\left(|u|^2 - |v|^2\right) = 0, \quad \Rightarrow \quad |u|^2 - |v|^2 = 0, \quad x \in \mathbb{R}. \tag{3.3.21}$$

Using polar form for $(u, v) \in \mathbb{C}^2$ with $|u| = |v|$, we write

$$\begin{cases} u(x) = \sqrt{Q(x)}e^{i\Theta(x)+i\Phi(x)}, \\ v(x) = \sqrt{Q(x)}e^{-i\Theta(x)+i\Phi(x)}, \end{cases} \tag{3.3.22}$$

where the three real-valued functions Q, Θ, and Φ satisfy the following system of differential equations

$$\begin{cases} iQ' - 2Q(\Theta' + \Phi') + 2\omega Q + 2Qe^{-2i\Theta} = 2\bar{u}\partial_{\bar{u}}W, \\ -iQ' - 2Q(\Theta' - \Phi') + 2\omega Q + 2Qe^{2i\Theta} = 2\bar{v}\partial_{\bar{v}}W. \end{cases} \tag{3.3.23}$$

Separating the real parts, we obtain

$$\begin{cases} Q(\cos(2\Theta) + \omega - \Theta' - \Phi') = \operatorname{Re}(\bar{u}\partial_{\bar{u}}W), \\ Q(\cos(2\Theta) + \omega - \Theta' + \Phi') = \operatorname{Re}(\bar{v}\partial_{\bar{v}}W), \\ Q' = 2Q\sin(2\Theta) + 2\operatorname{Im}(\bar{u}\partial_{\bar{u}}W), \\ Q' = 2Q\sin(2\Theta) - 2\operatorname{Im}(\bar{v}\partial_{\bar{v}}W), \end{cases} \tag{3.3.24}$$

where the last equation is redundant thanks to the gauge symmetry (3.3.20). On the other hand, it follows from representation (3.3.18) that

$$(u\partial_u + \bar{u}\partial_{\bar{u}} - v\partial_v - \bar{v}\partial_{\bar{v}}) W(|u|^2 + |v|^2, |u|^2|v|^2, u\bar{v} + v\bar{u})|_{|u|=|v|} = 0. \tag{3.3.25}$$

As a result, the first two equations of system (3.3.24) tell us that

$$\Phi'(x) = 0 \quad \Rightarrow \quad \Phi(x) = \Phi_0, \quad x \in \mathbb{R}.$$

If $\Phi_0 = 0$, the localized mode enjoys reduction (3.3.19), whereas if $\Phi_0 \in \mathbb{R}$, the reduction holds up to the multiplication of (u, v) by $e^{i\Phi_0}$. $\qquad \square$

If we set $\Phi_0 = 0$, then system (3.3.24) is rewritten in the equivalent form

$$\begin{cases} Q' = 2Q\sin(2\Theta) + 2\operatorname{Im}(\bar{u}\partial_{\bar{u}}W), \\ Q\Theta' = \omega Q + Q\cos(2\Theta) - \operatorname{Re}(\bar{u}\partial_{\bar{u}}W). \end{cases} \tag{3.3.26}$$

The following result shows that the Hamiltonian system (3.3.26) has the standard symplectic structure.

Corollary 3.1 *Let* $u = \bar{v} = \sqrt{Q}e^{i\Theta}$. *The stationary nonlinear Dirac equations* *(3.3.17) reduce to the planar Hamiltonian system,*

$$\frac{d}{dx} \begin{pmatrix} p \\ q \end{pmatrix} = \begin{pmatrix} 0 & -1 \\ +1 & 0 \end{pmatrix} \nabla h(p,q), \qquad (3.3.27)$$

where $p = 2\Theta$, $q = Q$, *and* $h(p,q) = \tilde{W}(p,q) - 2q\cos(p) - 2\omega q$ *with* $\tilde{W}(p,q) = W(2q, q^2, 2q\cos(p))$.

Proof System (3.3.26) is equivalent to the Hamiltonian system (3.3.27) if

$$\begin{cases} \partial_p \tilde{W}(p,q) = \mathrm{i}\,(u\partial_u - \bar{u}\partial_{\bar{u}})\,W(|u|^2 + |v|^2, |u|^2|v|^2, u\bar{v} + v\bar{u}), \\ q\partial_q \tilde{W}(p,q) = (u\partial_u + \bar{u}\partial_{\bar{u}})\,W(|u|^2 + |v|^2, |u|^2|v|^2, u\bar{v} + v\bar{u}), \end{cases} \qquad (3.3.28)$$

which follows from the chain rule. $\qquad\qquad\square$

The system of stationary equations (3.3.17) is of the fourth order. Although it also has the standard symplectic structure with a Hamiltonian function, it is not integrable unless another constant of motion exists. We are lucky to have this additional conserved quantity (3.3.21), which enables us to convert system (3.3.17) to the planar Hamiltonian system (3.3.27). The word "planar" means that the system can be analyzed on the phase plane (p,q), where we can draw all trajectories from the level set of the Hamiltonian function

$$h(p,q) = \tilde{W}(p,q) - 2q\cos(p) - 2\omega q = h_0.$$

The particular trajectory with $h(p,q) = 0$ gives the localized mode of the stationary nonlinear Dirac equations (3.3.17).

Exercise 3.30 Plot trajectories on the phase plane (p,q) for

$$W = \frac{1}{2}\left(|u|^4 + 4|u|^2|v|^2 + |v|^4\right)$$

and show the trajectory for the localized mode.

Explicit solutions of the planar Hamiltonian system (3.3.27) can be found if W is a homogeneous polynomial in its variables. These explicit solutions are obtained by Chugunova & Pelinovsky [35].

Exercise 3.31 Consider the stationary NLS equation with power nonlinearity,

$$\phi''(x) + (p+1)\phi^{2p+1}(x) + \omega\phi(x) = 0, \quad x \in \mathbb{R}, \quad p > 0,$$

and find the explicit solution for a localized mode $\phi(x)$.

Let W be a homogeneous polynomial of degree $2n \geq 4$. The function $\tilde{W}(p,q)$ can be parameterized in the form

$$\tilde{W}(p,q) = q^n \sum_{s=0}^{n} A_s \cos(sp), \qquad (3.3.29)$$

where $\{A_s\}_{s=0}^n$ is a set of real-valued coefficients. System (3.3.27) can be rewritten in the explicit form

$$\begin{cases} q' = 2q\sin(p) - q^n \sum_{s=0}^n sA_s \sin(sp), \\ p' = 2\omega + 2\cos(p) - nq^{n-1} \sum_{s=0}^n A_s \cos(sp). \end{cases} \quad (3.3.30)$$

The first integral of system (3.3.30) is now written in the form

$$h(p, q) = -2\omega q - 2\cos(p)q + q^n \sum_{s=0}^n A_s \cos(sp) = 0, \quad (3.3.31)$$

where the constant of integration is set to zero because $q(x) \to 0$ as $|x| \to \infty$. As a result, the second-order system (3.3.30) reduces to the first-order equation

$$p'(x) = -2(n-1)(\omega + \cos(p)), \quad x \in \mathbb{R}, \quad (3.3.32)$$

whereas the positive function $q(x)$ can be found from $p(x)$ by

$$q^{n-1} = \frac{(\cos(p) + \omega)}{\sum_{s=0}^n A_s \cos(sp)}. \quad (3.3.33)$$

As an example, let us consider the quartic function W,

$$W = \frac{a_1}{2}(|u|^4 + |v|^4) + a_2|u|^2|v|^2 + a_3(|u|^2 + |v|^2)(v\bar{u} + \bar{v}u) + \frac{a_4}{2}(v\bar{u} + \bar{v}u)^2, \quad (3.3.34)$$

where (a_1, a_2, a_3, a_4) are arbitrary parameters. Using (3.3.29) for $n = 2$, we obtain the correspondence

$$A_0 = a_1 + a_2 + a_4, \quad A_1 = 4a_3, \quad A_2 = a_4.$$

Define $t = \cos(p) \in [-1, 1]$ and

$$q(p) = \frac{t + \omega}{\phi(t)}, \quad \phi(t) = a_1 + a_2 + 4a_3 t + 2a_4 t^2.$$

Let us also introduce new parameters for any $\omega \in [-1, 1]$:

$$\beta = \sqrt{1 - \omega^2} \in [0, 1], \quad \mu = \frac{1 + \omega}{1 - \omega} \in \mathbb{R}_+.$$

The first-order equation (3.3.32) with $n = 2$ can be solved using the substitution

$$t = \frac{1 - z^2}{1 + z^2} \quad \Leftrightarrow \quad z^2 = \frac{1 - t}{1 + t}.$$

After integration with the symmetry constraint $p(0) = 0$, we obtain the solution

$$\left| \frac{z - \sqrt{\mu}}{z + \sqrt{\mu}} \right| = e^{2\beta x}, \quad x \in \mathbb{R}. \quad (3.3.35)$$

Two separate cases are considered:

(I) $\quad |z| \le \sqrt{\mu}: \quad z = -\sqrt{\mu}\frac{\sinh(\beta x)}{\cosh(\beta x)}, \quad t = \frac{\cosh^2(\beta x) - \mu\sinh^2(\beta x)}{\cosh^2(\beta x) + \mu\sinh^2(\beta x)}, \quad (3.3.36)$

and

(II) $\quad |z| \ge \sqrt{\mu}: \quad z = -\sqrt{\mu}\frac{\cosh(\beta x)}{\sinh(\beta x)}, \quad t = \frac{\sinh^2(\beta x) - \mu\cosh^2(\beta x)}{\sinh^2(\beta x) + \mu\cosh^2(\beta x)}. \quad (3.3.37)$

Let us introduce new parameters

$$A = a_1 + a_2 - 4a_3 + 2a_4,$$
$$B = 2(a_1 + a_2) - 4a_4,$$
$$C = a_1 + a_2 + 4a_3 + 2a_4.$$

It is clear that $A = \phi(-1)$ and $C = \phi(1)$.

In case (I), we have $t + \omega \geq 0$ and $\phi(t) \geq 0$, so that

$$q(x) = \frac{(1 + \omega)((\mu + 1) \cosh^2(\beta x) - \mu)}{(A\mu^2 + B\mu + C) \cosh^4(\beta x) - (B\mu + 2A\mu^2) \cosh^2(\beta x) + A\mu^2}. \tag{3.3.38}$$

In case (II), we have $t + \omega \leq 0$ and $\phi(t) \leq 0$, so that

$$q(x) = \frac{(1 + \omega)(1 - (\mu + 1) \cosh^2(\beta x))}{(A\mu^2 + B\mu + C) \cosh^4(\beta x) - (B\mu + 2C) \cosh^2(\beta x) + C}. \tag{3.3.39}$$

The asymptotic behavior of the function $q(x)$ as $|x| \to \infty$ depends on the location of the zeros of the function $\psi(\mu) = A\mu^2 + B\mu + C$.

Case $A < 0$, $C > 0$: The quadratic polynomial $\phi(t)$ has exactly one root at $t_1 \in (-1, 1)$. Two branches of decaying solutions with positive amplitude $q(x)$ exist. One branch occurs for $-1 \leq \omega \leq -t_1$ and the other one occurs for $-t_1 \leq \omega \leq 1$. At the point $\omega = -t_1$, the solution is bounded and decaying.

Case $A > 0$, $C > 0$: The quadratic polynomial $\phi(t)$ has no roots or two roots on $(-1, 1)$. If $\phi(t)$ does not have any roots on $(-1, 1)$, a decaying solution with positive amplitude $q(x)$ exists for any $\omega \in (-1, 1)$. If $\phi(t)$ has two roots at $t_1, t_2 \in (-1, 1)$, a decaying solution exists only on the interval $-1 \leq \omega \leq \max(t_1, t_2)$. At the point $\omega = -\max(t_1, t_2)$, the solution becomes bounded but non-decaying if $t_1 \neq t_2$ and unbounded if $t_1 = t_2$.

Case $A < 0$, $C < 0$: The quadratic polynomial $\phi(t)$ has no roots or two roots on $(-1, 1)$. If $\phi(t)$ does not have any roots on $(-1, 1)$, a decaying solution with positive amplitude $q(x)$ exists for any $\omega \in (-1, 1)$. If $\phi(t)$ has two roots at $t_1, t_2 \in (-1, 1)$, a decaying solution exists only on the interval $-\min(t_1, t_2) \leq \omega \leq 1$. At the point $\omega = -\min(t_1, t_2)$, the solution becomes bounded but non-decaying if $t_1 \neq t_2$ and unbounded if $t_1 = t_2$.

Case $A > 0$, $C < 0$: No decaying solutions with positive amplitude $q(x)$ exist.

If $a_1 = 1$, $a_2 = 2$, and $a_3 = a_4 = 0$, we obtain the explicit solution for the localized mode,

$$u(x) = \frac{\sqrt{2(1 + \omega)}}{\sqrt{3} \left[\cosh(\beta x) + i\sqrt{\mu} \sinh(\beta x)\right]}. \tag{3.3.40}$$

When $\omega \to -1$ (that is, $\mu \to 0$ and $\beta \to 0$), the localized mode (3.3.40) is small in amplitude and broad in width. In this case, $u(x) \sim \operatorname{sech}(\beta x)$, which shows the similarity of the localized mode with the soliton of the stationary NLS equation.

When $\omega \to 1$ (that is, $\mu \to \infty$ and $\beta \to 0$), the localized mode (3.3.40) is finite in amplitude and width and approaches the algebraically decaying solution,

$$\omega = 1: \quad u(x) = \frac{2}{\sqrt{3}(1 + 2ix)}.$$

The explicit solution (3.3.40) is shown on Figure 3.4 for three values of ω.

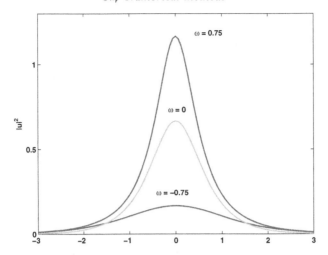

Figure 3.4 The graph of $|u|^2$ versus x in the explicit solution (3.3.40) for three values of parameter ω.

Exercise 3.32   Find the explicit solution for a localized mode if W is given by (3.3.34) with $a_1 = a_2 = a_4 = 0$ and $a_3 = 1$.

Exercise 3.33   Consider the stationary quintic nonlinear Dirac equations with the potential function

$$W = (|u|^2 + |v|^2)|u|^2|v|^2$$

and obtain the explicit solution for a localized mode.

3.4 Numerical methods

If none of the analytical methods help in the construction of localized modes, we have to rely upon numerical approximations. In one spatial dimension, a variety of numerical methods for finding localized modes are available, with the shooting and Newton's methods being the most widely used ones. However, in two and three spatial dimensions, iteration methods based on spectral (Fourier) representations are more desirable.

An effective iterative method for computing solitary wave solutions in a space of two dimensions was proposed by Petviashvili in the pioneer paper [168]. Although this method has been applied to numerous nonlinear problems in physics since the 1970s, it was only recently when this method got its second birth among applied mathematicians. First, the convergence analysis for the canonical Petviashvili method was developed by Pelinovsky & Stepanyants [160] and then used in the works of Demanet & Schlag [45] and Chugunova & Pelinovsky [34]. Second, various modifications of the Petviashvili and similar methods were proposed by Lakoba and Yang [125, 126, 216, 217].

We shall review the most important iterative procedures, spectral renormalization and imaginary-time integration, in the context of the nonlinear stationary equation

$$(c + L)\phi = \phi^p, \tag{3.4.1}$$

where parameters $p > 1$ and $c > 0$ are fixed and $L : \text{Dom}(L) \subseteq L^2 \to L^2$ is a non-negative self-adjoint operator. We are looking for strong solutions $\phi \in \text{Dom}(L)$ of equation (3.4.1).

In the context of the stationary focusing Gross–Pitaevskii equation,

$$-\Delta\phi + V\phi - |\phi|^2\phi = \omega\phi, \quad x \in \mathbb{R}^d, \tag{3.4.2}$$

we can think of the real-valued solutions $\phi(x) : \mathbb{R}^d \to \mathbb{R}$ for $\omega < \omega_0 := \inf \sigma(-\Delta + V)$. These solutions correspond to localized modes in the semi-infinite gap of $\sigma(-\Delta + V)$. Comparison with equation (3.4.1) gives $p = 3$, $c = \omega_0 - \omega > 0$, and $L := -\Delta + V - \omega_0$. Note that L is the Schrödinger operator with a periodic potential, which is self-adjoint in $L^2(\mathbb{R}^d)$ with $\text{Dom}(L) = H^2(\mathbb{R}^d)$ (Section 2.1). Thanks to the choice of ω_0, it is a non-negative operator in the sense of $\langle Lu, u \rangle_{L^2} \geq 0$.

3.4.1 Spectral renormalization

Since L is a non-negative operator, then $(c + L)$ is an invertible operator on L^2 for any fixed $c > 0$. As a result, the stationary equation (3.4.1) is equivalent to the fixed-point problem $\phi = T(\phi)$ for the nonlinear operator T given by

$$T(u) := (c + L)^{-1}u^p.$$

We would like to consider a sequence of functions $\{u_n\}_{n\in\mathbb{N}}$ in $\text{Dom}(L)$ and define the iterative rule $u_{n+1} = T(u_n)$, $n \in \mathbb{N}$ in a hope that the sequence converges to a fixed point ϕ of the nonlinear operator T in $\text{Dom}(L)$. In other words, after writing the stationary equation (3.4.1) as the fixed-point problem $\phi = T(\phi)$, we can propose a naive iterative algorithm,

$$u_{n+1} = (c + L)^{-1}u_n^p, \quad n \in \mathbb{N}. \tag{3.4.3}$$

However, this algorithm diverges for any $p > 1$ as can be seen from a simple sequence $u_n = a_n\phi$, where ϕ is a solution of the stationary equation (3.4.1) and the numerical sequence $\{a_n\}_{n\in\mathbb{N}}$ satisfies the power map,

$$a_{n+1} = a_n^p, \quad n \in \mathbb{N}.$$

The nonzero fixed point $a_* = 1$ of the power map is unstable for any $p > 1$.

Therefore, we need to renormalize the iterative rule T. Such renormalizations are sometimes referred to as the pre-conditioning of the iterative algorithms. Let us observe that if $\phi \in \text{Dom}(L)$ is a solution of the stationary equation (3.4.1), then it is true that

$$\langle (c + L)\phi, \phi \rangle_{L^2} = \langle \phi^p, \phi \rangle_{L^2}. \tag{3.4.4}$$

The quotient between the left and right sides of (3.4.4) is inversely proportional to the $(p-1)$th power of ϕ and this inverse power can compensate the divergence of the power map (3.4.3). As a result, let us define the new iterative rule by

$$u_{n+1} = M_n^\gamma \, (c+L)^{-1} \, u_n^p, \tag{3.4.5}$$

where $\gamma > 0$ is a parameter and

$$M_n \equiv M(u_n) := \frac{\langle (c+L)u_n, u_n \rangle_{L^2}}{\langle u_n^p, u_n \rangle_{L^2}}. \tag{3.4.6}$$

Since we modified the definition of the iterative rule, it is important to check that the fixed points of the iteration map (3.4.5)–(3.4.6) are still equivalent to the strong solutions of the stationary equation (3.4.1).

Lemma 3.14   *Assume $\gamma \neq 1 + 2n$ for any $n \in \mathbb{N}$. A set of fixed points of the iteration map (3.4.5)–(3.4.6) is equivalent to a set of strong solutions of the stationary equation (3.4.1).*

Proof   Let $u_n = \phi$ be a solution of the stationary equation (3.4.1) for some $n \geq 1$. Then, it follows from (3.4.6) that $M_n = 1$ and from (3.4.5) that $u_{n+1} = \phi$. Therefore, ϕ is a fixed point of the iteration map (3.4.5)–(3.4.6).

   In the other direction, let u_* be a fixed point of the iteration map (3.4.5)–(3.4.6). Multiplying (3.4.5) by $(c+L)u_*$ and integrating over $x \in \mathbb{R}^d$, we find that $M_* = M_*^\gamma$. When $\gamma \neq 1 + 2n$, $n \in \mathbb{N}$, there exist only two real-valued solutions of

$$M_*(1 - M_*^{\gamma-1}) = 0,$$

either $M_* = 0$ or $M_* = 1$. Since $(c+L)$ is strictly positive, the former solution is equivalent to a zero solution $u_* \equiv 0$. The nonzero fixed point of (3.4.5) with $M_* = 1$ satisfies the stationary equation (3.4.1), so that $u_* = \phi$. $\qquad\square$

   After we established that the iterative method (3.4.5)–(3.4.6) has the same set of fixed points as the stationary equation (3.4.1), we can now study if a sequence of iterations $\{u_n\}_{n \geq 1}$ converges to a nonzero fixed point ϕ in $\mathrm{Dom}(L)$. The following theorem gives the main result on convergence of the iterative method (3.4.5)–(3.4.6).

Theorem 3.13   *Let $L : \mathrm{Dom}(L) \subseteq L^2 \to L^2$ be a non-negative self-adjoint operator. Let $\phi \in \mathrm{Dom}(L)$ be a strong solution of the stationary equation (3.4.1). There exists an open neighborhood of $\phi \in \mathrm{Dom}(L)$, in which the sequence of iterations $\{u_n\}_{n \in \mathbb{N}}$ defined by the iterative method (3.4.5)–(3.4.6) with $\gamma \in (1, (p+1)/(p-1))$ converges to ϕ if $H = c + L - p\phi^{p-1} : \mathrm{Dom}(L) \to L^2$ has exactly one negative eigenvalue, no zero eigenvalue, and*

$$\text{either} \quad \phi^{p-1}(x) \geq 0 \quad \text{or} \quad \left| \inf_x \phi^{p-1}(x) \right| < \frac{c}{p}. \tag{3.4.7}$$

Remark 3.9   Because $H\phi = (c+L)\phi - p\phi^p = (1-p)\phi^p$ with $(1-p) < 0$, it is clear that H has at least one negative eigenvalue in L^2.

Remark 3.10 It will be clear from the proof that if H has two negative eigenvalues, the sequence of iterations $\{u_n\}_{n\in\mathbb{N}}$ diverges from the fixed point ϕ in $\mathrm{Dom}(L)$. Condition (3.4.7) gives a sufficient condition for convergence of the iterations and can be replaced by a sharper condition on eigenvalues of operator $(c+L)^{-1}H$.

Remark 3.11 If we formulate the stationary equation as the root of the nonlinear function

$$F(u) = (c+L)u - u^p : \mathrm{Dom}(L) \to L^2,$$

then $H = D_u F(\phi) = c + L - p\phi^{p-1}$ is the Jacobian operator for the nonlinear function $F(u)$.

Proof A nonlinear operator $\tilde{T}(u) : \mathrm{Dom}(L) \to \mathrm{Dom}(L)$ is defined by

$$\tilde{T}(u) := M^\gamma(u)(c+L)^{-1}u^p.$$

By the assumption of the theorem, we know that there exists a fixed point $\phi \in \mathrm{Dom}(L)$ of T such that $\phi = \tilde{T}(\phi)$. Moreover, $\tilde{T}$ is C^1 with respect to u in $\mathrm{Dom}(L)$. We need to prove that the iteration operator $\tilde{T}$ is a contraction in a small closed ball $B_\delta(\phi) \subset \mathrm{Dom}(L)$ of radius $\delta > 0$ centered at $\phi \in \mathrm{Dom}(L)$, that is, there is $Q \in (0,1)$ such that

$$\forall u, v \in B_\delta(\phi): \quad \|\tilde{T}(u) - \tilde{T}(v)\|_{\mathrm{Dom}(L)} \leq Q\|u - v\|_{\mathrm{Dom}(L)}. \tag{3.4.8}$$

If $v = \phi$ is taken in the contraction property (3.4.8), then $\tilde{T} : B_\delta(\phi) \to B_\delta(\phi)$. By the Banach Fixed-Point Theorem (Appendix B.2), iterations $\{u_n\}_{n\in\mathbb{N}}$ defined by $u_{n+1} = \tilde{T}(u_n)$ for $n \geq 1$ converge to the unique asymptotically stable fixed point ϕ in $B_\delta(\phi)$, that is, to ϕ.

Because $\tilde{T}(u)$ is C^1 with respect to u in $B_\delta(\phi)$, the continuity of $D_u\tilde{T}(u)$ near $u = \phi$ implies that if we can prove that the spectrum of $D_u\tilde{T}(\phi)$ in L^2 is confined inside the unit circle, then the operator $\tilde{T}(u)$ is necessarily a contraction in $B_\delta(\phi)$ for sufficiently small $\delta > 0$.

To study the spectrum of $D_u\tilde{T}(\phi)$ in L^2, we decompose the sequence $\{u_n\}_{n\in\mathbb{N}}$ near the fixed point ϕ in the form $u_n = \phi + v_n$ and neglect terms that are quadratic with respect to v_n. Then, we have

$$M(u_n) = 1 + \frac{(1-p)\langle \phi^p, v_n\rangle_{L^2}}{\langle \phi^p, \phi\rangle_{L^2}} + \mathcal{O}(\|v_n\|_{\mathrm{Dom}(L)}^2)$$

and

$$v_{n+1} = \frac{\gamma(1-p)\langle \phi^p, v_n\rangle_{L^2}}{\langle \phi^p, \phi\rangle_{L^2}}\phi + v_n - (c+L)^{-1}Hv_n + \mathcal{O}(\|v_n\|_{\mathrm{Dom}(L)}^2).$$

Operator $(c+L)^{-1}H = I - p(c+L)^{-1}\phi^{p-1} : \mathrm{Dom}(L) \to \mathrm{Dom}(L)$ is bounded. There is at least one negative eigenvalue $\lambda = 1 - p < 0$ of $(c+L)^{-1}H$ associated with the exact eigenfunction

$$(c+L)^{-1}H\phi = \phi - p(c+L)^{-1}\phi^p = (1-p)\phi.$$

Note that the adjoint operator $H(c+L)^{-1}$ has the exact eigenfunction for the same eigenvalue

$$H(c+L)^{-1}\phi^p = H\phi = (1-p)\phi^p.$$

Let us consider the orthogonal decomposition

$$v_n = \alpha_n \phi + w_n,$$

where α_n is the projection in the direction of the eigenvector ϕ, which is orthogonal to ϕ^p. Therefore, we have

$$\alpha_n = \frac{\langle \phi^p, v_n \rangle_{L^2}}{\langle \phi^p, \phi \rangle_{L^2}},$$

whereas w_n is uniquely defined in the constrained space

$$X_c = \{ w \in \text{Dom}(L) : \quad \langle \phi^p, w \rangle_{L^2} = \langle (c + L)\phi, w \rangle_{L^2} = 0 \}. \tag{3.4.9}$$

Substituting the decomposition for v_n into the linearized iteration equations, we obtain

$$\begin{cases} w_{n+1} = w_n - (c + L)^{-1} H w_n, \\ \alpha_{n+1} = (p - \gamma(p-1))\alpha_n, \end{cases} \quad n \in \mathbb{N}.$$

The sequence $\{\alpha_n\}_{n \in \mathbb{N}}$ converges to zero if and only if $\gamma \in (1, (p+1)/(p-1))$. The sequence $\{w_n\}_{n \in \mathbb{N}} \in \text{Dom}(L)$ converges to zero if and only if all eigenvalues of $(c + L)^{-1} H$ in the constrained space X_c are located in the interval $(0, 2)$.

We need to show that the operator $(c + L)^{-1} H$ in the constrained space X_c has no negative and zero eigenvalue if H in the unconstrained space L^2 has only one negative and no zero eigenvalues. This is done in two steps, which rely on the two abstract results from Section 4.1.

Let P_c be the orthogonal projection operator from $\text{Dom}(L)$ to X_c. We consider operator $H|_{X_c} = P_c H P_c$ restricted in the constrained space X_c.

First, it follows from Theorem 4.1 (Section 4.1.1) that if $\langle H^{-1} \phi^p, \phi^p \rangle_{L^2} < 0$, then $H|_{X_c}$ has one less negative eigenvalue compared to operator H in L^2. This condition is definitely true as $(c + L)$ is strictly positive and

$$\langle H^{-1} \phi^p, \phi^p \rangle_{L^2} = -\frac{1}{p-1} \langle \phi, \phi^p \rangle_{L^2} = -\frac{1}{p-1} \langle (c+L)\phi, \phi \rangle_{L^2} < 0.$$

Second, it follows from Theorem 4.2 (Section 4.1.2) that if $H|_{X_c}$ has no negative and zero eigenvalues, then $(c + L)^{-1} H|_{X_c}$ has no negative and zero eigenvalues for any positive operator $(c + L)^{-1}$.

We now need to eliminate eigenvalues of $(c + L)^{-1} H|_{X_c}$ at and above 2. To estimate the upper bound on the spectrum of $(c + L)^{-1} H|_{X_c}$, we note that

$$\sigma \left((c+L)^{-1} H|_{X_c} \right) - 1 \leq -p \inf_{\|u\|_{L^2}=1} \langle (c+L)^{-1} \phi^{p-1} u, u \rangle_{L^2}. \tag{3.4.10}$$

If $\phi^{p-1}(x) \geq 0$ for all $x \in \mathbb{R}^d$, the right-hand side of (3.4.10) is zero. Otherwise, the right-hand side of (3.4.10) is bounded from above by $(p/c) |\inf_{x \in \mathbb{R}^d} \phi^{p-1}(z)|$. To ensure that $\sigma \left((c+L)^{-1} H|_{X_c} \right) < 2$, we add the condition (3.4.7). $\square$

Remark 3.12 The optimal rate of convergence occurs if $\gamma = p/(p-1)$, in which case the sequence $\{\alpha_n\}_{n \in \mathbb{N}}$ reaches zero already at the first iteration. The convergence rate of the method is still linear since the sequence $\{w_n\}_{n \in \mathbb{N}}$ in X_c converges to zero only exponentially fast.

Remark 3.13 If the stationary equation (3.4.1) has translational symmetry and $\phi(x)$ is symmetric, then operator H has a zero eigenvalue with the eigenvectors given by $\{\partial_{x_j}\phi(x)\}_{j=1}^d$. However, the eigenvectors are antisymmetric and H has a trivial kernel in the space of symmetric functions in $\text{Dom}(L)$.

Exercise 3.34 Consider eigenvalues of the spectral problem

$$Hw = \lambda(c+L)w, \quad w \in X_c$$

where X_c is defined by (3.4.9) and assume that $\phi^{p-1}(x) \geq 0$ for all $x \in \mathbb{R}^d$ and that $(c+L)$ is a positive operator from $\text{Dom}(L)$ to L^2. Prove that there exists an infinite sequence of simple eigenvalues in the interval $\lambda \in (0,1)$, which accumulate to $\lambda = 1$ from the left.

Remark 3.14 The constrained space X_c defined by (3.4.9) appears naturally in the variational method of Section 3.1.1, where ϕ is a minimizer of

$$I(u) := \langle (c+L)u, u \rangle_{L^2}$$

subject to a fixed $J(u) = \|u\|_{L^{p+1}}^{p+1} > 0$. Moreover, the condition $M(\phi) = 1$ at the solution ϕ of the stationary equation (3.4.1) illustrates that $I(\phi) = J(\phi)$.

Exercise 3.35 Assume that $\phi(x)$ is a solution of the stationary equation (3.4.1), substitute $u_n(x) = a_n \phi(x)$ for a sequence $\{a_n\}_{n \in \mathbb{N}}$, and reduce the iteration map (3.4.5)–(3.4.6) to the power map,

$$a_{n+1} = a_n^{p-\gamma(p-1)}, \quad n \in \mathbb{N}.$$

Show that the fixed point $a_* = 1$ is stable if and only if $\gamma \in (1, (p+1)/(p-1))$.

Coming to the computational aspects of the spectral renormalization method (3.4.5)–(3.4.6), let us consider a localized mode in the focusing Gross–Pitaevskii equation (3.4.2) with $d = 1$. For cubic nonlinearity, we can fix $\gamma = \frac{3}{2}$ in (3.4.5).

Besides the theoretical convergence criterion described in Theorem 3.13, there are additional factors in the numerical approximation of $\phi(x)$ that come from the spatial discretization of the continuous derivatives, truncation of the computational interval, and termination of iterations within the given tolerance bound. There are several methods for computation of spatial discretizations of the continuous derivatives. One method is based on the truncation of the Fourier series for periodic functions. Another method is based on the finite-difference approximations of derivatives. Because operator L has x-dependent potential $V(x)$, the latter method is more robust in the context of localized modes in periodic potentials.

Let the periodic potential $V(x)$ have two centers of symmetry on $[0, 2\pi)$ at the minimum and maximum points at $x = 0$ and $x = \pi$ respectively. The Jacobian operator

$$H = c + L - 3\phi^2(x)$$

has exactly one negative eigenvalue if $\phi(x)$ is centered at the minimum of $V(x)$ and two negative eigenvalues if $\phi(x)$ is centered at the maximum of $V(x)$ at least for large $c > 0$ (Section 3.2.3). Both solutions are positive for all $x \in \mathbb{R}$. Therefore, the spectral renormalization method can be used for the former solution without

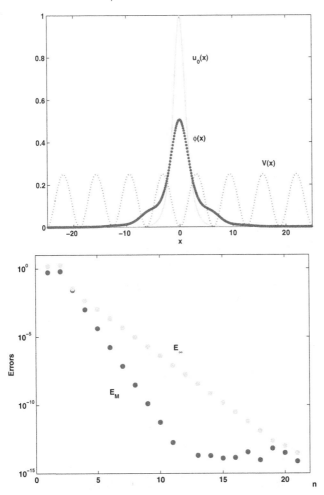

Figure 3.5 Top: the numerical approximation of the localized mode ϕ obtained in the spectral renormalization method. Bottom: convergence of the errors E_M and E_∞ with the number of iterations.

any modifications. For the latter solution, we can enforce the symmetry of solution $\phi(x)$ about the maximum point $x = \pi$. Because the corresponding eigenfunction for the second negative eigenvalue of H is odd, the operator H restricted to the space of even functions with respect to the point $x = \pi$ has only one negative eigenvalue.

Figure 3.5 (top) shows the numerical approximation of the localized mode $\phi(x)$ for $\omega = 0.1$ and $V(x) = 1 - \cos(x)$ starting with $u_0(x) = \text{sech}(x)$. Figure 3.5 (bottom) displays the convergence of the errors $E_M = |M_n - 1|$ and $E_\infty = \|u_{n+1} - u_n\|_{L^\infty}$ computed dynamically as n increases. We can see that the error E_M converges to zero faster than the error E_∞.

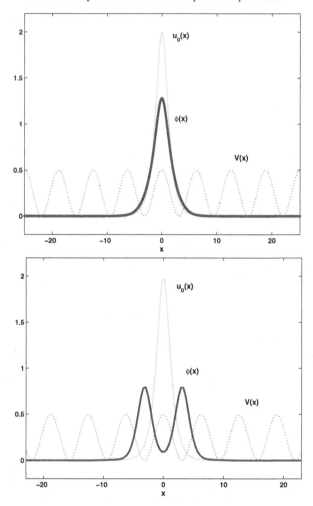

Figure 3.6 The numerical approximation of the localized mode ϕ obtained in the spectral renormalization method for $\omega = -0.5$ (top) and $\omega = 0.2$ (bottom).

Figure 3.6 shows the numerical approximation of the localized mode $\phi(x)$ for $V(x) = 1 + \cos(x)$ starting with $u_0(x) = 2\operatorname{sech}(x)$ for $\omega = -0.5$ (top) and $\omega = 0.2$ (bottom). The localized mode $\phi(x)$ is now centered at the point of the maximum of $V(x)$ and the symmetry of iterations $\{u_n(x)\}_{n \in \mathbb{N}}$ about this point is enforced for the convergence of the spectral renormalization method. We can see that the localized mode consists of one pulse for negative values of ω but it consists of two pulses for positive values of ω.

When the iteration method is applied to the two-pulse localized mode described in Theorem 3.6 (Section 3.2.4), we would run into a problem. The Jacobian operator H at the two-pulse localized mode $\phi(x)$ has two negative eigenvalues if each pulse is centered at the minimum point of $V(x)$ and four negative eigenvalues if each pulse is centered at the maximum of $V(x)$.

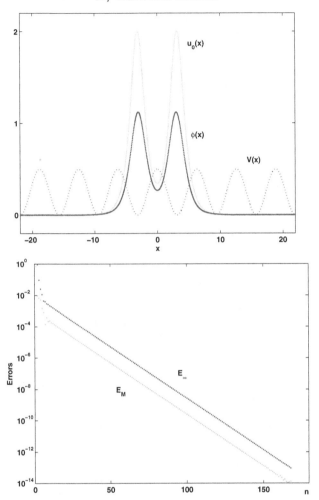

Figure 3.7 Top: the numerical approximation of the two-pulse in-phase localized mode ϕ obtained in the spectral renormalization method starting with (3.4.11). Bottom: convergence of the errors E_M and E_∞ with the number of iterations.

In the former case, only one negative eigenvalue of H is associated with the eigenfunction symmetric about the midpoint between the two pulses. Therefore, if the symmetry with respect to reflection about the midpoint is enforced, iterations of the spectral renormalization method converge for this two-pulse localized mode. In the latter case, however, two negative eigenvalues are associated with the eigenfunctions symmetric about the midpoint between the two pulses and this property causes divergence of iterations of the spectral renormalization method.

The top panels of Figures 3.7 and 3.8 illustrate the numerical approximations of the two-pulse localized mode for $\omega = -0.5$ and $V(x) = 1 + \cos(x)$ starting with

$$u_0(x) = 2\left(\operatorname{sech}(x - \pi) \pm \operatorname{sech}(x + \pi)\right), \tag{3.4.11}$$

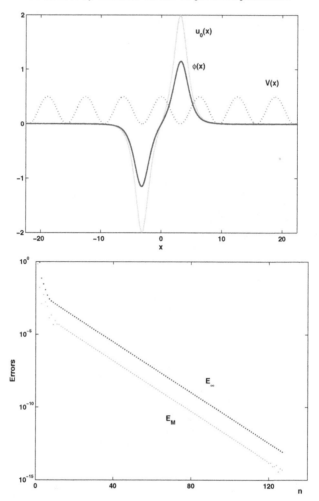

Figure 3.8 Top: the numerical approximation of the two-pulse out-of-phase localized mode ϕ obtained in the spectral renormalization method starting with
(3.4.11). Bottom: convergence of the errors E_M and E_∞ with the number of
iterations.

where the in-phase mode occurs for the plus sign and the out-of-phase mode occurs
for the minus sign. The bottom panels of Figures 3.7 and 3.8 display the convergence
of the errors E_M and E_∞ computed dynamically as n increases. Because both pulses
are localized near the minima of the periodic potential, there is no problem with
convergence of iterations when the even or odd symmetry of $u_n(x)$ is enforced.

Figures 3.9 and 3.10 display the final outputs (top) and the errors E_M and E_∞
(bottom) for $\omega = -1$ and $V(x) = 1 - \cos(x)$ starting with

$$u_0(x) = 2\left(\operatorname{sech}(x - \pi - a) + \operatorname{sech}(x + \pi + a)\right) \qquad (3.4.12)$$

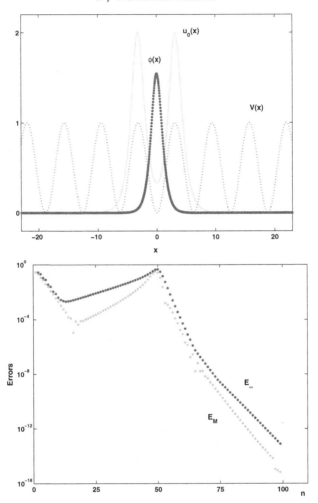

Figure 3.9 Top: the numerical approximation of the localized mode ϕ obtained in the spectral renormalization method starting with (3.4.12) for $a = 0.035$. Bottom: convergence of the errors E_M and E_∞.

for $a = 0.035$ and $a = 0.040$, respectively. In both cases, the iterations of the spectral renormalization method involve three stages. At the first stage, the initial condition relaxes to two uncoupled pulses centered at the maximum of the periodic potential $V(x)$. However, this relaxation is weakly unstable due to the second negative eigenvalue of H in the space of even functions with respect to the midpoint between the two pulses. The instability of iterations develops at the second stage leading to the convergence of iterations to other stable fixed points at the third stage. For $a = 0.035$, the iterations converge to the one-pulse localized mode centered at the point $x = 0$ (Figure 3.9). For $a = 0.040$, the iterations converge to

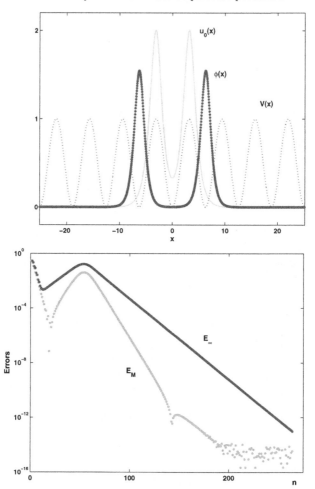

Figure 3.10 Top: the numerical approximation of the localized mode ϕ obtained in the spectral renormalization method starting with (3.4.12) for $a = 0.040$. Bottom: convergence of the errors E_M and E_∞.

the two-pulse localized mode centered at the points $x = 2\pi$ and $x = -2\pi$ (Figure 3.10). In all these cases, the localized modes are centered at the points of minimum of $V(x)$. Similar results are valid for out-of-phase configurations of the initial two-pulse approximation.

To capture the two-pulse localized mode in a similar situation, Chugunova & Pelinovsky [34] used a root finding procedure for the minimum of the error E_∞ with respect to different values of a in the starting approximation u_0. This numerical subroutine captures the unstable fixed points of the iterative method, when the instability is induced by the second (small) negative eigenvalue of the linearized operator H.

The spectral renormalization method cannot be useful for approximations of three-pulse and multi-pulse localized modes because the number of negative eigenvalues in the space of even functions exceeds one and violates the contraction criterion in Theorem 3.13. In one dimension, one can apply a shooting method or Newton iterations, whereas approximations of multi-pulse solutions in higher dimensions is a more challenging problem.

A number of modifications of the spectral renormalization method have appeared recently in the literature. Lakoba & Yang [125] considered non-power nonlinear functions in the boundary-value problem

$$(c + L)\phi = f(\phi).$$

In this case, the negative eigenvalue of the operator $T = (c + L)^{-1}f'(\phi)$ still exists but ϕ is no longer an exact eigenvector for this eigenvalue. As a result, it is more difficult to prove that the constraint $\langle f(\phi), w \rangle_{L^2}$ removes the negative eigenvalue of the linearized iteration operator in the constrained space X_c. Other modifications are also reported in [125] concerning the choice of the constraint in the stabilizing factor M_n and the choice of the iterative algorithm.

Exercise 3.36 Consider another iteration method for (3.4.1),

$$u_{n+1} = M_n^\gamma (c + L)^{-1} u_n^p, \quad n \in \mathbb{N},$$

where

$$M_n = \frac{\langle u_n, u_n \rangle_{L^2}}{\langle u_n^p, (c + L)^{-1} u_n \rangle_{L^2}},$$

and prove that the convergence criterion of this method is the same as in Theorem 3.13.

Exercise 3.37 Consider another iteration method for (3.4.1),

$$u_{n+1} = u_n + h \left(-u_n + (c + L)^{-1} u_n^p + \gamma \frac{\langle u_n, (c + L)u_n - u_n^p \rangle_{L^2}}{\langle u_n, (c + L)u_n \rangle_{L^2}} u_n \right), \quad n \in \mathbb{N},$$

and prove that there is a choice of $h > 0$ and $\gamma > 0$ such that the linearized iteration operator is a contraction, no matter if the condition (3.4.7) is satisfied or not.

3.4.2 Imaginary-time integration

Solutions of the stationary equation (3.4.1) are also equivalent to the time-independent solutions of the evolution equation

$$u_\tau = -(c + L)u + u^p, \quad \tau > 0, \tag{3.4.13}$$

where τ is an artificial time variable. Compared to the Gross–Pitaevskii equation,

$$i\psi_t = (c + L)\psi - |\psi|^2\psi, \quad t > 0, \tag{3.4.14}$$

where $c + L = -\Delta + V(x) - \omega$ for a fixed $\omega \in \mathbb{R}$, the solution $u(x, \tau)$ is a real-valued function and the time $\tau \in \mathbb{R}$ is related to the imaginary time $t \in i\mathbb{R}$ by $\tau = it$. This correspondence suggests the name of the numerical method as *imaginary-time integration*.

The imaginary-time integration method is inspired by an old idea that the iterative algorithm for solutions of an elliptic equation can be obtained from a numerical solution of a parabolic equation, which has the same set of stationary solutions as the elliptic equation does. When the stationary solution is asymptotically stable with respect to the time evolution of a parabolic equation, these iterations recover solutions of the elliptic equation as $\tau \to \infty$. If we look for strong solutions $u(\tau) \in C(\mathbb{R}_+, \mathrm{Dom}(L))$ of the parabolic equation (3.4.13), then the iterations recover strong solutions $\phi \in \mathrm{Dom}(L)$ of the stationary equation (3.4.1).

There are multiple ways to discretize the time evolution equation (3.4.13) to replace a continuous function $u(\cdot, \tau)$ in $\mathrm{Dom}(L)$ by a sequence of functions $\{u_n\}_{n \geq 1}$ in $\mathrm{Dom}(L)$ at $\tau = \tau_n = nh$ with a small time step $h > 0$. One of the simplest methods is the explicit Euler method, which takes the form

$$u_{n+1} = u_n + h\left(-(c+L)u_n + u_n^p\right), \quad n \in \mathbb{N}. \tag{3.4.15}$$

However, there are two difficulties associated with the explicit Euler method (3.4.15).

First, since $(c + L)$ is a strictly positive self-adjoint operator in L^2, the fundamental evolution operator $e^{-\tau(c+L)}$ of the linearized equation

$$u_\tau = -(c+L)u$$

forms a contraction semi-group and converges to zero as $\tau \to +\infty$ at an exponential rate. As a result, the solution of the initial-value problem associated with the time evolution equation (3.4.13) for a small initial data is expected to converge to the zero solution of the stationary equation (3.4.1).

Second, if we look at the fundamental evolution operator of the discretized linearized equation

$$u_{n+1} = u_n - h(c+L)u_n,$$

then we realize that the evolution operator is not a contraction and may lead to artificial instabilities if L is unbounded from above, no matter how small $h > 0$ is! These instabilities have a numerical origin since the unstable spectrum of the discretized evolution operator extends to the negative domain beyond multiplier -1, whereas the spectrum of the continuous evolution operator is always confined in the interval $[0, 1]$ for $\tau \geq 0$.

The first problem is fixed if we add a constraint on the iterative method (3.4.15) that would allow us to obtain a nonzero stationary solution ϕ of the stationary equation (3.4.1). For example, we can preserve the L^2 norm of functions in the sequence $\{u_n\}_{n \geq 1}$ obtaining thus the iterative method

$$\begin{cases} v_{n+1} = u_n(1 - ch) + h(-Lu_n + u_n^p), \\ u_{n+1} = Q^{1/2}v_{n+1}/\|v_{n+1}\|_{L^2}, \end{cases} \quad n \in \mathbb{N}, \tag{3.4.16}$$

where $Q = \|u_n\|_{L^2}^2$ is fixed for any $n \in \mathbb{N}$. The second problem can be fixed by working with the implicit Euler or Crank–Nicholson methods or by using the pre-conditioning operator $(I + L)^{-1}$, which transforms the fundamental evolution operator of the iterative method to a bounded operator. To simplify our presentation, we shall assume that L is a bounded operator, but more robust treatments of unbounded operators can be found in Exercises 3.38 and 3.39.

In the context of the stationary Gross–Pitaevskii equation (3.4.2), parameter c is related to parameter ω of the localized mode. Recall from Section 3.2 that parameters ω and Q are related along the solution family. If we fix ω, then Q is determined along the solution family, and vice versa. This observation implies that the explicit iterative method (3.4.16) with two parameters c and Q (we exclude the time step parameter $h > 0$ from the count) is over-determined and only one parameter (namely, the fixed power Q) plays a role in the limiting stationary solution

$$\phi = \lim_{\tau \to \infty} u(\cdot, \tau).$$

The other parameter (namely, c) can be set to any number, for instance, to zero. To confirm this claim, we prove the following lemma about the fixed points of the iterative method (3.4.16).

Lemma 3.15 *For sufficiently small $h > 0$, the set of fixed points of the iterative method (3.4.16) in $\mathrm{Dom}(L)$ is equivalent to the set of solutions of the stationary equation*

$$(\tilde{c} + L)\phi = \phi^p, \tag{3.4.17}$$

where $\tilde{c}$ is uniquely defined from ϕ by the Rayleigh quotient,

$$\tilde{c} = \frac{\langle (-L\phi + \phi^p), \phi \rangle_{L^2}}{\|\phi\|_{L^2}^2}. \tag{3.4.18}$$

Proof Let ϕ be a solution of the stationary equation (3.4.17) for some $\tilde{c}$. Then, if $u_n = \phi$, we obtain $Q = \|\phi\|_{L^2}^2$ and

$$v_{n+1} = (1 - hc + h\tilde{c})\phi.$$

For sufficiently small $h > 0$, we have $1 - hc + h\tilde{c} > 0$. Hence, $u_{n+1} = \phi$ and ϕ is a fixed point of the iterative method (3.4.16).

In the opposite direction, let u_* be the fixed point of (3.4.16). Then, we obtain

$$u_* = \|u_*\|_{L^2} \frac{u_*(1 - ch) + h(-Lu_* + u_*^p)}{\|u_*(1 - ch) + h(-Lu_* + u_*^p)\|_{L^2}}.$$

Therefore, there is a constant $\tilde{c} \in \mathbb{R}$ such that $-Lu_* + u_*^p = \tilde{c}u_*$. In other words, $u_* = \phi$ is a solution of the stationary equation (3.4.17).

In both cases, $\tilde{c}$ is found from the Rayleigh quotient (3.4.18) for the stationary equation (3.4.17). $\qquad\square$

As a consequence of Lemma 3.15, it is sufficient to set $c = 0$ in the time evolution equation (3.4.13) and in the iterative method (3.4.16). Thus, we consider the iterative method

$$\begin{cases} v_{n+1} = u_n + h(-Lu_n + u_n^p), \\ u_{n+1} = Q^{1/2}v_{n+1}/\|v_{n+1}\|_{L^2}, \end{cases} \quad n \in \mathbb{N}. \tag{3.4.19}$$

On the other hand, it would be more natural to denote $\tilde{c} = c$ in the stationary equation (3.4.17) for the fixed points of the iterative method (3.4.19). The following theorem gives the main result on convergence of the iterative method (3.4.19).

Theorem 3.14   *Let $L : \mathrm{Dom}(L) \subseteq L^2 \to L^2$ be a non-negative bounded self-adjoint operator. Given $Q > 0$, let $\phi \in \mathrm{Dom}(L)$ be a strong solution of the stationary equation (3.4.1) for some $c \in \mathbb{R}$ and assume that the map $c \mapsto Q$ is C^1. For sufficiently small $h > 0$, there exists an open neighborhood of ϕ in $\mathrm{Dom}(L)$, in which the sequence of iterations $\{u_n\}_{n\geq 1}$ defined by the iterative method (3.4.19) for the same Q converges to ϕ if and only if $H = c + L - p\phi^{p-1} : \mathrm{Dom}(L) \to L^2$ has exactly one negative eigenvalue, no zero eigenvalues, and $dQ/dc > 0$.*

Proof   The proof of this theorem follows the same steps as the proof of Theorem 3.13 with the only difference being the computation of the linearized iteration equation. We need to show that the linearized iteration operator is a contraction operator. Let us substitute $u_n = \phi + v_n$ into the iteration method (3.4.19) and neglect all terms quadratic in $\|v_n\|_{\mathrm{Dom}(L)}$. After some straightforward computations, we obtain

$$v_{n+1} = v_n - \frac{h}{1 + hc}\left(Hv_n - \frac{\langle Hv_n, \phi\rangle_{L^2}}{\|\phi\|_{L^2}^2}\phi\right), \qquad (3.4.20)$$

where c appears in the stationary equation (3.4.17) for $\phi \in \mathrm{Dom}(L)$.

It follows from (3.4.20) that

$$\langle v_{n+1}, \phi\rangle_{L^2} = \langle v_n, \phi\rangle_{L^2}, \quad n \in \mathbb{N}.$$

The projection of $\{v_n\}_{n\in\mathbb{N}}$ in the direction of ϕ is constant in $n \in \mathbb{N}$. Convergence to zero can only be achieved in the constrained L^2 space,

$$X_c = \{v \in \mathrm{Dom}(L) : \quad \langle v, \phi\rangle_{L^2} = 0\}, \qquad (3.4.21)$$

and we should require that $\langle v_1, \phi\rangle_{L^2} = 0$ initially. The constrained L^2 space reflects the conservation of power

$$Q = \|u_n\|_{L^2}^2 = \|\phi\|_{L^2}^2 + 2\langle v_n, \phi\rangle_{L^2} + \|v_n\|_{L^2}^2, \quad n \in \mathbb{N}.$$

If $v_n \in X_c$, then $Q = \|\phi\|_{L^2}^2$ up to the quadratic terms in v_n. Let P_c be the orthogonal projection operator from $\mathrm{Dom}(L)$ to X_c. The linearized iteration equation (3.4.20) can be rewritten in the form

$$v_{n+1} = v_n - \frac{h}{1 + hc}P_c H P_c v_n, \quad n \in \mathbb{N}. \qquad (3.4.22)$$

We need to consider the spectrum of $H|_{X_c} = P_c H P_c$ restricted in the constrained space X_c. In particular, any negative eigenvalue of $H|_{X_c}$ leads to the divergence of the linearized iteration equation (3.4.22). Thanks to Theorem 4.1 (Section 4.1.1), operator $H|_{X_c}$ has one less negative eigenvalue compared to operator H if $\langle H^{-1}\phi, \phi\rangle_{L^2} < 0$ and has the same number of negative eigenvalues if

$$\langle H^{-1}\phi, \phi\rangle_{L^2} > 0.$$

Since $H\partial_c\phi = -\phi$, we obtain

$$\langle H^{-1}\phi, \phi\rangle_{L^2} = -\langle \partial_c\phi, \phi\rangle_{L^2} = -\frac{1}{2}\frac{dQ}{dc},$$

where the map $c \mapsto \phi \mapsto Q$ is assumed to be C^1 in c. If H has exactly one negative eigenvalue and $Q'(c) > 0$, then $H|_{X_c}$ has no negative eigenvalues. The linearized iteration operator is a contraction if additionally

$$\frac{h}{1 + hc} \sigma(H|_{X_c}) < 2. \tag{3.4.23}$$

Since L is a bounded operator, $\sigma(H|_{X_c})$ is bounded from above. As a result, there exists a small $h_0 > 0$ such that condition (3.4.23) is satisfied for any $h \in (0, h_0)$ and the iterative method (3.4.19) is a contraction in a local neighborhood of the fixed point $\phi \in \text{Dom}(L)$. If H has more than one negative eigenvalue or if $Q'(c) < 0$, then $H|_{X_c}$ has at least one negative eigenvalue and the iterative method (3.4.19) diverges near the fixed point ϕ. □

If the unbounded differential operator $L = -\partial_x^2 + V(x)$ is replaced by the finite-difference operator $L = -\Delta + V$, where

$$(\Delta u)_n = u_{n+1} - 2u_n + u_{n-1}, \quad (Vu)_n = V_n u_n, \quad n \in \mathbb{Z},$$

then the conditions of Theorem 3.14 on L are satisfied.

The conditions in Theorem 3.14 on H and dQ/dc coincide with the stability condition of a stationary solution ϕ with respect to the time evolution of the parabolic equation (3.4.13). This condition coincides with the condition for orbital stability of the *positive* stationary solution ϕ with respect to the time evolution of the Gross–Pitaevskii equation (3.4.14) (Section 4.4.2). For a *non-positive* stationary solution ϕ, the relationship becomes less explicit since another operator $L_- = c + L - \phi^{p-1}$ needs to be considered in addition to the Jacobian operator $L_+ = c + L - p\phi^{p-1} \equiv H$ (Section 4.3).

The constrained space X_c given by (3.4.21) appears naturally in the variational formalism of the stationary solution ϕ as the critical point of the Hamiltonian

$$H(u) = \langle (Lu - u^p), u \rangle_{L^2}$$

under a fixed $Q(u) = \|u\|_{L^2}^2$, where c is a Lagrange multiplier of the energy functional $E_c(u) = H(u) - cQ(u)$. Equation (3.4.13) is a gradient flow which realizes a continuous minimization of the Hamiltonian $H(u)$ under fixed power $Q(u)$. The concentration compactness principle of Lions [134] allows one to predict when the minimum can be achieved. Various discretizations of the minimization procedure under a Q-normalized gradient flow were discussed by Bao & Du [16].

To discretize the continuous imaginary-time evolution equation (3.4.13), we applied the explicit Euler method (3.4.16). One can consider other discretizations such as implicit methods which would leave out the requirement of Theorem 3.14 that the linear operator L is bounded.

Exercise 3.38 Consider the implicit Euler method,

$$\begin{cases} v_{n+1} = (I + h(c + L))^{-1}(u_n + hu_n^p), \\ u_{n+1} = Q^{1/2} v_{n+1} / \|v_{n+1}\|_{L^2}, \end{cases} \quad n \in \mathbb{N}, \tag{3.4.24}$$

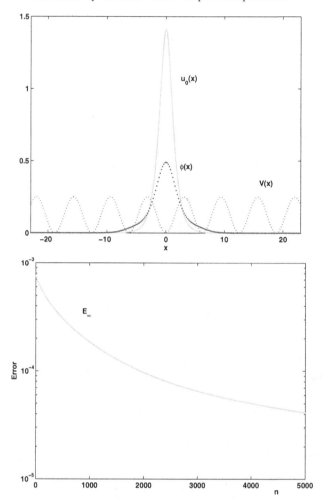

Figure 3.11 Top: the numerical approximation of the localized mode ϕ obtained in the explicit imaginary-time integration method. Bottom: convergence of the error E_∞ with the number of iterations.

and prove that this algorithm converges for small $h > 0$ if H and dQ/dc satisfy the same conditions as in Theorem 3.14, no matter whether L is a bounded or unbounded operator.

Exercise 3.39 Consider the implicit Crank–Nicholson method for the imaginary-time evolution problem (3.4.13) and prove that it converges for small $h > 0$ if H and dQ/dc satisfy the same conditions as in Theorem 3.14, no matter whether L is a bounded or unbounded operator.

For an illustration on the convergence of the imaginary-time integration method (3.4.19), let us consider again a localized mode in the focusing Gross–Pitaevskii equation (3.4.2) with $d = 1$. Figure 3.11 (top) shows the numerical approximation

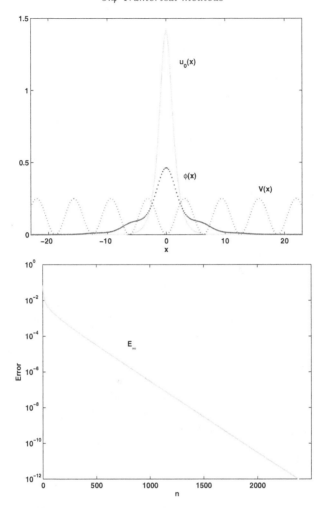

Figure 3.12 Top: the numerical approximation of the localized mode ϕ obtained in the implicit imaginary-time integration method. Bottom: convergence of the error E_∞ with the number of iterations.

of the localized mode for $Q = 4$ and $V(x) = 1 - \cos(x)$ starting with $u_0(x) = \sqrt{2}\,\mathrm{sech}(x)$. Figure 3.11 (bottom) displays the convergence of the error $E_\infty = \|u_{n+1} - u_n\|_{L^\infty}$ versus n. We can see that the error E_∞ converges very slowly because the time step $h = 0.001$ is too small. The second-order difference approximation is used for the second derivatives with the step size 0.2 and the time step h has to be small to preserve stability of iterations of the explicit Euler method.

To avoid the stability constraint on h, we can implement the implicit Euler method (3.4.24), where the parameter c is dynamically approximated at each iteration from the Rayleigh quotient, with

$$c_n = \frac{\langle (-Lu_n + u_n^p), u_n \rangle_{L^2}}{Q}.$$

Figure 3.12 (top) shows the numerical approximation of the localized mode for $Q = 4$ and $V(x) = 1 - \cos(x)$ starting with $u_0(x) = \sqrt{2}\,\mathrm{sech}(x)$. Figure 3.12 (bottom) displays the convergence of the error E_∞ versus n. We can see that the error E_∞ converges faster because the time step $\tau = 0.1$ is bigger. The same level of accuracy, 10^{-13}, as on Figure 3.5 is now reached in 2630 iterations compared to 25 iterations of the spectral renormalization method.

There are several ways to accelerate convergence of the imaginary-time integration method (3.4.16). García-Ripoll & Pérez-García [64] used Sobolev gradients to accelerate minimization of the square energy function

$$E(u) = \|(Lu - u^p)\|_{L^2}^2$$

under the fixed $Q(u)$. Yang & Lakoba [216] introduced a similar square operator method, written in the explicit form by

$$u_{n+1} = u_n + h\left(-L(-Lu_n + u_n^p) + p(-Lu_n + u_n^p)^{p-1}\right), \quad n \in \mathbb{N},$$

for a time step $h > 0$. The latter method was applied to more general nonlinear evolution equations and was modified using "pre-conditioned" factors to accelerate iterations. As a result, localized modes were approximated numerically even in finite band gaps of the spectrum of $L = -\Delta + V(x)$ with $V \in L_{\mathrm{per}}^\infty(\mathbb{R}^d)$ for $d = 2$.

In their follow-up works, Yang & Lakoba also discussed acceleration of the spectral normalization and imaginary-time integration methods using a mode elimination technique [126] and a "pre-conditioned" positive operator [217], which are introduced to improve the contraction properties of the linearized iteration operator. Lakoba [127] also considered the conjugate gradient method as yet another iteration method to reduce the number of iterations in the numerical approximation of stationary localized modes.

4

Stability of localized modes

The only reason for time is so that everything doesn't happen at once.
– Albert Einstein.

Be wise and linearize!
– Mathematical proverb.

Localized modes mean nothing to physicists if they are not observed in real experiments. No physicists would waste their time talking about unstable localized modes, which are destroyed by a small disturbance in the initial data. To sort out different families of localized modes in nonlinear evolution equations, the existence of which was studied in Chapter 3, we introduce here the concept of their stability with respect to time evolution and describe robust methods of stability analysis.

Many nonlinear evolution equations with localized modes can be written in the Hamiltonian form (Section 1.2),

$$i\frac{d\psi}{dt} = \nabla_{\bar{\psi}}H(\psi, \bar{\psi}),$$

where $H(\psi, \bar{\psi}) : X \times X \to \mathbb{R}$ is the Hamiltonian functional defined on some Hilbert space $X \subset L^2$ and ψ is the complex-valued vector function. Because of the time translational invariance, the value of $H(\psi, \bar{\psi})$ is constant in $t \in \mathbb{R}$ if $\psi(t)$ is a solution of the Hamiltonian system. Additionally, there is a conserved gauge functional $Q(\psi, \bar{\psi}) : X \times X \to \mathbb{R}$, which is related to the gauge transformation of the Hamiltonian system.

If a localized mode $\phi \in X$ is a stationary solution of the Hamiltonian system, then ϕ is a critical point of the energy functional (Section 3.1)

$$E_\omega(\psi, \bar{\psi}) = H(\psi, \bar{\psi}) - \omega Q(\psi, \bar{\psi})$$

in the sense

$$\forall \mathbf{v} \in X : \quad \frac{d}{d\epsilon}E_\omega(\phi + \epsilon\mathbf{v}, \bar{\phi} + \epsilon\bar{\mathbf{v}})\bigg|_{\epsilon=0} = 0.$$

Here ω is the Lagrange multiplier and, simultaneously, a parameter of the localized mode ϕ. Using the gauge transformation, the Hamiltonian system can be rewritten in the equivalent form

$$i\frac{d\mathbf{u}}{dt} = \nabla_{\bar{\mathbf{u}}}E_\omega(\mathbf{u}, \bar{\mathbf{u}}),$$

where $\mathbf{u}$ is a new variable expressed from ψ via the linear transformation (Section 1.2). Now $\mathbf{u} = \phi$ becomes a time-independent solution of the Hamiltonian system in the new coordinates.

Let us expand the energy functional $E_\omega(\mathbf{u}, \bar{\mathbf{u}})$ in the neighborhood of the critical point ϕ,

$$\forall \mathbf{v} \in X : \quad E_\omega(\phi + \mathbf{v}, \bar{\phi} + \bar{\mathbf{v}}) = E_\omega(\phi, \bar{\phi}) + \frac{1}{2} E_\omega''(\mathbf{v}, \bar{\mathbf{v}}) + \mathcal{O}(\|\mathbf{v}\|_X^3),$$

where $E_\omega''(\mathbf{v}, \bar{\mathbf{v}})$ is the quadratic part of the energy functional evaluated at the critical point ϕ for a given value of ω. The quadratic part can be expressed in terms of the Hessian operator $H_\omega : X \times X \to L^2 \times L^2$, which acts on the vector $\mathbf{V} = (\mathbf{v}, \bar{\mathbf{v}})$,

$$E_\omega''(\mathbf{v}, \bar{\mathbf{v}}) = \langle H_\omega \mathbf{V}, \mathbf{V} \rangle_{L^2}.$$

Because $E_\omega(\mathbf{u}, \bar{\mathbf{u}}) : X \times X \to \mathbb{R}$, the Hessian operator H_ω is necessarily self-adjoint in $L^2 \times L^2$. Nevertheless, it is not a block-diagonal operator in variables $(\mathbf{v}, \bar{\mathbf{v}})$ in the general case. Expanding solutions of the Hamiltonian system as $\mathbf{u} = \phi + \mathbf{v}$ and truncating by the linear terms in $\mathbf{v}$, we obtain the linearized Hamiltonian system

$$i \frac{d\mathbf{v}}{dt} = (H_\omega)_{11} \mathbf{v} + (H_\omega)_{12} \bar{\mathbf{v}},$$

which is coupled with the complex conjugate equation

$$-i \frac{d\bar{\mathbf{v}}}{dt} = (H_\omega)_{21} \mathbf{v} + (H_\omega)_{22} \bar{\mathbf{v}},$$

where $(H_\omega)_{11} = (\bar{H}_\omega)_{22}$ and $(H_\omega)_{12} = (\bar{H}_\omega)_{21}$. Since H_ω does not depend on time t, we can separate the time and space variables and reduce the linearized Hamiltonian system to the spectral problem. This separation of variables is given by the substitution

$$\begin{cases} \mathbf{v} = \mathbf{v}_1 e^{\lambda t} + \bar{\mathbf{v}}_2 e^{\bar{\lambda} t}, \\ \bar{\mathbf{v}} = \mathbf{v}_2 e^{\lambda t} + \bar{\mathbf{v}}_1 e^{\bar{\lambda} t}, \end{cases}$$

The spectral problem now takes the form

$$\begin{cases} i\lambda \mathbf{v}_1 = (H_\omega)_{11} \mathbf{v}_1 + (H_\omega)_{12} \mathbf{v}_2, \\ -i\lambda \mathbf{v}_2 = (H_\omega)_{21} \mathbf{v}_1 + (H_\omega)_{22} \mathbf{v}_2, \end{cases}$$

or, simply,

$$JH_\omega \mathbf{V} = \lambda \mathbf{V}, \quad \mathbf{V} \in X \times X, \quad \lambda \in \mathbb{C},$$

where

$$J = \begin{bmatrix} -i & 0 \\ 0 & i \end{bmatrix} \quad \text{and} \quad \mathbf{V} = \begin{bmatrix} \mathbf{v}_1 \\ \mathbf{v}_2 \end{bmatrix}.$$

The symplectic operator J destroys the self-adjoint structure of the Hessian operator H_ω and the resulting linearized operator JH_ω is neither self-adjoint nor skew-adjoint in $L^2 \times L^2$. Nevertheless, the spectrum of JH_ω is related to the spectrum of H_ω and this relationship between the spectra of these two operators is the central part of this chapter.

It is now time to consider an example of the linearized Hamiltonian system in the context of the Gross–Pitaevskii equation,

$$i\psi_t = -\Delta\psi + V(x)\psi + \sigma|\psi|^2\psi, \quad x \in \mathbb{R}^d,$$

where $\psi(x,t) : \mathbb{R}^d \times \mathbb{R} \to \mathbb{C}$, $\sigma \in \{1,-1\}$, and $V(x) : \mathbb{R}^d \to \mathbb{R}$. If $V \in L^\infty(\mathbb{R}^d)$, weak solutions of the Gross–Pitaevskii equation in $X = H^1(\mathbb{R}^d)$ are equivalent to strong solutions in $X = H^2(\mathbb{R}^d)$. The energy functional is given by

$$E_\omega(u,\bar{u}) = \int_{\mathbb{R}^d} \left(|\nabla u|^2 + V|u|^2 - \omega|u|^2 + \frac{1}{2}\sigma|u|^4\right) dx,$$

and the gauge transformation is given by $\psi(x,t) = e^{-i\omega t}u(x,t)$. After the expansion near $\phi \in H^2(\mathbb{R}^d)$, the quadratic part of the energy functional is given by

$$E_\omega''(v,\bar{v}) = \int_{\mathbb{R}^d} \left(|\nabla v|^2 + V|v|^2 - \omega|v|^2 + \frac{1}{2}\sigma(\phi^2\bar{v}^2 + 4|\phi|^2|v|^2 + \bar{\phi}^2v^2)\right) dx$$

and it is expressed in terms of the Hessian operator,

$$H_\omega = \begin{bmatrix} -\Delta + V - \omega + 2\sigma|\phi|^2 & \sigma\phi^2 \\ \sigma\bar{\phi}^2 & -\Delta + V - \omega + 2\sigma|\phi|^2 \end{bmatrix}.$$

The spectral problem takes the explicit form

$$\begin{cases} \left(-\nabla^2 + V - \omega + 2\sigma|\phi|^2\right)v_1 + \phi^2 v_2 = i\lambda v_1, \\ \bar{\phi}^2 v_1 + \left(-\nabla^2 + V - \omega + 2\sigma|\phi|^2\right)v_2 = -i\lambda v_2, \end{cases}$$

where $(v_1, v_2) \in H^2(\mathbb{R}^d) \times H^2(\mathbb{R}^d)$ and $\lambda \in \mathbb{C}$.

There are four main definitions of stability of localized modes in nonlinear evolution equations, which we rank here from the weakest to the strongest: (i) spectral stability; (ii) linearized stability; (iii) orbital stability; and (iv) asymptotic stability. The *spectral stability* of a localized mode ϕ is defined by the spectral problem,

$$JH_\omega\mathbf{V} = \lambda\mathbf{V}, \quad \mathbf{V} \in X \times X, \ \lambda \in \mathbb{C},$$

where JH_ω is neither self-adjoint nor skew-adjoint in $L^2 \times L^2$.

Definition 4.1 We say that the localized mode ϕ is *spectrally unstable* if the spectral problem for JH_ω has at least one eigenvalue λ with $\mathrm{Re}\,\lambda > 0$ and $\mathbf{V} \in X \times X$. Otherwise, it is *(weakly) spectrally stable*.

The *linearized stability* of a localized mode ϕ relies on the time evolution problem associated with the spectral problem above,

$$\frac{d\mathbf{V}}{dt} = JH_\omega\mathbf{V}, \quad t \in \mathbb{R}_+.$$

The gauge invariance of the Hamiltonian system leads to a weak growth of the solution $\mathbf{V}(t)$ as $t \to \infty$. To remove this weak growth, we shall add the constraint that the perturbation $\mathbf{V}(t)$ does not affect the first variation of the conserved quantity $Q(\mathbf{u},\bar{\mathbf{u}})$ in the sense that

$$X_c = \left\{\mathbf{v} \in X : \ \langle\nabla_{\bar{\mathbf{u}}}Q(\phi,\bar{\phi}),\mathbf{v}\rangle_X + \langle\nabla_{\mathbf{u}}Q(\phi,\bar{\phi}),\bar{\mathbf{v}}\rangle_X = 0\right\}.$$

If $\mathbf{V}(t) \in X_c \times X_c$ for all $t \in \mathbb{R}_+$, we can study the evolution of perturbations $\mathbf{V}(t)$ in the linearized Hamiltonian system.

Definition 4.2   We say that the localized mode ϕ is *stable with respect to the linearization* if there exists the limit

$$\limsup_{t \to \infty} \|\mathbf{v}(t)\|_X < \infty,$$

for the solution $\mathbf{v}(t)$ of the linearized Hamiltonian system in X_c. Otherwise, it is *linearly unstable*.

To define *orbital stability*, we need to consider a family of localized modes $T(\theta)\phi$ (called an *orbit*), where parameter θ belongs to an open set and $T(\theta)$ is a generator of the group related to the gauge transformation of the Hamiltonian system. This symmetry leads to the conservation of the power $Q(\mathbf{u}, \bar{\mathbf{u}})$ (Section 1.2.1). We say that the localized mode ϕ is orbitally stable in the sense of Lyapunov if the solution $\mathbf{u}(t)$ of the time evolution problem that starts in a local neighborhood of ϕ remains in a local neighborhood of the orbit $T(\theta)\phi$ for all $t \in \mathbb{R}_+$. More precisely, we have the following definition.

Definition 4.3   The localized mode ϕ is said to be *orbitally stable* if for any $\epsilon > 0$ there is a $\delta > 0$, such that if $\|\mathbf{u}(0) - \phi\|_X \leq \delta$ then

$$\inf_{\theta \in \mathbb{R}} \|\mathbf{u}(t) - T(\theta)\phi\|_X \leq \epsilon,$$

for all $t \in \mathbb{R}_+$. Otherwise, it is *orbitally unstable*.

Finally, *asymptotic stability* is the strongest concept of stability since the solution $\mathbf{u}(t)$ of the nonlinear evolution equation is required not only to remain in the neighborhood of the orbit $T(\theta)\phi$ but also to approach a particular orbit $T(\theta)\phi_\infty$ in the limit $t \to \infty$, where ϕ_∞ is generally different from ϕ by the value of the parameter ω.

Definition 4.4   The localized mode ϕ is said to be *asymptotically stable* if it is orbitally stable and for any $\mathbf{u}(0)$ near ϕ there is ϕ_∞ near ϕ such that

$$\lim_{t \to \infty} \inf_{\theta \in \mathbb{R}} \|\mathbf{u}(t) - T(\theta)\phi_\infty\|_X = 0.$$

In the context of the Gross–Pitaevskii equation, the spectral problem can be simplified if ϕ is real-valued as in most of Chapter 3. If we consider again the spectral problem

$$\begin{cases} \left(-\nabla^2 + V - \omega + 2\sigma|\phi|^2\right) v_1 + \phi^2 v_2 = \mathrm{i}\lambda v_1, \\ \bar{\phi}^2 v_1 + \left(-\nabla^2 + V - \omega + 2\sigma|\phi|^2\right) v_2 = -\mathrm{i}\lambda v_2, \end{cases}$$

and assume that $\phi \in \mathbb{R}$, then substitution

$$v_1 = u + \mathrm{i}w, \quad v_2 = u - \mathrm{i}w$$

gives a new spectral problem

$$L_+ u = -\lambda w, \quad L_- w = \lambda u,$$

where L_+ and L_- are linear Schrödinger operator in the form

$$\begin{cases} L_+ = -\Delta + V - \omega + 3\sigma\phi^2, \\ L_- = -\Delta + V - \omega + \sigma\phi^2. \end{cases}$$

In other words, the Hessian operator H_ω becomes block-diagonal in variables (u, w) and the quadratic form for the energy functional becomes

$$E''_\omega(u + iw, u - iw) = \langle L_+ u, u \rangle_{L^2} + \langle L_- w, w \rangle_{L^2}.$$

Although L_+ and L_- are self-adjoint operators in L^2, the Hamiltonian structure associated with the canonical symplectic matrix

$$\begin{bmatrix} 0 & -1 \\ 1 & 0 \end{bmatrix}$$

still makes the spectral problem for (u, w) neither self-adjoint nor skew-adjoint.

It turns out that the spectral problem

$$L_+ \mathbf{u} = -\lambda \mathbf{w}, \quad L_- \mathbf{w} = \lambda \mathbf{u}, \quad \mathbf{u} \in \mathrm{Dom}(L_+), \quad \mathbf{w} \in \mathrm{Dom}(L_-), \quad \lambda \in \mathbb{C},$$

where L_+ and L_- are self-adjoint operators in L^2, becomes relevant for other nonlinear evolution equations associated with localized modes even if ϕ is no longer real-valued.

If λ is an eigenvalue of the spectral problem above with the eigenvector $(\mathbf{u}, \mathbf{w})$, so are eigenvalues $(-\lambda)$, $\bar{\lambda}$, and $(-\bar{\lambda})$ with eigenvectors $(\mathbf{u}, -\mathbf{w})$, $(\bar{\mathbf{u}}, \bar{\mathbf{w}})$, and $(\bar{\mathbf{u}}, -\bar{\mathbf{w}})$, respectively. This property is general for linearized Hamiltonian systems and it implies that eigenvalues λ occur only as real or purely imaginary pairs or as quartets in the complex plane. If one of the operators is invertible (say, L_-), the spectral problem can be rewritten in the form of the generalized eigenvalue problem,

$$L_+ \mathbf{u} = \gamma (L_-)^{-1} \mathbf{u}, \quad \mathbf{u} \in \mathrm{Dom}(L_+), \quad \gamma \in \mathbb{C},$$

where $\gamma = -\lambda^2$ is a new eigenvalue. Real and purely imaginary pairs of λ become negative and positive eigenvalues of γ respectively, while quartets of complex eigenvalues of λ become pairs of complex conjugate eigenvalues of γ.

The generalized eigenvalue problem is a classical problem of simultaneous block-diagonalization of two self-adjoint operators L_+ and L_-^{-1}. Since both L_+ and L_- are self-adjoint operators in L^2, each operator can be orthogonally diagonalized by the Spectral Theorem (Appendix B.11). However, the orthogonal diagonalization of each operator is relevant for the problem of their simultaneous diagonalization only if the two operators commute. In the latter case, there exists a common basis of eigenvectors for both operators. Since operators L_+ and L_- do not commute in general, eigenvectors of each operator are not related to eigenvectors of the generalized eigenvalue problem. Moreover, complex and multiple eigenvalues may generally occur in the generalized eigenvalue problem if either operator is not sign-definite.

Analysis of spectral stability of equilibrium configurations in a Hamiltonian system of finitely many interacting particles relies on the same generalized eigenvalue problem, where L_+ and L_- are symmetric matrices for kinetic and potential energies and $(\mathbf{u}, \mathbf{w}) \in \mathbb{R}^n \times \mathbb{R}^n$. In this context, it follows from the Sylvester Law of Inertia [200] that if one matrix (say, L_-) is positive definite and thus invertible, the numbers of positive, zero and negative eigenvalues γ in the generalized eigenvalue problem *equals* the numbers of positive, zero and negative eigenvalues of the other matrix (L_+). If L_- is not positive definite (but still invertible), a complete classification of eigenvalues γ of the generalized eigenvalue problem in terms of real eigenvalues of L_+ and L_- can be developed with the use of the Pontryagin Invariant Subspace Theorem, originally proved for a single negative eigenvalue of L_- by Pontryagin [173] and then extended to finitely many negative eigenvalues of L_- by Krein and his students [12, 70, 91].

In the context of solitary waves in nonlinear evolution equations, L_+ and L_- are self-adjoint differential operators in a Hilbert space. There has recently been a rapidly growing sequence of publications about characterization of unstable eigenvalues of the generalized eigenvalue problem in terms of real eigenvalues of the operators L_+ and L_-. An incomplete list of literature is papers [32, 33, 43, 84, 105, 119, 122, 153, 181, 208].

We shall describe the spectral stability theory for the nonlinear Schrödinger equations. We also show how the spectral stability theory can be useful for analysis of linearized, orbital, and asymptotic stability of localized modes. Besides the nonlinear Schrödinger equation, we also discuss the nonlinear Dirac equations, where the spectral stability of localized modes cannot be studied using the count of eigenvalues of operators L_+ and L_-.

4.1 Schrödinger operators with decaying potentials

Since spectral stability of a localized mode in a nonlinear evolution equation relies on analysis of self-adjoint differential operators L_+ and L_-, let us start with the properties of a one-dimensional Schrödinger operator,

$$L = -\partial_x^2 + c + V(x),$$

where $c > 0$ is a parameter and $V(x) : \mathbb{R} \to \mathbb{R}$ is a bounded potential. We assume the exponential decay of the potential $V(x)$ as $|x| \to \infty$, that is, we assume that there exist $\kappa > 0$ and $C > 0$ such that $V \in L^\infty(\mathbb{R})$ satisfies

$$\lim_{x \to \pm\infty} e^{\kappa|x|} |V(x)| = C. \tag{4.1.1}$$

This exponential decay of the potential $V(x)$ is related to the exponential decay of the localized mode that determines $V(x)$.

If $V \in L^\infty(\mathbb{R})$, the arguments of Section 2.1 can be used to state that L maps continuously $H^2(\mathbb{R})$ to $L^2(\mathbb{R})$. Since L is self-adjoint with respect to $\langle \cdot, \cdot \rangle_{L^2}$, the Spectral Theorem (Appendix B.11) states that the spectrum of L is a subset of the real axis and its eigenfunctions form an orthonormal basis in $L^2(\mathbb{R})$. We will show that the spectrum $\sigma(L)$ of the self-adjoint operator with the exponentially decaying potential is divided into two disjoint parts: the *point spectrum of eigenvalues* denoted by $\sigma_p(L)$ and the *continuous spectrum* denoted by $\sigma_c(L)$.

We will give general definitions of the point and continuous spectra of linear operators, which are also useful for analysis of a non-self-adjoint spectral problem.

Definition 4.5 We say that λ is an *eigenvalue* of an operator

$$L : \mathrm{Dom}(L) \subseteq L^2 \to L^2$$

of geometric multiplicity $m \geq 1$ and algebraic multiplicity $k \geq m$ if

$$\dim \mathrm{Ker}(L - \lambda I) = m \quad \text{and} \quad \dim \bigcup_{n \in \mathbb{N}} \mathrm{Ker}(L - \lambda I)^n = k.$$

The eigenvalue is said to be *semi-simple* if $k = m$ and *multiple* if $k > m$.

Exercise 4.1 Show that if L is a self-adjoint operator in L^2, then all eigenvalues are semi-simple.

Definition 4.6 We say that λ is a *point of the continuous spectrum* of an operator

$$L : \mathrm{Dom}(L) \subseteq L^2 \to L^2$$

if

$$\mathrm{Ker}(L - \lambda I) = \varnothing \quad \text{and} \quad \mathrm{Ran}(L - \lambda I) \neq L^2.$$

Exercise 4.2 Show with the help of the Fourier transform that if $L = -\partial_x^2 + c$, then $\sigma_c(L) = [c, \infty)$.

If λ is an eigenvalue of $\sigma_p(L)$, we say that the eigenvalue is *isolated* from the continuous spectrum if the distance between λ and points in $\sigma_c(L)$ is bounded away from zero and *embedded* into the continuous spectrum if the distance between λ and points in $\sigma_c(L)$ is zero. Continuous spectrum, embedded eigenvalues, and eigenvalues of infinite multiplicities are sometimes combined together under the name of *essential spectrum*, while isolated eigenvalues of finite multiplicities are singled out under the name of *discrete spectrum*. The great simplification that comes from the assumption of the exponential decay of V is that the two subdivisions become identical and the embedded eigenvalues or eigenvalues of infinite multiplicities do not occur in the spectrum of L. As a result, we have

$$\sigma(L) = \sigma_p(L) \cup \sigma_c(L) \quad \text{and} \quad \sigma_p(L) \cap \sigma_c(L) = \varnothing.$$

This partition of $\sigma(L)$ is based on the following lemma.

Lemma 4.1 *Let $L = -\partial_x^2 + c + V(x)$ for a fixed $c > 0$ and let $V(x)$ satisfy (4.1.1). If $\lambda \in \sigma_p(L)$, then $\lambda < c$, whereas $\sigma_c(L) = [c, \infty)$.*

Proof Let us parameterize $\lambda \geq c$ by $\lambda = c + k^2$ for $k = \sqrt{\lambda - c} \geq 0$. If $\lambda = c + k^2$ is an eigenvalue of L, there exists a solution $\varphi \in H^2(\mathbb{R})$ of the differential equation

$$\varphi''(x) + k^2 \varphi(x) = V(x)\varphi(x), \quad x \in \mathbb{R}. \tag{4.1.2}$$

Assuming decay of $V(x)$ as $x \to -\infty$ and using the Green's function for $\partial_x^2 + k^2$, the differential equation (4.1.2) can be rewritten in the equivalent integral form,

$$\varphi(x) = c_1 \cos(kx) + c_2 \sin(kx) + k^{-1} \int_{-\infty}^{x} \sin[k(x - x')]V(x')\varphi(x')dx', \quad (4.1.3)$$

where c_1 and c_2 are arbitrary constants. Since the integral decays to zero as $x \to -\infty$, we need to set $c_1 = c_2 = 0$ if $\varphi \in H^2(\mathbb{R})$. Therefore, $\varphi(x)$ solves the homogeneous integral equation (4.1.3) for $c_1 = c_2 = 0$ and we need to show that the homogeneous equation admits no solutions in $H^2(\mathbb{R})$ if $V(x)$ satisfies the exponential decay (4.1.1).

Let $\varphi_0 = \sup_{x \in (-\infty, x_0]} |\varphi(x)|$ for a fixed $x_0 < 0$ and observe that

$$k^{-1} \sin[k(x - x')] \le (x - x'), \quad x - x' \ge 0, \quad k \in \mathbb{R}.$$

Estimating the integral, we obtain

$$\varphi_0 \le \sup_{x \in (-\infty, x_0]} k^{-1} \int_{-\infty}^{x} |\sin[k(x - x')]V(x')\varphi(x')| \, dx',$$

$$\le \varphi_0 \sup_{x \in (-\infty, x_0]} \int_{-\infty}^{x} |(x - x')V(x')| \, dx'. \quad (4.1.4)$$

If $V \in L^\infty(\mathbb{R})$ satisfies the exponential decay (4.1.1), then there are $\kappa > 0$ and $C > 0$ such that

$$|V(x)| \le Ce^{\kappa x}, \quad \forall x \in (-\infty, x_0],$$

for any $x_0 < 0$. If x_0 is a sufficiently large negative number, then

$$\sup_{x \in (-\infty, x_0]} \int_{-\infty}^{x} |(x - x')V(x')| \, dx' \le C \int_{-\infty}^{x} (x - x')e^{\kappa x'} \, dx' = C\kappa^{-2}e^{\kappa x_0} < 1.$$

Bound (4.1.4) implies that $\varphi_0 = 0$ and the integral equation for $\varphi(x)$ admits zero solution on $(-\infty, x_0]$. Unique continuation of the zero solution of the differential equation (4.1.2) gives $\varphi(x) = 0$ for all $x \in \mathbb{R}$, so that no eigenvalues of $\sigma_p(L)$ with $\lambda \ge c$ exist.

On the other hand, for $\lambda \ge c$ ($k \in \mathbb{R}$) we can also consider a solution of the inhomogeneous integral equation

$$\varphi(x; k) = e^{ikx} + k^{-1} \int_{-\infty}^{x} \sin[k(x - x')]V(x')\varphi(x'; k)dx'. \quad (4.1.5)$$

Using the same proof as above, a unique solution $\varphi(\cdot; k) \in L^\infty((-\infty, x_0])$ of the integral equation (4.1.5) exists for sufficiently large negative x_0. From the linear differential equation (4.1.2), this solution is uniquely continued for all $x \in \mathbb{R}$. The solution $\varphi(x; k)$ is bounded as $x \to \infty$ if $k \ne 0$. Analyzing limits $x \to \pm\infty$ from the integral equation (4.1.5), we find the limiting representation for the unique solution $\varphi(\cdot; k) \in L^\infty(\mathbb{R})$ for any $k \ne 0$:

$$\varphi(x; k) \to \begin{cases} e^{ikx} & \text{as} \quad x \to -\infty, \\ a(k)e^{ikx} + b(k)e^{-ikx} & \text{as} \quad x \to +\infty, \end{cases}$$

where $a(k)$ and $b(k)$ (referred to as the *scattering data*) are given by

$$a(k) = 1 + \frac{1}{2ik} \int_{\mathbb{R}} e^{-ikx} V(x) \varphi(x; k) dx,$$

$$b(k) = -\frac{1}{2ik} \int_{\mathbb{R}} e^{ikx} V(x) \varphi(x; k) dx.$$

Using Weyl's Theorem (Appendix B.15), we can construct wave packets $\{u_n\}_{n \geq 1} \in H^2(\mathbb{R})$ from the eigenfunctions $\varphi(x; k)$ compactly supported on $[k_0 - n^{-1}, k_0 + n^{-1}]$ for any $k_0 > 0$. Under the normalization $\|u_n\|_{L^2} = 1$ for all $n \geq 1$, one can show that

$$\lim_{n \to \infty} \|(L - \lambda_0 I) u_n\|_{L^2} = 0, \quad \lambda_0 = c + k_0^2,$$

which confirm that $\lambda_0 = c + k_0^2 \in \sigma_c(L)$. Taking the closure of all such points λ_0, we conclude that the continuous spectrum of L is located at $\sigma_c(L) = [c, \infty)$, that is, for all $\lambda \geq c$. $\qquad\Box$

Remark 4.1 Embedded eigenvalues do occur in the spectrum of $L = -\partial_x^2 + c + V(x)$ for $\lambda \geq c$ if the potential $V(x)$ decays slower than the exponential rate (4.1.1). For instance, if $V(x) = \mathcal{O}(x^{-2})$ as $|x| \to \infty$, an embedded eigenvalue may occur at the end point of the continuous spectrum at $\lambda = c$, whereas if $V(x) = \mathcal{O}(|x|^{-1})$ as $|x| \to \infty$, an embedded eigenvalue may also occur in the spectrum of L for $\lambda > c$. Examples of these algebraically decaying potentials and their embedded eigenvalues are given by Klaus *et al.* [115].

Isolated eigenvalues λ of $\sigma_p(L)$ located for $\lambda < c$ play an important role in analysis of spectral stability of localized modes. These eigenvalues and the corresponding eigenfunctions are characterized by the following lemma.

Lemma 4.2 *Let $L = -\partial_x^2 + c + V(x)$ for a fixed $c > 0$ and let $V(x)$ satisfy (4.1.1). There exists a unique solution of the differential equation $Lu_0 = \lambda_0 u_0$ for a fixed $\lambda_0 < c$ such that $u_0 \in H^2((-\infty, x_0])$ for any $x_0 < \infty$ and*

$$\lim_{x \to -\infty} e^{-\nu_0 x} u_0(x) = 1, \quad \text{with} \quad \nu_0 = \sqrt{c - \lambda_0} > 0.$$

If $u_0(x)$ has $n(\lambda_0)$ zeros on $\mathbb{R}$ ($u_0(x)$ may diverge as $x \to \infty$), then there exists exactly $n(\lambda_0)$ eigenvalues of $\sigma_p(L)$ for $\lambda < \lambda_0$.

Proof Let us now parameterize $\lambda < c$ by $\lambda = c - \nu^2$ for fixed $\nu = \sqrt{c - \lambda} > 0$. We consider a solution of the differential equation

$$u''(x) - \nu^2 u(x) = V(x) u(x), \quad x \in \mathbb{R}. \tag{4.1.6}$$

The potential-free part has two particular solutions $e^{\nu x}$ and $e^{-\nu x}$, one of which corresponds to the exponential decay and the other one to the exponential growth in either limit $x \to -\infty$ or $x \to +\infty$. Let us consider a particular solution $u(x; \lambda)$ such that

$$\lim_{x \to -\infty} e^{-\nu x} u(x; \lambda) = 1.$$

In other words, $u(x, \lambda)$ decays to zero as $x \to -\infty$ and this decay is uniquely normalized. Using the Green's function, we infer that the particular solution $u(x; \lambda)$ satisfies the inhomogeneous integral equation

$$u(x; \lambda) = e^{\nu x} + \nu^{-1} \int_{-\infty}^{x} \sinh[\nu(x - x')] V(x') u(x'; \lambda) dx'. \qquad (4.1.7)$$

If $u(x; \lambda) = e^{\nu x} w(x; \lambda)$, then $w(x; \lambda)$ satisfies the equivalent integral equation

$$w(x; \lambda) = 1 + \nu^{-1} \int_{-\infty}^{x} \sinh[\nu(x - x')] e^{-\nu(x-x')} V(x') w(x'; \lambda) dx'. \qquad (4.1.8)$$

We note that

$$\nu^{-1} \sinh[\nu(x - x')] e^{-\nu(x-x')} = \frac{1}{2\nu} \left[1 - e^{-2\nu(x-x')} \right]$$

is bounded for any $x \geq x'$. Using the same technique as in the proof of Lemma 4.1, it is easy to prove that there exists a unique solution $w \in L^\infty((-\infty, x_0])$ of the inhomogeneous integral equation (4.1.8) for sufficiently large negative x_0. From the linear differential equation (4.1.6), this solution is uniquely continued for all $x \in \mathbb{R}$. Moreover, it has the bounded limit as $x \to \infty$:

$$A(\lambda) = \lim_{x \to \infty} w(x; \lambda) = 1 + \frac{1}{2\nu} \int_{\mathbb{R}} V(x) w(x; \lambda) dx.$$

The unique solution $w \in L^\infty(\mathbb{R})$ gives a unique solution $u \in H^2((-\infty, x_0])$ of the integral equation (4.1.7) for any $x_0 < \infty$. If $A(\lambda) \neq 0$, then $u(x; \lambda)$ diverges exponentially as $x \to \infty$.

On the other hand, if $A(\lambda) = 0$, then $w(x; \lambda) = \mathcal{O}(e^{-2\nu x})$ as $|x| \to \infty$ so that $u(x; \lambda)$ decays to zero as $x \to \infty$. The functions $w(x; \lambda)$ and $A(\lambda)$ are analytic for all $\lambda < c$. Because L is a self-adjoint operator with only semi-simple eigenvalues (Exercise 4.1), if $A(\lambda_0) = 0$ for some $\lambda_0 < c$, then $A'(\lambda_0) \neq 0$.

If $\lambda \leq \inf_{x \in \mathbb{R}} V(x)$ and $u(x) > 0$ for large negative x, then equation (4.1.8) implies that

$$u''(x) = (c + V(x) - \lambda) u(x) \geq 0, \quad x \in \mathbb{R}.$$

Therefore, $u(x; \lambda)$ remains strictly positive on $\mathbb{R}$ for this (sufficiently large negative) λ, hence $A(\lambda) > 0$. If λ increases, either $u(x; \lambda)$ remains positive for all $x \in \mathbb{R}$ (and then the lemma is proved for this λ) or $u(x; \lambda)$ acquires a new zero for $x > -\infty$ for some $\lambda_1 > \inf_{x \in \mathbb{R}} V(x)$.

For any $\lambda < c$, a zero of $u(x; \lambda)$ at any $x \in \mathbb{R}$ is simple (if the derivative is also zero at this x, then $u(x; \lambda) = 0$ for all $x \in \mathbb{R}$). Therefore, a continuous deformation of $u(x; \lambda)$ in λ may generate a new zero on $\mathbb{R}$ for $\lambda = \lambda_1$ if and only if $A(\lambda_1) = 0$. Therefore, the new zero of $u(x; \lambda)$ always occurs at $x = \infty$ if $\lambda_1 \in \sigma_p(L)$. Since $A'(\lambda_1) \neq 0$, $A(\lambda)$ changes sign for λ slightly larger than λ_1 so that $u(x; \lambda)$ has exactly one zero on $\mathbb{R}$ for λ slightly larger than λ_1. Therefore, the lemma is proved for this λ and we can continue increasing λ for $\lambda > \lambda_1$ and counting zeros of $u(x; \lambda)$ on $\mathbb{R}$ and eigenvalues of $\sigma_p(L)$ below this λ. The rest of the proof goes by induction. $\qquad \square$

Corollary 4.1 *If $\lambda \in \sigma_p(L)$, then λ is simple. Moreover, if λ is the nth eigenvalue sorted in increasing order, then the eigenfunction $u \in H^2(\mathbb{R})$ has exactly $n-1$ zeros on $\mathbb{R}$.*

Proof The subspace of all solutions of equation (4.1.6) that decay to zero as $x \to -\infty$ is spanned by solution $u(x; \lambda)$ of Lemma 4.2. Therefore, there is at most one linearly independent eigenfunction $u \in H^2(\mathbb{R})$ for any $\lambda \in \sigma_p(L)$. By the Sturm Comparison Theorem (Appendix B.12), the eigenfunction for the $(n+1)$th eigenvalue has at least one more zero on $\mathbb{R}$ compared to the eigenfunction for the nth eigenvalue. Since the eigenfunction for the first eigenvalue has no zeros on the real line, the eigenfunction for the nth eigenvalue has exactly $n-1$ zeros on $\mathbb{R}$. $\square$

We shall consider three results for an abstract self-adjoint operator L with $\sigma_c(L) \geq c > 0$ and $\dim \sigma_p(L) < \infty$. These results hold for the Schrödinger operator $L = -\partial_x^2 + c + V(x)$ with the exponentially decaying potential $V(x)$. In particular, we describe eigenvalues of a self-adjoint operator in constrained L^2 spaces (Section 4.1.1), Sylvester's Law of Inertia for differential operators (Section 4.1.2), and Pontryagin's invariant subspaces for L-symmetric operators (Section 4.1.3). The three results cover the main aspects in the modern analysis of spectral stability of localized modes in the nonlinear Schrödinger equations.

4.1.1 Eigenvalues in constrained L^2 spaces

Since the spectral stability problem involves inversions of differential operators L_+ and L_-, we need to add constraints on L^2 spaces if L_+ or L_- have non-trivial kernels and the rest of their spectrum is bounded away from zero. These constraints remove the zero eigenvalue from the spectrum of the stability problem and allow inversion of L_+ and L_-. From a geometric point of view, zero eigenvalues of L_+ and L_- are induced by the symmetries of the nonlinear evolution equation with respect to translational or gauge invariance.

Let L be a self-adjoint operator in L^2 with $\sigma_c(L) \geq c > 0$. Let $n(L)$ and $z(L)$ denote the finite numbers of negative and zero eigenvalues of $\sigma_p(L)$. Given a linearly independent set of vectors $\{v_j\}_{j=1}^N \in L^2$, define a subspace of L^2 by

$$L_c^2 = \left\{ u \in L^2 : \{\langle u, v_j \rangle_{L^2} = 0\}_{j=1}^N \right\}. \tag{4.1.9}$$

Since L_c^2 is determined by constraints, we say that L_c^2 is a *constrained L^2 space*. Using Lagrange multipliers $\{\nu_j\}_{j=1}^N$, eigenvalues of operator L restricted on L_c^2 are determined by the constrained spectral problem

$$Lu = \mu u - \sum_{j=1}^N \nu_j v_j, \quad u \in L^2, \tag{4.1.10}$$

subject to the conditions that $\langle u, v_j \rangle_{L^2} = 0$ for all $j \in \{1, 2, ..., N\}$. To be precise, we can find Lagrange multipliers $\{\nu_j\}_{j=1}^N$ by

$$\nu_j = -\frac{\langle Lu, v_j \rangle_{L^2}}{\|v_j\|_{L^2}^2}, \quad j \in \{1, 2, ..., N\},$$

if the vectors $\{v_j\}_{j=1}^N$ are mutually orthogonal.

We shall consider how the negative and zero eigenvalues of operator L change as a result of the constraints (4.1.9). The numbers of the corresponding eigenvalues of $L|_{L_c^2}$ are denoted by $n_c(L)$ and $z_c(L)$. Although the main result may have a much older history, the first count of eigenvalues of abstract self-adjoint operators in constrained L^2 spaces in the context of stability theory of solitary waves was developed by Grillakis *et al.* [75]. More recently, the same result was recovered with a simpler technique of matrix algebra by Pelinovsky [153] and Cuccagna *et al.* [43].

Theorem 4.1   *Let L be a self-adjoint operator in L^2 with $\sigma_c(L) \geq c > 0$ and $\dim \sigma_p(L) < \infty$. Let $A(\mu)$ be the matrix-valued function defined by*

$$\forall \mu \notin \sigma(L): \quad A_{i,j}(\mu) = \langle (\mu - L)^{-1} v_i, v_j \rangle_{L^2}, \quad i, j \in \{1, 2, ..., N\}. \qquad (4.1.11)$$

Let n_0, z_0, and p_0 be the numbers of negative, zero, and positive eigenvalues of $\lim_{\mu \uparrow 0} A(\mu)$ and denote $z_\infty = N - n_0 - z_0 - p_0$. Then,

$$n_c(L) = n(L) - p_0 - z_0, \quad z_c(L) = z(L) + z_0 - z_\infty. \qquad (4.1.12)$$

Remark 4.2   If $z(L) \geq 1$, then L is not invertible and some eigenvalues of $A(\mu)$ as $\mu \uparrow 0$ can be unbounded unless $\mathrm{Ker}(L) \subset L_c^2$.

Exercise 4.3   Let $\{v_j\}_{j=1}^N$ be a set of orthogonal eigenvectors of a self-adjoint operator L for negative eigenvalues. Show that $A(\mu)$ is a diagonal matrix and

$$n_c(L) = n(L) - N, \quad z_c = z(L), \quad p_0 = N, \quad z_0 = n_0 = z_\infty = 0.$$

Proof of Theorem 4.1 We represent $A(\mu)$ by

$$A_{i,j}(\mu) = \int_{\mu_1}^{\infty} \frac{\langle dE_\lambda v_i, v_j \rangle_{L^2}}{\mu - \lambda}, \quad \mu \notin \sigma(L), \qquad (4.1.13)$$

where E_λ is the spectral family associated with the operator L, and $\mu_1 \in \sigma_p(L)$ is the smallest eigenvalue of L. An easy calculation yields

$$A'_{i,j}(\mu) = -\int_{\mu_1}^{\infty} \frac{\langle dE_\lambda v_i, v_j \rangle_{L^2}}{(\mu - \lambda)^2} = -\langle (\mu - L)^{-2} v_i, v_j \rangle_{L^2}, \quad \mu \notin \sigma(L). \qquad (4.1.14)$$

Note that $B = -(\mu - L)^{-2}$ is a negative definite operator in L^2 in the sense

$$\langle Bu, u \rangle_{L^2} < 0 \quad \text{for all nonzero } u \in L^2.$$

Because the set $\{v_j\}_{j=1}^N \in L^2$ is linearly independent, the matrix representation of operator B with elements $[M(B)]_{i,j} = \langle Bv_i, v_j \rangle_{L^2}$ for all $i, j \in \{1, 2, ..., N\}$ is negative definite in $\mathbb{C}^N$ in the sense

$$\langle M(B)\mathbf{x}, \mathbf{x} \rangle_{\mathbb{C}^N} = \left\langle B \sum_{j=1}^N x_j v_j, \sum_{j=1}^N x_j v_j \right\rangle_{L^2} < 0 \quad \text{for all nonzero } \mathbf{x} \in \mathbb{C}^N.$$

Therefore, the matrix $A'(\mu)$ is negative definite on $\mathbb{C}^N$.

Note that the matrix $A(\mu)$ is symmetric. Let $\{\alpha_i(\mu)\}_{i=1}^N$ be a set of real eigenvalues of $A(\mu)$ and $\{\mathbf{x}_i(\mu)\}_{i=1}^N$ be a set of orthonormal eigenvectors of $A(\mu)$ in $\mathbb{C}^N$. Since

$$\alpha_i'(\mu) = \langle A'(\mu)\mathbf{x}_i(\mu), \mathbf{x}_j(\mu) \rangle_{\mathbb{C}^N}$$

and $A'(\mu)$ is negative definite, eigenvalues $\{\alpha_i(\mu)\}_{i=1}^N$ are monotonically decreasing functions for all $\mu \notin \sigma(L)$.

Let μ_k be an isolated eigenvalue of operator L of multiplicity m_k and $\{u_k^j\}_{j=1}^{m_k} \in L^2$ be an orthonormal set of eigenvectors of L for μ_k. The eigenvectors of L are orthonormal because L is self-adjoint. Denote the projection operator to the eigenspace of $\operatorname{Ker}(L - \mu_k I)$ by $P_k : L^2 \to U_k \subset L^2$, where $U_k = \operatorname{span}\{u_k^1, ..., u_k^{m_k}\}$. Via spectral calculus, we have

$$A(\mu) = \frac{1}{\mu - \mu_k} A_k + B_k(\mu) , \qquad (4.1.15)$$

where

$$[A_k]_{i,j} = \langle P_k v_i, P_k v_j \rangle_{L^2}, \quad [B_k]_{i,j}(\mu) = \int_{[\mu_1,\infty)\setminus(\mu_k-\delta,\mu_k+\delta)} \frac{\langle dE_\lambda v_i, v_j \rangle_{L^2}}{\mu - \lambda},$$

for all $i, j \in \{1, 2, ..., N\}$ and some small $\delta > 0$. Using Cauchy estimates, we obtain

$$\left| \frac{d^n}{d\mu^n} [B_k]_{i,j}(\mu) \right| \le n! \left(\frac{2}{\delta} \right)^{n+1} \|v_i\|_{L^2} \|v_j\|_{L^2}, \quad \mu \in \left(\mu_k - \frac{\delta}{2}, \mu_k + \frac{\delta}{2} \right).$$

By comparison with geometric series, $B_k(\mu)$ is analytic in the neighborhood of $\mu = \mu_k$. Therefore, the eigenvalue problem for $A(\mu)$ can be written as

$$(A_k + B_k(\mu)(\mu - \mu_k)) \mathbf{x}_i(\mu) = \alpha_i(\mu)(\mu - \mu_k)\mathbf{x}_i(\mu), \quad i \in \{1, 2, ..., N\}.$$

Eigenvalues of the analytic Hermitian matrix $A_k + B_k(\mu)(\mu - \mu_k)$ are analytic in the neighborhood of $\mu = \mu_k$, hence we obtain

$$\alpha_i(\mu)(\mu - \mu_k) = \alpha_i^0 + (\mu - \mu_k)\beta_i(\mu), \quad i \in \{1, 2, ..., N\},$$

where α_i^0 are eigenvalues of A_k and $\beta_i(\mu)$ are analytic near $\mu = \mu_k$. Therefore,

$$\alpha_i(\mu) = \frac{\alpha_i^0}{\mu - \mu_k} + \beta_i(\mu), \quad i \in \{1, 2, ..., N\} \qquad (4.1.16)$$

near $\mu = \mu_k$, which means that the behavior of $\alpha_i(\mu)$ in the neighborhood of $\mu = \mu_k$ depends on the rank of the matrix A_k. We note that the matrix A_k is non-negative for all $\mathbf{x} \in \mathbb{C}^N$ in the sense

$$\langle A_k \mathbf{x}, \mathbf{x} \rangle_{\mathbb{C}^N} = \left\| P_k \sum_{i=1}^N x_i v_i \right\|_{L^2}^2 \ge 0.$$

Therefore, $\alpha_i^0 \ge 0$, $1 \le i \le N$. Given $N_k = \operatorname{rank}(A_k)$, where $0 \le N_k \le \min(m_k, N)$, there are precisely N_k linearly independent $\{P_k v_i\}_{i=1}^{N_k}$. We can hence construct an orthonormal set of eigenvectors $\{u_k^j\}_{j=1}^{m_k}$ corresponding to μ_k, such that $u_k^j \notin L_c^2$, $1 \le j \le N_k$ and $u_k^j \in L_c^2$, $N_k + 1 \le j \le m_k$. In other words, $\alpha_i^0 > 0$, $1 \le i \le N_k$ and these $\alpha_i(\mu)$ diverge at μ_k as follows:

$$\lim_{\mu \uparrow \mu_k} \alpha_i(\mu) = -\infty, \qquad \lim_{\mu \downarrow \mu_k} \alpha_i(\mu) = +\infty.$$

On the other hand, $\alpha_i^0 = 0$, $N_k + 1 \le i \le N$ and these $\alpha_i(\mu)$ are continuous at μ_k. Assume now that the ith eigenvalue $\alpha_i(\mu)$ has vertical asymptotes at

$$\mu_{i_1} < \mu_{i_2} < \cdots < \mu_{i_{K_i}} < 0.$$

Using Schwarz inequality and spectral calculus for $\mu < \mu_1$, we obtain

$$|A_{ij}(\mu)| \le \left(\int_{\mu_1}^{\infty} \frac{\langle dE_\lambda v_i, v_i \rangle_{L^2}}{(\mu - \lambda)^2} \right)^{1/2} \|v_j\|_{L^2} \le \frac{\|v_i\|_{L^2} \|v_j\|_{L^2}}{|\mu| - |\mu_1|}, \quad \mu < \mu_1.$$

Therefore, $\lim_{\mu \to -\infty} \alpha_i(\mu) = -0$ and $\alpha_i(\mu)$ is monotonically decreasing on $(-\infty, \mu_{i_1})$.

On the interval $\mu \in (\mu_{i_l}, \mu_{i_{l+1}})$, $1 \le l \le K_i - 1$, the eigenvalue $\alpha_i(\mu)$ is continuous, monotonic, and has a simple zero at $\mu_l^* \in (\mu_{i_l}, \mu_{i_{l+1}})$ with the unique solution of the constrained eigenvalue problem (4.1.10) given by

$$u_l^* = \sum_{j=1}^{N} x_j (\mu_l^* - L)^{-1} v_j \in L_c^2,$$

where $A(\mu_l^*)\mathbf{x} = \mathbf{0}$ and $\|u_l^*\|_{L^2} < \infty$ since for each $i \in \{1, 2, ..., N\}$,

$$\|(\mu_l^* - L)^{-1} v_i\|_{L^2}^2 = \int_{\mu_1}^{\infty} \frac{\langle dE_\lambda v_i, v_i \rangle_{L^2}}{(\mu_l^* - \lambda)^2} \le \frac{\|v_i\|_{L^2}^2}{\min\{|\mu_l^* - \lambda|^2, \lambda \in \sigma(L)\}} < \infty.$$

Therefore, $(K_i - 1)$ negative eigenvalues of operator $L|_{L_c^2}$ are located at $\mu_l^* \in (\mu_{i_l}, \mu_{i_{l+1}})$, $1 \le l \le K_i - 1$. Due to the monotonicity, $\alpha_i(\mu) > 0$ on $\mu \in (\mu_{i_{K_i}}, 0)$ if $\alpha_i(0) > 0$ or has precisely one zero at $\mu_{K_i}^* \in (\mu_{i_{K_i}}, 0)$ if $\alpha_j(0) < 0$ or $\lim_{\mu \uparrow 0} \alpha_i(\mu) = -\infty$.

It is now time to count the numbers of negative and zero eigenvalues of operator $L|_{L_c^2}$. Sorting the negative eigenvalues of L in increasing order

$$\mu_1 < \mu_2 < \cdots < \mu_K < 0,$$

we have

$$n(L) = \sum_{k=1}^{K} m_k, \quad z(L) = m_{K+1}.$$

Let $\Theta(x)$ be the Heaviside step function such that $\Theta(x) = 1$ for $x \ge 0$ and $\Theta(x) = 0$ for $x < 0$. Each eigenvalue $\alpha_i(\mu)$ has K_i jump discontinuities and $K_i - \Theta(\alpha_i(0))$ zeros on $(-\infty, 0)$. On the other hand, for each eigenvalue $\mu_k < 0$, only $N_k \le \min(m_k, N)$ eigenvalues of $\{\alpha_i(\mu)\}_{i=1}^{N}$ diverge at μ_k and there exist $m_k - N_k$ eigenvectors $\{u_k^j\}_{j=1}^{m_k}$ that belong to L_c^2. Summing all contributions to the index $n_c(L)$ together and noting that $\sum_{i=1}^{N} K_i = \sum_{k=1}^{K} N_k$, we obtain

$$n_c(L) = \sum_{i=1}^{N} [K_i - \Theta(\alpha_i(0))] + \sum_{k=1}^{K} [m_k - N_k] = n(L) - (p_0 + z_0),$$

where $p_0 + z_0$ is the total number of non-negative eigenvalues of $\lim_{\mu \uparrow 0} A(\mu)$. Counting what is left over at $\mu = 0$, we obtain

$$z_c(L) = z_0 + m_{K+1} - N_{K+1} = z(L) + z_0 - z_\infty,$$

where $z_\infty = N_{K+1}$ is the number of unbounded eigenvalues of $A(\mu)$ as $\mu \uparrow 0$. $\square$

Exercise 4.4 Let ϕ be a localized mode of the stationary NLS equation with the power nonlinearity,

$$-\phi''(x) + c\phi(x) - (p+1)\phi^{2p+1}(x) = 0, \quad x \in \mathbb{R},$$

where $c > 0$ and $p > 0$ are parameters. Let

$$L = -\partial_x^2 + c - (2p+1)(p+1)\phi^{2p}(x)$$

and consider the constrained L^2 space given by

$$L_c^2(\mathbb{R}) = \left\{ u \in L^2(\mathbb{R}) : \langle u, \phi \rangle_{L^2} = 0 \right\}.$$

Show that

$$n(L) = 1, \quad z(L) = 1, \quad n_c(L) = \begin{cases} 1, & p > 2, \\ 0, & p < 2, \end{cases} \quad z_c(L) = 1.$$

4.1.2 Sylvester's Law of Inertia

Spectral stability problems for localized modes of the nonlinear Schrödinger equations can be written in many cases as the generalized eigenvalue problem,

$$Lu = \gamma M^{-1}u, \tag{4.1.17}$$

where L and M are unbounded differential operators. We assume that M is a strictly positive self-adjoint operator such that $M^{-1} : L^2 \to L^2$ is a bounded non-negative operator, whereas L is a self-adjoint operator in L^2 with $\sigma_c(L) \geq c > 0$.

Let $n(L)$ and $z(L)$ denote finite numbers of negative and zero eigenvalues of operator L. Under this condition, we would like to ask if the numbers of negative and zero eigenvalues of the generalized eigenvalue problem (4.1.17) can be counted from the numbers $n(L)$ and $z(L)$. The answer is given by a generalization of Sylvester's Law of Inertia [200] from matrices to differential operators obtained by Pelinovsky [153].

Let $\{u_k\}_{k=1}^{n(L)}$ be a set of orthonormal eigenvectors of L for negative eigenvalues $\{\lambda_k\}_{k=1}^{n(L)}$ and let $\{u_k\}_{k=n(L)+1}^{n(L)+z(L)}$ be an orthonormal system of eigenvectors of L for the zero eigenvalue if it exists. We denote $U = \text{Span}\{u_1, ..., u_K\} \subset L^2$ with $K = n(L) + z(L)$ and consider the orthogonal decomposition of L^2 into the direct sum $L^2 = U \oplus U^\perp$, where $U^\perp$ is the orthogonal complement of U in L^2. In the coordinate representation, we have

$$\forall u \in L^2 : \quad u = \sum_{k=1}^{K} \alpha_k u_k + u^\perp, \quad \alpha_k = \langle u, u_k \rangle_{L^2}, \quad u^\perp \in U^\perp. \tag{4.1.18}$$

Let us now consider the *quadratic form* $Q_L(u)$ associated with the self-adjoint operator L given by

$$\forall u \in \text{Dom}(L) : \quad Q_L(u) = \langle Lu, u \rangle_{L^2}.$$

Let $P_c : L^2 \to U^\perp \subset L^2$ be the orthogonal projection operator. The quadratic form $Q_L(u)$ becomes diagonal in the coordinate representation (4.1.18) according to the sum of squares

$$Q_L(u) = \sum_{k=1}^{K} \lambda_k |\alpha_k|^2 + Q_L(P_c u), \quad Q_L(P_c u) = \langle P_c L P_c u, u \rangle_{L^2}. \tag{4.1.19}$$

Since P_cLP_c is strictly positive, the last term of the sum of squares (4.1.19) is positive for $u \neq 0$, whereas the first sum contains $n(L)$ negative and $z(L)$ zero terms.

Let us represent the positive self-adjoint operator M in the form $M := SS^*$, where $S : \text{Dom}(S) \to L^2$ is an invertible operator and S^* is the adjoint operator in the sense

$$\forall f \in \text{Dom}(S), \quad \forall g \in \text{Dom}(S^*) : \quad \langle f, S^* g \rangle_{L^2} = \langle Sf, g \rangle_{L^2}.$$

For all nonzero $u \in \text{Dom}(M)$, we have $\langle Mu, u \rangle_{L^2} = \|S^* u\|_{L^2}^2 > 0$.

Exercise 4.5 Let ϕ be a positive localized mode of the stationary NLS equation with the power nonlinearity,

$$-\phi''(x) + c\phi(x) - (p+1)\phi^{2p+1}(x) = 0, \quad x \in \mathbb{R},$$

where $c > 0$ and $p > 0$ are parameters. Show that

$$L = -\partial_x^2 + c - (p+1)\phi^{2p}(x) = S^* S,$$

where

$$S = -\partial_x + \frac{\phi'(x)}{\phi(x)},$$

and prove that $\text{Ker}(L) = \text{Ker}(S) = \text{Span}\{\phi\}$.

Let us define the *congruent* operator to L by

$$\tilde{L} = S^* L S. \tag{4.1.20}$$

The quadratic forms $Q_{\tilde{L}}(v)$ and $Q_L(u)$ are related by the transformation

$$u = Sv, \quad v = S^{-1} u,$$

thanks to the explicit computation

$$Q_{\tilde{L}}(v) = \langle S^* LSv, v \rangle_{L^2} = \langle L(Sv), (Sv) \rangle_{L^2} = Q_L(Sv) = Q_L(u).$$

The spectrum of $\tilde{L}$ is not, however, equivalent to the spectrum of L since

$$\tilde{L}v = \lambda v \;\Rightarrow\; S^* LSv = \lambda v \;\Rightarrow\; L(Sv) = \lambda (S^*)^{-1} S^{-1}(Sv) \;\Rightarrow\; Lu = \lambda M^{-1} u.$$

Therefore, if (v, λ) is an eigenvector–eigenvalue pair for $\tilde{L}$, then the generalized eigenvalue problem (4.1.17) with $M = SS^*$ has eigenvector $u = Sv$ for eigenvalue λ.

If L is self-adjoint, then $\tilde{L}$ is self-adjoint and the eigenvalues of the generalized eigenvalue problem (4.1.17) are real and semi-simple. Therefore, the multiplicities of the zero eigenvalue in $\sigma_p(L)$ and $\sigma_p(\tilde{L})$ coincide and $z(\tilde{L}) = z(L)$. We shall now consider the number of negative eigenvalues of $\tilde{L}$ and compare $n(\tilde{L})$ and $n(L)$.

Let $\{v_k\}_{k=1}^{\tilde{K}}$ with $\tilde{K} = n(\tilde{L}) + z(\tilde{L})$ be a set of orthonormal eigenvectors of $\tilde{L}$ for negative and zero eigenvalues $\{\gamma_k\}_{k=1}^{\tilde{K}}$ of $\sigma_p(\tilde{L})$. Eigenvectors $\{\tilde{u}_k\}_{k=1}^{\tilde{K}}$ with $\tilde{u}_k = Sv_k$ of the generalized eigenvalue problem (4.1.17) inherit orthonormality with respect to the weight M^{-1} since

$$\langle M^{-1}\tilde{u}_j, \tilde{u}_k \rangle_{L^2} = \langle (S^*)^{-1} S^{-1} Sv_j, Sv_k \rangle_{L^2} = \langle v_j, v_k \rangle_{L^2}.$$

Let $\tilde{U} = \text{span}\{\tilde{u}_1, \tilde{u}_2, ..., \tilde{u}_{\tilde{K}}\}$ and $\tilde{P}_c : L^2 \to \tilde{U}^\perp \subset L^2$ be the orthogonal projection operator with respect to the weight M^{-1} in the sense

$$\forall \tilde{u}^\perp \in \tilde{U}^\perp : \quad \langle M^{-1}\tilde{u}^\perp, \tilde{u}_k \rangle_{L^2} = 0, \quad k \in \{1, 2, ..., \tilde{K}\}.$$

In the coordinate representation, we have

$$\forall u \in L^2 : \quad u = \sum_{k=1}^{\tilde{K}} \beta_k \tilde{u}_k + \tilde{u}^\perp, \quad \beta_k = \langle M^{-1}u, \tilde{u}_k \rangle_{L^2}, \quad \tilde{u}^\perp \in \tilde{U}^\perp. \quad (4.1.21)$$

The quadratic form $Q_L(u) = Q_{\tilde{L}}(S^{-1}u)$ becomes diagonal in the coordinate representation (4.1.21) according to the sum of squares

$$Q_L(u) = \sum_{k=1}^{\tilde{K}} \gamma_k |\beta_k|^2 + Q_L(\tilde{P}_c u), \quad Q_L(\tilde{P}_c u) = \langle \tilde{P}_c L \tilde{P}_c u, u \rangle_{L^2}. \quad (4.1.22)$$

Again operator $\tilde{P}_c L \tilde{P}_c$ is strictly positive because the spectrum of the generalized eigenvalue problem (4.1.17) includes only $n(\tilde{L})$ negative and $z(\tilde{L})$ zero eigenvalues and the rest of the spectrum γ is strictly positive. As a result, the last term in the last sum of squares (4.1.22) is positive for $u \neq 0$, whereas the first sum contains $n(\tilde{L})$ negative and $z(\tilde{L})$ zero terms.

We shall now prove that if the quadratic form $Q_L(u)$ is diagonalized to the sum of squares in two different ways (4.1.19) and (4.1.22), then the numbers of negative and zero terms in each sum of squares are equal.

Theorem 4.2 *Let L be a self-adjoint operator in L^2 with $\sigma_c(L) \geq c > 0$ and $\dim \sigma_p(L) < \infty$. Let $\tilde{L}$ be a congruent operator (4.1.20) associated with an invertible operator S. Then, the congruent operators have equal numbers of negative and zero eigenvalues, that is, $n(\tilde{L}) = n(L)$ and $z(\tilde{L}) = z(L)$.*

Proof We only need to prove that $n(\tilde{L}) = n(L)$ because the kernels of L and $\tilde{L}$ are related by the transformation $u = Sv$ and $v = S^{-1}u$ for solutions of $Lu = 0$ and $\tilde{L}v = 0$ in L^2.

Let us consider negative eigenvalues of L and $\tilde{L}$. By contradiction, we assume that $n(\tilde{L}) > n(L)$ and show that this is false. The case $n(\tilde{L}) < n(L)$ is treated similarly. To find a contradiction, we construct a vector $u \in L^2$ by the following linear combination

$$u = \sum_{k=1}^{n(\tilde{L})} c_k \tilde{u}_k + u^\perp, \quad u^\perp \in U^\perp.$$

Expand $\tilde{u}_k$ into similar linear combinations

$$\tilde{u}_k = \sum_{j=1}^{n(L)} \alpha_{k,j} u_j + u_k^\perp, \quad k \in \{1, 2, ..., n(\tilde{L})\},$$

so that

$$u = \sum_{j=1}^{n(L)} \left(\sum_{k=1}^{n(\tilde{L})} \alpha_{k,j} c_k \right) u_j + \left(u^\perp + \sum_{k=1}^{n(\tilde{L})} c_k u_k^\perp \right).$$

Since the set of eigenvectors of $\sigma_p(L)$ is linearly independent, setting $u = 0$ gives an under-determined system of linear equations on $\{c_1, c_2, ..., c_{n(\tilde{L})}\}$,

$$\sum_{k=1}^{n(\tilde{L})} \alpha_{k,j} c_k = 0, \quad j \in \{1, ..., n(L)\}, \qquad (4.1.23)$$

and the residual equation

$$u^\perp + \sum_{k=1}^{n(\tilde{L})} c_k u_k^\perp = 0.$$

If $n(L) < n(\tilde{L})$, the linear system (4.1.23) is under-determined and always admits a nonzero solution for $\{c_k\}_{k=1}^{n(\tilde{L})}$ that generates a nonzero vector $u_0 \in L^2$ given by

$$u_0 = \sum_{k=1}^{n(\tilde{L})} c_k \tilde{u}_k = -u^\perp \neq 0.$$

The quadratic form $Q_L(u_0)$ is bounded in two contradictory ways as follows:

$$Q_L(u_0) = \sum_{k=1}^{n(\tilde{L})} \sum_{j=1}^{n(\tilde{L})} c_k c_j \langle L\tilde{u}_k, \tilde{u}_j \rangle_{L^2} = \sum_{k=1}^{n(\tilde{L})} \gamma_k |c_k|^2 < 0,$$

$$Q_L(u_0) = \langle Lu^\perp, u^\perp \rangle_{L^2} = \langle P_c L P_c u, u \rangle_{L^2} > 0,$$

since $P_c L P_c$ is strictly positive. The contradiction is resolved if and only if $n(\tilde{L}) = n(L)$. $\qquad \square$

If operator M^{-1} in the generalized eigenvalue problem (4.1.17) is invertible but non-positive, eigenvalues γ can be complex-valued and multiple. In this case, the factorization $M = SS^*$ and the congruent self-adjoint operator $\tilde{L} = S^*LS$ do not exist and it is hard to prove that $Q_L(\tilde{P}_c u) > 0$ for all nonzero $u \in L^2$, where $\tilde{P}_c$ is now the projection operator to the orthogonal complement of all eigenvectors of the generalized eigenvalue problem (4.1.17).

One way around this obstacle is to develop scattering theory for solutions of $Lu = \gamma M^{-1}u$ under some restrictive assumptions – see Buslaev & Perelman [25] and Pelinovsky [153]. The other way out is to use the wave operator theory as in Weder [210] and Cuccagna *et al.* [43]. In both methods, the count of complex and multiple eigenvalues in the generalized eigenvalue problem (4.1.17) in terms of the negative eigenvalues of L is performed using the count of negative eigenvalues in the Jordan canonical blocks for eigenspaces related to the complex and multiple eigenvalues. Construction of Jordan blocks for multiple eigenvalues of the generalized eigenvalue problem (4.1.17) is discussed by Vougalter & Pelinovsky [209].

Exercise 4.6 Consider the generalized eigenvalue problem (4.1.17) and assume that M is invertible but non-positive. Also assume that there is a simple pair of complex conjugate eigenvalues γ_0 and $\bar{\gamma}_0$ with complex-valued eigenvectors $u_0, \bar{u}_0 \in L^2$ and that all other isolated eigenvalues are real and simple. Prove that

$$n(L) = N_r + 1,$$

where N_r is the number of real eigenvalues γ with eigenvectors u such that $\langle Lu, u \rangle_{L^2} < 0$.

Exercise 4.7 Consider the generalized eigenvalue problem (4.1.17) and assume that M is invertible but non-positive. Also assume that there is a double eigenvalue $\gamma_0 \in \mathbb{R}$, $\gamma \neq 0$ with the generalized eigenvectors $u_0, u_1 \in L^2$ defined by

$$Lu_0 = \gamma_0 M^{-1} u_0, \quad Lu_1 = \gamma_0 M^{-1} u_1 + M^{-1} u_0,$$

under the conditions that

$$\langle M^{-1} u_0, u_0 \rangle_{L^2} = 0 \quad \text{and} \quad \langle M^{-1} u_1, u_0 \rangle_{L^2} \neq 0.$$

Assume that all other isolated eigenvalues are real and simple. Prove that

$$n(L) = N_r + 1,$$

where N_r is the number of real eigenvalues γ with eigenvectors u such that $\langle Lu, u \rangle_{L^2} < 0$.

4.1.3 Pontryagin's invariant subspaces

A generalization of Sylvester's Law of Inertia is useful to count eigenvalues of the generalized eigenvalue problem (4.1.17) for sign-indefinite operators L and M with a finite number of negative and zero eigenvalues. This generalization is referred to as Pontryagin's Invariant Subspace Theorem.

Let $L : \text{Dom}(L) \subseteq L^2 \to L^2$ be a self-adjoint differential operator with $\sigma_c(L) \geq c > 0$. We assume that L has κ negative eigenvalues and no zero eigenvalues, that is, $n(L) = \kappa < \infty$ and $z(L) = 0$.

Definition 4.7 We say that $\text{Dom}(L) \subseteq L^2$ is *Pontryagin space* (denoted as Π_κ) if it can be decomposed into the direct orthogonal sum

$$\Pi_\kappa = \Pi_+ \oplus \Pi_-, \quad \Pi_+ \cap \Pi_- = \varnothing, \tag{4.1.24}$$

in the sense

$$\forall u \in \Pi_+ : \langle Lu, u \rangle_{L^2} > 0, \quad \forall u \in \Pi_- : \langle Lu, u \rangle_{L^2} < 0,$$

and

$$\forall u_+ \in \Pi_+, \forall u_- \in \Pi_- : \langle Lu_+, u_- \rangle_{L^2} = 0.$$

The direct orthogonal sum (4.1.24) implies that any nonzero $u \in \Pi_\kappa$ can be represented in the form $u = u_+ + u_-$, where $u_\pm \in \Pi_\pm$. We shall use the notation $u = \{u_-, u_+\}$ for this representation and the notation $[\cdot, \cdot] = \langle L\cdot, \cdot \rangle_{L^2}$ for the quadratic form associated with the operator L.

Definition 4.8 We say that Π is a *non-positive subspace* of Π_κ if $\langle Lu, u \rangle_{L^2} \leq 0$ for all $u \in \Pi$. We say that Π is a *maximal non-positive subspace* if any subspace of Π_κ of dimension higher than $\dim(\Pi)$ is not a non-positive subspace of Π_κ.

Lemma 4.3 *The dimension of the maximal non-positive subspace of* Π_κ *is* κ.

Proof By contradiction, we assume that there exists a $(\kappa + 1)$-dimensional non-positive subspace $\tilde{\Pi}$ of Π_κ. Let $\{e_1, e_2, ..., e_\kappa\}$ be a basis in Π_-. We fix two elements $u_1, u_2 \in \tilde{\Pi}$ with the same projections to $\{e_1, e_2, ..., e_\kappa\}$, so that

$$u_1 = \alpha_1 e_1 + \alpha_2 e_2 + \cdots + \alpha_\kappa e_\kappa + u_1^+,$$
$$u_2 = \alpha_1 e_1 + \alpha_2 e_2 + \cdots + \alpha_\kappa e_\kappa + u_2^+,$$

where $u_1^+, u_2^+ \in \Pi_+$. It is clear that $u_1 - u_2 = u_1^+ - u_2^+ \in \Pi_+$, so that

$$[u_1^+ - u_2^+, u_1^+ - u_2^+] \geq 0.$$

On the other hand, since $\tilde{\Pi}$ is a subspace, $u_1 - u_2 \in \tilde{\Pi}$, so that

$$[u_1^+ - u_2^+, u_1^+ - u_2^+] \leq 0.$$

Hence, $u_1^+ = u_2^+$ and $u_1 = u_2$. Therefore, u_1^+ is uniquely determined by coordinates $\{\alpha_1, \alpha_2, ..., \alpha_\kappa\}$ and $\dim \tilde{\Pi} = \kappa$. $\qquad\square$

Exercise 4.8 Prove the Cauchy–Schwarz inequality in Pontryagin spaces:

$$\forall f, g \in \Pi : \quad |[f, g]|^2 \leq [f, f][g, g],$$

where Π is a non-positive subspace of Π_κ.

Exercise 4.9 Prove that if Π is a T-invariant subspace of Π_κ and $\Pi^\perp$ is the orthogonal complement of Π in Π_κ with respect to $[\cdot, \cdot]$, then $\Pi^\perp$ is also invariant with respect to T.

The main application of the Pontryagin space is the Pontryagin Invariant Subspace Theorem, which is formulated for the L-symmetric operators below.

Theorem 4.3 *Let T be an L-symmetric bounded operator in Π_κ such that*

$$\forall f, g \in \mathrm{Dom}(T) : \quad \langle LTf, g \rangle_{L^2} = \langle Lf, Tg \rangle_{L^2}. \qquad (4.1.25)$$

There exists a κ-dimensional, maximal non-positive, T-invariant subspace of Π_κ.

The first proof of this theorem was given by L.S. Pontryagin in [173] using the theory of analytic functions. A more general proof based on angular operators was developed by M.G. Krein and his students (see books [12, 70, 91]). The same theorem was rediscovered by M. Grillakis in [77] with the use of topology. In the usual twist between Russian and American literature, the Pontryagin Invariant Subspace Theorem is often referred to as the Grillakis Theorem by readers who are unfamiliar with the history of the subject.

In what follows, we follow the work of Chugunova & Pelinovsky [33] and give a geometric proof of Theorem 4.3 based on the Schauder Fixed-Point Theorem (Appendix B.2). The proof uses the Cayley transformation of a self-adjoint operator in Π_κ to a unitary operator in Π_κ (Lemma 4.4) and the Krein representation of the maximal non-positive subspace of Π_κ in terms of a graph of the contraction map (Lemma 4.5).

Lemma 4.4 *Let T be a linear operator in Π_κ and $z \in \mathbb{C}$, $\mathrm{Im}(z) > 0$ be a regular point of the operator T, such that $z \notin \sigma(T)$. Let U be the Cayley transform of T defined by*

$$U = (T - \bar{z})(T - z)^{-1}.$$

The operators T and U have the same invariant subspaces in Π_κ.

Proof Let Π be a finite-dimensional invariant subspace of the operator T in Π_κ. If $z \notin \sigma(T)$, then $(T - z)\Pi = \Pi$, $(T - z)^{-1}\Pi = \Pi$ and $(T - \bar{z})(T - z)^{-1}\Pi \subseteq \Pi$, that is $U\Pi \subseteq \Pi$. Conversely, let Π be an invariant subspace of the operator U. It follows from $U - I = (z - \bar{z})(T - z)^{-1}$ that $1 \notin \sigma(U)$, therefore $\Pi = (U - I)\Pi = (T - z)^{-1}\Pi$. From there, $\Pi \subseteq \mathrm{Dom}(T)$ and $(T - z)\Pi = \Pi$ so that $T\Pi \subseteq \Pi$. $\qquad\square$

Corollary 4.2 *If T is a self-adjoint operator in Π_κ, then U is a unitary operator in Π_κ.*

Proof We shall prove that

$$\forall g \in \mathrm{Dom}(U) \subset \Pi_\kappa : \quad \langle LUg, Ug \rangle_{L^2} = \langle Lg, g \rangle_{L^2}.$$

Using the notation $[\cdot, \cdot] := \langle L\cdot, \cdot \rangle_{L^2}$, we proceed by the explicit computation

$$[Ug, Ug] = [(T - \bar{z})f, (T - \bar{z})f] = [Tf, Tf] - \bar{z}[f, Tf] - z[Tf, f] + |z|^2[f, f],$$
$$[g, g] = [(T - z)f, (T - z)f] = [Tf, Tf] - \bar{z}[f, Tf] - z[Tf, f] + |z|^2[f, f],$$

where we have introduced $f \in \mathrm{Dom}(T)$ such that $f = (T - z)^{-1}g$. $\qquad\square$

Lemma 4.5 *A linear subspace $\Pi \subseteq \Pi_\kappa$ is a κ-dimensional non-positive subspace of Π_κ if and only if it is a graph of the contraction map $K : \Pi_- \to \Pi_+$, such that*

$$\forall u \in \Pi : \quad u = \{u_-, Ku_-\} \quad \text{and} \quad \|Ku_-\| \le \|u_-\|,$$

where $\|u_\pm\| := |\langle Lu_\pm, u_\pm \rangle_{L^2}|^{1/2}$, $u_\pm \in \Pi_\pm$.

Proof Let Π be a κ-dimensional non-positive subspace of Π_κ. We will show that there exists a contraction map $K : \Pi_- \to \Pi_+$ such that Π is a graph of K. Indeed, the subspace Π is a graph of a linear operator K if and only if it follows from $u = \{0, u_+\} \in \Pi$ that $u_+ = 0$. Since Π is non-positive with respect to $\langle L\cdot, \cdot \rangle_{L^2}$, then

$$\forall u = \{u_-, u_+\} \in \Pi : \quad \langle Lu, u \rangle_{L^2} = \langle Lu_+, u_+ \rangle_{L^2} + \langle Lu_-, u_- \rangle_{L^2}$$
$$\equiv \|u_+\|^2 - \|u_-\|^2 \le 0.$$

Therefore, $0 \le \|u_+\| \le \|u_-\|$ and if $u_- = 0$ then $u_+ = 0$. Moreover, for any $u_- \in \Pi_-$, it is true that $\|Ku_-\| \le \|u_-\|$, hence K is a contraction map. Conversely, let K be a contraction map $K : \Pi_- \to \Pi_+$. Then, we have

$$\forall u = \{u_-, u_+\} \in \Pi : \quad \langle Lu, u \rangle_{L^2} = \|Ku_-\|^2 - \|u_-\|^2 \le 0.$$

Therefore, the graph of K belongs to the non-positive subspace of Π_κ and $\dim(\Pi) = \dim(\Pi_\kappa) = \kappa$. $\qquad\square$

Proof of Theorem 4.3 Let $z \in \mathbb{C}$, $\mathrm{Im}(z) > 0$ be a regular point of the self-adjoint operator T in Π_κ. Let $U = (T - \bar{z})(T - z)^{-1}$ be the Cayley transform of T. By Corollary 4.2, U is a unitary operator in Π_κ. By Lemma 4.4, T and U have the same invariant subspaces in Π_κ. Therefore, the existence of the maximal non-positive invariant subspace for the self-adjoint operator T can be proved from the existence of such a subspace for the unitary operator U. Let $u = \{u_-, u_+\}$ and let

$$U = \left[\begin{array}{cc} U_{11} & U_{12} \\ U_{21} & U_{22} \end{array} \right]$$

be the matrix representation of the operator U with respect to the decomposition (4.1.24). Let Π denote a κ-dimensional non-positive subspace in Π_κ. Since U has a trivial kernel in Π_κ and U is unitary in Π_κ, then

$$\forall u \in \Pi : \quad \langle LUu, Uu \rangle_{L^2} = \langle Lu, u \rangle_{L^2} \leq 0.$$

Therefore, $\tilde{\Pi} = U\Pi$ is also a κ-dimensional non-positive subspace of Π_κ. By Lemma 4.5, there exist two contraction mappings K and $\tilde{K}$ for subspaces Π and $\tilde{\Pi}$, respectively. As a result, $\tilde{\Pi} = U\Pi$ is written in the form,

$$\left(\begin{array}{c} \tilde{u}_- \\ \tilde{K}\tilde{u}_- \end{array} \right) = \left[\begin{array}{cc} U_{11} & U_{12} \\ U_{21} & U_{22} \end{array} \right] \left(\begin{array}{c} u_- \\ Ku_- \end{array} \right) = \left(\begin{array}{c} (U_{11} + U_{12}K)u_- \\ (U_{21} + U_{22}K)u_- \end{array} \right),$$

which is equivalent to the scalar equation,

$$U_{21} + U_{22}K = \tilde{K}(U_{11} + U_{12}K).$$

We shall prove that the operator $(U_{11} + U_{12}K)$ is invertible. By contradiction, we assume that there exists $u_- \neq 0$ such that $\tilde{u}_- = (U_{11} + U_{12}K)u_- = 0$. If $\tilde{u}_- = 0$, then $\tilde{u}_+ = \tilde{K}\tilde{u}_- = 0$ and we obtain that $\{u_-, Ku_-\}$ is an eigenvector in the kernel of U. However, U has a trivial kernel in Π_κ, hence $u_- = 0$. Let $F(K)$ be an operator-valued function on the space of contraction operators given by

$$F(K) = (U_{21} + U_{22}K)(U_{11} + U_{12}K)^{-1},$$

and rewrite the equation above in the form $\tilde{K} = F(K)$. By Lemma 4.5, the operator $F(K)$ maps the operator unit ball $\|K\| \leq 1$ to itself. Since U is a continuous operator and U_{12} is a finite-dimensional operator, then U_{12} is a compact operator. Hence the operator ball $\|K\| \leq 1$ is a weakly compact set and the function $F(K)$ is continuous with respect to weak topology. By the Schauder Fixed-Point Theorem (Appendix B.2), there exists a fixed point K_0 such that $F(K_0) = K_0$ and $\|K_0\| \leq 1$. By Lemma 4.5, the graph of K_0 defines the κ-dimensional non-positive subspace Π, which is invariant with respect to U. By Lemma 4.3, the κ-dimensional non-positive subspace Π is the maximal non-positive subspace of Π_κ. $\square$

Exercise 4.10   Consider a generalized eigenvalue problem $Lu = \lambda M^{-1}u$, where L and M are self-adjoint bounded invertible operators in L^2 and show that $T = L^{-1}M^{-1}$ is symmetric with respect to operator M^{-1} in the sense of (4.1.25). Assume that $\kappa = n(M) = 0$ and derive the Sylvester Law of Inertia (Theorem 4.2) from the Pontryagin Invariant Subspace Theorem (Theorem 4.3).

Exercise 4.11 Under the same conditions as in Exercise 4.10, show that $T = L^{-1}M^{-1}$ is also symmetric with respect to operator L in Pontryagin space Π_κ with $\kappa = n(L)$.

4.2 Unstable and stable eigenvalues in the spectral stability problem

Consider the generalized nonlinear Schrödinger equation,

$$i\psi_t = -\psi_{xx} - f(|\psi|^2)\psi, \tag{4.2.1}$$

where $\psi(x,t) : \mathbb{R} \times \mathbb{R}_+ \to \mathbb{C}$ is the wave function and $f(|\psi|^2) : \mathbb{R}_+ \to \mathbb{R}$ is the nonlinear function such that $f(0) = 0$ and $f'(0) > 0$.

Let $\phi(x) : \mathbb{R} \to \mathbb{R}$ be a localized mode of the stationary generalized NLS equation,

$$-\phi''(x) - f(\phi^2)\phi(x) = \omega\phi(x), \quad x \in \mathbb{R}. \tag{4.2.2}$$

When we substitute

$$\psi(x,t) = \left(\phi(x) + [u(x) + iw(x)]e^{\lambda t} + [\bar{u}(x) + i\bar{w}(x)]e^{\bar{\lambda}t}\right)e^{-i\omega t} \tag{4.2.3}$$

and neglect quadratic terms in u and w, we obtain the spectral stability problem,

$$L_+u = -\lambda w, \quad L_-w = \lambda u, \quad \lambda \in \mathbb{C} \tag{4.2.4}$$

associated with the Schrödinger operators $L_\pm$,

$$\begin{cases} L_+ = -\partial_x^2 - \omega - f(\phi^2) - 2\phi^2 f'(\phi^2), \\ L_- = -\partial_x^2 - \omega - f(\phi^2). \end{cases}$$

If we consider the case of the power NLS equation with $f(\phi^2) = (p+1)\phi^{2p}$ for $p > 0$, then the localized mode ϕ exists in explicit form for any $\omega < 0$ (Exercise 3.31 in Section 3.3.4),

$$\phi(x) = |\omega|^{1/2p} \operatorname{sech}^{1/p}\left(p\sqrt{|\omega|}(x-s)\right), \quad x \in \mathbb{R}, \tag{4.2.5}$$

where $s \in \mathbb{R}$ is an arbitrary parameter of the spatial translation. Note that $\phi(x) > 0$ for all $x \in \mathbb{R}$. The following theorem gives sufficient conditions for stability and instability of the localized mode ϕ in the generalized NLS equation (4.2.1).

Theorem 4.4 *Assume that a localized mode ϕ exists in the stationary NLS equation (4.2.2) such that $\phi(x) > 0$ for all $x \in \mathbb{R}$ and the map $\mathbb{R} \ni \omega \mapsto \phi \in H^2(\mathbb{R})$ is C^1. The localized mode ϕ is spectrally unstable if $\partial_\omega \|\phi\|_{L^2}^2 > 0$ and is spectrally stable if $\partial_\omega \|\phi\|_{L^2}^2 < 0$.*

Proof By Lemma 4.1, because $\phi(x) \to 0$ as $|x| \to \infty$ exponentially fast, we have

$$\sigma_c(L_\pm) = \sigma(-\partial_x^2 - \omega) = [|\omega|, \infty).$$

Lemma 4.2 implies now that $\sigma_p(L_\pm)$ includes finitely many non-positive eigenvalues. The number of negative eigenvalues of $L_\pm$ is denoted by $n(L_\pm)$.

It is clear that $L_-\phi = 0$ is equivalent to the stationary NLS equation (4.2.2) thanks to the gauge translation. Because $\phi(x) > 0$ for all $x \in \mathbb{R}$, Corollary 4.1 implies that

$$n(L_-) = 0, \quad \mathrm{Ker}(L_-) = \mathrm{Span}\{\phi\}.$$

Furthermore, $L_+\phi' = 0$ thanks to the spatial translation of $\phi(x)$. Using the same argument, since $\phi'(x)$ has only one zero on $\mathbb{R}$, we obtain

$$n(L_+) = 1, \quad \mathrm{Ker}(L_+) = \mathrm{Span}\{\phi'\} \perp \mathrm{Ker}(L_-).$$

To count the unstable eigenvalues in the spectral stability problem (4.2.4) with $\mathrm{Re}(\lambda) > 0$, we note that if $(u, w) \in H^2(\mathbb{R}) \times H^2(\mathbb{R})$ is an eigenvector for $\lambda \neq 0$, then u belongs to the constrained L^2 space,

$$L_c^2 = \left\{ u \in L^2(\mathbb{R}) : \quad \langle u, \phi \rangle_{L^2} = 0 \right\}. \tag{4.2.6}$$

Let $P_c : L^2 \to L_c^2 \subset L^2$ be the orthogonal projection operator and note that $L_c^2 = \mathrm{Ran}(L_-)$. Inverting L_- for $u \in L_c^2$, we eliminate w from the second equation of system (4.2.4) by

$$w = c\phi + \lambda P_c L_-^{-1} P_c u,$$

where $c \in \mathbb{R}$ is uniquely found from the projection of the first equation of system (4.2.4) to ϕ for $\lambda \neq 0$:

$$c = -\frac{\langle (L_+ + \lambda^2 P_c L_-^{-1} P_c)u, \phi \rangle_{L^2}}{\lambda \|\phi\|_{L^2}^2}.$$

Using the projection operator P_c for the first equation of system (4.2.4), we obtain the generalized eigenvalue problem

$$P_c L_+ P_c u = -\lambda P_c w = \gamma P_c L_-^{-1} P_c u, \quad \gamma = -\lambda^2. \tag{4.2.7}$$

It is clear that $P_c L_-^{-1} P_c$ is a strictly positive bounded operator and that the generalized eigenvalue problem (4.2.7) has only real semi-simple eigenvalues γ. We are particularly interested in negative eigenvalues γ since these eigenvalues correspond to pairs of real eigenvalues λ in the spectral stability problem (4.2.4), which are located symmetrically about 0 and include one unstable eigenvalue (Definition 4.1).

We note that $n(L_+) = 1$, $\mathrm{Ker}(L_+) \in L_c^2$, and $L_+\partial_\omega \phi = \phi$, so that

$$\langle L_+^{-1}\phi, \phi \rangle_{L^2} = \langle \partial_\omega \phi, \phi \rangle_{L^2} = \frac{1}{2}\frac{dQ}{d\omega},$$

where $Q(\omega) = \|\phi\|_{L^2}^2$. By Theorem 4.1, $P_c L_+ P_c$ has no negative eigenvalues if $Q'(\omega) < 0$ and has exactly one negative eigenvalue if $Q'(\omega) > 0$. By Theorem 4.2, there is exactly one negative eigenvalue γ (real positive λ) if $Q'(\omega) > 0$ and no negative eigenvalues γ if $Q'(\omega) < 0$. $\square$

Corollary 4.3 *Let ϕ be the localized mode (4.2.5) of the power NLS equation (4.2.1) with $f(\phi^2) = (p + 1)\phi^{2p}$ for $p > 0$. The localized mode ϕ is spectrally unstable if $p > 2$ and it is spectrally stable if $0 < p < 2$.*

Proof   Using the exact solution (4.2.5), we compute the dependence $Q(\omega)$ explicitly as

$$Q(\omega) = |\omega|^{\frac{1}{p}-\frac{1}{2}} \int_{\mathbb{R}} \text{sech}^{2/p}(pz)dz, \quad \omega < 0. \tag{4.2.8}$$

Therefore, $Q'(\omega) < 0$ for $p < 2$ and the localized mode ϕ is spectrally stable in the power NLS equation for $p < 2$. On the other hand, $Q'(\omega) > 0$ for $p > 2$ and the localized mode ϕ is spectrally unstable for $p > 2$. $\square$

Remark 4.3   The case $p = 2$ is critical since it follows from (4.2.8) that $Q'(\omega) = 0$ for $p = 2$. The localized mode ϕ is still unstable in this case because of the highly degenerate zero eigenvalue of the spectral stability problem (4.2.4) and the associated perturbations that grow algebraically in time. Perelman [167] and Comech & Pelinovsky [40] discuss this case in more detail.

Similar results on the count of the unstable eigenvalues for a *positive* localized mode ϕ hold for $d \geq 1$. If $n(L_+) = 1$ and $Q'(\omega) < 0$, the localized mode ϕ is spectrally (and orbitally) stable. This result is usually quoted as the Stability Theorem of Grillakis, Shatah and Strauss [74], although it appeared in the Russian literature 10 years earlier in the work of Vakhitov & Kolokolov [207]. The spectral stability condition $Q'(\omega) < 0$ is commonly known in the physics literature as the *Vakhitov–Kolokolov* criterion, although physicists often forget to check the assumptions $n(L_-) = 0$ and $n(L_+) = 1$, which are necessary for the validity of this criterion.

If ϕ is *not sign-definite* but $n(L_+)$ and $n(L_-)$ are still finite, the Sylvester Law of Inertia (Theorem 4.2) is not applicable and we need the Pontryagin Invariant Subspace Theorem (Theorem 4.3) in an appropriately defined Pontryagin space. Application of this theory for abstract self-adjoint operators L_+ and L_- is the goal of this section.

From a historical perspective, more general theorems on spectral stability of a sign-indefinite localized mode ϕ in the Gross–Pitaevskii equation (4.2.1) can be found in the second paper of Grillakis *et al.* [75] as well as in the works of Jones [99, 100] and Grillakis [76, 77] published at the end of the 1980s. While the second paper of Grillakis [77] has almost all the ingredients of the modern stability analysis, it took 20 more years to obtain the most general count of all unstable (real and complex) eigenvalues in the generalized eigenvalue problem for operators L_+ and L_- in a constrained L^2 space [33, 43, 105, 153].

4.2.1 Count of eigenvalues of the generalized eigenvalue problem

Let L_+ and L_- be real-valued self-adjoint operators in L^2. Throughout this section, we assume that $\sigma_c(L_\pm) \geq c_\pm$ for some $c_\pm > 0$ and $\sigma_p(L_\pm)$ includes finitely many non-positive eigenvalues of finite multiplicities. The negative and zero indices of $L_\pm$ are denoted by $n(L_\pm)$ and $z(L_\pm)$.

We write an abstract spectral stability problem in the form

$$L_+ u = -\lambda w, \quad L_- w = \lambda u, \quad \lambda \in \mathbb{C}. \tag{4.2.9}$$

Let $P_c : L^2 \to L_c^2 \subset L^2$ be the projection operator to the orthogonal complement of $\mathrm{Ker}(L_-)$ given by

$$L_c^2 = \left\{ u \in L^2 : \ \langle u, u_0 \rangle_{L^2} = 0 \ \text{for all} \ u_0 \in \mathrm{Ker}(L_-) \right\}. \qquad (4.2.10)$$

We are only interested in nonzero eigenvalues of the spectral problem (4.2.9) because only nonzero eigenvalues λ determine the spectral stability or instability of the underlying solution. Since $\sigma_c(L_-) \geq c_- > 0$ and $L^2 = \mathrm{Ker}(L_-) \oplus \mathrm{Ran}(L_-)$ for the self-adjoint operator L_-, we have $L_c^2 \equiv \mathrm{Ran}(L_-)$. If $\lambda \neq 0$ and $u \in L^2$, then $u \in L_c^2$. As a result, we express w from the second equation of system (4.2.9)

$$w = \lambda P_c L_-^{-1} P_c u + w_0, \quad w_0 \in \mathrm{Ker}(L_-). \qquad (4.2.11)$$

Substituting w into the first equation of system (4.2.9) and using the projection operator P_c, we obtain a closed equation for u,

$$P_c L_+ P_c u = -\lambda^2 P_c L_-^{-1} P_c u, \quad u \in \mathcal{H} \qquad (4.2.12)$$

and a unique expression for w_0,

$$w_0 = -\frac{1}{\lambda}(I - P_c)(L_+ + \lambda^2 P_c L_-^{-1} P_c)u, \qquad (4.2.13)$$

where $\lambda \neq 0$ and $(I - P_c)$ is the orthogonal projection from L^2 to $\mathrm{Ker}(L_-)$.

Equation (4.2.12) shows that the linear eigenvalue problem (4.2.9) for nonzero λ is equivalent to the generalized eigenvalue problem for nonzero γ,

$$Lu = \gamma M^{-1} u, \quad \gamma \in \mathbb{C}, \qquad (4.2.14)$$

where $L = P_c L_+ P_c$, $M^{-1} = P_c L_-^{-1} P_c$, and $\gamma = -\lambda^2$. The following lemma states the equivalence of quadratic forms $\|u\|_{L^2}^2$ and $\langle M^{-1} u, u \rangle_{L^2}$ for solutions of the generalized eigenvalue problem (4.2.14).

Lemma 4.6   *The generalized eigenvalue problem*

$$Lu = \gamma M^{-1} u \quad \text{with} \quad \|u\|_{L^2} < \infty$$

is equivalent to the generalized eigenvalue problem

$$Lu = \gamma M^{-1} u \quad \text{with} \quad |\langle M^{-1} u, u \rangle_{L^2}| < \infty.$$

Proof   Since M^{-1} is a bounded invertible self-adjoint operator with $\mathrm{Dom}(M^{-1}) \equiv L^2$, there exists $C > 0$ such that

$$\forall u \in L_c^2 : \quad |\langle M^{-1} u, u \rangle_{L^2}| \leq C\|u\|_{L^2}^2. \qquad (4.2.15)$$

Therefore, if u is an eigenvector of $Lu = \gamma M^{-1} u$ and $\|u\|_{L^2} < \infty$, then $|\langle M^{-1} u, u \rangle_{L^2}| < \infty$. On the other hand, L is generally an unbounded non-invertible self-adjoint operator with $\mathrm{Dom}(L) \subseteq L_c^2$. If u is the eigenvector of $Lu = \gamma M^{-1} u$ with $|\langle M^{-1} u, u \rangle_{L^2}| < \infty$, then $u \in \mathrm{Dom}(L) \subseteq L_c^2$, so that $\|u\|_{L^2} < \infty$. $\square$

Operators L and M have finitely many non-positive eigenvalues in L_c^2. By the spectral theory of self-adjoint operators, the constrained space L_c^2 can be equivalently decomposed into the direct orthogonal sums

$$L_c^2 = \mathcal{H}_M^- \oplus \mathcal{H}_M^+, \tag{4.2.16}$$

$$L_c^2 = \mathcal{H}_L^- \oplus \mathcal{H}_L^0 \oplus \mathcal{H}_L^+, \tag{4.2.17}$$

where $\mathcal{H}_L^-$ denotes the eigenspace for negative eigenvalues of operator L, $\mathcal{H}_L^0$ denotes the kernel of L, and $\mathcal{H}_L^+$ denotes the eigenspace for positive eigenvalues and the continuous spectrum.

Since P_c is a projection operator defined by the kernel of L_- and $M^{-1} = P_c L_-^{-1} P_c$, it is obvious that

$$\dim(\mathcal{H}_M^-) = n(L_-). \tag{4.2.18}$$

Let $\mathrm{Ker}(L_-) = \mathrm{Span}\{v_1, v_2, ..., v_N\}$ with $N = z(L_-)$ and define a matrix-valued function $A(\mu)$ by

$$\forall \mu \notin \sigma(L_+): \quad A_{i,j}(\mu) = \langle (\mu - L_+)^{-1} v_i, v_j \rangle_{L^2}, \quad i, j \in \{1, 2, ..., N\}.$$

The eigenvalues of L are related to the eigenvalues of L_+ by Theorem 4.1, which gives

$$\dim(\mathcal{H}_L^-) = n(L_+) - p_0 - z_0 \tag{4.2.19}$$

and

$$\dim(\mathcal{H}_L^0) = z(L_+) + z_0 - z_\infty, \tag{4.2.20}$$

where n_0, z_0, and p_0 are the numbers of negative, zero, and positive eigenvalues of $\lim_{\mu \uparrow 0} A(\mu)$ and $z_\infty = N - n_0 - z_0 - p_0$.

If $\mathcal{H}_L^0$ is trivial (that is, $\dim(\mathcal{H}_L^0) = 0$), operator L is invertible and we can proceed with analysis of the generalized eigenvalue problem (4.2.14). However, if L is not invertible, we would like to shift this problem and reduce it to an invertible operator.

Let γ_{-1} be the smallest (in absolute value) negative eigenvalue of ML. There exists a small number $\delta \in (0, |\gamma_{-1}|)$ such that the operator $L + \delta M^{-1}$ is continuously invertible and the generalized eigenvalue problem (4.2.14) is rewritten in the form

$$(L + \delta M^{-1})u = (\gamma + \delta)M^{-1}u, \quad \gamma \in \mathbb{C}. \tag{4.2.21}$$

By the spectral theory, for any fixed $\delta \in (0, |\gamma_{-1}|)$, we have

$$L_c^2 = \mathcal{H}_{L+\delta M^{-1}}^- \oplus \mathcal{H}_{L+\delta M^{-1}}^+. \tag{4.2.22}$$

The following lemma characterizes the dimension of $\mathcal{H}_{L+\delta M^{-1}}^-$.

Lemma 4.7 *Assume that* $\dim(\mathcal{H}_L^0) = 1$ *and let* $\{u_1, ..., u_n\}$ *be the Jordan chain of* L^2*-normalized eigenvectors of the generalized eigenvalue problem (4.2.14) given* by

$$\begin{cases} Lu_1 = 0, \\ Lu_2 = M^{-1}u_1, \\ \quad \cdots \\ Lu_n = M^{-1}u_{n-1}, \end{cases}$$

which is truncated if $\langle M^{-1}u_n, u_1 \rangle_{L^2} \neq 0$. Fix $\delta \in (0, |\gamma_{-1}|)$, where γ_{-1} is the smallest (in absolute value) negative eigenvalue of (4.2.14). Then

$$\dim(\mathcal{H}^-_{L+\delta M^{-1}}) = \dim(\mathcal{H}^-_L) + N_0, \qquad (4.2.23)$$

where

$$N_0 = \begin{cases} 0 \text{ if } n \text{ is odd and } \langle M^{-1}u_n, u_1 \rangle_{L^2} > 0 \text{ or if } n \text{ is even and } \langle M^{-1}u_n, u_1 \rangle_{L^2} < 0, \\ 1 \text{ if } n \text{ is odd and } \langle M^{-1}u_n, u_1 \rangle_{L^2} < 0 \text{ or if } n \text{ is even and } \langle M^{-1}u_n, u_1 \rangle_{L^2} > 0. \end{cases}$$

Proof Since we shift a self-adjoint operator L to a self-adjoint operator $L + \delta M^{-1}$ for a sufficiently small $\delta > 0$, the zero eigenvalue of operator L becomes a small real eigenvalue $\mu(\delta)$ of operator $L + \delta M^{-1}$. By perturbation theory for an isolated eigenvalue of a self-adjoint operator (Chapter VII.3 in [108]), eigenvalue $\mu(\delta)$ is a continuous function of δ and

$$\lim_{\delta \downarrow 0} \frac{\mu(\delta)}{\delta^n} = (-1)^{n+1} \langle M^{-1}u_n, u_1 \rangle_{L^2}. \qquad (4.2.24)$$

The assertion of the lemma follows from the limiting relation (4.2.24). Since no eigenvalues of (4.2.14) exist in $(-|\gamma_{-1}|, 0)$, the eigenvalue $\mu(\delta)$ remains sign-definite for $\delta \in (0, |\gamma_{-1}|)$. $\qquad \square$

Remark 4.4 Assumption $\dim(\mathcal{H}^0_L) = 1$ of Lemma 4.7 can be removed by considering the Jordan block decomposition for the zero eigenvalue and by summing contributions from all Jordan blocks.

If 0 is a semi-simple eigenvalue of the generalized eigenvalue problem (4.2.14), the following lemma gives an analogue of Lemma 4.7.

Lemma 4.8 *Let* $\mathrm{Ker}(L) = \mathrm{span}\{u_1, u_2, ..., u_n\}$ *and* $M_K \in \mathbb{R}^{n \times n}$ *be the matrix with elements*

$$(M_K)_{ij} = \langle M^{-1}u_i, u_j \rangle_{L^2}, \quad 1 \leq i, j \leq n.$$

Then for small $\delta > 0$

$$\dim(\mathcal{H}^-_{L+\delta M^{-1}}) = \dim(\mathcal{H}^-_L) + \dim(\mathcal{H}^-_{M_K}). \qquad (4.2.25)$$

Proof Let $\{u_1, u_2, ..., u_n\}$ be a basis for $\mathrm{Ker}(L)$, which is orthogonal with respect to $\langle M^{-1}\cdot, \cdot \rangle_{L^2}$ (such a basis always exists if 0 is a semi-simple eigenvalue of $Lu = \gamma M^{-1}u$). Then, for the jth Jordan block, the result of Lemma 4.7 with $n = 1$ shows that $\dim(\mathcal{H}^-_{L+\delta M^{-1}}) = \dim(\mathcal{H}^-_L) + 1$ if $\langle M^{-1}u_j, u_j \rangle_{L^2} < 0$. Equality (4.2.25) holds after summing contributions from all Jordan blocks for this basis. The number of negative eigenvalues of M_K is invariant with respect to the choice of basis in $\mathrm{Ker}(L)$. $\qquad \square$

We shall now introduce notation for particular eigenvalues of the generalized eigenvalue problem (4.2.14) and formulate our main result on the count of unstable eigenvalues in the spectral stability problem (4.2.9). Recall from Definition 4.1 that we are looking for unstable eigenvalues $\lambda \in \mathbb{C}$ with $\mathrm{Re}(\lambda) > 0$. By the symmetries of the spectral stability problem (4.2.9), pairs of real eigenvalues λ correspond to negative eigenvalues γ, pairs of purely imaginary eigenvalues λ correspond to positive eigenvalues γ, and quartets of complex eigenvalues λ correspond to pairs of complex conjugate eigenvalues γ. Therefore, the unstable eigenvalues of the spectral stability problem (4.2.9) correspond to negative and complex eigenvalues of the generalized eigenvalue problem (4.2.14).

Let N_p^- (N_n^-), N_p^0 (N_n^0), and N_p^+ (N_n^+) be the numbers of negative, zero, and positive eigenvalues γ of the generalized eigenvalue problem (4.2.14) with the account of their algebraic multiplicities whose generalized eigenvectors are associated to the non-negative (non-positive) values of the quadratic form $\langle M^{-1}\cdot, \cdot \rangle_{L^2}$. Numbers N_p^+ (N_n^+) include both isolated and embedded eigenvalues with respect to the continuous spectrum of the generalized eigenvalue problem (4.2.14). Let N_{c+} (N_{c-}) be the number of complex eigenvalues in the upper (lower) half-plane $\gamma \in \mathbb{C}$ for $\mathrm{Im}(\gamma) > 0$ ($\mathrm{Im}(\gamma) < 0$). Because L and M are real-valued operators, it is obvious that $N_{c+} = N_{c-}$. Our main result is the following theorem.

Theorem 4.5 *Assume that $\sigma_c(L_\pm) \geq c_\pm > 0$ and $\sigma_p(L_\pm)$ includes finitely many non-positive eigenvalues of finite multiplicities. Let $L = P_c L_+ P_c$, $M^{-1} = P_c L_-^{-1} P_c$, and $\delta \in (0, |\gamma_{-1}|)$, where $P_c : L^2 \to [\mathrm{Ker}(L_-)]^\perp$ and γ_{-1} is the smallest (in absolute value) negative eigenvalue of (4.2.14). Eigenvalues of the generalized eigenvalue problem (4.2.14) satisfy the following identities:*

$$N_p^- + N_n^0 + N_n^+ + N_{c+} = \dim(\mathcal{H}_{L+\delta M^{-1}}^-), \qquad (4.2.26)$$

$$N_n^- + N_n^0 + N_n^+ + N_{c+} = \dim(\mathcal{H}_M^-), \qquad (4.2.27)$$

where $\dim(\mathcal{H}_M^-)$ and $\dim(\mathcal{H}_{L+\delta M^{-1}}^-)$ are defined by (4.2.18), (4.2.19), (4.2.23), and (4.2.25).

Remark 4.5 If L is invertible, then $N_n^0 = 0$ and $\dim(\mathcal{H}_{L+\delta M^{-1}}^-) = \dim(\mathcal{H}_L^-)$.

Exercise 4.12 Let L and M be invertible self-adjoint operators with finite numbers $n(L)$ and $n(M)$ of negative eigenvalues. Assume that L and M commute and hence they have a common set of eigenvectors. Show that

$$N_p^- + N_n^+ = n(L), \quad N_n^- + N_n^+ = n(M), \quad N_{c+} = 0,$$

which agrees with (4.2.26) and (4.2.27).

A direct consequence of Theorem 4.5 is the fact that the total number of unstable and potentially unstable eigenvalues of the generalized eigenvalue problem (4.2.14) *equals* the total number of negative eigenvalues of the self-adjoint operators $L + \delta M^{-1}$ and M, while the number of negative eigenvalues of the generalized eigenvalue problem (4.2.14) is bounded from below by the difference between negative eigenvalues of $L + \delta M^{-1}$ and M.

Corollary 4.4 *Let* $N_{\text{neg}} = \dim(\mathcal{H}^-_{L+\delta M^{-1}}) + \dim(\mathcal{H}^-_M)$, $N_{\text{unst}} = N^-_p + N^-_n + 2N_{c+}$, *and* $N^{\text{pot}}_{\text{unst}} = 2N^+_n + 2N^0_n$. *Then,*

$$N_{\text{unst}} + N^{\text{pot}}_{\text{unst}} = N_{\text{neg}}. \qquad (4.2.28)$$

Proof Equality (4.2.28) follows by the sum of (4.2.26) and (4.2.27). □

Corollary 4.5 *Let* $N^- = N^-_p + N^-_n$. *Then,*

$$N^- \geq \left| \dim(\mathcal{H}^-_{L+\delta M^{-1}}) - \dim(\mathcal{H}^-_M) \right|. \qquad (4.2.29)$$

Proof Inequality (4.2.29) follows by the difference between (4.2.26) and (4.2.27). □

We note that the number of unstable eigenvalues N_{unst} includes $N^- = N^-_p + N^-_n$ negative eigenvalues $\gamma < 0$ and $N_c = N_{c+} + N_{c-} = 2N_{c+}$ complex eigenvalues with $\text{Im}(\gamma) \neq 0$. While eigenvalues $N^{\text{pot}}_{\text{unst}} = 2N^+_n + 2N^0_n$ are neutrally stable, they are potentially unstable, since they can bifurcate to the complex domain in parameter continuation at the localized modes (Sections 4.3.1 and 4.3.3).

Exercise 4.13 Assume that $\dim(\mathcal{H}^0_L) = 0$ (so that δ can be set to 0) and show that Theorem 4.5 and Corollary 4.4 recover the following theorem from Ref. [75]: "The localized mode is spectrally stable if $n(L_+) + n(L_-) = p_0$ and spectrally unstable if the number $n(L_+) + n(L_-) - p_0$ is odd."

Exercise 4.14 Assume that $\dim(\mathcal{H}^0_L) = 0$ (so that δ can be set to 0) and show that Theorem 4.5 and Corollary 4.5 recover the following theorem from Ref. [76]: "There exist at least $|n(L_+) - n(L_-) - p_0|$ real positive eigenvalues λ in the spectral stability problem (4.2.9). If $n(L_-) = 0$, there are exactly $n(L_+) - p_0$ real positive eigenvalues λ."

Theorem 4.5 appeared in [33, 43, 105, 153], although the methods of proof presented therein were quite different. The method of [43] relies on the sequence of constrained spaces that reduces L^2 to the invariant subspace for the continuous spectrum of the generalized eigenvalue problem. The method of [105] applies the Grillakis Theorem from [77]. The method of [153] is based on an application of the Sylvester Law of Inertia extended to non-positive operators M. The proof we give below follows the method of [33] and it is based on an application of the Pontryagin Invariant Subspace Theorem.

Proof of Theorem 4.5 Consider the generalized eigenvalue problem (4.2.21) for a fixed $\delta \in (0, |\gamma_1|)$ and define a bounded invertible operator

$$T = (L + \delta M^{-1})^{-1} M^{-1}.$$

Operator T is self-adjoint with respect to $\langle M^{-1} \cdot, \cdot \rangle_{L^2}$. By Theorem 4.3, it has a κ-dimensional maximal non-positive invariant subspace in Pontryagin space Π_κ, where $\kappa = \dim(\mathcal{H}^-_M)$. Identity (4.2.27) follows from the count of eigenvalues of the generalized eigenvalue problem (4.2.21), which relies on the analysis of T-invariant subspaces in Pontryagin space (Section 4.2.2).

On the other hand, let us define another bounded invertible operator

$$\tilde{T} = M^{-1}(L + \delta M^{-1})^{-1},$$

which is self-adjoint with respect to $\langle (L+\delta M^{-1})^{-1}\cdot, \cdot \rangle_{L^2}$. By Theorem 4.3 again, it has a $\tilde{\kappa}$-dimensional maximal non-positive invariant subspace in Pontryagin space $\tilde{\Pi}_{\tilde{\kappa}}$, where $\tilde{\kappa} = \dim(\mathcal{H}^-_{L+\delta M^{-1}})$.

Let $\mathcal{H}_{\gamma_0}$ be the eigenspace of L^2_c associated with an eigenvalue γ_0 of operator $\tilde{T}$. It follows from the generalized eigenvalue problem (4.2.21) that

$$\forall f, g \in \text{Dom}(L): \quad \langle (L+\delta M^{-1})f, g \rangle_{L^2} = (\gamma_0 + \delta)\langle M^{-1}f, g \rangle_{L^2}.$$

Therefore, for $\tilde{f} = (L+\delta M^{-1})f$ and $\tilde{g} = (L+\delta M^{-1})g$, we have

$$\forall \tilde{f}, \tilde{g} \in L^2_c: \quad \langle (L+\delta M^{-1})^{-1}\tilde{f}, \tilde{g} \rangle_{L^2} = (\gamma_0 + \delta)\langle M^{-1}f, g \rangle_{L^2}.$$

If $\gamma_0 \geq 0$ or $\text{Im}(\gamma_0) \neq 0$, the maximal non-positive eigenspace of $\tilde{T}$ in $\tilde{\Pi}_{\tilde{\kappa}}$ associated with γ_0 coincides with the maximal non-positive eigenspace of T in Π_κ. If $\gamma_0 < 0$, the maximal non-positive eigenspace of $\tilde{T}$ in $\tilde{\Pi}_{\tilde{\kappa}}$ coincides with the maximal non-negative eigenspace of T in Π_κ. Identity (4.2.26) follows from the same count of eigenvalues after N_n^- and κ are replaced by N_p^- and $\tilde{\kappa}$ respectively. $\square$

4.2.2 Analysis of invariant subspaces in Pontryagin space

The proof of Theorem 4.5 uses the count of eigenvalues of the generalized eigenvalue problem (4.2.21) that contribute to the maximal non-positive invariant subspace in Pontryagin space Π_κ. Here we count these eigenvalues in the main result, Theorem 4.7 below.

Let us define the *sesquilinear form* $[\cdot, \cdot]$ and the symmetric operator T by

$$[\cdot, \cdot] := \langle M^{-1}\cdot, \cdot \rangle_{L^2}, \quad T := (L + \delta M^{-1})^{-1}M^{-1}. \tag{4.2.30}$$

We shall consider various sign-definite T-invariant subspaces of Π_κ. In general, these subspaces do not provide a canonical decomposition of Π_κ compared with the direct orthogonal sum (4.2.16). Recall from Definition 4.7 the notation $\Pi_+ = \mathcal{H}_M^-$ and $\Pi_- = \mathcal{H}_M^+$ in the canonical decomposition

$$\Pi_\kappa = \Pi_+ \oplus \Pi_-, \quad \Pi_+ \cap \Pi_- = \varnothing.$$

Let us denote the eigenvalue of T by $\lambda := 1/(\gamma + \delta)$. Let $\mathcal{H}_{c+}$ ($\mathcal{H}_{c-}$) denote the T-invariant subspace associated with a complex eigenvalue λ in the upper (lower) half-plane and $\mathcal{H}_n (\mathcal{H}_p)$ denote the non-positive (non-negative) T-invariant subspace associated with a real eigenvalue λ.

The spectrum of the generalized eigenvalue problem (4.2.21) consists of not only eigenvalues but also the continuous spectrum (Definitions 4.5 and 4.6). Nevertheless, we show that the maximal non-positive T-invariant subspace does not include the continuous spectrum and only includes isolated and embedded eigenvalues of finite multiplicities.

Lemma 4.9 *The continuous spectrum of T is real.*

Proof Let P^+ and P^- be projection operators orthogonal to Π^+ and Π^- respectively. The self-adjoint operator M^{-1} admits the polar decomposition $M^{-1} = J|M^{-1}|$, where $J = P^+ - P^-$ and $|M^{-1}|$ is a positive operator. Since $J^2 = I$ and M^{-1} is self-adjoint, we have

$$J|M^{-1}|J = |M^{-1}| \quad \text{and} \quad J|M^{-1}|^{1/2}J = |M^{-1}|^{1/2}.$$

Operator $T = BM^{-1}$ with $B = (L + \delta M^{-1})^{-1}$ is similar to the operator

$$
\begin{aligned}
|M^{-1}|^{1/2}BJ|M^{-1}|^{1/2} &= |M^{-1}|^{1/2}BJ|M^{-1}|^{1/2}(J + 2P^-) \\
&= |M^{-1}|^{1/2}B|M^{-1}|^{1/2} + 2|M^{-1}|^{1/2}BJ|M^{-1}|^{1/2}P^-.
\end{aligned}
$$

Since P^- is a projection to a finite-dimensional subspace, the second operator in the sum is a finite-rank perturbation of the first operator in the sum. By Theorem 18 of Glazman [69], the continuous part of the self-adjoint operator $|M^{-1}|^{1/2}B|M^{-1}|^{1/2}$ is the same as that of $|M^{-1}|^{1/2}BJ|M^{-1}|^{1/2}$, which is the same as that of T by the similarity transformation. $\qquad\square$

Theorem 4.6 *Let Π_c be an invariant subspace associated with the continuous spectrum of T. Then, $[f, f] > 0$ for any nonzero $f \in \Pi_c$.*

Proof By Theorem 4.3, the operator T has a κ-dimensional maximal non-positive invariant subspace of Π_κ. Let us denote this subspace by Π. Because any finite-dimensional invariant subspace of T cannot be a part of Π_c, Π and Π_c intersect trivially. Assume now that there exists a nonzero $f_0 \in \Pi_c$ such that $[f_0, f_0] \leq 0$. Since $f_0 \notin \Pi$, the subspace spanned by f_0 and the basis vectors in Π is a $(\kappa + 1)$-dimensional non-positive subspace of Π_κ. However, by Lemma 4.3, the maximal dimension of any non-positive subspace of Π_κ is κ. Therefore, $[f_0, f_0] > 0$ for any nonzero $f_0 \in \Pi_c$. $\qquad\square$

Note that Theorem 4.6 states that the quadratic form associated with the self-adjoint operator M^{-1} is *strictly positive* on the subspace related to the continuous spectrum of the generalized eigenvalue problem (4.2.21). This result has a technical significance since it establishes a similarity between the spectral stability analysis based on Sylvester's Law of Inertia with $Q_L(\tilde{P}_c u) > 0$ (Section 4.1.2) and the one based on Pontryagin's invariant subspaces (Section 4.1.3).

It remains to count isolated and embedded eigenvalues of finite multiplicities for operator T. Note that T is self-adjoint in the Pontryagin space Π_κ. Let $\mathcal{H}_\lambda$ denote the eigenspace of Π_κ associated with the eigenvalue λ of T, so that

$$\text{Ker}(T - \lambda I) \neq \varnothing, \qquad \mathcal{H}_\lambda = \cap_{k \in \mathbb{N}}\text{Ker}(T - \lambda I)^k.$$

As previously, the eigenvalue λ is said to be *semi-simple* if

$$\dim \text{Ker}(T - \lambda I) = \dim(\mathcal{H}_\lambda)$$

and *multiple* if

$$\dim \text{Ker}(T - \lambda I) < \dim(\mathcal{H}_\lambda).$$

In the latter case, the eigenspace $\mathcal{H}_\lambda$ can be represented by the union of Jordan blocks and the canonical basis for each Jordan block is built by the generalized eigenvectors,

$$f_j \in \Pi_\kappa : \quad Tf_j = \lambda f_j + f_{j-1}, \qquad j \in \{1, ..., n\}, \tag{4.2.31}$$

where $f_0 = 0$. Each Jordan block of generalized eigenvectors (4.2.31) is associated with a single eigenvector of T. We start with an elementary result about the generalization of the Fredholm theory for a symmetric operator T in the Pontryagin space Π_κ.

Lemma 4.10 *Let λ be an isolated eigenvalue of T associated with a one-dimensional eigenspace* $\mathrm{Ker}(T - \lambda I) = \mathrm{Span}\{f_1\}$. *Then, $\lambda \in \mathbb{R}$ is algebraically simple if and only if $[f_1, f_1] \neq 0$ and $\lambda \in \mathbb{C}$ is algebraically simple if and only if $[f_1, \bar{f}_1] \neq 0$.*

Proof Since $(L + \delta M^{-1})^{-1}$ and M^{-1} are bounded invertible self-adjoint operators, the eigenvalue problem $Tf = \lambda f$ in the Pontryagin space Π_κ is rewritten as the generalized eigenvalue problem

$$M^{-1}f = \lambda(L + \delta M^{-1})f \tag{4.2.32}$$

in the Hilbert space L_c^2. Since λ is an isolated eigenvalue, the Fredholm theory for the generalized eigenvalue problem (4.2.32) implies that $\lambda \in \mathbb{R}$ is algebraically simple if and only if

$$\langle (L + \delta M^{-1})f_1, f_1 \rangle_{L^2} \neq 0,$$

while $\lambda \in \mathbb{C}$ is algebraically simple if and only if

$$\langle (L + \delta M^{-1})f_1, \bar{f}_1 \rangle_{L^2} \neq 0.$$

Since $\lambda \neq 0$ (otherwise, M^{-1} is not invertible), the condition of the Fredholm theory is equivalent to the condition that $\langle M^{-1}f_1, f_1 \rangle_{L^2} \neq 0$ and $\langle M^{-1}f_1, \bar{f}_1 \rangle_{L^2} \neq 0$, respectively. The assertion of the lemma is proved when definition (4.2.30) of the sesquilinear form is used. □

Lemma 4.11 *Let $\mathcal{H}_\lambda$ and $\mathcal{H}_\mu$ be eigenspaces associated with eigenvalues λ and μ of the operator T and $\lambda \neq \bar{\mu}$. Then $\mathcal{H}_\lambda$ is orthogonal to $\mathcal{H}_\mu$ with respect to $[\cdot, \cdot]$.*

Proof Let $n \geq 1$ and $m \geq 1$ be the dimensions of $\mathcal{H}_\lambda$ and $\mathcal{H}_\mu$, respectively. Using the Jordan chain (4.2.31), we write

$$f \in \mathcal{H}_\lambda \Longleftrightarrow (T - \lambda I)^n f = 0,$$

$$g \in \mathcal{H}_\mu \Longleftrightarrow (T - \mu I)^m g = 0.$$

We should prove that $[f, g] = 0$ by induction for $n + m \geq 2$. If $n + m = 2$ ($n = m = 1$), then it is found directly that

$$(\bar{\mu} - \lambda)[f, g] = 0, \qquad f \in \mathcal{H}_\lambda, \quad g \in \mathcal{H}_\mu,$$

so that $[f, g] = 0$ for $\lambda \neq \bar{\mu}$. Let us assume that subspaces $\mathcal{H}_\lambda$ and $\mathcal{H}_\mu$ are orthogonal for $2 \leq n + m \leq k$ and prove that an extended subspace $\tilde{\mathcal{H}}_\lambda$ with $\tilde{n} = n + 1$ remains orthogonal to $\mathcal{H}_\mu$. To do so, we define $\tilde{f} = (T - \lambda I)f$, $\tilde{g} = (T - \lambda I)g$ and verify that

$$f \in \tilde{\mathcal{H}}_\lambda \iff (T - \lambda I)^{\tilde{n}}f = (T - \lambda I)^n \tilde{f} = 0,$$

$$g \in \tilde{\mathcal{H}}_\mu \iff (T - \mu I)^m f = (T - \mu I)^{m-1}\tilde{g} = 0.$$

By the inductive assumption, we have $[\tilde{f}, g] = [f, \tilde{g}] = 0$, so that

$$0 = [\tilde{f}, g] = [(T - \lambda I)f, g] = (\bar{\mu} - \lambda)[f, g] + [f, \tilde{g}] = (\bar{\mu} - \lambda)[f, g].$$

Therefore, $[f, g] = 0$ for $\lambda \neq \bar{\mu}$. Similarly, an extended subspace $\tilde{\mathcal{H}}_\mu$ with $\tilde{m} = m+1$ remains orthogonal to $\mathcal{H}_\lambda$. The assertion of the lemma follows by the induction method. $\qquad\square$

Lemma 4.12 *Let $\mathcal{H}_\lambda$ be an eigenspace associated with an isolated eigenvalue $\lambda \in \mathbb{R}$ of T and $\{f_1, f_2, ..., f_n\}$ be the Jordan chain of eigenvectors. Let*

$$\mathcal{H}_0 = \mathrm{Span}\{f_1, f_2, ..., f_k\} \subset \mathcal{H}_\lambda$$

and

$$\tilde{\mathcal{H}}_0 = \mathrm{Span}\{f_1, f_2, ..., f_k, f_{k+1}\} \subset \mathcal{H}_\lambda,$$

where $k = \frac{1}{2}n$ if n is even and $k = \frac{1}{2}(n - 1)$ if n is odd.

(i) *If n is even $(n = 2k)$, the neutral subspace $\mathcal{H}_0$ is the maximal sign-definite subspace of $\mathcal{H}_\lambda$.*

(ii) *If n is odd $(n = 2k + 1)$, the subspace $\tilde{\mathcal{H}}_0$ is the maximal non-negative subspace of $\mathcal{H}_\lambda$ if $[f_1, f_n] > 0$ and the maximal non-positive subspace of $\mathcal{H}_\lambda$ if $[f_1, f_n] < 0$, while the neutral subspace $\mathcal{H}_0$ is the maximal non-positive subspace of $\mathcal{H}_\lambda$ if $[f_1, f_n] > 0$ and the maximal non-negative subspace of $\mathcal{H}_\lambda$ if $[f_1, f_n] < 0$.*

Proof Without loss of generality we will consider the case $\lambda = 0$ (the same argument is applied to the operator $\tilde{T} = T - \lambda I$ if $\lambda \neq 0$). We will show that $[f, f] = 0$ for any $f \in \mathcal{H}_0$. Using a decomposition over the basis in $\mathcal{H}_0$, we obtain

$$\forall f = \sum_{i=1}^k \alpha_i f_i : \quad [f, f] = \sum_{i=1}^k \sum_{j=1}^k \alpha_i \bar{\alpha}_j [f_i, f_j]. \qquad (4.2.33)$$

For any $1 \leq i, j \leq k$, we note

$$[f_i, f_j] = [Tf_{i+1}, Tf_{j+1}] = ... = [T^k f_{i+k}, T^k f_{j+k}] = [T^{2k} f_{i+k}, f_{j+k}].$$

In the case of even $n = 2k$, we have

$$[f_i, f_j] = [T^n f_{i+k}, f_{j+k}] = 0, \quad 1 \leq i, j \leq k.$$

In the case of odd $n = 2k + 1$, we have

$$[f_i, f_j] = [T^{n+1} f_{i+k+1}, f_{j+k+1}] = 0, \quad 1 \leq i, j \leq k.$$

Therefore, $\mathcal{H}_0$ is a neutral subspace of $\mathcal{H}_{\lambda_0}$. To show that it is the maximal neutral subspace of $\mathcal{H}_{\lambda_0}$, let $\mathcal{H}_0' = \mathrm{Span}\{f_1, f_2, ..., f_k, f_{k_0}\}$, where $k+1 \leq k_0 \leq n$. Since f_{n+1} does not exist in the Jordan chain (4.2.31) (otherwise, the algebraic multiplicity is

$n + 1$) and λ_0 is an isolated eigenvalue, then $[f_1, f_n] \neq 0$ by Lemma 4.10. It follows from the Jordan chain (4.2.31) that

$$[f_1, f_n] = [T^{m-1} f_m, f_n] = [f_m, T^{m-1} f_n] = [f_m, f_{n-m+1}] \neq 0. \tag{4.2.34}$$

When $n = 2k$, we have $1 \leq n - k_0 + 1 \leq k$, so that $[f_{k_0}, f_{n-k_0+1}] \neq 0$ and the subspace $\mathcal{H}'_0$ is sign-indefinite in the decomposition (4.2.33). When $n = 2k + 1$, we have $1 \leq n - k_0 + 1 \leq k$ for $k_0 \geq k+2$ and $n - k_0 + 1 = k+1$ for $k_0 = k+1$. In either case, $[f_{k_0}, f_{n-k_0+1}] \neq 0$ and the subspace $\mathcal{H}'_0$ is sign-indefinite in the decomposition (4.2.33) unless $k_0 = k + 1$. In the latter case, we have

$$[f_{k+1}, f_{k+1}] = [f_1, f_n] \neq 0 \quad \text{and} \quad [f_j, f_{k+1}] = [T^{2k} f_{j+k}, f_n] = 0, \quad 1 \leq j \leq k.$$

Therefore, the subspace $\mathcal{H}'_0$ with $k_0 = k+1$, that is $\tilde{\mathcal{H}}_0$, is non-negative for $[f_1, f_n] > 0$ and non-positive for $[f_1, f_n] < 0$. $\qquad\square$

Lemma 4.13 *If $\lambda \in \mathbb{R}$ is an algebraically simple embedded eigenvalue, then the corresponding eigenspace $\mathcal{H}_\lambda = \mathrm{Span}\{f_1\}$ is either positive or negative or neutral depending on the value of $[f_1, f_1]$.*

Proof Because the embedded eigenvalue is assumed to be algebraically simple, the assertion of the lemma is trivial. $\qquad\square$

There is a technical problem to extend the result of Lemma 4.13 to multiple embedded eigenvalues. If $\lambda \in \mathbb{R}$ is an embedded eigenvalue of T, the Jordan chain (4.2.31) can terminate at f_n even if $[f_1, f_n] = 0$. Indeed, the Fredholm theory for the generalized eigenvalue problem in Lemma 4.10 gives a necessary but not a sufficient condition for existence of the solution f_{n+1} in the Jordan chain (4.2.31) if the eigenvalue λ is embedded into the continuous spectrum. If $[f_1, f_n] = 0$ but f_{n+1} does not exist in Π_κ, the neutral subspaces $\mathcal{H}_0$ for $n = 2k$ and $\tilde{\mathcal{H}}_0$ for $n = 2k + 1$ in Lemma 4.12 may not be the maximal non-positive or non-negative subspaces.

Lemma 4.14 *Let $\lambda \in \mathbb{C}$, $\mathrm{Im}(\lambda) > 0$ be an eigenvalue of T, $\mathcal{H}_\lambda$ be the corresponding eigenspace, and $\tilde{\mathcal{H}}_\lambda = \mathcal{H}_\lambda \cup \mathcal{H}_{\bar\lambda}$. Then, the neutral subspace $\mathcal{H}_\lambda$ is the maximal sign-definite subspace of $\tilde{\mathcal{H}}_\lambda$.*

Proof By Lemma 4.11 with $\lambda = \mu$, the eigenspace $\mathcal{H}_\lambda$ is orthogonal to itself with respect to $[\cdot, \cdot]$, so that $\mathcal{H}_\lambda$ is a neutral subspace. It remains to prove that $\mathcal{H}_\lambda$ is the maximal sign-definite subspace of $\tilde{\mathcal{H}}_\lambda$. Let $\mathcal{H}_\lambda = \mathrm{Span}\{f_1, f_2, ..., f_n\}$, where $\{f_1, f_2, ..., f_n\}$ is the Jordan chain of eigenvectors (4.2.31). Consider a subspace $\mathcal{H}'_{\lambda_0} = \mathrm{Span}\{f_1, f_2, ..., f_n, \bar{f}_j\}$ for any $1 \leq j \leq n$ and construct a linear combination of f_{n+1-j} and $\bar{f}_j$:

$$\forall \alpha \in \mathbb{C}: \quad [f_{n+1-j} + \alpha \bar{f}_j, f_{n+1-j} + \alpha \bar{f}_j] = 2 \, \mathrm{Re}\left(\alpha [\bar{f}_j, f_{n+1-j}]\right) = 2 \, \mathrm{Re}\left(\alpha [\bar{f}_1, f_n]\right).$$

By Lemma 4.10, we have $[f_n, \bar{f}_1] \neq 0$, so that the linear combination $f_{n+1-j} + \alpha \bar{f}_j$ is sign-indefinite with respect to $[\cdot, \cdot]$. $\qquad\square$

Exercise 4.15 Let $\gamma = \gamma_R + i\gamma_I$ be a complex eigenvalue of $Lu = \gamma M^{-1}u$ with a complex-valued eigenvector $u = u_R + iu_I$. Let $u = c_R u_R + c_I u_I$, where $\mathbf{c} = (c_R, c_I) \in \mathbb{R}^2$ and show that

$$\langle Lu, u \rangle_{L^2} - \langle \hat{L}\mathbf{c}, \mathbf{c} \rangle_{\mathbb{R}^2}, \quad \langle M^{-1}u, u \rangle_{L^2} = \langle \hat{M}\mathbf{c}, \mathbf{c} \rangle_{\mathbb{R}^2},$$

where matrices $\hat{L}$ and $\hat{M}$ have exactly one positive and one negative eigenvalue.

Exercise 4.16 Let $\gamma_0 \in \mathbb{R}$ be a double eigenvalue of $Lu = \gamma M^{-1}u$ with an eigenvector u_0 and a generalized eigenvector u_1, according to equations

$$\begin{cases} Lu_0 = \gamma_0 M^{-1}u_0, \\ Lu_1 = \gamma_0 M^{-1}u_1 + M^{-1}u_0, \end{cases}$$

which terminates at u_1 if $\langle M^{-1}u_1, u_0 \rangle_{L^2} \neq 0$. Show that the maximal non-negative subspace of $\mathcal{H}_{\gamma_0} = \mathrm{Span}\{u_0, u_1\}$ with respect to both $\langle L\cdot, \cdot \rangle$ and $\langle M^{-1}\cdot, \cdot \rangle_{L^2}$ is one-dimensional.

Exercise 4.17 Let $\gamma_0 > 0$ be an eigenvalue of $Lu = \gamma M^{-1}u$ of algebraic multiplicity three with an eigenvector u_0 and the generalized eigenvectors u_1 and u_2, according to equations

$$\begin{cases} Lu_0 = \gamma_0 M^{-1}u_0, \\ Lu_1 = \gamma_0 M^{-1}u_1 + M^{-1}u_0, \\ Lu_2 = \gamma_0 M^{-1}u_2 + M^{-1}u_1, \end{cases}$$

which terminates at u_2 if $\langle M^{-1}u_2, u_0 \rangle_{L^2} \neq 0$. Show that the maximal non-negative subspace of $\mathcal{H}_{\gamma_0} = \mathrm{Span}\{u_0, u_1, u_2\}$ with respect to both $\langle L\cdot, \cdot \rangle$ and $\langle M^{-1}\cdot, \cdot \rangle_{L^2}$ is two-dimensional if $\langle M^{-1}u_2, u_0 \rangle_{L^2} > 0$ and one-dimensional if $\langle M^{-1}u_2, u_0 \rangle_{L^2} < 0$.

The following theorem summarizes the count of the dimensions of the maximal non-positive and non-negative subspaces associated with eigenspaces of T in Π_κ.

Theorem 4.7 *Let $N_n(\lambda)$ ($N_p(\lambda)$) denote the dimension of the maximal non-positive (non-negative) subspace of Π_κ corresponding to the eigenspace $\mathcal{H}_\lambda$ of operator T for an eigenvalue λ.*

If $\lambda \in \mathbb{R}$ is isolated from the continuous spectrum, then

$$\dim(\mathcal{H}_\lambda) = N_p(\lambda) + N_n(\lambda) \tag{4.2.35}$$

and, for each Jordan block of generalized eigenvectors, we have the following:

(1) *If $n = 2k$, then $N_p(\lambda) = N_n(\lambda) = k$.*
(2) *If $n = 2k + 1$ and $[f_1, f_n] > 0$, then $N_p(\lambda) = k + 1$ and $N_n(\lambda) = k$.*
(3) *If $n = 2k + 1$ and $[f_1, f_n] < 0$, then $N_p(\lambda) = k$ and $N_n(\lambda) = k + 1$.*

If $\lambda \in \mathbb{R}$ is a simple embedded eigenvalue, then we have the following:

(1) *If $[f_1, f_1] > 0$, then $N_p(\lambda) = 1$, $N_n(\lambda) = 0$.*
(2) *If $[f_1, f_1] < 0$, then $N_p(\lambda) = 0$, $N_n(\lambda) = 1$.*
(3) *If $[f_1, f_1] = 0$, then $N_p(\lambda) = N_n(\lambda) = 1$.*

If $\lambda \notin \mathbb{R}$, then $\dim(\mathcal{H}_\lambda) = N_p(\lambda) = N_n(\lambda)$.

Proof The assertions of the theorem follow from Lemmas 4.12, 4.13, and 4.14. □

Remark 4.6 For a simple embedded eigenvalue $\lambda \in \mathbb{R}$, equality (4.2.35) does not hold in case (3) of Theorem 4.7 as

$$1 = \dim(\mathcal{H}_\lambda) < N_p(\lambda) + N_n(\lambda) = 2.$$

If $\lambda \in \mathbb{R}$ is a multiple embedded eigenvalue, computation of the projection matrix $[f_i, f_j]$ is needed in order to find the dimensions $N_p(\lambda)$ and $N_n(\lambda)$.

If the number of negative eigenvalues of operators $L + \delta M^{-1}$ and M equal the number of particular eigenvalues of the generalized eigenvalue problem (4.2.21), it is natural to ask if the total number of isolated eigenvalues of $L + \delta M^{-1}$ and M is related to the total number of isolated eigenvalues of the generalized eigenvalue problem (4.2.14). The following exercise shows that the latter is *bounded from above* by the former under the assumption of no *embedded* eigenvalues in the generalized eigenvalue problem [33].

Exercise 4.18 Assume that no embedded eigenvalues occur in the generalized eigenvalue problem (4.2.21) and prove that the eigenvalues satisfy the inequality

$$N_p^- + N_p^0 + N_p^+ + N_{c^+} \leq N_L + N_M, \qquad (4.2.36)$$

where N_L and N_M are the total numbers of isolated eigenvalues of L and M.

4.3 Spectral stability of localized modes

Let us consider the Gross–Pitaevskii equation,

$$i\psi_t = -\psi_{xx} + V(x)\psi + \sigma|\psi|^2\psi, \qquad (4.3.1)$$

where $\psi(x,t) : \mathbb{R} \times \mathbb{R}_+ \to \mathbb{C}$, $\sigma \in \{1, -1\}$, and $V(x) : \mathbb{R} \to \mathbb{R}$ is a 2π-periodic bounded potential.

The stationary Gross–Pitaevskii equation in the focusing case $\sigma = -1$,

$$-\phi''(x) + V(x)\phi(x) - \phi^3(x) = \omega\phi(x), \quad x \in \mathbb{R}, \qquad (4.3.2)$$

admits a strong localized mode $\phi \in H^2(\mathbb{R})$ for any $\omega < \omega_0 = \inf \sigma(L)$, where $L = -\partial_x^2 + V(x)$ (Section 3.1). The localized mode $\phi(x)$ decays to zero as $|x| \to \infty$ exponentially fast.

Using the same linearization technique as in Section 4.2, we obtain the spectral stability problem

$$\mathcal{L}\mathbf{u} = \lambda\mathbf{u}, \quad \mathcal{L} = \begin{bmatrix} 0 & L_- \\ -L_+ & 0 \end{bmatrix}, \quad \mathbf{u} = \begin{bmatrix} u \\ w \end{bmatrix}, \qquad (4.3.3)$$

associated with the Schrödinger operators $L_\pm$,

$$\begin{cases} L_+ = -\partial_x^2 + V(x) - \omega - 3\phi^2(x), \\ L_- = -\partial_x^2 + V(x) - \omega - \phi^2(x). \end{cases}$$

We note that $L_\pm$ has both periodic potentials $V(x)$ and exponentially decaying potentials $\phi^2(x)$. The following theorem describes the spectral stability of the localized mode ϕ.

Theorem 4.8 *Assume that the map $\mathbb{R} \ni \omega \mapsto \phi \in H^2(\mathbb{R})$ is C^1 near a fixed ω, $0 \notin \sigma(L_+)$ for this ω, and $\phi(x) > 0$ for all $x \in \mathbb{R}$. The localized mode ϕ is spectrally unstable if $n(L_+) \geq 2$ or $n(L_+) = 1$ and $\frac{d}{d\omega}\|\phi\|_{L^2}^2 > 0$ and it is spectrally stable if $n(L_+) = 1$ and $\frac{d}{d\omega}\|\phi\|_{L^2}^2 < 0$.*

Proof Because $\phi(x)$ decays to zero exponentially fast as $|x| \to \infty$, Lemma 4.1 extended to the case of both periodic and decaying potentials implies that

$$\sigma_c(L_\pm) = \sigma(-\partial_x^2 + V(x) - \omega) \geq \omega_0 - \omega > 0.$$

The point spectrum of operators $L_\pm$ is isolated from the continuous spectrum $\sigma_c(L_\pm)$ and it includes finitely many negative eigenvalues and, possibly, a zero eigenvalue.

To compute the location of $\sigma(\mathcal{L})$, we consider the spectrum of the limiting problem

$$(-\partial_x^2 + V(x) - \omega)u = -\lambda w, \quad (-\partial_x^2 + V(x) - \omega)w = \lambda u.$$

This problem becomes diagonal in variables $v_\pm = u \pm iw$,

$$(-\partial_x^2 + V(x) - \omega)v_\pm = \pm i\lambda v_\pm.$$

Therefore, $\sigma_c(\mathcal{L}) = \pm i\sigma(-\partial_x^2 + V(x) - \omega)$, that is, $\sigma_c(\mathcal{L}) \subset i\mathbb{R}$ with the gap $(-ic, ic)$ near the origin, where $c = \omega_0 - \omega > 0$.

Because $L_-\phi = 0$ is equivalent to the stationary equation (4.3.2), we note that $0 \in \sigma_p(L_-)$. By the assumption that $\phi(x) > 0$ for all $x \in \mathbb{R}$. Lemma 4.2 and Corollary 4.1 extended to the case of both periodic and decaying potentials imply that 0 is a simple eigenvalue of L_- and no negative eigenvalues of L_- exist.

By the assumption of the theorem, $0 \notin \sigma_p(L_+)$. There are still negative eigenvalues of L_+, which follows from the negativity of the quadratic form

$$\langle L_+\phi, \phi \rangle_{L^2} = -2\phi^2 < 0.$$

Hence, we have $n(L_+) \geq 1$.

Because L_+ is invertible, $L_+^{-1}\phi$ exists. Taking the derivative of the stationary equation (4.3.2) in ω under the assumption that the map $\mathbb{R} \ni \omega \mapsto \phi \in H^2(\mathbb{R})$ is C^1 shows that $L_+^{-1}\phi = \partial_\omega\phi$. Algorithmic computations then show that

$$\text{Ker}(\mathcal{L}) = \text{span}\left\{ \begin{bmatrix} 0 \\ \phi \end{bmatrix} \right\}, \quad \bigcup_{n\geq 1}\text{Ker}(\mathcal{L}^n) = \text{span}\left\{ \begin{bmatrix} 0 \\ \phi \end{bmatrix}, \begin{bmatrix} \partial_\omega\phi \\ 0 \end{bmatrix} \right\},$$

under the condition that $Q'(\omega) \neq 0$, where $Q(\omega) = \|\phi\|_{L^2}^2$. Indeed, $\mathcal{L}u = 0$ gives

$$L_+u = 0, \quad L_-w = 0,$$

with the only solution $u = 0$ and $w = C_1\phi$ for $C_1 \in \mathbb{R}$. If $\mathcal{L}^2u = 0$, then $\mathcal{L}u \in \text{Ker}(\mathcal{L})$, which gives the system

$$L_+u = C_1\phi, \quad L_-w = 0,$$

with the only solution $u = C_1 \partial_\omega \phi$ and $w = C_2 \phi$ for $(C_1, C_2) \in \mathbb{R}^2$. Finally, if $\mathcal{L}^3 \mathbf{u} = \mathbf{0}$, then $\mathcal{L} \mathbf{u} \in \text{Ker}(\mathcal{L}^2)$, which gives the system

$$L_+ u = C_1 \phi, \quad L_- w = C_2 \partial_\omega \phi,$$

which does not generate new solutions if

$$\langle \partial_\omega \phi, \phi \rangle_{L^2} = \frac{1}{2} \frac{dQ}{d\omega} \neq 0.$$

Thus, $0 \in \sigma_p(\mathcal{L})$ is an eigenvalue of geometric multiplicity *one* and algebraic multiplicity *two*.

We can now consider nonzero isolated eigenvalues of $\sigma_p(\mathcal{L})$. If $\mathbf{u} = (u, w) \in H^2(\mathbb{R}) \times H^2(\mathbb{R})$ is an eigenvector of $\mathcal{L}$ for $\lambda \neq 0$, then

$$\langle u, \phi \rangle_{L^2} = 0, \quad \langle w, \partial_\omega \phi \rangle_{L^2} = 0. \tag{4.3.4}$$

Let us consider the constrained L^2 space,

$$L_c^2 = \left\{ u \in L^2(\mathbb{R}) : \quad \langle u, \phi \rangle_{L^2} = 0 \right\}$$

and let $P_c : L^2 \to L_c^2 \subset L^2$ be the orthogonal projection operator. If $u \in L_c^2$, then we can now eliminate the component w from system (4.3.3) by

$$w = C\phi + \lambda P_c L_-^{-1} P_c u,$$

where $C \in \mathbb{R}$ is uniquely found from the projection to ϕ for $\lambda \neq 0$:

$$C = -\frac{\langle (L_+ + \lambda^2 P_c L_-^{-1} P_c) u, \phi \rangle_{L^2}}{\lambda \|\phi\|_{L^2}^2}.$$

Using the projection operator P_c, we can close system (4.3.3) at the generalized eigenvalue problem,

$$P_c L_+ P_c u = -\lambda P_c w = \gamma P_c L_-^{-1} P_c u, \quad \gamma = -\lambda^2. \tag{4.3.5}$$

Because operator $P_c L_-^{-1} P_c$ is strictly positive, all eigenvalues γ are real and simple and the number of negative eigenvalues γ equals the number of negative eigenvalues of $P_c L_+ P_c$ (Theorem 4.2).

By Theorem 4.1, we know that $n(P_c L_+ P_c) = n(L_+) - 1$ if $\langle L_+^{-1} \phi, \phi \rangle_{L^2} < 0$, which gives $Q'(\omega) < 0$, whereas $n(P_c L_+ P_c) = n(L_+)$ if $\langle L_+^{-1} \phi, \phi \rangle_{L^2} > 0$, which gives $Q'(\omega) > 0$. Spectral stability corresponds to the case of strictly positive eigenvalues γ, which may only happen if $n(L_+) = 1$ and $Q'(\omega) < 0$. In the case $n(L_+) \geq 2$ or $n(L_+) = 1$ and $Q'(\omega) > 0$, there is at least one negative eigenvalue $\gamma < 0$, which gives at least one unstable eigenvalue $\lambda > 0$. $\qquad \square$

Two distinct branches of localized modes ϕ exist with $Q'(\omega) < 0$ in the limit of large negative ω (Section 3.2.3). Exercise 3.10 shows that $n(L_+) = 1$ for the localized mode ϕ which is centered at the minimum of $V(x)$, and $n(L_+) = 2$ for the localized mode ϕ which is centered at the maximum of $V(x)$. Therefore, the former localized mode is spectrally stable, whereas the latter localized mode is spectrally unstable with exactly one real unstable eigenvalue.

If $V(x) \equiv 0$, the two distinct localized modes belong to the family of translationally invariant localized modes $\phi(x - s)$, $s \in \mathbb{R}$ of the stationary cubic NLS equation,

$$-\phi''(x) - \phi^3(x) = \omega\phi(x), \quad x \in \mathbb{R}.$$

In this case, the localized mode exists in the explicit form,

$$\phi(x) = \sqrt{2|\omega|} \operatorname{sech}(\sqrt{|\omega|}x), \quad \omega < 0.$$

Direct computation gives $Q'(\omega) < 0$. Because $L_+\phi'(x) = 0$ and $\phi'(x)$ has exactly one zero on $\mathbb{R}$, we have $n(L_+) = 1$ and $z(L_+) = 1$, so that $\mathcal{L}$ has additional degeneracy at 0 but no unstable eigenvalues with $\operatorname{Re}(\lambda) > 0$. As a result, the localized mode $\phi(x - s)$ is *spectrally stable* for all $s \in \mathbb{R}$ in the case of the cubic NLS equation (Corollary 4.3 for $p = 1$).

If the localized mode ϕ is sign-indefinite on $\mathbb{R}$, the count of unstable eigenvalues is developed from Theorem 4.5. However, this construction only works if the number of zeros of ϕ is finite on $\mathbb{R}$, that is, if ω is fixed in the semi-infinite gap with $\omega < \omega_0$. Unfortunately, no information on stability of localized modes ϕ can be extracted for ω in a finite gap of the spectrum of $L = -\partial_x^2 + V(x)$.

To overcome this limitation, we shall apply reductions of the Gross–Pitaevskii equation with a periodic potential to the nonlinear evolution equations with constant coefficients. We restrict ourselves to a number of prototypical examples that include solitons in the coupled NLS equation (Section 4.3.1), vortices in the NLS equation (Section 4.3.2), soliton configurations in the DNLS equation (Section 4.3.3), vortex configurations in the DNLS equation (Section 4.3.4), and gap solitons in the nonlinear Dirac equations (Section 4.3.5).

Other examples of localized modes in nonlinear evolution equations will remain outside the scope of this book. Readers can find stability analysis of solitons of the generalized Korteweg–de Vries equation in [34], solitons of the Boussinesq equation in [132], solitons of the Maxwell–Bloch equations in [128], among many other examples.

4.3.1 Solitons in the coupled NLS equations

Let us consider the coupled cubic NLS equations,

$$\begin{cases} i\partial_t\psi_1 + \partial_x^2\psi_1 + \left(|\psi_1|^2 + \chi|\psi_2|^2\right)\psi_1 = 0, \\ i\partial_t\psi_2 + \partial_x^2\psi_2 + \left(\chi|\psi_1|^2 + |\psi_2|^2\right)\psi_2 = 0, \end{cases} \tag{4.3.6}$$

where $\psi_{1,2}(x,t) : \mathbb{R} \times \mathbb{R}_+ \to \mathbb{C}$ are envelope amplitudes and $\chi > 0$ is the coupling constant. This system coincides with system (1.2.11) from Section 1.2.1, where it is shown that the system has the standard complex-valued Hamiltonian structure with the conserved Hamiltonian,

$$H = \int_{\mathbb{R}} \left(|\partial_x\psi_1|^2 + |\partial_x\psi_2|^2 - \frac{1}{2}\left(|\psi_1|^4 + 2\chi|\psi_1|^2|\psi_2|^2 + |\psi_2|^4\right)\right) dx,$$

the conserved powers,

$$Q_1 = \int_{\mathbb{R}} |\psi_1|^2 dx, \quad Q_2 = \int_{\mathbb{R}} |\psi_2|^2 dx,$$

and the conserved momentum,

$$P = \mathrm{i} \int_{\mathbb{R}} \left(\psi_1 \partial_x \bar{\psi}_1 - \bar{\psi}_1 \partial_x \psi_1 + \psi_2 \partial_x \bar{\psi}_2 - \bar{\psi}_2 \partial_x \psi_2 \right) dx.$$

Stationary localized modes are given by

$$\psi_1(x,t) = \phi_1(x)e^{-\mathrm{i}\omega_1 t}, \quad \psi_2(x,t) = \phi_2(x)e^{-\mathrm{i}\omega_2 t},$$

where (ϕ_1, ϕ_2) are real-valued functions and (ω_1, ω_2) belong to an open domain $D \subset \mathbb{R}^2$. Weak stationary solutions $(\phi_1, \phi_2) \in H^1(\mathbb{R}) \times H^1(\mathbb{R})$ are critical points of the energy functional

$$E_\omega(u_1, u_2) = H(u_1, u_2) - \omega_1 Q_1(u_1) - \omega_2 Q_2(u_2). \tag{4.3.7}$$

Exercise 4.19 Assume that the map $\mathbb{R}^2 \ni (\omega_1, \omega_2) \mapsto (\phi_1, \phi_2) \in H^1(\mathbb{R}) \times H^1(\mathbb{R})$ is C^1 and show that the Hessian matrix of the energy surface $\mathcal{D}(\omega_1, \omega_2) = E_\omega(\phi_1, \phi_2)$ is symmetric with

$$U_{i,j} := \frac{\partial^2 \mathcal{D}}{\partial \omega_i \partial \omega_j} = -\frac{\partial Q_i}{\partial \omega_j} = -\frac{\partial Q_j}{\partial \omega_i}, \quad i,j \in \{1,2\}. \tag{4.3.8}$$

The linearization of the coupled NLS equations (4.3.6) leads to the spectral stability problem

$$L_+ \mathbf{u} = -\lambda \mathbf{w}, \quad L_- \mathbf{w} = \lambda \mathbf{u}, \quad \lambda \in \mathbb{C}, \tag{4.3.9}$$

where $L_\pm$ are 2×2 matrix Schrödinger operators,

$$(L_-)_{i,j} = \left(-\partial_x^2 + \omega_j\right)\delta_{i,j} - V_{i,j}(x),$$

$$(L_+)_{i,j} = \left(-\partial_x^2 + \omega_j\right)\delta_{i,j} - V_{i,j}(x) - W_{i,j}(x),$$

associated with the decaying potentials

$$V(x) = \begin{bmatrix} \phi_1^2(x) + \chi\phi_2^2(x) & 0 \\ 0 & \chi\phi_1^2(x) + \phi_2^2(x) \end{bmatrix},$$

$$W(x) = 2 \begin{bmatrix} \phi_1^2(x) & \chi\phi_1(x)\phi_2(x) \\ \chi\phi_1(x)\phi_2(x) & \phi_2^2(x) \end{bmatrix}.$$

Exercise 4.20 Assume that $\phi_1(x)$ and $\phi_2(x)$ are not identically zero for all $x \in \mathbb{R}$ and show that

$$\mathrm{Ker}(L_-) = \mathrm{Span}\left\{\phi_1 \mathbf{e}_1, \phi_2 \mathbf{e}_2\right\},$$

where $\{\mathbf{e}_1, \mathbf{e}_2\}$ is a standard basis in $\mathbb{R}^2$. Show that $\phi' \in \mathrm{Ker}(L_+)$, where $\phi' = (\phi_1'(x), \phi_2'(x))$.

To invert the operator L_-, we define the constrained L^2 space,

$$L_c^2 = \left\{ \mathbf{u} = (u_1, u_2) \in L^2(\mathbb{R}) \times L^2(\mathbb{R}) : \quad \langle u_1, \phi_1 \rangle_{L^2} = \langle u_2, \phi_2 \rangle_{L^2} = 0 \right\}.$$

There are also constraints on $\mathbf{w} = (w_1, w_2)$, but, as we did earlier, the eigenvalue problem (4.3.9) for $\lambda \neq 0$ can be closed at the generalized eigenvalue problem,

$$P_c L_+ P_c \mathbf{u} = \gamma P_c L_-^{-1} P_c \mathbf{u}, \quad \gamma = -\lambda^2, \tag{4.3.10}$$

where $P_c : L^2 \to L_c^2 \subset L^2$ is an orthogonal projection operator.

Exercise 4.21 Assume that $\mathrm{Ker}(L_+) \perp \mathrm{Ker}(L_-)$ and show that

$$n(P_c L_+ P_c) = n(L_+) - p(U) - z(U), \quad z(P_c L_+ P_c) = z(L_+) + z(U),$$

where $p(U)$ and $z(U)$ are the numbers of positive and zero eigenvalues of the Hessian matrix U of the energy surface $\mathcal{D}(\omega_1, \omega_2)$.

Exercise 4.22 Consider a traveling localized mode of the coupled NLS equations (4.3.6),

$$\psi_1(x, t) = \tilde{\phi}_1(x + 2ct)e^{-i\tilde{\omega}_1 t}, \quad \psi_2(x, t) = \tilde{\phi}_2(x + 2ct)e^{-i\tilde{\omega}_2 t},$$

where $(\tilde{\phi}_1, \tilde{\phi}_2)$ is a critical point of the energy functional

$$\tilde{E}_\omega(u_1, u_2) = H(u_1, u_2) - \tilde{\omega}_1 Q_1(u_1) - \tilde{\omega}_2 Q_2(u_2) - cP(u_1, u_2).$$

Define the Hessian matrix $\tilde{U}$ of the extended energy surface $\tilde{\mathcal{D}}(\omega_1, \omega_2, c) = \tilde{E}_\omega(\tilde{\phi}_1, \tilde{\phi}_2)$ and use Galileo invariance (1.2.19) to show that $p(\tilde{U}) = p(U)$ and $z(\tilde{U}) = z(U)$.

The following theorem determines the spectral stability of localized modes in the coupled NLS equations (4.3.6) based on Theorem 4.5.

Theorem 4.9 *Assume that $z(L_+) = 1$ and $z(U) = 0$ for a point (ω_1, ω_2) in the domain D. The numbers of eigenvalues γ of the generalized eigenvalue problem (4.3.10) satisfy*

$$N_p^- + N_n^+ + N_{c^+} = n(L_+) - p(U), \tag{4.3.11}$$

$$N_n^- + N_n^+ + N_{c^+} = n(L_-). \tag{4.3.12}$$

Proof The assumptions of Theorem 4.5 are satisfied and equalities (4.3.11) and (4.3.12) follow from equalities (4.2.26) and (4.2.27) if $N_n^0 = 0$ and

$$\dim(\mathcal{H}_{L+\delta M^{-1}}^-) = n(L_+) - p(U).$$

From the assumptions of the theorem, we have $\mathrm{Ker}(L_+) = \mathrm{Span}\{\phi'\}$, where $\phi' = (\phi_1'(x), \phi_2'(x))$. Let $\phi = (\phi_1(x), \phi_2(x))$ and observe that

$$L_-(x\phi) = -2\phi'.$$

Since $\mathrm{Ker}(L_+) \perp \mathrm{Ker}(L_-)$, we have

$$\langle L_-^{-1}\phi', \phi' \rangle_{L^2} = -\frac{1}{2}\langle x\phi, \phi' \rangle_{L^2} = \frac{1}{4}(Q_1 + Q_2) > 0.$$

By the definition of N_n^0 and N_p^0, we have $N_n^0 = 0$ and $N_p^0 = z(L_+) = 1$. By Lemma 4.7, we obtain $\dim(\mathcal{H}_{L+\delta M^{-1}}^-) = \dim(\mathcal{H}_L^-)$. On the other hand, since

$$L_+\partial_{\omega_1}\boldsymbol{\phi} = \phi_1\mathbf{e}_1, \quad L_+\partial_{\omega_2}\boldsymbol{\phi} = \phi_2\mathbf{e}_2,$$

we obtain

$$\lim_{\mu\uparrow 0} A_{i,j}(\mu) = -\langle L_+^{-1}\phi_i\mathbf{e}_i, \phi_j\mathbf{e}_j\rangle_{L^2} = -\langle \partial_{\omega_i}\boldsymbol{\phi}, \phi_j\mathbf{e}_j\rangle_{L^2} = \frac{1}{2}U_{i,j}, \quad i,j \in \{1,2\}.$$

Therefore, $p_0 = p(U)$ and $z_0 = z(U) = 0$. Using (4.2.19), we obtain $\dim(\mathcal{H}_L^-) = n(L_+) - p(U)$. The assertion of the theorem is proved. $\qquad\square$

Equalities (4.3.11) and (4.3.12) are not precise as they do not specify if the localized mode $\boldsymbol{\phi}$ is spectrally stable or spectrally unstable. To make the conclusion more precise, we consider a particular family of localized modes that bifurcates from the solution of the stationary cubic NLS equation,

$$-\phi_0''(x) - \phi_0^3 = \omega\phi_0, \quad x \in \mathbb{R},$$

for any fixed $\omega < 0$. The exact localized solution for $\phi_0(x)$ is available in the analytic form,

$$\phi_0(x) = \sqrt{2|\omega|}\operatorname{sech}(\sqrt{|\omega|}x),$$

where it is chosen to be centered at $x = 0$ thanks to the translational invariance of the NLS equation.

If $\phi_1 = \phi_0$ and $\omega_1 = \omega$, then substituting ϕ_0 into the second equation of system (4.3.6) and linearizing it around the zero solution, we obtain the linear stationary Schrödinger equation,

$$-\psi''(x) - 2\chi|\omega|\operatorname{sech}^2(\sqrt{|\omega|}x)\psi(x) = \lambda\psi(x), \quad x \in \mathbb{R}. \tag{4.3.13}$$

The second-order differential equation (4.3.13) can be reduced to the hypergeometric equation, from which exact expressions for the eigenvalues and eigenfunctions can be obtained [213]. As a summary of this theory, we state that for $\chi \in (\chi_n, \chi_{n+1})$, where

$$\chi_n = \frac{n(n+1)}{2}, \quad n \in \mathbb{N}_0,$$

there exist exactly $n+1$ negative eigenvalues of the stationary Schrödinger equation (4.3.13) at $\lambda \in \{-\lambda_0^2|\omega|, -\lambda_1^2|\omega|, ..., -\lambda_n^2|\omega|\}$, where

$$\lambda_k = \frac{\sqrt{1+8\chi} - (2k+1)}{2}, \quad k \in \{0, 1, ..., n\}.$$

The eigenfunction $\psi_0(x)$ for $\lambda = -\lambda_0^2|\omega|$ is

$$\psi_0(x) = \operatorname{sech}^{\lambda_0}(\sqrt{|\omega|}x) > 0, \quad x \in \mathbb{R}.$$

The eigenfunction $\psi_k(x)$ for $\lambda = -\lambda_k^2|\omega|$ has exactly k zeros on $\mathbb{R}$, in agreement with Corollary 4.1.

We shall now consider the family of localized modes $\boldsymbol{\phi} = (\phi_1, \phi_2)$ of the coupled NLS equations (4.3.6) such that the first component $\phi_1(x)$ is positive for all $x \in \mathbb{R}$ and the second component $\phi_2(x)$ has exactly n zeros on $\mathbb{R}$. The following local bifurcation result shows that this family exists for $\chi > \chi_n$ in a neighborhood of the solution $(\phi_1, \phi_2) = (\phi_0, 0)$, which corresponds to the bifurcation curve $\omega_2 = \lambda_n^2 \omega_1$. Thanks to the scaling invariance, we can set $\omega_1 = -1$ for convenience.

Lemma 4.15 *Fix $n \in \mathbb{N}_0$, $\chi > \chi_n$ and assume that $\delta_n \neq 0$, where*

$$\delta_n = -2\chi^2 \langle \phi_0 \psi_n^2, L_1^{-1} \phi_0 \psi_n^2 \rangle_{L^2} - \|\psi_n\|_{L^4}^4. \tag{4.3.14}$$

There exists an $\epsilon_n > 0$ such that the coupled NLS equations (4.3.6) admit a localized mode (ϕ_1, ϕ_2) near $(\phi_0, 0)$ for $\omega_1 = -1$ and $\omega_2 \in B_n$, where

$$B_n = (-\lambda_n^2, -\lambda_n^2 + \epsilon_n) \quad \text{if} \quad \delta_n > 0, \quad B_n = (-\lambda_n^2 - \epsilon_n, -\lambda_n^2) \quad \text{if} \quad \delta_n < 0.$$

Proof The proof follows the method of Lyapunov–Schmidt reductions (Section 3.2.1). To simplify details, let us expand the solution (ϕ_1, ϕ_2) near $(\phi_0, 0)$ as

$$\phi_1(x) = \phi_0(x) + \epsilon^2 \tilde{\phi}_1(x) + \mathcal{O}(\epsilon^4),$$

$$\phi_2(x) = \epsilon \psi_n(x) + \epsilon^3 \tilde{\phi}_2(x) + \mathcal{O}(\epsilon^5),$$

$$\omega_2 = -\lambda_n^2 + \epsilon^2 \delta_n + \mathcal{O}(\epsilon^4),$$

where $\phi_0(x) = \sqrt{2}\,\mathrm{sech}(x)$, $\psi_n(x)$ is the L^2-normalized eigenfunction of the linear Schrödinger equation (4.3.13) for $\omega = -1$ and $\lambda = -\lambda_n^2$, and ϵ is a formal small parameter.

The first two corrections of the perturbation series satisfy the linear inhomogeneous equations

$$L_1 \tilde{\phi}_1 = \chi \phi_0 \psi_n^2,$$

$$L_2 \tilde{\phi}_2 = \delta_n \psi_n + 2\chi \phi_0 \psi_n \tilde{\phi}_1 + \psi_n^3,$$

where

$$L_1 = -\partial_x^2 + 1 - 6\,\mathrm{sech}^2(x), \quad L_2 = -\partial_x^2 + \lambda_n^2 - 2\chi\,\mathrm{sech}^2(x).$$

We note that $\mathrm{Ker}(L_1) = \mathrm{Span}\{\phi_0'\}$ and $\mathrm{Ker}(L_2) = \mathrm{Span}\{\psi_n\}$. Since $\phi_0'(x)$ is odd on $\mathbb{R}$, the operator L_1 is invertible in the space of even functions and a unique even solution $\tilde{\phi}_1 \in H^2(\mathbb{R})$ exists. On the other hand, the solution $\tilde{\phi}_2 \in H^2(\mathbb{R})$ exists if and only if δ_n is given by

$$\delta_n = -\langle \psi_n^2, 2\chi \phi_0 \tilde{\phi}_1 + \psi_n^2 \rangle_{L^2},$$

which is equivalent to (4.3.14). Since $L_+ = \mathrm{diag}(L_1, L_2)$ at $\epsilon = 0$, existence and uniqueness of a localized mode (ϕ_1, ϕ_2) near $(\phi_0, 0)$ follow by the Lyapunov–Schmidt reduction algorithm, provided $\delta_n \neq 0$. □

By Corollary 4.1, we infer that $z(L_-) = 2$ and $n(L_-) = n$ are preserved for any $\epsilon > 0$ if the localized mode has positive $\phi_1(x)$ and n zeros for $\phi_2(x)$ on $\mathbb{R}$. The following lemma computes the numbers $n(L_+)$ and $p(U)$ for the localized mode near the bifurcation curve $\omega_2 = \lambda_n^2 \omega_1$.

Lemma 4.16 *Let $\phi = (\phi_1, \phi_2)$ be a localized mode in Lemma 4.15 for $\omega_1 = -1$ and $\omega_2 \in B_n$. Then, we have $z(L_+) = 1$, $z(U) = 0$, and*

$$n(L_+) = \begin{cases} 1 + n, & \delta_n > 0, \\ 2 + n, & \delta_n < 0, \end{cases} \qquad p(U) = \begin{cases} 1, & \delta_n > 0, \\ 2, & \delta_n < 0, \end{cases}$$

where δ_n is given by (4.3.14).

Proof From the proof of Lemma 4.15, we know that $n(L_+) = 1 + n$ and $z(L_+) = 1 + 1 = 2$ at $\epsilon = 0$. When $\omega_2 \in B_n$, the degeneracy of $\mathrm{Ker}(L_+)$ is broken since only a simple zero eigenvalue with the eigenvector $\phi' = (\phi_1'(x), \phi_2'(x))$ is preserved by the translational symmetry of the coupled NLS equations (4.3.6).

Let $(\lambda, \mathbf{u})$ be the eigenvalue–eigenvector pair for $L_+ \mathbf{u} = \lambda \mathbf{u}$. The bifurcating eigenvalue of L_+ can be traced by the regular perturbation series,

$$\mathbf{u}(x) = \begin{bmatrix} 0 \\ \psi_n(x) \end{bmatrix} + \epsilon \begin{bmatrix} u_1(x) \\ 0 \end{bmatrix} + \epsilon^2 \begin{bmatrix} 0 \\ u_2(x) \end{bmatrix} + \mathcal{O}(\epsilon^3),$$

$$\lambda = \epsilon^2 \lambda_2 + \mathcal{O}(\epsilon^4).$$

Corrections of the perturbation series satisfy linear inhomogeneous equations

$$L_1 u_1 = 2\chi \phi_0 \psi_n^2,$$

$$L_2 u_2 = (\lambda_2 + \delta_n)\psi_n + 2\chi \phi_0 \psi_n \left(\tilde{\phi}_1 + u_1 \right) + 3\psi_n^3.$$

From the linear inhomogeneous equations in Lemma 4.15, we infer that

$$u_1 = 2\tilde{\phi}_1, \quad u_2 = 3\tilde{\phi}_2, \quad \text{and} \quad \lambda_2 = 2\delta_n.$$

Adding this perturbation result for $\epsilon > 0$ to the previous count at $\epsilon = 0$, we obtain

$$n(L_+) = \begin{cases} 1 + n, & \delta_n > 0, \\ 2 + n, & \delta_n < 0, \end{cases} \qquad z(L_+) = 1.$$

To trace the zero eigenvalue of U, we compute perturbation expansions

$$U_{1,1} = -\frac{\partial Q_1}{\partial \omega_1}\bigg|_{\omega_1 = -1} = 2 - 2\langle \phi_0, \tilde{\phi}_1 \rangle_{L^2} \frac{\partial \epsilon^2}{\partial \omega_1}\bigg|_{\omega_1 = -1} + \mathcal{O}(\epsilon^2),$$

$$U_{1,2} = -\frac{\partial Q_1}{\partial \omega_2}\bigg|_{\omega_1 = -1} = -2\langle \phi_0, \tilde{\phi}_1 \rangle_{L^2} \frac{\partial \epsilon^2}{\partial \omega_2}\bigg|_{\omega_1 = -1} + \mathcal{O}(\epsilon^2),$$

$$U_{2,1} = -\frac{\partial Q_2}{\partial \omega_1}\bigg|_{\omega_1 = -1} = -\frac{\partial \epsilon^2}{\partial \omega_1}\bigg|_{\omega_1 = -1} + \mathcal{O}(\epsilon^2),$$

$$U_{2,2} = -\frac{\partial Q_2}{\partial \omega_2}\bigg|_{\omega_1 = -1} = -\frac{\partial \epsilon^2}{\partial \omega_2}\bigg|_{\omega_1 = -1} + \mathcal{O}(\epsilon^2).$$

From Lemma 4.15, we know that

$$\frac{\partial \epsilon^2}{\partial \omega_2}\bigg|_{\omega_1 = -1} = \frac{1}{\delta_n} + \mathcal{O}(\epsilon^2),$$

which gives $\det(U) = -2/\delta_n + \mathcal{O}(\epsilon^2)$. Since $\delta_n \neq 0$, we obtain

$$p(U) = \begin{cases} 1, & \delta_n > 0, \\ 2, & \delta_n < 0, \end{cases} \qquad z(U) = 0,$$

which completes the proof of the lemma. $\qquad\qquad\qquad\qquad\qquad\square$

Substituting the result of Lemma 4.16 into equations (4.3.11) and (4.3.12), we obtain

$$\begin{cases} N_p^- + N_n^+ + N_{c^+} = n, \\ N_n^- + N_n^+ + N_{c^+} = n. \end{cases}$$

In particular, the positive localized mode ϕ is spectrally stable for $n = 0$ and the sign-indefinite localized mode ϕ may have at most $(2n)$ unstable eigenvalues for $n \geq 1$. If $\omega_1 = -1$ and $\omega_2 = -\lambda_n^2$ ($\epsilon = 0$), the spectral stability problem (4.3.9) becomes block-diagonal with two blocks,

$$L_1 u_1 = -\lambda w_1, \quad L_0 w_1 = \lambda u_1 \qquad\qquad (4.3.15)$$

and

$$L_2(u_2 \pm i w_2) = \pm i\lambda(u_2 \pm i w_2), \qquad\qquad (4.3.16)$$

where $L_0 = -\partial_x^2 + 1 - 2\,\mathrm{sech}^2(x)$.

The first problem (4.3.15) is a linearization of the cubic NLS equation. By Corollary 4.3 for $p = 1$, this problem gives no contribution to the count of N_p^-, N_n^-, N_n^+, and N_{c^+}.

The second problem (4.3.16) is given by an uncoupled pair of linear Schrödinger equations (4.3.13) with $\omega = -1$. From analysis of these equations, we know that there exist $2n$ isolated eigenvalues at

$$\lambda = \pm i\left(\lambda_k^2 - \lambda_n^2\right), \quad k \in \{0, 1, ..., n-1\}.$$

Moreover, since

$$\langle L_+ \mathbf{u}, \mathbf{u} \rangle_{L^2} = -(\lambda_k^2 - \lambda_n^2)\|\psi_k\|_{L^2}^2 < 0, \quad k \in \{0, 1, ..., n-1\}, \qquad (4.3.17)$$

we obtain

$$\omega_2 = -\lambda_n^2: \quad N_p^- = N_n^- = N_{c^+} = 0, \quad N_n^+ = n. \qquad (4.3.18)$$

We note that

$$\lambda_k^2 - \lambda_n^2 = (n-k)\,(2\lambda_n + (n-k)) > (n-k)^2 \geq 1, \quad k \in \{0, 1, ..., n-1\}.$$

Because system (4.3.15) has the continuous spectrum for $\mathrm{Re}(\lambda) = 0$ and $|\mathrm{Im}(\lambda)| \geq 1$, pairs of imaginary eigenvalues of negative energy (4.3.17) in the count N_n^+ are embedded into the continuous spectrum of the spectral stability problem (4.3.9).

By the analysis developed by Grillakis [77] and Cuccagna *et al.* [43], embedded eigenvalues of negative energy bifurcate generally to complex unstable eigenvalues.

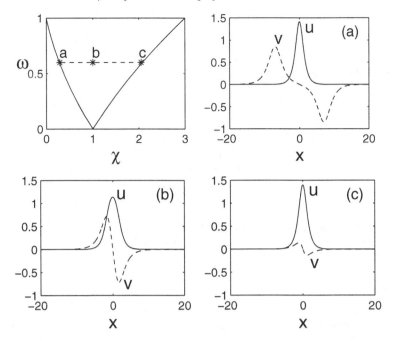

Figure 4.1 The existence domain of the first family of localized modes with $\omega_1 = -1$ and $\omega_2 = -\omega$ on the plane (ω, χ) and the three particular localized modes for three values of χ at $\omega = 0.6$. Reproduced from [166].

Therefore, we should expect that the count of eigenvalues for $\omega_2 \neq -\lambda_n^2$ becomes

$$\omega_2 \in B_n : \quad N_p^- = N_n^- = N_n^+ = 0, \quad N_{c+} = n, \qquad (4.3.19)$$

hence $2n$ complex eigenvalues in the count N_{c+} are unstable.

To confirm this result, we approximate numerically the first family of localized modes in the coupled NLS equations (4.3.6). These numerical approximations are discussed by Yang [213], Champneys & Yang [31], and Pelinovsky & Yang [166].

Figure 4.1 shows the localized mode with $n = 1$ for $\omega_1 = -1$, $\omega_2 = -0.6$, and $\chi \in (\chi_1, \chi_2)$, where χ_2 is the local bifurcation boundary in Lemma 4.15 and χ_1 is the non-local bifurcation boundary studied in [31]. The one-sided domain B_1 near χ_2 is located to the left of the local bifurcation curve, hence $\delta_1 < 0$ in Lemma 4.15.

Figure 4.2 shows the unstable eigenvalues in the interval (χ_1, χ_2). Near the local bifurcation boundary $\chi = \chi_2$, there is a pair of unstable complex eigenvalues $\lambda = \operatorname{Re}(\sigma_2) \pm \operatorname{i} \operatorname{Im}(\sigma_2)$, which bifurcate from the embedded eigenvalues $\lambda = \pm \operatorname{i}(\lambda_0^2 - \lambda_1^2)$ according to the above analysis.

No unstable eigenvalues exist at $\chi = 1$ since $N_n^+ = 1$ and $N_{c+} = 0$. However, unstable eigenvalues exist for both $\chi > 1$ and $\chi < 1$ with $N_n^+ = 0$ and $N_{c+} = 1$ since the embedded eigenvalues with negative energy are structurally unstable with respect to parameter continuation. An additional pair of real unstable eigenvalues $\lambda = \pm \sigma_1$ exists for $\chi < 1$. It bifurcates from the pair of purely imaginary eigenvalues

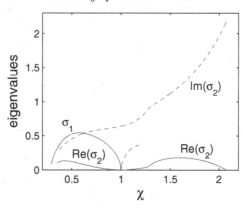

Figure 4.2 Eigenvalues of the first family of localized modes for $\chi \in (\chi_1, \chi_2)$ at $\omega = 0.6$. The solid (dashed) lines are the real (imaginary) parts of the eigenvalues. Reproduced from [166].

$\lambda = \pm i\sigma_1$ for $\chi > 1$. The latter pair merges to the end points $\lambda = \pm i\omega^2$ of the continuous spectrum at $\chi = 1.185$ and coalesces at the origin at $\chi = 1$.

Let us summarize the numerical computations of the unstable eigenvalues and indices $n(L_+)$ and $p(U)$,

$$n(L_+) = \begin{cases} 4, & \chi_1 < \chi < 1, \\ 3, & 1 < \chi < \chi_2, \end{cases} \qquad p(U) = 2, \text{ for all } \chi_1 < \chi < \chi_2,$$

so that

$$n(L_+) - p(U) = \begin{cases} 2, & \chi_1 < \chi < 1, \\ 1, & 1 < \chi < \chi_2. \end{cases}$$

On the other hand, $n(L_-) = 1$ for all $\chi_1 < \chi < \chi_2$ and

$$N_{c+} = \begin{cases} 1, & \chi_1 < \chi < 1, \\ 1, & 1 < \chi < \chi_2 \end{cases} \qquad N_p^- = \begin{cases} 1, & \chi_1 < \chi < 1, \\ 0, & 1 < \chi < \chi_2. \end{cases}$$

The count of unstable eigenvalues agrees with the result of Theorem 4.9 and it is more precise than equalities (4.3.11) and (4.3.12).

The spectral stability of the localized mode ϕ with $n = 1$ at $\chi = 1$ is not a coincidence. The system of coupled NLS equations (4.3.6) with $\chi = 1$ is the Manakov system (1.4.3), which belongs to the list of integrable evolution equations (Section 1.4). In this case, there exists an exact localized mode $\phi = (\phi_1, \phi_2)$ with $n = 1$ for $\omega_1 = -1$ and $\omega_2 = -\omega$ with $\omega \in (0, 1)$ in the analytic form,

$$\begin{cases} \phi_1(x) = \dfrac{\sqrt{1 - \omega^2} \cosh(\omega x)}{\cosh(x) \cosh(\omega x) - \omega \sinh(x) \sinh(\omega x)}, \\[3mm] \phi_2(x) = -\dfrac{\omega\sqrt{1 - \omega^2} \sinh(x)}{\cosh(x) \cosh(\omega x) - \omega \sinh(x) \sinh(\omega x)}, \end{cases} \qquad (4.3.20)$$

The Manakov system has an additional rotational symmetry (Section 1.2.1) in the nonlinear potential

$$W(|\psi_1|^2, |\psi_2|^2) = \frac{1}{2}(|\psi_1|^2 + |\psi_2|^2)^2.$$

As a result, this symmetry generates a pair of imaginary eigenvalues of negative energy in the count $N_n^+ = 1$ for $\gamma = 1$, as per the following exercise.

Exercise 4.23 Show that the spectral stability problem (4.3.9) for $\chi = 1$ admits the exact solution

$$\mathbf{u} = \begin{bmatrix} -\phi_2 \\ \phi_1 \end{bmatrix}, \quad \mathbf{w} = \mp\mathrm{i}\begin{bmatrix} \phi_2 \\ \phi_1 \end{bmatrix}, \quad \lambda = \pm\mathrm{i}(1 - \omega^2). \qquad (4.3.21)$$

Furthermore, show that these eigenvalues contribute to the count N_n^+ since

$$\langle L_+\mathbf{u}, \mathbf{u}\rangle_{L^2} = -(1 - \omega^2)\left(\|\phi_1\|_{L^2}^2 - \|\phi_2\|_{L^2}^2\right) < 0, \qquad (4.3.22)$$

where (ϕ_1, ϕ_2) are given by (4.3.20).

We also note that the Manakov system admits another family of exact asymmetric localized modes with an additional free parameter that includes the localized mode (4.3.20) [213]. Because of the presence of this additional parameter, $z(L_+) = 2$ at $\chi = 1$ and this additional zero eigenvalue of L_+ leads to a bifurcation of an unstable eigenvalue in the count N_p^- for $\chi < 1$.

The rotational symmetry in the nonlinear potential $W(|\psi_1|^2, |\psi_2|^2)$ may become useful to stabilize a localized mode ϕ with $n = 1$ in the physically relevant situations, which are not modeled by the integrable coupled NLS equations. Pelinovsky & Yang [166] also considered the system of coupled saturable NLS equations,

$$\begin{cases} \mathrm{i}\partial_t\psi_1 + \partial_x^2\psi_1 + \dfrac{|\psi_1|^2 + |\psi_2|^2}{1 + s(|\psi_1|^2 + |\psi_2|^2)}\psi_1 = 0, \\[3mm] \mathrm{i}\partial_t\psi_2 + \partial_x^2\psi_2 + \dfrac{|\psi_1|^2 + |\psi_2|^2}{1 + s(|\psi_1|^2 + |\psi_2|^2)}\psi_2 = 0, \end{cases}$$

where $s > 0$ is an arbitrary parameter. The exact solution (4.3.21) also exists in the associated spectral stability problem for the pair of imaginary eigenvalues of negative energy. The embedded eigenvalue persists in the continuous spectrum of the spectral stability problem. As a result, the localized mode ϕ with $n = 1$ is spectrally stable at least near the local bifurcation point.

Exercise 4.24 Consider the Gross–Pitaevskii equation,

$$\mathrm{i}\psi_t = -\psi_{xx} + V(x)\psi - |\psi|^2\psi,$$

where $V(x)$ is bounded and decaying to zero exponentially fast as $|x| \to \infty$. Assume that the operator $L = -\partial_x^2 + V(x)$ has $(n + 1)$ negative eigenvalues $\{-\lambda_0^2, ..., -\lambda_n^2\}$ and show that a localized mode $\phi(x)$ with n nodes on $\mathbb{R}$ bifurcates near $\omega = -\lambda_n^2$. Show that if

$$|\lambda_{n-1}^2 - \lambda_n^2| > \lambda_n^2,$$

then the other n eigenvalues of L contribute generally to the count of unstable eigenvalues $N_{c+} = n$ with $N_n^- = N_p^- = N_n^+ = N_n^0 = 0$.

4.3.2 Vortices in the NLS equation

Let us consider the generalized NLS equation in polar coordinates (r, θ) on the plane $\mathbb{R}^2$,

$$i\psi_t = -\Delta_r \psi - f(|\psi|^2)\psi, \quad \Delta_r := \partial_r^2 + \frac{1}{r}\partial_r + \frac{1}{r^2}\partial_\theta^2, \qquad (4.3.23)$$

where $\psi(r, \theta, t) : \mathbb{R}_+ \times [0, 2\pi] \times \mathbb{R}_+ \to \mathbb{C}$ is the amplitude function and $f(|\psi|^2) : \mathbb{R}_+ \to \mathbb{R}$ is a C^1 nonlinear function. Assume that the NLS equation (4.3.23) admits a vortex of charge m,

$$\psi(r, \theta, t) = \phi(r)e^{im\theta - i\omega t},$$

where $\omega \in \mathbb{R}$, $m \in \mathbb{N}$, and $\phi(r) : \mathbb{R}_+ \to \mathbb{R}$ is a solution of the stationary generalized NLS equation,

$$-\Delta_m \phi - f(\phi^2)\phi = \omega\phi, \quad \Delta_m := \partial_r^2 + \frac{1}{r}\partial_r - \frac{m^2}{r^2}. \qquad (4.3.24)$$

A *fundamental* vortex of charge m is defined by the C^1 function $\phi(r)$ such that $\phi(0) = 0$, $\phi(r) > 0$ for $r \in \mathbb{R}_+$, and $\phi(r) \to 0$ as $r \to \infty$ exponentially fast. Such solutions exist only for $\omega < 0$. Existence results for vortices of charge m in the cubic–quintic NLS equation with $f(|\psi|^2) = |\psi|^2 - |\psi|^4$ were obtained by Pego & Warchall [151].

Linearization of the NLS equation (4.3.23) with the substitution

$$\psi = \left[\phi(r)e^{im\theta} + \varphi_+(r, \theta)e^{\lambda t} + \bar{\varphi}_-(r, \theta)e^{\bar{\lambda} t} \right] e^{-i\omega t}$$

results in the spectral stability problem

$$\sigma_3 H\boldsymbol{\varphi} = i\lambda\boldsymbol{\varphi}, \quad \lambda \in \mathbb{C}, \qquad (4.3.25)$$

where $\boldsymbol{\varphi} = (\varphi_+, \varphi_-)$ and

$$H = \begin{bmatrix} -\Delta_r - \omega - f(\phi^2) - \phi^2 f'(\phi^2) & -\phi^2 f'(\phi^2)e^{2im\theta} \\ -\phi^2 f'(\phi^2)e^{-2im\theta} & -\Delta_r - \omega - f(\phi^2) - \phi^2 f'(\phi^2) \end{bmatrix}.$$

The spectral stability problem (4.3.25) involves a partial differential operator Δ_r. If $\boldsymbol{\varphi}(r, \theta)$ is 2π-periodic in θ, we can separate variables r and θ and reduce the eigenvalue problem to an infinite set of uncoupled ordinary differential equations. To do this, let us expand $\boldsymbol{\varphi}(r, \theta)$ in a Fourier series,

$$\boldsymbol{\varphi}(r, \theta) = \sum_{n \in \mathbb{Z}} \boldsymbol{\varphi}^{(n)}(r)e^{in\theta},$$

and substitute it into equation (4.3.25). Equations for $\varphi_+^{(n+m)}$ and $\varphi_-^{(n-m)}$ detach from the other equations and become a closed eigenvalue problem,

$$\sigma_3 H_n \boldsymbol{\varphi}_n = i\lambda\boldsymbol{\varphi}_n, \qquad n \in \mathbb{Z}, \qquad (4.3.26)$$

where $\boldsymbol{\varphi}_n = (\varphi_+^{(n+m)}, \varphi_-^{(n-m)})$ and

$$H_n := \begin{bmatrix} -\Delta_{n+m} - \omega - f(\phi^2) - \phi^2 f'(\phi^2) & -\phi^2 f'(\phi^2) \\ -\phi^2 f'(\phi^2) & -\Delta_{n-m} - \omega - f(\phi^2) - \phi^2 f'(\phi^2) \end{bmatrix}.$$

The eigenvalue problems (4.3.26) are different between cases $n = 0$ and $n \neq 0$. Both cases can be transformed, however, to the standard form

$$L_+ \mathbf{u} = -\lambda \mathbf{w}, \quad L_- \mathbf{w} = \lambda \mathbf{u}, \qquad (4.3.27)$$

for suitably defined operators $L_\pm$ and vectors $(\mathbf{u}, \mathbf{w})$.

When $n = 0$, the eigenvalue problem (4.3.26) transforms to the form (4.3.27) for vectors

$$\mathbf{u} = \varphi_+^{(m)} + \varphi_-^{(-m)} \in \mathbb{C}, \quad \mathbf{w} = -\mathrm{i}(\varphi_+^{(m)} - \varphi_-^{(-m)}) \in \mathbb{C}$$

and operators

$$L_+ = -\Delta_m - \omega - f(\phi^2) - 2\phi^2 f'(\phi^2), \quad L_- = -\Delta_m - \omega - f(\phi^2).$$

When $n \in \mathbb{N}$, the eigenvalue problem (4.3.26) transforms to the form (4.3.27) for vectors

$$\mathbf{u} = \boldsymbol{\varphi}_n \in \mathbb{C}^2, \quad \mathbf{w} = -\mathrm{i}\sigma_3 \boldsymbol{\varphi}_n \in \mathbb{C}^2$$

and operators

$$L_+ = H_n, \quad L_- = \sigma_3 H_n \sigma_3.$$

Note that $H_{-n} = \sigma_1 H_n \sigma_1$ and $\sigma_3 \sigma_1 = -\sigma_1 \sigma_3$. As a result of this symmetry, the eigenvalue problem (4.3.26) with $-n \in \mathbb{N}$ is equivalent to that with $n \in \mathbb{N}$.

Let us consider the Hilbert space $L_r^2(\mathbb{R}_+)$ equipped with the weighted inner product,

$$\forall f, g \in L_r^2(\mathbb{R}_+) : \quad \langle f, g \rangle_{L_r^2} := \int_0^\infty f(r)g(r)r\,dr.$$

Because functions $f(\phi^2)$ and $\phi^2 f'(\phi^2)$ decay to zero as $r \to \infty$ exponentially fast, Lemma 4.1 gives $\sigma_c(H_n) = [c, \infty)$ with $c = -\omega > 0$ for any $n \in \mathbb{Z}$. The number of negative and zero isolated eigenvalues of H_n is finite.

Thanks to the gauge and translational symmetry of the NLS equation, there are at least three eigenvectors in the non-trivial kernels of H_{-1}, H_0, and H_1 given by

$$n = \pm 1 : \quad \boldsymbol{\varphi}_{\pm 1} = \phi'(r)\mathbf{1} \mp \frac{m}{r}\phi(r)\sigma_3 \mathbf{1}, \quad n = 0 : \quad \boldsymbol{\varphi}_0 = \phi(r)\sigma_3 \mathbf{1},$$

where $\mathbf{1} = (1, 1) \in \mathbb{R}^2$.

The case $n = 0$ is similar to the case of localized modes of the one-dimensional NLS equation. If $\phi(r) > 0$ for all $r > 0$, then L_- is non-negative and L_+ has $n(L_+) - 1$ negative eigenvalues in a constrained subspace of $L_r^2(\mathbb{R}_+)$ if

$$\langle L_+^{-1}\phi, \phi \rangle_{L^2} = \frac{1}{2}\frac{d}{d\omega}\|\phi\|_{L_r^2}^2 < 0. \qquad (4.3.28)$$

By Theorem 4.8, the spectral stability problem (4.3.27) for $n = 0$ has no unstable eigenvalues if $n(L_+) = 1$ and the constraint (4.3.28) is met.

We shall now consider the spectral stability problem (4.3.27) for $n \neq 0$. Combining these equations for $\pm n$, we write

$$\begin{cases} \sigma_3 H_n \boldsymbol{\varphi}_n = \mathrm{i}\lambda\boldsymbol{\varphi}_n, \\ \sigma_3 H_{-n} \boldsymbol{\varphi}_{-n} = \mathrm{i}\lambda\boldsymbol{\varphi}_{-n}, \end{cases} \quad n \in \mathbb{N}. \qquad (4.3.29)$$

As mentioned above, the second equation of system (4.3.29) is equivalent to the first one and both equations can be reduced to the standard form (4.3.27).

Exercise 4.25 Let λ be an eigenvalue of the spectral stability problem (4.3.29) with the eigenvector $(\varphi_n, \mathbf{0})$. Show that there exists another eigenvalue $-\lambda$ with the eigenvector $(\mathbf{0}, \sigma_1 \varphi_n)$. Moreover, show that if $\mathrm{Re}(\lambda) > 0$, there exist two more eigenvalues $\bar{\lambda}$ and $-\bar{\lambda}$ with the eigenvectors $(\mathbf{0}, \sigma_1 \bar{\varphi}_n)$ and $(\bar{\varphi}_n, \mathbf{0})$.

The following theorem gives the main result on the count of unstable eigenvalues in the spectral stability problem (4.3.29).

Theorem 4.10 *Assume that* $\mathrm{Ker}(H_{\pm 1}) = \mathrm{Span}\{\varphi_{\pm 1}\}$ *and* $\mathrm{Ker}(H_n) = \varnothing$ *for* $n \geq 2$. *Let* N_{real} *be the number of real positive eigenvalues in the stability problem (4.3.29),* N_{comp} *be the number of complex eigenvalues in the first quadrant, and* N_{imag}^{-} *be the number of purely imaginary eigenvalues with* $\mathrm{Im}(\lambda) > 0$ *and* $\langle H_n \varphi_n, \varphi_n \rangle_{L_r^2} \leq 0$. *Then, these numbers satisfy*

$$N_{\mathrm{real}} + 2N_{\mathrm{comp}} + 2N_{\mathrm{imag}}^{-} = 2n(H_n). \tag{4.3.30}$$

Proof For $n \geq 2$, $L_- = \sigma_3 H_n \sigma_3$ is invertible on $L_r^2(\mathbb{R}_+)$, so that we can rewrite the eigenvalue problem (4.3.29) as the generalized eigenvalue problem,

$$L_+ \mathbf{u} = \gamma L_-^{-1} \mathbf{u}, \quad \gamma = -\lambda^2. \tag{4.3.31}$$

Because the spectral stability problem (4.3.29) has a double size, let us abuse the notation and extend the size of the generalized eigenvalue problem (4.3.31) for

$$\mathbf{u} = (\varphi_n, \varphi_{-n}) \in \mathbb{C}^4, \quad \mathbf{w} = -\mathrm{i}(\sigma_3 \varphi_n, \sigma_3 \varphi_n) \in \mathbb{C}^4.$$

Therefore, from now on, L_+ is a diagonal composition of $[H_n, H_{-n}]$ and L_- is a diagonal composition of $[\sigma_3 H_n \sigma_3, \sigma_3 H_{-n} \sigma_3]$.

For each $\lambda \in \mathbb{R}$ ($\gamma < 0$), we have

$$\mathbb{R} \ni \langle H_n \varphi_n, \varphi_n \rangle_{L_r^2} = \mathrm{i}\lambda \langle \sigma_3 \varphi_n, \varphi_n \rangle_{L_r^2} \in \mathrm{i}\mathbb{R} \quad \Rightarrow \quad \langle H_n \varphi_n, \varphi_n \rangle_{L_r^2} = 0.$$

Similarly, $\langle H_n \bar{\varphi}_n, \bar{\varphi}_n \rangle_{L_r^2} = 0$. By Theorem 4.5, we have $N_n^- = N_p^- = N_{\mathrm{real}}$.

From symmetries of eigenvalues (Exercise 4.25) we conclude that

$$N_{c^+} = 2N_{\mathrm{comp}}, \quad N_n^+ = 2N_{\mathrm{imag}}^{-}.$$

Since the spectra of H_n, $H_{-n} = \sigma_1 H_n \sigma_1$, and $\sigma_3 H_n \sigma_3$ coincide, we have

$$n(L_-) = n(L_+) = 2n(H_n).$$

Since $N_n^0 = 0$ because $\mathrm{Ker}(H_n) = \varnothing$, equality (4.3.30) follows from either equality (4.2.26) or equality (4.2.27) in Theorem 4.5.

For $n = 1$, we need a constrained L_r^2 space because $\mathrm{Ker}(\sigma_3 H_{\pm 1} \sigma_3) = \mathrm{span}\{\sigma_3 \varphi_{\pm 1}\}$. Therefore, we define

$$L_{\pm}^2 = \left\{ \varphi \in L_r^2(\mathbb{R}_+) : \quad \langle \sigma_3 \varphi_{\pm 1}, \varphi \rangle_{L_r^2} = 0 \right\}.$$

The generalized eigenvalue problem (4.3.31) is now defined in the constrained space $\mathbf{u} \in L_+^2 \times L_-^2$.

Since $\langle \sigma_3 \varphi_{\pm 1}, \varphi_{\pm 1} \rangle_{L_r^2} = 0$, we know that $\varphi_{\pm 1} \perp \mathrm{Ker}(\sigma_3 H_{\pm 1} \sigma_3)$. By direct computation, we obtain $(\sigma_3 H_{\pm 1} \sigma_3)^{-1} \varphi_{\pm 1} = -\frac{1}{2} r \phi(r) \mathbf{1}$, which gives

$$\langle (\sigma_3 H_{\pm 1} \sigma_3)^{-1} \varphi_{\pm 1}, \varphi_{\pm 1} \rangle_{L_r^2} = \int_0^\infty \phi^2(r) r\, dr > 0.$$

By Theorem 4.1, we still have $n(L_+) = 2n(H_1)$, $N_n^0 = 0$, and $N_p^0 = 1$. By Lemma 4.7, we also have $\dim(\mathcal{H}_{L+\delta M^{-1}}^-) = \dim(\mathcal{H}_L^-) = n(L_+)$. Again, for $n = 1$, equality (4.3.30) follows from either equality (4.2.26) or equality (4.2.27) in Theorem 4.5. $\quad\square$

In the case of the defocusing NLS equation (such as the cubic–quintic NLS equation [151] or the cubic NLS equation with a harmonic potential [118]), it is possible to show that $n(H_n) = 0$ for $|n| \geq 2m$. By Theorem 4.10, this useful result shows that the unstable and potentially unstable eigenvalues only appear for $|n| < 2m$, which limits the numerical analysis of the spectral stability problem (4.3.29) by a finite number of differential equations. Numerical approximations of unstable eigenvalues with the use of Evans functions are reported in [118, 151].

Unfortunately, no bounds on $n(H_n)$ exist in the case of a general nonlinear function $f(|\psi|^2)$, unless the localized mode ϕ is small. Yang & Pelinovsky [215] used arguments similar to the ones in Section 4.3.1 and showed for the coupled NLS equations with the saturable nonlinear function $f(|\psi|^2)$ that the vortex of charge $m = 1$ may become spectrally unstable for perturbations with $|n| = 2$ because $n(H_2) = 1$. From the numerical point of view, Theorem 4.10 remains a useful tool to trace unstable and potentially unstable eigenvalues of the spectral stability problem (4.3.29) after the index $n(H_n)$ is computed numerically or analytically.

4.3.3 Soliton configurations in the DNLS equation

Let us consider the one-dimensional cubic DNLS equation,

$$i\dot{\psi}_n + \epsilon\,(\psi_{n+1} - 2\psi_n + \psi_{n-1}) + |\psi_n|^2 \psi_n = 0, \tag{4.3.32}$$

where $\psi_n(t) : \mathbb{R}_+ \to \mathbb{C}$, $n \in \mathbb{Z}$, and $\epsilon > 0$ is a coupling parameter. Because of the scaling transformation, we can normalize the frequency of the time-periodic localized mode by $\omega = -1$.

All localized modes are real-valued on the lattice of one dimension, up to a multiplication by $e^{i\theta}$ for $\theta \in \mathbb{R}$ (Section 3.2.5). Therefore, these localized modes are the time-periodic solutions $\psi_n(t) = \phi_n e^{it}$, where the real-valued sequence $\{\phi_n\}_{n \in \mathbb{Z}}$ is found from the second-order difference equation

$$(1 - \phi_n^2)\phi_n = \epsilon\,(\phi_{n+1} - 2\phi_n + \phi_{n-1}), \quad n \in \mathbb{Z}. \tag{4.3.33}$$

By Theorem 3.8, localized modes for small $\epsilon \in \mathbb{R}$ can be characterized by the limiting configuration

$$\phi_n^{(0)} = \begin{cases} \pm 1, & n \in U_\pm, \\ 0, & n \in U_0, \end{cases} \tag{4.3.34}$$

where $\mathbb{Z}$ is decomposed into subsets $U_+ \cup U_- \cup U_0$ and $N := \dim(U_+ \cup U_-) < \infty$. To be more precise, for each limiting solution (4.3.34) there is a unique solution $\{\phi_n\}_{n\in\mathbb{Z}}$ in $l^2(\mathbb{Z})$ of the difference equation (4.3.33) for a small $\epsilon \in \mathbb{R}$ such that $\{\phi_n\}_{n\in\mathbb{Z}}$ depends analytically on ϵ near $\epsilon = 0$ and

$$\lim_{\epsilon\to 0} \phi_n = \phi_n^{(0)}, \quad n \in \mathbb{Z}.$$

The linearization of the DNLS equation (4.3.32) at the localized mode $\{\phi_n\}_{n\in\mathbb{Z}}$ results in the spectral stability problem

$$(L_+u)_n = -\lambda w_n, \quad (L_-w)_n = \lambda u_n, \tag{4.3.35}$$

where $L_\pm$ are the discrete Schrödinger operators,

$$(L_+u)_n = \left(1 - 3\phi_n^2\right) u_n - \epsilon \left(u_{n+1} - 2u_n + u_{n-1}\right),$$

$$(L_-w)_n = \left(1 - \phi_n^2\right) w_n - \epsilon \left(w_{n+1} - 2w_n + w_{n-1}\right).$$

Operators $L_\pm$ are bounded in $l^2(\mathbb{Z})$. We are looking for eigenvectors $(\mathbf{u}, \mathbf{w}) \in l^2(\mathbb{Z}) \times l^2(\mathbb{Z})$ and eigenvalues $\lambda \in \mathbb{C}$ of the non-self-adjoint spectral problem

$$\mathcal{L} \begin{bmatrix} \mathbf{u} \\ \mathbf{w} \end{bmatrix} = \lambda \begin{bmatrix} \mathbf{u} \\ \mathbf{w} \end{bmatrix}, \quad \mathcal{L} = \begin{bmatrix} 0 & L_- \\ -L_+ & 0 \end{bmatrix}. \tag{4.3.36}$$

Orbital stability of the fundamental localized mode with $U_+ = \{0\}$ and $U_- = \varnothing$ was proved with the variational technique by Weinstein [212]. Stability of this mode close to the continuum limit $\epsilon \to \infty$ was considered by Kapitula & Kevrekidis [102]. Spectral instabilities of localized modes with $\dim(U_+ \cup U_-) > 1$ were considered numerically in [142] and analytically in [155].

For $\epsilon = 0$ and $\phi = \phi^{(0)}$, operators $L_\pm$ are diagonal and the spectra of $L_\pm$ and $\mathcal{L}$ are known explicitly.

- The spectrum of L_+ consists of the semi-simple eigenvalue -2 of multiplicity N and the semi-simple eigenvalue 1 of an infinite multiplicity.
- The spectrum of L_- consists of the semi-simple eigenvalue 0 of multiplicity N and the semi-simple eigenvalue 1 of an infinite multiplicity.
- The spectrum of $\mathcal{L}$ consists of the eigenvalue 0 of geometric multiplicity N and algebraic multiplicity $2N$ and the semi-simple eigenvalues $\pm i$ of an infinite multiplicity. The subspace of $l^2(\mathbb{Z}) \times l^2(\mathbb{Z})$ corresponding to the zero eigenvalue is associated with the set $U_+ \cup U_-$ and that for the pair of eigenvalues $\pm i$ is associated with the set U_0.

For small $\epsilon > 0$, small nonzero eigenvalues λ in the spectral stability problem (4.3.35) bifurcate from the zero eigenvalue. Spectral instabilities may arise from the zero eigenvalue and the following theorem controls the number of unstable eigenvalues near 0. Bifurcations of the continuous spectrum from points $\pm i$ were considered recently by Pelinovsky & Sakovich [159]. No instabilities arise from points $\pm i$.

Theorem 4.11 *Let n_0 be the number of sign changes in $\{\phi_n^{(0)}\}_{n\in U_+\cup U_-}$ and $N := \dim(U_+ \cup U_-)$. For sufficiently small $\epsilon > 0$, the localized mode ϕ is spectrally stable if $n_0 = N - 1$ and spectrally unstable if $n_0 < N - 1$ with exactly $N - 1 - n_0$ real positive eigenvalues λ in the spectral stability problem (4.3.36).*

Before discussing the proof of Theorem 4.11, let us consider the immediate consequence of the theorem.

Corollary 4.6 *Given the set of nodes $U_+ \cup U_-$, there exist only two configurations of the limiting solution (4.3.34) which are spectrally stable for small $\epsilon > 0$.*

Proof The spectrally stable configurations must have $n_0 = N - 1$, which means that the distributions of $\{+1, -1\}$ in the set of nodes $U_+ \cup U_-$ must be sign-alternating. One configuration has $+1$ on the left-most node $n \in U_+ \cup U_-$ and the other one has -1 on that node. $\square$

Remark 4.7 The fundamental localized mode with $U_+ = \{0\}$ and $U_- = \varnothing$ has no nonzero eigenvalues of the spectral stability problem (4.3.36) near 0 for small $\epsilon > 0$. The fundamental localized mode is stable not only spectrally but also orbitally (Section 4.4.2).

We shall break the proof of Theorem 4.11 into a number of elementary results. In particular, we shall characterize for small $\epsilon > 0$ the continuous spectrum of $\mathcal{L}$ (Lemma 4.17), the negative and zero eigenvalues of $L_\pm$ (Lemma 4.18), and the small nonzero eigenvalues of $\mathcal{L}$ in a constrained l^2 space (Lemma 4.19).

Lemma 4.17 *For any $\epsilon > 0$, for which ϕ exists,*

$$\sigma_c(\mathcal{L}) = [-(1 + 4\epsilon)i, i] \cup [i, (1 + 4\epsilon)i].$$

Proof Since $I - \epsilon\Delta : l^2(\mathbb{Z}) \to l^2(\mathbb{Z})$ is a strictly positive operator, the localized mode ϕ decays to zero as $|n| \to \infty$ exponentially fast for any $\epsilon > 0$, for which it exists. By the discrete version of the Weyl Theorem (Appendix B.15), $\sigma_c(\mathcal{L}) = \sigma(\mathcal{L}_0)$, where

$$\mathcal{L}_0 = \begin{bmatrix} 0 & I - \epsilon\Delta \\ -(I - \epsilon\Delta) & 0 \end{bmatrix}.$$

Diagonalizing the eigenvalue problem for $\mathcal{L}_0$ in variables $\mathbf{v}_\pm = \mathbf{u} \pm i\mathbf{w}$, we obtain

$$(I - \epsilon\Delta)\mathbf{v}_\pm = \pm i\lambda\mathbf{v}_\pm,$$

from which it follows that if $\lambda \in \sigma(\mathcal{L}_0)$, then $\pm i\lambda \in [1, 1 + 4\epsilon]$. $\square$

Lemma 4.18 *Let n_0 be the number of sign changes in $\{\phi_n^{(0)}\}_{n \in U_+ \cup U_-}$. For sufficiently small $\epsilon > 0$, we have*

$$n(L_-) = n_0, \quad z(L_-) = 1, \quad n(L_+) = N, \quad z(L_+) = 0,$$

where $n(L)$ and $z(L)$ are the number of negative and zero eigenvalues of operator L with the account of their multiplicities.

Proof By the discrete analogue of Corollary 4.1 for the second-order discrete Schrödinger operators $L_\pm$, if $\lambda \in \sigma_p(L_\pm)$ for $\epsilon \neq 0$, then λ is a simple eigenvalue.

For $\epsilon = 0$, the diagonal operator L_+ has the semi-simple negative eigenvalue -2 of multiplicity N and no zero eigenvalues. For $\epsilon > 0$, $I - \epsilon\Delta$ is strictly positive and ϕ is analytic in ϵ. Analytic perturbation theory for isolated eigenvalues of the self-adjoint operators implies that there are exactly N negative eigenvalues of L_+

with the account of their multiplicities and no zero eigenvalues for sufficiently small $\epsilon > 0$.

For any $\epsilon > 0$, for which ϕ exists, equation $L_-\phi = \mathbf{0}$ is equivalent to the stationary DNLS equation (4.3.33). Therefore, $z(L_-) = 1$ for $\epsilon > 0$. By the discrete analogue of Lemma 4.2, there exist n_0 negative eigenvalues of L_-, where n_0 is the number of sign changes in $\{\phi_n\}_{n\in\mathbb{Z}}$. These negative eigenvalues and the remaining $N - 1 - n_0$ positive eigenvalues of L_- for $\epsilon > 0$ are small as they bifurcate from the zero eigenvalue of L_- of algebraic multiplicity N that exists for $\epsilon = 0$. It remains to prove that the number of sign changes in $\{\phi_n\}_{n\in\mathbb{Z}}$ is semi-continuous in $\epsilon \geq 0$. Hence it equals the number of sign changes in $\{\phi_n^{(0)}\}_{n\in U_+\cup U_-}$.

Consider two adjacent nodes $n_1, n_2 \in U_+ \cup U_-$ separated by M nodes in U_0 on $\mathbb{Z}$, so that $n_2 - n_1 = 1 + M$. The difference equation (4.3.33) on $[n_1 + 1, n_2 - 1]$ can be rewritten as the linear system

$$\mathcal{A}_M \boldsymbol{\phi}_M = \epsilon \mathbf{b}_M, \qquad (4.3.37)$$

where $\boldsymbol{\phi}_M = (\phi_{n_1+1}, ..., \phi_{n_2-1}) \in \mathbb{R}^{n_2-n_1-1}$, $\mathbf{b} = (\phi_{n_1}, 0, ..., 0, \phi_{n_2}) \in \mathbb{R}^{n_2-n_1-1}$, and

$$\mathcal{A}_M = \begin{bmatrix} 1+2\epsilon - \phi_{n_1+1}^2 & -\epsilon & 0 & ... & 0 \\ -\epsilon & 1+2\epsilon - \phi_{n_1+2}^2 & -\epsilon & ... & 0 \\ \vdots & \vdots & \vdots & ... & \vdots \\ 0 & 0 & 0 & ... & 1+2\epsilon - \phi_{n_2-1}^2 \end{bmatrix}.$$

Let $D_{I,J}$, $1 \leq I \leq J \leq N$ be the determinant of the block of the matrix $\mathcal{A}_M$ between the Ith and Jth rows and columns. By Cramer's rule, we have

$$\phi_{n_1+j} = \frac{\epsilon^j \phi_{n_1} D_{j+1,N} + \epsilon^{N-j+1}\phi_{n_2} D_{1,N-j}}{D_{1,N}}. \qquad (4.3.38)$$

Since $\lim_{\epsilon\to 0} D_{I,J} = 1$ for all $1 \leq I \leq J \leq N$, we have

$$\lim_{\epsilon\to 0} \epsilon^{-j}\phi_{n_1+j} = \phi_{n_1}, \qquad 1 \leq j < \frac{N+1}{2},$$

$$\lim_{\epsilon\to 0} \epsilon^{-j}\phi_{n_1+j} = \phi_{n_1} + \phi_{n_2}, \qquad j = \frac{N+1}{2},$$

$$\lim_{\epsilon\to 0} \epsilon^{j-1-N}\phi_{n_1+j} = \phi_{n_2}, \qquad \frac{N+1}{2} < j \leq N.$$

Therefore, the number of sign changes in $\{\phi_n\}_{n\in[n_1,n_2]}$ for small $\epsilon > 0$ is exactly one if $\text{sign}(\phi_{n_2}) = -\text{sign}(\phi_{n_1})$ and zero if $\text{sign}(\phi_{n_2}) = \text{sign}(\phi_{n_1})$. Combining with the sign changes on other segments between two adjacent nodes in $U_+ \cup U_-$, we obtain the semi-continuity of the sign changes in the sequence $\{\phi_n\}_{n\in\mathbb{Z}}$ in $\epsilon \geq 0$. □

Remark 4.8 The semi-continuity of the sign changes in the sequence $\{\phi_n\}_{n\in\mathbb{Z}}$ can be viewed as the discrete version of elliptic smoothing of the operator $I - \epsilon\Delta$ for $\epsilon > 0$. This result is not valid for $\epsilon < 0$ and the number of sign changes in $\{\phi_n\}_{n\in\mathbb{Z}}$ for $\epsilon < 0$ exceeds generally the number of sign changes in $\{\phi_n^{(0)}\}_{n\in U_+\cup U_-}$.

Let us extend the family of localized modes to $\phi(\omega)$ for a fixed $\epsilon > 0$, where $\{\phi_n(\omega)\}_{n\in\mathbb{Z}}$ satisfies

$$(|\omega| - \phi_n^2(\omega))\phi_n(\omega) = \epsilon(\phi_{n+1}(\omega) - 2\phi_n(\omega) + \phi_{n-1}(\omega)), \quad n \in \mathbb{Z},$$

and ω is defined near $\omega = -1$. The limiting configuration (4.3.34) becomes now

$$\phi_n^{(0)}(\omega) = \begin{cases} \pm\sqrt{|\omega|}, & n \in U_\pm, \\ 0, & n \in U_0. \end{cases}$$

For small $\epsilon > 0$, we obtain

$$\|\phi(\omega)\|_{l^2}^2 = N|\omega| + \mathcal{O}(\epsilon),$$

which hence gives

$$\frac{d}{d\omega}\|\phi(\omega)\|_{l^2}^2 = -N + \mathcal{O}(\epsilon) < 0.$$

Under this condition, 0 is a double eigenvalue of $\mathcal{L}$ for any $\epsilon \neq 0$ and

$$\text{Ker}(\mathcal{L}) = \text{span}\left\{ \begin{bmatrix} \mathbf{0} \\ \phi \end{bmatrix} \right\}, \quad \text{Ker}(\mathcal{L}^2) = \text{span}\left\{ \begin{bmatrix} \mathbf{0} \\ \phi \end{bmatrix}, \begin{bmatrix} \partial_\omega \phi \\ \mathbf{0} \end{bmatrix} \right\}, \quad (4.3.39)$$

where $\partial_\omega \phi \equiv \partial_\omega \phi(\omega)|_{\omega=-1}$. Since the zero eigenvalue of $\mathcal{L}$ for $\epsilon = 0$ has algebraic multiplicity $2N$, this computation shows for $N \geq 2$ that $2N-2$ eigenvalues bifurcate from 0 for $\epsilon > 0$. If $\lambda \in \sigma_p(\mathcal{L})$ and $\lambda \neq 0$, then $(\mathbf{u}, \mathbf{w}) \in l^2(\mathbb{Z}) \times l^2(\mathbb{Z})$ satisfy the constraints

$$\langle \mathbf{u}, \phi \rangle_{l^2} = 0, \quad \langle \mathbf{w}, \partial_\omega \phi \rangle_{l^2} = 0. \quad (4.3.40)$$

If we let

$$l_c^2 = \{\mathbf{u} \in l^2(\mathbb{Z}) : \quad \langle \mathbf{u}, \phi \rangle_{l^2} = 0\}, \quad (4.3.41)$$

and introduce the orthogonal projection operator $P_c : l^2 \to l_c^2 \subset l^2$, then the non-self-adjoint spectral problem (4.3.36) can be closed as the generalized eigenvalue problem

$$(P_c L_+ P_c)\mathbf{u} = \gamma (P_c L_-^{-1} P_c)\mathbf{u}, \quad \gamma = -\lambda^2, \quad (4.3.42)$$

similarly to the continuous case.

We recall that L_+ is invertible and $n(P_c L_+ P_c) = n(L_+) - 1$ since for small $\epsilon > 0$, we have

$$\langle L_+^{-1}\phi, \phi \rangle_{l^2} = \langle \partial_\omega \phi, \phi \rangle_{l^2} = \frac{1}{2}\frac{d}{d\omega}\|\phi(\omega)\|_{l^2}^2 \bigg|_{\omega=-1} < 0.$$

By equalities (4.2.26) and (4.2.27) in Theorem 4.5, we thus obtain

$$\begin{cases} N_p^- + N_n^+ + N_{c^+} = n(L_+) - 1 = N - 1, \\ N_n^- + N_n^+ + N_{c^+} = n(L_-) = n_0. \end{cases} \quad (4.3.43)$$

If $n_0 = 0$, the count of eigenvalues in (4.3.43) gives the result of Theorem 4.11 as

$$N_p^- = N - 1, \quad N_n^- = N_n^+ = N_{c^+} = 0.$$

In other words, if $n_0 = 0$ and $N = 1$, no small nonzero eigenvalues γ exist, whereas if $n_0 = 0$ and $N \geq 2$, exactly $N-1$ small negative eigenvalues γ exist in the generalized eigenvalue problem (4.3.42), which give $N - 1$ small positive eigenvalues λ in the spectral stability problem (4.3.36).

If $n_0 \geq 1$, we have to admit that the count of eigenvalues in (4.3.43) does not provide a proof of Theorem 4.11. In addition, the operator norm of $P_c L_-^{-1} P_c$ diverges as $\epsilon \to 0$ because $N - 1$ small nonzero eigenvalues of L_- converge to 0 as $\epsilon \to 0$. On the other hand, L_+ is invertible for any $\epsilon \geq 0$. Therefore, we may think about the other constrained l^2 space for the vector $\mathbf{w}$ of the spectral stability problem (4.3.36). Instead of the constrained space (4.3.41), let us define

$$l_c^2 = \{\mathbf{w} \in l^2(\mathbb{Z}) : \quad \langle \mathbf{w}, \partial_\omega \phi \rangle_{l^2} = 0\}. \tag{4.3.44}$$

In all previous examples, the component $\mathbf{w}$ was eliminated from system (4.3.36) and the constraint on $\mathbf{w}$ was neglected. Let us now eliminate $\mathbf{u}$ from the system and neglect the constraint on $\mathbf{u}$. Since L_+ is invertible, we do not need any projection operator to find that

$$\mathbf{u} = -\lambda L_+^{-1} \mathbf{w}.$$

The non-self-adjoint spectral problem (4.3.36) is now replaced by the generalized eigenvalue problem,

$$L_- \mathbf{w} = \gamma L_+^{-1} \mathbf{w}, \quad \gamma = -\lambda^2. \tag{4.3.45}$$

Because

$$0 = \langle L_- \mathbf{w}, \phi \rangle_{l^2} = \gamma \langle L_+^{-1} \mathbf{w}, \phi \rangle_{l^2} = \gamma \langle \mathbf{w}, \partial_\omega \phi \rangle_{l^2},$$

we realize that if $\mathbf{w} \in l^2(\mathbb{Z})$ is an eigenvector of the generalized eigenvalue problem (4.3.45) for $\gamma \neq 0$, then $\mathbf{w} \in l_c^2$, where l_c^2 is defined by (4.3.44). Therefore, we do not need to recompute the numbers of negative eigenvalues of operators L_- and L_+ in the new constrained l^2 space.

Lemma 4.19 *For sufficiently small $\epsilon > 0$, there exist n_0 small positive eigenvalues γ and $(N-1-n_0)$ small negative eigenvalues γ in the generalized eigenvalue problem (4.3.45).*

Proof Let $\mathbf{w}(\epsilon) \in l^2(\mathbb{Z})$ be an eigenvector of the generalized eigenvalue problem (4.3.45) for a small eigenvalue $\gamma(\epsilon)$ such that $\gamma(\epsilon) \to 0$ as $\epsilon \to 0$. Thanks to analyticity of the coefficients of $L_\pm$ in ϵ, $\mathbf{w}(\epsilon)$ is continuous as $\epsilon \to 0$. If we normalize $\|\mathbf{w}(\epsilon)\|_{l^2} = 1$ for all $\epsilon \geq 0$, there exists a nonzero vector $\mathbf{w}(0) = \lim_{\epsilon \to 0} \mathbf{w}(\epsilon) \in l^2(\mathbb{Z})$. Because $\lim_{\epsilon \to 0} \gamma(\epsilon) = 0$, the vector $\mathbf{w}(0)$ is compactly supported on nodes of $U_+ \cup U_-$, hence

$$L_+^{-1} \mathbf{w}(0) = -\frac{1}{2} \mathbf{w}(0).$$

By continuity in ϵ, $\langle L_+^{-1} \mathbf{w}(\epsilon), \mathbf{w}(\epsilon) \rangle_{l^2} < 0$ for small $\epsilon \geq 0$. Therefore, we can think about application of Sylvester's Law of Inertia (Theorem 4.2).

Let U_0 be the subspace of $l^2(\mathbb{Z})$ associated with eigenvectors for N small or zero eigenvalues of L_- for a fixed $\epsilon \geq 0$. Let $P_0 : l^2(\mathbb{Z}) \to U_0$ and $P_0^\perp : l^2(\mathbb{Z}) \to U_0^\perp$ be the orthogonal projection operators to the invariant subspaces with respect to self-adjoint operator L_- so that $P_0 + P_0^\perp = I$. The generalized eigenvalue problem

(4.3.45) can be written in the form

$$\begin{cases} P_0 L_- P_0 w = \gamma \left(P_0 L_+^{-1} P_0 w + P_0 L_+^{-1} P_0^\perp w \right), \\ P_0^\perp L_- P_0^\perp w = \gamma \left(P_0^\perp L_+^{-1} P_0 w + P_0^\perp L_+^{-1} P_0^\perp w \right). \end{cases} \tag{4.3.46}$$

The spectrum of $P_0^\perp L_- P_0^\perp$ is bounded away from zero and is strictly positive. If γ is small, there exists a unique solution of the second equation of system (4.3.46),

$$P_0^\perp w = \gamma P_0^\perp \left(P_0^\perp L_- P_0^\perp - \gamma P_0^\perp L_+^{-1} P_0^\perp \right)^{-1} P_0^\perp L_+^{-1} P_0 w.$$

Substituting this expression into the first equation of system (4.3.46), we obtain the finite-dimensional eigenvalue problem for small γ,

$$P_0 L_- P_0 w = \gamma \left(P_0 L_+^{-1} P_0 + \gamma P_0 L_+^{-1} P_0^\perp \left(P_0^\perp L_- P_0^\perp - \gamma P_0^\perp L_+^{-1} P_0^\perp \right)^{-1} P_0^\perp L_+^{-1} P_0 \right) P_0 w,$$

which can be rewritten as the nonlinear eigenvalue equation

$$L \mathbf{w}_0 = \gamma M(\gamma) \mathbf{w}_0, \tag{4.3.47}$$

for $\mathbf{w}_0 \in \mathbb{R}^N$ and $N \times N$ matrices L and $M(\gamma)$.

The matrix $L = P_0 L_- P_0$ is diagonal with n_0 negative, one zero, and $N - 1 - n_0$ positive diagonal elements for small $\epsilon > 0$. On the other hand, matrix $M(0) = P_0 L_+^{-1} P_0$ is a strictly negative matrix for small $\epsilon \geq 0$. Consider the truncated problem $L \mathbf{w}_0 = \gamma M(0) \mathbf{w}_0$. By Sylvester's Law of Inertia (Theorem 4.2), there exist n_0 negative eigenvalues $-\gamma$ and $N - 1 - p_0$ positive eigenvalues $-\gamma$ for small $\epsilon > 0$. Moreover, for each eigenvalue, $\gamma \to 0$ as $\epsilon \to 0$. Because $M(\gamma)$ is a self-adjoint operator for real γ and γ is small, negative and positive eigenvalues persist in the nonlinear eigenvalue problem (4.3.47) for sufficiently small $\epsilon > 0$. $\qquad \square$

Corollary 4.7 *Let $\mu(\epsilon)$ be a small eigenvalue of L_- such that $\mu(\epsilon) \to 0$ as $\epsilon \to 0$. The spectral stability problem (4.3.36) has a pair of small eigenvalues $\lambda(\epsilon)$ and $-\lambda(\epsilon)$ such that $\lambda(\epsilon) \to 0$ as $\epsilon \to 0$ and*

$$\lim_{\epsilon \to 0} \frac{\lambda^2(\epsilon)}{\mu(\epsilon)} = 2. \tag{4.3.48}$$

Moreover, if $\mu(\epsilon) < 0$ and $(\mathbf{u}, \mathbf{w})$ is the eigenvector of the spectral stability problem (4.3.36), then

$$\langle L_+ \mathbf{u}, \mathbf{u} \rangle_{l^2} < 0 \quad \text{and} \quad \langle L_- \mathbf{w}, \mathbf{w} \rangle_{l^2} < 0. \tag{4.3.49}$$

Proof The first statement (4.3.48) follows from the Rayleigh quotient for the generalized eigenvalue problem (4.3.45),

$$\lambda^2 = -\frac{\langle L_- \mathbf{w}, \mathbf{w} \rangle_{l^2}}{\langle L_+^{-1} \mathbf{w}, \mathbf{w} \rangle_{l^2}}$$

and the limit

$$\lim_{\epsilon \to 0} \langle L_+^{-1} \mathbf{w}, \mathbf{w} \rangle_{l^2} = -\frac{1}{2},$$

which is obtained in the proof of Lemma 4.19.

The second statement (4.3.49) follows from the relationship

$$\langle L_+ \mathbf{u}, \mathbf{u} \rangle_{l^2} = -\lambda \langle \mathbf{w}, \mathbf{u} \rangle_{l^2} = \langle \mathbf{w}, \lambda \mathbf{u} \rangle_{l^2} = \langle L_- \mathbf{w}, \mathbf{w} \rangle_{l^2},$$

which is obtained from the spectral stability problem (4.3.36) for $\lambda \in i\mathbb{R}$. $\qquad \square$

The proof of Theorem 4.11 relies on Lemma 4.19. It follows from equalities (4.3.43) that

$$N_{\text{real}} + 2N_{\text{imag}}^- + 2N_{\text{comp}} = n(L_+) + n(L_-) - 1, \qquad (4.3.50)$$

where $N_{\text{real}} = N_p^- + N_n^-$ is the number of positive real eigenvalues λ, $N_{\text{comp}} = 2N_{c+}$ is the number of complex eigenvalues λ in the first quadrant, and $N_{\text{imag}}^- = 2N_n^+$ is the number of purely imaginary eigenvalues λ with $\text{Im}(\lambda) > 0$ and the eigenvector $(\mathbf{u}, \mathbf{w})$ such that $\langle L_+ \mathbf{u}, \mathbf{u} \rangle_{l^2} < 0$.

By Lemma 4.19 and Corollary 4.7 for a small $\epsilon > 0$, we have obtained a more precise count of eigenvalues,

$$N_{\text{real}} = N_p^- = N - 1 - n_0, \quad N_{\text{imag}}^- = N_n^+ = n_0, \quad N_{\text{comp}} = N_{c+} = N_n^- = N_p^+ = 0.$$

Even if eigenvalues $\lambda \in i\mathbb{R}$ in N_{imag}^- and $\lambda \in \mathbb{R}$ in N_{real} are semi-simple, they are structurally stable on $i\mathbb{R}$ and $\mathbb{R}$ with respect to parameter continuation in ϵ.

Exercise 4.26   Consider the set $U_+ \cup U_- = \{1, 2, ..., N\} \subset \mathbb{Z}$ and prove that the set of N small eigenvalues $\{\mu_j(\epsilon)\}_{1 \le j \le N}$ of operator L_- is given asymptotically by

$$\mu_j(\epsilon) = \nu_j \epsilon + \mathcal{O}(\epsilon^2),$$

where $\{\nu_j\}_{j=1}^N$ is a set of eigenvalues of the $N \times N$ matrix with the elements

$$M_{i,j} = \begin{cases} s_{i+1} + s_i, & j = i, \\ -s_{i+1}, & j = i+1, \\ -s_i, & j = i-1, \\ 0, & \text{otherwise,} \end{cases} \quad 1 \le i, j \le N,$$

where

$$s_i = \begin{cases} +1 \ \text{if} \ \phi_j^{(0)} \phi_{j-1}^{(0)} > 0, \\ -1 \ \text{if} \ \phi_j^{(0)} \phi_{j-1}^{(0)} < 0, \end{cases} \quad i \in \{2, ..., N\},$$

and $s_1 = s_{N+1} = 0$.

As a numerical example, we consider two localized modes for $N = 2$ with $U_+ = \{1, 2\}$, $U_- = \varnothing$ and $U_+ = \{1\}$, $U_- = \{2\}$. The first configuration corresponds to the *bond-symmetric* soliton ϕ and it is spectrally unstable with exactly one real positive eigenvalue since $n_0 = 0 < N - 1 = 1$. The second configuration corresponds to a *twisted* soliton ϕ and it is spectrally stable with a pair of purely imaginary eigenvalues of negative energy because $n_0 = 1 = N - 1$.

From Corollary 4.7 and Exercise 4.26, the small positive eigenvalue of the spectral stability problem (4.3.35) for the bond-symmetric localized mode is expanded by

$$\lambda = 2\sqrt{\epsilon}(1 + \mathcal{O}(\epsilon)) \quad \text{as} \quad \epsilon \to 0$$

whereas the small imaginary eigenvalues for the twisted localized mode are expanded by

$$\lambda = \pm 2i\sqrt{\epsilon}(1 + \mathcal{O}(\epsilon)) \quad \text{as} \quad \epsilon \to 0.$$

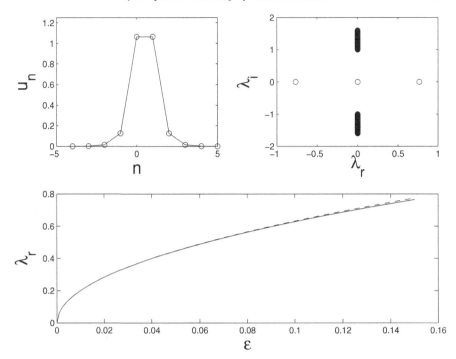

Figure 4.3 Top: the bond-symmetric localized mode (left) and the corresponding spectral plane λ (right) for $\epsilon = 0.15$. Bottom: the real positive eigenvalue λ versus ϵ from asymptotic (dashed line) and numerical (solid line) approximations. Reprinted from [155].

These results are illustrated in Figures 4.3 and 4.4. The top panels of each figure show the mode profiles (left) and the spectral plane of λ (right) for $\epsilon = 0.15$. The bottom panels indicate the corresponding asymptotic approximations of the small eigenvalues (dashed line) versus the numerical approximations (solid line). We observe for the twisted localized mode that the purely imaginary eigenvalues coalesce at $\epsilon = \epsilon_c \approx 0.146$ with the continuous spectrum at $\pm i$. As a result, the complex unstable eigenvalues bifurcate for $\epsilon > \epsilon_c$. Therefore, $N_{\text{comp}} = 1$ and $N_{\text{imag}}^- = 0$ for $\epsilon > \epsilon_c$, in agreement with equality (4.3.50).

4.3.4 Vortex configurations in the DNLS equation

Let us consider the two-dimensional cubic DNLS equation,

$$i\dot{\psi}_n + \epsilon(\Delta\psi)_n + |\psi_n|^2\psi_n = 0, \qquad (4.3.51)$$

where $\psi_n(t) : \mathbb{R}_+ \to \mathbb{C}$, $n \in \mathbb{Z}^2$, $\epsilon > 0$ is a small parameter, and Δ is the two-dimensional discrete Laplacian,

$$(\Delta\psi)_n := \psi_{n+e_1} + \psi_{n-e_1} + \psi_{n+e_2} + \psi_{n-e_2} - 4\psi_n, \qquad n \in \mathbb{Z}^2,$$

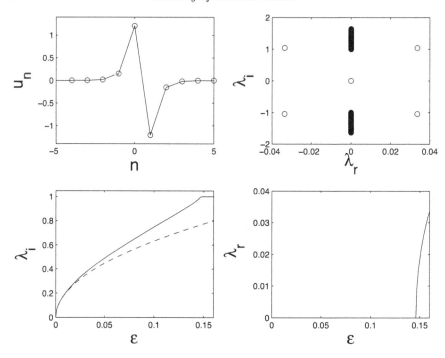

Figure 4.4 Top: the twisted localized mode (left) and the corresponding spectral plane λ (right) for $\epsilon = 0.15$. Bottom: the imaginary (left) and real (right) parts of the small eigenvalue λ versus ϵ from asymptotic (dashed line) and numerical (solid line) approximations. Reprinted from [155].

with $\{e_1, e_2\}$ being unit vectors on $\mathbb{Z}^2$. For small $\epsilon > 0$, we consider vortex solutions of the DNLS equation (4.3.51),

$$\psi_n(t) = \phi_n e^{it}, \quad n \in \mathbb{Z}^2, \tag{4.3.52}$$

where amplitudes $\{\phi_n\}_{n \in \mathbb{Z}^2}$ satisfy the stationary DNLS equation

$$(1 - |\phi_n|^2)\phi_n = \epsilon(\Delta\phi)_n, \quad n \in \mathbb{Z}^2. \tag{4.3.53}$$

For $\epsilon = 0$, the limiting vortex configuration is defined by

$$\phi_n^{(0)} = \begin{cases} e^{i\theta_n}, & n \in S, \\ 0, & n \in S^\perp, \end{cases} \tag{4.3.54}$$

where $S \subset \mathbb{Z}^2$ with $N = \dim(S) < \infty$, $S^\perp = \mathbb{Z}^2 \backslash S$, and $\boldsymbol{\theta} \in \mathbb{T}^N$. Recall from the Lyapunov–Schmidt reduction method (Section 3.2.5) that the vortex solution persists for $\epsilon \neq 0$ only if $\{\theta_n\}_{n \in S}$ is a root of the function

$$\mathbf{g}(\boldsymbol{\theta}, \epsilon) = \sum_{k=1}^{\infty} \epsilon^k \mathbf{g}^{(k)}(\boldsymbol{\theta}) = 0. \tag{4.3.55}$$

For the vortex configurations, we require that if $\theta_{n_0} = 0$ for one $n_0 \in S$, then $\{\theta_n\}_{n \in S} \notin \{0, \pi\}$. Under this constraint, the localized mode $\{\phi_n\}_{n \in \mathbb{Z}^2}$ is not equivalent to the real-valued solution up to the multiplication by $e^{i\alpha}$ with $\alpha \in \mathbb{R}$.

Theorems 3.9 and 3.10 give us the conditions (and the computational algorithm), under which the limiting configuration $\phi^{(0)}$ is uniquely continued into the vortex solution ϕ for small $\epsilon > 0$. Moreover, both $\phi \in l^2(\mathbb{Z})$ and $\theta \in \mathbb{T}^N$ can be expanded in power series

$$\phi = \phi^{(0)}(\theta_0) + \sum_{k=1}^{\infty} \epsilon^k \phi^{(k)}, \quad \theta = \theta_0 + \sum_{k=1}^{\infty} \epsilon^k \theta_k. \quad (4.3.56)$$

Linearization of the DNLS equation (4.3.51) at the vortex solution (4.3.52) is defined by the substitution

$$\psi_n(t) = \left[\phi_n + a_n e^{\lambda t} + \bar{b}_n e^{\bar{\lambda} t} \right] e^{it}, \quad n \in \mathbb{Z}^2,$$

which results in the spectral stability problem

$$\sigma \mathcal{H} \mathbf{v} = i\lambda \mathbf{v}, \quad (4.3.57)$$

where vector $\mathbf{v}$ has components (a_n, b_n) for all $n \in \mathbb{Z}^2$, σ is a diagonal matrix of $\{+1, -1\}$, and $\mathcal{H}$ is the Jacobian operator (Section 3.2.5),

$$\mathcal{H}_n = \begin{bmatrix} 1 - 2|\phi_n|^2 & -\phi_n^2 \\ -\bar{\phi}_n^2 & 1 - 2|\phi_n|^2 \end{bmatrix} - \epsilon \left(\delta_{+e_1} + \delta_{-e_1} + \delta_{+e_2} + \delta_{-e_2} - 4 \right) \begin{bmatrix} 1 & 0 \\ 0 & 1 \end{bmatrix}.$$

For $\epsilon = 0$, the spectrum of $\mathcal{H}$ includes the semi-simple eigenvalue -2 of multiplicity $N = \dim(S)$, the semi-simple eigenvalue 0 of multiplicity N, and the semi-simple eigenvalue 1 of infinite multiplicity. On the other hand, the spectrum of $i\sigma \mathcal{H}$ includes the zero eigenvalue of geometric multiplicity N and algebraic multiplicity $2N$ and the semi-simple eigenvalues $\pm i$ of infinite multiplicity.

Using power series expansion (4.3.56), operator $\mathcal{H}$ is expanded into the power series

$$\mathcal{H} = \mathcal{H}^{(0)} + \sum_{k=1}^{\infty} \epsilon^k \mathcal{H}^{(k)}. \quad (4.3.58)$$

Using the power series (4.3.55) and (4.3.56), the Jacobian $N \times N$ matrices $\mathcal{M}^{(k)} = D_\theta g^{(k)}(\theta_0)$ are obtained. These matrices determine persistence of the limiting vortex configuration (Theorem 3.10). In particular, if $g^{(k)} \equiv \mathbf{0}$ for $k = 1, 2, ..., k_0 - 1$ for some $k_0 \geq 1$, $g^{(k_0)} \neq \mathbf{0}$, and $\mathcal{M}^{(k_0)}$ has a simple zero eigenvalue with the eigenvector $\mathbf{p}_0 = (1, 1, ..., 1) \in \mathbb{R}^N$ (thanks to the gauge invariance), then the limiting vortex configuration (4.3.54) persists for small $\epsilon > 0$.

It is natural to expect that the eigenvalues of matrices $\{\mathcal{M}^{(k)}\}_{k \in \mathbb{N}}$ determine the small eigenvalues of $\mathcal{H}$ for small $\epsilon > 0$. Note that $0 \in \sigma(\mathcal{H})$ for any $\epsilon \in \mathbb{R}$ as long as the vortex solution exists, thanks to the gauge invariance of the DNLS equation (4.3.51). The following theorem establishes the relation between eigenvalues of $\{\mathcal{M}^{(k)}\}_{k \geq 1}$ and small eigenvalues of $\mathcal{H}$.

Theorem 4.12 *Let ϕ be the vortex solution of the stationary equation (4.3.53) in the power series form (4.3.56). Let $P^{(k-1)} : \mathbb{R}^N \to \mathrm{Ker}(\mathcal{M}^{(k-1)}) \subset \mathbb{R}^N$ be the orthogonal projection operator and let μ_0 be a nonzero eigenvalue of the matrix $P^{(k-1)} \mathcal{M}^{(k)} P^{(k-1)}$ for any $k \geq 1$. There exists an eigenvalue μ of operator $\mathcal{H}$ such that $\mu = \epsilon^k \mu_0 + \mathcal{O}(\epsilon^{k+1})$ as $\epsilon \to 0$.*

Proof Let μ be an eigenvalue of $\mathcal{H}$ with the eigenvector $\mathbf{v}$. From the power series (4.3.58), we write

$$\left[\mathcal{H}^{(0)} + \epsilon \mathcal{H}^{(1)} + \cdots + \epsilon^{k-1} \mathcal{H}^{(k-1)} + \epsilon^k \mathcal{H}^{(k)} + \mathcal{O}(\epsilon^{k+1}) \right] \mathbf{v} = \mu \mathbf{v}.$$

Let $\{\mathcal{M}^{(m)}\}_{m \in \mathbb{N}}$ be Jacobian matrices such that

$$\mathrm{Ker}(\mathcal{M}^{(m+1)}) \subset \mathrm{Ker}(\mathcal{M}^{(m)}) \subset \mathbb{R}^N \quad \text{for any} \quad m \geq 1.$$

Let $\boldsymbol{\alpha}$ be an element in the intersection of

$$\mathrm{Ker}(\mathcal{M}^{(1)}) \cap \mathrm{Ker}(\mathcal{M}^{(2)}) \cap \ldots \cap \mathrm{Ker}(\mathcal{M}^{(k-1)}) \subset \mathbb{R}^N$$

but $\boldsymbol{\alpha} \notin \mathrm{Ker}(\mathcal{M}^{(k)})$ for some $k \geq 1$.

Recall the projection operator $\mathcal{P} : l^2(\mathbb{Z}^2) \times l^2(\mathbb{Z}^2) \to \mathrm{Ker}(\mathcal{H}^{(0)}) \subset l^2(\mathbb{Z}^2) \times l^2(\mathbb{Z}^2)$ given by

$$\forall \mathbf{f} \in l^2(\mathbb{Z}^2) : \quad (\mathcal{P}\mathbf{f})_n = \frac{1}{2i} \begin{cases} \left(e^{-i\theta_n}(\mathbf{f})_n - e^{i\theta_n}(\bar{\mathbf{f}})_n \right), \\ - \left(e^{i\theta_n}(\bar{\mathbf{f}})_n - e^{-i\theta_n}(\mathbf{f})_n \right), \end{cases} \quad n \in S, \qquad (4.3.59)$$

and the obvious relation

$$\forall \boldsymbol{\theta}_0 \in \mathcal{T} : \quad \boldsymbol{\phi}^{(0)}(\boldsymbol{\theta}_0) + \sum_{n \in S} \alpha_n \mathbf{e}_n = \boldsymbol{\phi}^{(0)}(\boldsymbol{\theta}_0 + \boldsymbol{\alpha}) + \mathcal{O}(\|\boldsymbol{\alpha}\|^2). \qquad (4.3.60)$$

Hence we obtain

$$\boldsymbol{\alpha} = \mathcal{P} \left(\sum_{n \in S} \alpha_n \mathbf{e}_n \right), \qquad \left(\sum_{n \in S} \alpha_n \mathbf{e}_n \right) = D_{\boldsymbol{\theta}} \boldsymbol{\phi}^{(0)}(\boldsymbol{\theta}_0) \boldsymbol{\alpha},$$

where $D_{\boldsymbol{\theta}} \boldsymbol{\phi}^{(0)}(\boldsymbol{\theta}_0)$ is the Jacobian matrix of the infinite-dimensional vector $\boldsymbol{\phi}^{(0)}(\boldsymbol{\theta})$ with respect to the N-dimensional vector $\boldsymbol{\theta}$. It is clear that

$$\mathcal{H}^{(0)} \mathbf{v}^{(0)} = 0, \quad \text{where} \quad \mathbf{v}^{(0)} = \sum_{n \in S} \alpha_n \mathbf{e}_n = D_{\boldsymbol{\theta}} \boldsymbol{\phi}^{(0)}(\boldsymbol{\theta}_0) \boldsymbol{\alpha}.$$

Furthermore, the partial $(k-1)$th sum of the power series (4.3.56) satisfies the stationary equation (4.3.53) up to the order of $\mathcal{O}(\epsilon^k)$.

If $\dim \mathrm{Ker}(\mathcal{M}^{(k-1)}) = d_{k-1} + 1$, then the partial $(k-1)$th sum of the power series (4.3.56) has $(d_{k-1}+1)$ arbitrary parameters when $\boldsymbol{\theta}_0$ is shifted in the direction of the vector $\boldsymbol{\alpha}$. In the tangent space of the stationary equation (4.3.53) in the direction of $\boldsymbol{\alpha}$, the linear inhomogeneous system

$$\mathcal{H}^{(0)} \mathbf{v}^{(m)} + \mathcal{H}^{(1)} \mathbf{v}^{(m-1)} + \ldots + \mathcal{H}^{(m)} \mathbf{v}^{(0)} = 0, \quad m = 1, 2, \ldots, k-1$$

has a particular solution in the form $\mathbf{v}^{(m)} = D_{\boldsymbol{\theta}} \boldsymbol{\phi}^{(m)}(\boldsymbol{\theta}_0) \boldsymbol{\alpha}$ for $m = 1, 2, \ldots, k-1$.

Using the regular perturbation series for isolated zero eigenvalues of $\mathcal{H}^{(0)}$,

$$\mathbf{v} = \mathbf{v}^{(0)} + \epsilon \mathbf{v}^{(1)} + \ldots + \epsilon^k \mathbf{v}^{(k)} + \mathcal{O}(\epsilon^{k+1}), \quad \mu = \mu_k \epsilon^k + \mathcal{O}(\epsilon^{k+1}),$$

we obtain the linear inhomogeneous equation

$$\mathcal{H}^{(0)}\mathbf{v}^{(k)} + \mathcal{H}^{(1)}\mathbf{v}^{(k-1)} + \cdots + \mathcal{H}^{(k)}\mathbf{v}^{(0)} = \mu_k \mathbf{v}^{(0)}. \tag{4.3.61}$$

Applying the projection operator $\mathcal{P}$ and recalling the definition of the vector $\mathbf{g}(\boldsymbol{\theta}, \epsilon) = \mathcal{P}\mathbf{F}(\boldsymbol{\phi}, \epsilon)$, we find that the left-hand side of the inhomogeneous equation (4.3.61) reduces to the form

$$\mathcal{P}\left[\mathcal{H}^{(1)} D_{\boldsymbol{\theta}}\boldsymbol{\phi}^{(k-1)}(\boldsymbol{\theta}_0) + \cdots + \mathcal{H}^{(k)} D_{\boldsymbol{\theta}}\boldsymbol{\phi}^{(0)}(\boldsymbol{\theta}_0)\right]\boldsymbol{\alpha} = D_{\boldsymbol{\theta}}\mathbf{g}^{(k)}(\boldsymbol{\theta}_0)\boldsymbol{\alpha} = \mathcal{M}^{(k)}\boldsymbol{\alpha},$$

whereas the right-hand side becomes $\mu_k \mathcal{P}\mathbf{v}^{(0)} = \mu_k \boldsymbol{\alpha}$. Therefore, μ_k is an eigenvalue of the Jacobian matrix $\mathcal{M}^{(k)}$ and $\boldsymbol{\alpha}$ is the corresponding eigenvector. $\square$

Small eigenvalues of the self-adjoint operator $\mathcal{H}$ are expected to be related to the small eigenvalues of the spectral stability problem (4.3.57), similarly to the one-dimensional DNLS equation (Section 4.3.3). We will show that the relationship between these small eigenvalues for the vortex solutions of the two-dimensional DNLS equation may be more complicated than that for solitons in the one-dimensional DNLS equation. The following theorem gives the main result on the relationship between nonzero eigenvalues of the Jacobian matrices $\{\mathcal{M}^{(k)}\}_{k\geq 1}$ and small eigenvalues of the spectral stability problem (4.3.57).

Theorem 4.13 *Under the conditions of Theorem 4.12, there exists a pair of small eigenvalues λ of the spectral stability problem (4.3.57) such that*

$$\lambda = \epsilon^{k/2}\lambda_{k/2} + \mathcal{O}(\epsilon^{k/2+1}),$$

where $\lambda_{k/2}$ is found from the quadratic eigenvalue problems,

odd k:
$$\mathcal{M}^{(k)}\boldsymbol{\alpha} = \frac{1}{2}\lambda_{k/2}^2\boldsymbol{\alpha}, \tag{4.3.62}$$

even k:
$$\mathcal{M}^{(k)}\boldsymbol{\alpha} + \frac{1}{2}\lambda_{k/2}\mathcal{L}^{(k)}\boldsymbol{\alpha} = \frac{1}{2}\lambda_{k/2}^2\boldsymbol{\alpha}. \tag{4.3.63}$$

Here $\mathcal{L}^{(k)}$ is a skew-symmetric matrix defined by (4.3.66) below.

Proof Similarly to the computations of small nonzero eigenvalues of H in Theorem 4.12, we write the spectral stability problem (4.3.57) in the form

$$\sigma\left[\mathcal{H}^{(0)} + \epsilon\mathcal{H}^{(1)} + \cdots + \epsilon^{k-1}\mathcal{H}^{(k-1)} + \epsilon^k\mathcal{H}^{(k)} + \mathcal{O}(\epsilon^{k+1})\right]\mathbf{v} = \mathrm{i}\lambda\mathbf{v}.$$

Using relations

$$\hat{\mathbf{e}}_n = -\mathrm{i}\sigma\mathbf{e}_n, \quad \mathcal{H}^{(0)}\hat{\mathbf{e}}_n = -2\hat{\mathbf{e}}_n, \quad n \in S,$$

we find a solution of the linear inhomogeneous equation

$$\mathcal{H}^{(0)}\boldsymbol{\varphi}^{(0)} = 2\mathrm{i}\sigma D_{\boldsymbol{\theta}}\boldsymbol{\phi}^{(0)}(\boldsymbol{\theta}_0)\boldsymbol{\alpha}$$

in the form

$$\boldsymbol{\varphi}^{(0)} = \sum_{n\in S}\alpha_n\hat{\mathbf{e}}_n = \boldsymbol{\Phi}^{(0)}(\boldsymbol{\theta}_0)\boldsymbol{\alpha},$$

where $\boldsymbol{\Phi}^{(0)}(\boldsymbol{\theta}_0)$ is the matrix extension of $\boldsymbol{\phi}^{(0)}(\boldsymbol{\theta}_0)$, which consists of vector columns $\hat{\mathbf{e}}_n$, $n \in S$.

Similarly, there exists a particular solution of the inhomogeneous problem

$$\mathcal{H}^{(0)}\boldsymbol{\varphi}^{(m)} + \mathcal{H}^{(1)}\boldsymbol{\varphi}^{(m-1)} + \cdots + \mathcal{H}^{(m)}\boldsymbol{\varphi}^{(0)} = 2i\sigma D_{\boldsymbol{\theta}}^{\mathrm{T}}\boldsymbol{\phi}^{(m)}(\boldsymbol{\theta}_0)\boldsymbol{\alpha}, \quad m = 1, 2, \cdots, k',$$

in the form $\boldsymbol{\varphi}^{(m)} = \boldsymbol{\Phi}^{(m)}(\boldsymbol{\theta}_0)\boldsymbol{\alpha}$, where

$$k' = \frac{k-1}{2} \text{ if } k \text{ is odd} \quad \text{and} \quad k' = \frac{k}{2} - 1 \text{ if } k \text{ is even.}$$

Using the regular perturbation series for an isolated zero eigenvalue of $\sigma\mathcal{H}^{(0)}$,

$$\mathbf{v} = \mathbf{v}^{(0)} + \epsilon\mathbf{v}^{(1)} + \cdots + \epsilon^{k-1}\mathbf{v}^{(k-1)} + \frac{1}{2}\lambda\left(\boldsymbol{\varphi}^{(0)} + \epsilon\boldsymbol{\varphi}^{(1)} + \cdots + \epsilon^{k'}\boldsymbol{\varphi}^{(k')}\right)$$
$$+ \epsilon^k\mathbf{v}^{(k)} + \mathcal{O}(\epsilon^{k+1}),$$

and

$$\lambda = \epsilon^{k/2}\lambda_{k/2} + \mathcal{O}(\epsilon^{k/2+1}),$$

where

$$\mathbf{v}^{(m)} = D_{\boldsymbol{\theta}}\boldsymbol{\phi}^{(m)}(\boldsymbol{\theta}_0)\boldsymbol{\alpha} \quad \text{for} \quad m = 0, 1, \cdots, k-1,$$

$$\boldsymbol{\varphi}^{(m)} = \boldsymbol{\Phi}^{(m)}(\boldsymbol{\theta}_0)\boldsymbol{\alpha} \quad \text{for} \quad m = 0, 1, .., k',$$

we obtain a linear inhomogeneous equation for $\mathbf{v}^{(k)}$. When k is odd, the linear equation for $\mathbf{v}^{(k)}$ takes the form

$$\mathcal{H}^{(0)}\mathbf{v}^{(k)} + \mathcal{H}^{(1)}\mathbf{v}^{(k-1)} + \cdots + \mathcal{H}^{(k)}\mathbf{v}^{(0)} = \frac{i}{2}\lambda_{k/2}^2\sigma\boldsymbol{\varphi}^{(0)}. \tag{4.3.64}$$

When k is even, the linear equation for $\mathbf{v}^{(k)}$ takes the form

$$\mathcal{H}^{(0)}\mathbf{v}^{(k)} + \mathcal{H}^{(1)}\mathbf{v}^{(k-1)} + \cdots + \mathcal{H}^{(k)}\mathbf{v}^{(0)}$$
$$+ \frac{1}{2}\lambda_{k/2}\left(\mathcal{H}^{(1)}\boldsymbol{\varphi}^{(k')} + \cdots + \mathcal{H}^{(k'+1)}\boldsymbol{\varphi}^{(0)}\right) = \frac{i}{2}\lambda_{k/2}^2\sigma\boldsymbol{\varphi}^{(0)}. \tag{4.3.65}$$

Using the projection operator $\mathcal{P}$ and the definition of the new matrix

$$\mathcal{L}^{(k)} = \mathcal{P}\left[\mathcal{H}^{(1)}\boldsymbol{\Phi}^{(k')}(\boldsymbol{\theta}_0) + \cdots + \mathcal{H}^{(k'+1)}\boldsymbol{\Phi}^{(0)}(\boldsymbol{\theta}_0)\right], \tag{4.3.66}$$

we obtain reduced problems (4.3.62) and (4.3.63) from the solvability conditions for the linear inhomogeneous problems (4.3.64) and (4.3.65). $\qquad\square$

Exercise 4.27 Assume that matrix $\mathcal{L}^{(k)}$ is skew-symmetric and prove that eigenvalues of the quadratic eigenvalue problem (4.3.63) occur in pairs $\lambda_{k/2}$ and $-\lambda_{k/2}$.

A general count of unstable eigenvalues in the quadratic eigenvalue problem (4.3.63) was developed by Chugunova & Pelinovsky [37] and Kollar [119].

Let us apply Theorems 4.12 and 4.13 to the localized modes $\boldsymbol{\phi}$ associated with the vortex cell,

$$S = \{(1, 1), (2, 1), (2, 2), (1, 2)\} \subset \mathbb{Z}^2.$$

The persistent localized modes (Section 3.2.5) include eight soliton configurations and one vortex configuration of charge $L = 1$ with the phase distribution

$$\theta_j = \frac{\pi(j-1)}{2}, \quad j = 1, 2, 3, 4. \tag{4.3.67}$$

The Jacobian matrix $\mathcal{M}^{(1)}$ is given by

$$\mathcal{M}^{(1)} = \begin{bmatrix} s_1 + s_4 & -s_1 & 0 & -s_4 \\ -s_1 & s_1 + s_2 & -s_2 & 0 \\ 0 & -s_2 & s_2 + s_3 & -s_3 \\ -s_4 & 0 & -s_3 & s_3 + s_4 \end{bmatrix}, \qquad (4.3.68)$$

where

$$s_1 = \cos(\theta_2 - \theta_1), \quad s_2 = \cos(\theta_3 - \theta_2), \quad s_3 = \cos(\theta_4 - \theta_3), \quad s_4 = \cos(\theta_1 - \theta_4).$$

For the soliton configurations, $\mathcal{M}^{(1)}$ is generally nonzero. Every positive eigenvalue of $\mathcal{M}^{(1)}$ generates a pair of real eigenvalues of the quadratic problem (4.3.62) and every negative eigenvalue of $\mathcal{M}^{(1)}$ generates a pair of purely imaginary eigenvalues of the quadratic problem (4.3.62). These results are similar to the analysis of stability of localized modes in one dimension (Section 4.3.3).

For the vortex configuration (4.3.67), $\mathcal{M}^{(1)} \equiv 0$ is zero. Therefore, we proceed with the Jacobian matrix $\mathcal{M}^{(2)}$ at the order $k = 2$,

$$\mathcal{M}^{(2)} = \begin{bmatrix} 1 & 0 & -1 & 0 \\ 0 & 1 & 0 & -1 \\ -1 & 0 & 1 & 0 \\ 0 & -1 & 0 & 1 \end{bmatrix}.$$

Using the projection formula (4.3.66), we obtain

$$\mathcal{L}^{(2)} = 2 \begin{bmatrix} 0 & 1 & 0 & -1 \\ -1 & 0 & 1 & 0 \\ 0 & -1 & 0 & 1 \\ 1 & 0 & -1 & 0 \end{bmatrix}$$

and consider eigenvalues in the quadratic problem (4.3.63). Note that the eigenvalue problem is equivalent to the difference equation with constant coefficients

$$-\alpha_{j+2} + 2\alpha_j - \alpha_{j-2} = \lambda_1^2 \alpha_j + 2\lambda_1 (\alpha_{j+1} - \alpha_{j-1}), \quad j = 1, 2, 3, 4,$$

subject to periodic boundary conditions. Using the discrete Fourier transform, the difference equation reduces to the characteristic equation:

$$\left(\lambda_1 + 2i \sin \frac{\pi n}{2}\right)^2 = 0, \quad n = 1, 2, 3, 4.$$

There exist two eigenvalues of algebraic multiplicity two at $\lambda_1 = -2i$ and $\lambda_1 = 2i$ and a zero eigenvalue of algebraic multiplicity four.

Exercise 4.28 Show that if $\mathbf{v}$ is given by the perturbation series in Theorem 4.13, then

$$\langle \mathcal{H}\mathbf{v}, \mathbf{v} \rangle_{l^2} = 2\epsilon^2 \lambda_1 \langle (\mathcal{L}^{(1)} + \lambda_1 I)\boldsymbol{\alpha}, \boldsymbol{\alpha} \rangle + \mathcal{O}(\epsilon^3),$$

hence the double eigenvalues $\lambda_1 = \pm 2i$ have zero energy $\langle \mathcal{H}\mathbf{v}, \mathbf{v} \rangle_{l^2}$ at the order $k = 2$.

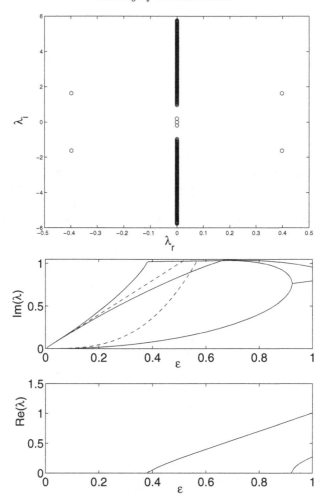

Figure 4.5 Top: the spectral plane $\lambda = \lambda_r + i\lambda_i$ for $\epsilon = 0.6$. Bottom: imaginary and real parts of small eigenvalues versus ϵ using asymptotic (dashed lines) and numerical (solid lines) approximations. Reprinted from [156].

To detect the splitting of the zero eigenvalue of multiplicity four, one needs to extend computations up to the order $k = 6$. This was performed in [156], where it was found that $\lambda_3 = \pm 4\sqrt{2}i$. Taking into account that the double eigenvalues at $\lambda_1 = \pm 2i$ split along the imaginary axis at the order $k = 4$ for $\epsilon > 0$, we conclude that the vortex configuration (4.3.67) is spectrally *stable* for small $\epsilon > 0$.

Figure 4.5 shows numerical results for the vortex configuration (4.3.67). The linearization spectrum of the vortex solution at the order $k = 2$ has a zero eigenvalue of algebraic multiplicity four and two pairs of imaginary eigenvalues $\lambda \approx \pm 2\epsilon i$. These pairs split along the imaginary axis at the order $k = 4$. The pair of larger purely imaginary eigenvalues of negative energy undertakes a bifurcation to a quartet of complex eigenvalues for a larger value of ϵ upon collision with the continuous spectrum. The pair of smaller purely imaginary eigenvalues of positive energy merges with the continuous spectrum avoiding a bifurcation to a quartet of complex eigenvalues. The zero eigenvalue produces another pair of purely imaginary eigenvalues $\lambda \approx \pm 4\sqrt{2}\epsilon^3 i$ at the order $k = 6$. This pair has negative energy and it bifurcates to a quartet of complex eigenvalues upon collision with another pair of purely imaginary eigenvalues of positive energy for larger values of ϵ.

Exercise 4.29 Compute matrices $\mathcal{M}^{(1)}$, $\mathcal{M}^{(2)}$, and $\mathcal{L}^{(2)}$ and the eigenvalues of the quadratic eigenvalue problems (4.3.62) and (4.3.63) for the three vortex configurations

$$\theta_j = \frac{\pi(j-1)L}{4}, \quad j \in \{1,...,8\}, \quad L \in \{1,2,3\}$$

associated with a larger vortex cell,

$$S = \{(1,1),(2,1),(3,1),(3,2),(3,3),(2,3),(1,3),(1,2)\} \subset \mathbb{Z}^2.$$

Prove that the configuration with charge $L = 1$ is stable, whereas the configurations with charges $L = 2$ and $L = 3$ are unstable. Show that if the node $(2,2)$ is removed from $\mathbb{Z}^2$, the vortex configuration with charge $L = 2$ becomes spectrally stable.

Exercise 4.30 Study stability of the vortex configuration,

$$\theta_j = \frac{\pi(j-1)}{2}, \quad j \in \{1,...,4\}$$

associated with the vortex cross,

$$S = \{(0,-1),(1,0),(0,1),(-1,0)\} \subset \mathbb{Z}^2$$

and show that the quadratic eigenvalue problem (4.3.63) at $k = 2$ has zero matrix $\mathcal{L}^{(2)} \equiv 0$. Prove that the vortex cross configuration is spectrally stable for small $\epsilon > 0$.

4.3.5 Gap solitons in the nonlinear Dirac equations

Let us consider the nonlinear Dirac equations,

$$\begin{cases} i(u_t + u_x) + v = \partial_{\bar{u}} W(u,\bar{u},v,\bar{v}), \\ i(v_t - v_x) + u = \partial_{\bar{v}} W(u,\bar{u},v,\bar{v}), \end{cases} \quad (4.3.69)$$

where $(u,v) \in \mathbb{C}^2$, $(x,t) \in \mathbb{R}^2$, and

$$W(u,\bar{u},v,\bar{v}) \equiv W(|u|^2 + |v|^2, |u|^2|v|^2, u\bar{v} + v\bar{u}).$$

We assume the existence of the localized mode,

$$u(x,t) = u_0(x)e^{-i\omega t}, \quad v(x,t) = v_0(x)e^{-i\omega t}, \quad u_0(x) = \bar{v}_0(x),$$

for some $\omega \in (-1,1)$ (Section 3.3.4). This localized mode is known as the *gap soliton* because $(-1,1)$ corresponds to the gap in the continuous spectrum of the nonlinear Dirac equations (4.3.69).

Linearization of the nonlinear Dirac equations (4.3.69) with the substitution

$$\begin{cases} u(x,t) = \left[u_0(x) + V_1(x)e^{\lambda t}\right] e^{-i\omega t}, \\ \bar{u}(x,t) = \left[\bar{u}_0(x) + V_2(x)e^{\lambda t}\right] e^{i\omega t}, \\ v(x,t) = \left[v_0(x) + V_3(x)e^{\lambda t}\right] e^{-i\omega t}, \\ \bar{v}(x,t) = \left[\bar{v}_0(x) + V_4(x)e^{\lambda t}\right] e^{i\omega t} \end{cases}$$

results in the spectral stability problem

$$\sigma H \mathbf{V} = i\lambda \mathbf{V}, \tag{4.3.70}$$

where $\mathbf{V} = (V_1, V_2, V_3, V_4) \in \mathbb{C}^4$, σ is a diagonal 4×4 matrix of $\{1, -1, 1, -1\}$, and $H = L + V(x)$ is a sum of the differential operator with constant coefficients

$$L = \begin{bmatrix} -\omega - i\partial_x & 0 & -1 & 0 \\ 0 & -\omega + i\partial_x & 0 & -1 \\ -1 & 0 & -\omega + i\partial_x & 0 \\ 0 & -1 & 0 & -\omega - i\partial_x \end{bmatrix} \tag{4.3.71}$$

and the matrix potential function

$$V(x) = \begin{bmatrix} \partial^2_{\bar{u}_0 u_0} & \partial^2_{\bar{u}_0^2} & \partial^2_{\bar{u}_0 v_0} & \partial^2_{\bar{u}_0 \bar{v}_0} \\ \partial^2_{u_0^2} & \partial^2_{u_0 \bar{u}_0} & \partial^2_{u_0 v_0} & \partial^2_{u_0 \bar{v}_0} \\ \partial^2_{\bar{v}_0 u_0} & \partial^2_{\bar{v}_0 \bar{u}_0} & \partial^2_{\bar{v}_0 v_0} & \partial^2_{\bar{v}_0^2} \\ \partial^2_{v_0 u_0} & \partial^2_{v_0 \bar{u}_0} & \partial^2_{v_0^2} & \partial^2_{v_0 \bar{v}_0} \end{bmatrix} W(u_0, \bar{u}_0, v_0, \bar{v}_0). \tag{4.3.72}$$

We note that

$$H : H^1(\mathbb{R}) \times H^1(\mathbb{R}) \times H^1(\mathbb{R}) \times H^1(\mathbb{R}) \to L^2(\mathbb{R}) \times L^2(\mathbb{R}) \times L^2(\mathbb{R}) \times L^2(\mathbb{R})$$

is a self-adjoint operator. It arises in the quadratic form associated with the energy functional $E_\omega(u, v) = H(u, v) - \omega Q(u, v)$ after it is expanded near the critical point (u_0, v_0) (Exercise 3.29).

Exercise 4.31 Use the gauge and translational symmetries of the nonlinear Dirac equations (4.3.69) and show that $\mathrm{Ker}(H_\omega) = \mathrm{Span}\{\mathbf{V}_1, \mathbf{V}_2\}$, where

$$\mathbf{V}_1 = \sigma \mathbf{u}_0, \quad \mathbf{V}_2 = \partial_x \mathbf{u}_0,$$

where $\mathbf{u}_0 = (u_0, \bar{u}_0, v_0, \bar{v}_0)$.

When $u_0(x) = \bar{v}_0(x)$ for all $x \in \mathbb{R}$, elements of $V(x)$ in (4.3.72) enjoy additional symmetry relations

$$\partial^2_{u_0 \bar{u}_0} W = \partial^2_{v_0 \bar{v}_0} W, \quad \partial^2_{\bar{u}_0^2} W = \partial^2_{v_0^2} W, \quad \partial^2_{u_0 v_0} W = \partial^2_{\bar{u}_0 \bar{v}_0} W.$$

In this case, the eigenvalue problem $H\mathbf{V} = \mu\mathbf{V}$ admits two particular reductions:

$$\text{(i) } V_1 = V_4, \ V_2 = V_3, \quad \text{(ii) } V_1 = -V_4, \ V_2 = -V_3.$$

These reductions allow us to block-diagonalize the spectral stability problem (4.3.70), after which the spectral stability problem can be transformed to the conventional form

$$L_+ \mathbf{U} = -\lambda \mathbf{W}, \quad L_- \mathbf{W} = \lambda \mathbf{U}, \tag{4.3.73}$$

where $\mathbf{U} \in \mathbb{C}^2$, $\mathbf{W} \in \mathbb{C}^2$, and $L_\pm$ are 2×2 matrix differential operators. The block-diagonalization of the spectral stability problem (4.3.70) was discovered by Chugunova & Pelinovsky [35].

Lemma 4.20 *Assume $u_0(x) = \bar{v}_0(x)$ for all $x \in \mathbb{R}$. There exists an orthogonal similarity transformation with an orthogonal matrix*

$$S = \frac{1}{\sqrt{2}} \begin{bmatrix} 1 & 0 & 1 & 0 \\ 0 & 1 & 0 & 1 \\ 0 & 1 & 0 & -1 \\ 1 & 0 & -1 & 0 \end{bmatrix}$$

that simultaneously block-diagonalizes the self-adjoint operator H,

$$S^{\mathrm{T}} H S = \begin{bmatrix} H_+ & 0 \\ 0 & H_- \end{bmatrix}, \tag{4.3.74}$$

and the linearized operator σH

$$S^{\mathrm{T}} \sigma H S = \sigma \begin{bmatrix} 0 & H_- \\ H_+ & 0 \end{bmatrix}, \tag{4.3.75}$$

where $H_\pm$ are 2×2 Dirac operators

$$H_\pm = \begin{bmatrix} -\omega - i\partial_x & \mp 1 \\ \mp 1 & -\omega + i\partial_x \end{bmatrix} + V_\pm(x),$$

with the new potential functions

$$V_\pm(x) = \begin{bmatrix} \partial^2_{\bar{u}_0 u_0} \pm \partial^2_{\bar{u}_0 \bar{v}_0} & \partial^2_{u_0^2} \pm \partial^2_{\bar{u}_0 v_0} \\ \partial^2_{u_0^2} \pm \partial^2_{u_0 \bar{v}_0} & \partial^2_{\bar{u}_0 u_0} \pm \partial^2_{u_0 v_0} \end{bmatrix} W(u_0, \bar{u}_0, v_0, \bar{v}_0).$$

Proof Block-diagonalizations (4.3.74) and (4.3.75) are checked by explicit computations. □

Corollary 4.8 *Let*

$$\mathbf{V} = S \begin{bmatrix} \mathbf{U} \\ i\sigma_3 \mathbf{W} \end{bmatrix}, \quad \mathbf{U}, \mathbf{W} \in \mathbb{C}^2.$$

The spectral stability problem (4.3.70) is equivalent to the standard form (4.3.73) with

$$L_+ = H_+ \quad \text{and} \quad L_- = \sigma_3 H_- \sigma_3.$$

Let us abuse the notation and denote $\mathbf{u}_0 = (u_0, \bar{u}_0) \in \mathbb{C}^2$. The kernel of H (Exercise 4.31) is now decomposed into

$$\mathrm{Ker}(L_+) = \mathrm{Span}\{\partial_x \mathbf{u}_0\} \quad \text{and} \quad \mathrm{Ker}(L_-) = \mathrm{Span}\{\mathbf{u}_0\}.$$

Define the constrained L^2 space,

$$L_c^2 = \left\{ \mathbf{U} \in L^2(\mathbb{R}) : \quad \langle \mathbf{u}_0, \mathbf{U} \rangle_{L^2} = 0 \right\}. \tag{4.3.76}$$

Thanks to the exponential decay of $V_{\pm}(x)$ as $|x| \to \infty$, we find

$$\sigma_c(L_{\pm}) = (-\infty, -(1+\omega)] \cup [1 - \omega, \infty)$$

for any fixed $\omega \in (-1, 1)$. Therefore, the zero eigenvalue of $L_{\pm}$ is bounded away from the continuous spectrum. Using the standard algorithm, the spectral stability problem (4.3.73) can be reduced to the generalized eigenvalue problem in the constrained L^2 space,

$$L\mathbf{U} = \gamma M^{-1} \mathbf{U}, \quad \gamma = -\lambda^2, \tag{4.3.77}$$

where $L = P_c L_+ P_c$, $M^{-1} = P_c L_-^{-1} P_c$, and $P_c : L^2(\mathbb{R}) \to L_c^2 \subset L^2(\mathbb{R})$ is an orthogonal projection operator.

Although we are now in the formalism of Section 4.2, the assumptions of Theorem 4.5 are not satisfied since $\sigma_c(L_{\pm})$ is not bounded from below by a positive constant.

No analytic theory of the spectral stability of localized modes in the nonlinear Dirac equations (4.3.69) has been developed so far. Numerical approximations of eigenvalues were constructed using the Evans function computations [46] and the Chebyshev interpolation [35].

For illustrations, let us consider the nonlinear Dirac equations (4.3.69) with

$$W = \frac{1}{2}(|u|^4 + |v|^4)$$

when the potential functions $V_{\pm}(x)$ are computed explicitly by

$$V_+ = V_- = \begin{pmatrix} 2|u_0|^2 & u_0^2 \\ \bar{u}_0^2 & 2|u_0|^2 \end{pmatrix}.$$

Figure 4.6 displays the spectra of operators H_+, H_-, and $L := i\sigma H$ for six values of parameter ω in the interval $(-1, 1)$. When ω is close to 1, there exists a single nonzero eigenvalue for H_+ and H_- and a single pair of purely imaginary eigenvalues of L (top left panel). The first set of arrays on the figure indicates that the pair of eigenvalues of L becomes visible at the same value of ω as the eigenvalue of H_+. This correlation between eigenvalues of L and H_+ can be traced throughout the entire parameter domain.

When ω decreases, the operator H_- acquires another nonzero eigenvalue from the continuous spectrum, with no changes in the number of isolated eigenvalues of L (top right panel). The first complex instability occurs near $\omega \approx -0.18$, when the pair of purely imaginary eigenvalues of L collides with the continuous spectrum and emerge as a quartet of complex eigenvalues, with no changes in the number of isolated eigenvalues for H_+ and H_- (middle left panel).

The second complex instability occurs at $\omega \approx -0.54$, when the operator H_- acquires a third nonzero eigenvalue and the linearized operator L acquires another quartet of complex eigenvalues (middle right panel). The second set of arrays on the figure indicates a correlation between these eigenvalues of L and H_-.

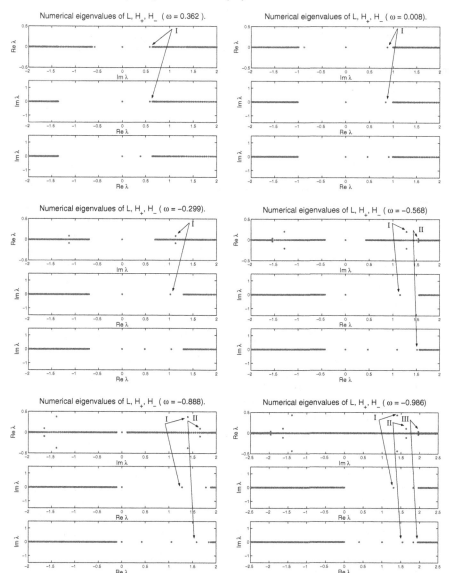

Figure 4.6 Eigenvalues and instability bifurcations for the nonlinear Dirac equations. Reprinted from [35].

When ω decreases further, the operators H_+ and H_- acquire one more isolated eigenvalue, with no change in the spectrum of L (bottom left panel). Finally, when ω is close to -1, the third complex instability occurs, correlated with another bifurcation from the continuous spectrum in the operator H_- (bottom right panel). The third set of arrays on the figure indicates this correlation. In a narrow domain near $\omega = -1$, the operator H_+ has two nonzero eigenvalues, the operator H_- has five nonzero eigenvalues and the operator L has three quartets of complex eigenvalues.

These numerical results imply that there exists a correlation between bifurcations of isolated eigenvalues in the operator L and those in the Dirac operators H_+ and H_-. However, analysis of this correlation remains open for further studies.

4.4 Other methods in stability analysis

After the spectral stability of localized modes is proved, the spectral information can be used to study the nonlinear stability of localized modes. If the spectral stability problem is defined by a non-self-adjoint linearized operator, the nonlinear stability problem involves analysis of the time evolution of the original equation.

Analysis of the linearized stability in the sense of Definition 4.2 can already be difficult compared to the spectral stability analysis. If the localized mode is unstable spectrally, it is unstable linearly (because we can choose initial data of the linearized evolution in the direction of an unstable eigenvector). However, what if the localized mode is spectrally stable and no perturbations grow exponentially fast? Several subtle issues may be obstacles for the linearized stability of the spectrally stable localized modes:

- Nonzero eigenvalues on the imaginary axis may be multiple and lead to the polynomial growth of solutions of the linearized equation.
- The zero eigenvalue may have a higher algebraic multiplicity than the one prescribed by the symmetries of the original equation, leading again to the polynomial growth of solutions of the linearized equation.
- The eigenvectors of the spectral problem may not form a basis in L^2 space, leading to no semi-group properties of the linearized evolution operator.

The first two issues can occur also in finite-dimensional evolution problems, but the third issue can only arise in infinite-dimensional evolution equations. A subtle phenomenon of ill-posedness of linearized evolution equations was studied recently by Chugunova & Pelinovsky [36] in the context of the advection–diffusion equation with a sign-varying dispersion. Good news is that the initial-value problem for linearized Gross–Pitaevskii and nonlinear Schrödinger equations is well-posed and eigenvectors of the spectral problem do form a basis in L^2 space.

If the spectral information is relatively simple and the linearized evolution is stable, we can then ask if the localized wave is orbitally stable in the sense of Definition 4.3. The orbital stability is studied as a variant of the classical stability of critical points in the sense of Lyapunov. Recall that the localized wave ϕ is a critical point of the energy functional $E_\omega(\mathbf{u}) = H(\mathbf{u}) - \omega Q(\mathbf{u})$ in a function space X. If we can prove that ϕ is a local minimizer of $E_\omega(\mathbf{u})$ in X and the minimum is non-degenerate in the sense of

$$\exists C > 0 : \quad E_\omega(\mathbf{u}) - E_\omega(\phi) \geq C \|\mathbf{u} - \phi\|_X^2,$$

for all $\mathbf{u}$ sufficiently close to ϕ in X, then ϕ is stable in the sense of Lyapunov. In other words, for $\epsilon > 0$, there exists $\delta > 0$ such that if $\|\mathbf{u}(0) - \phi\|_X \leq \delta$ then

$\|\mathbf{u}(t) - \boldsymbol{\phi}\|_X \leq \epsilon$ for all $t \in \mathbb{R}_+$. Indeed, continuity of $E_\omega(\mathbf{u})$ with respect to $\mathbf{u}$, conservation of $E_\omega(\mathbf{u}(t))$ for a solution $\mathbf{u}(t) : \mathbb{R}_+ \mapsto X$, and the convexity of $E_\omega(\mathbf{u})$ at $\boldsymbol{\phi}$ imply the stability of $\boldsymbol{\phi}$ in the sense of Lyapunov.

The classical method of Lyapunov needs to be revised, however, because the symmetries of the nonlinear evolution equations result in the existence of a continuous orbit $T(\theta)\boldsymbol{\phi}$ near the localized mode $\boldsymbol{\phi}$, where θ is an arbitrary parameter along the soliton orbit. We would not say that the solution $\boldsymbol{\phi}$ is unstable if $\mathbf{u}$ simply shifts along the orbit $T(\theta)\boldsymbol{\phi}$. Therefore, only the relative distance from $\mathbf{u}$ to the orbit $T(\theta)\boldsymbol{\phi}$ matters for the stability analysis and one needs to specify how to identify the parameter θ to ensure that $\|\mathbf{u} - T(\theta)\boldsymbol{\phi}\|_X$ is minimal in θ. A sufficient condition for orbital stability of localized modes in the context of nonlinear Schrödinger and Klein–Gordon equations was obtained by Grillakis *et al.* [74, 75].

If the localized mode $\boldsymbol{\phi}$ passes through the test of orbital stability, one can finally consider the strongest asymptotic stability in the sense of Definition 4.4. Asymptotic stability would guarantee that the dynamics near the orbit $T(\theta)\boldsymbol{\phi}$ approaches the orbit $T(\theta)\boldsymbol{\phi}_\infty$ as $t \to \infty$ and the remainder term disperses away in a local norm on X. The solution $\boldsymbol{\phi}_\infty$ near $\boldsymbol{\phi}$ corresponds to another value ω_∞ of parameter ω. To put it in a different way, asymptotic stability guarantees that the localized mode $\boldsymbol{\phi}$ is an attractor in the time evolution of the nonlinear equation in the sense that initial data close to a localized mode $\boldsymbol{\phi}$ resolves into a localized mode $\boldsymbol{\phi}_\infty$ and a dispersive remainder. In this case, dispersion-induced radiation arises in the Hamiltonian dynamical systems with conserved energy and power.

A combination of dispersive decay estimates following from the linearized time evolution, the modulation equations for varying parameters (ω, θ), and continuation arguments for the dispersive remainder is needed for the proof of asymptotic stability of localized modes.

For all three topics, the central element of analysis is the conditions of symplectic orthogonality on the dispersive remainder. These symmetry constraints have been used already in the formalism of the spectral stability problem, when the non-self-adjoint eigenvalue problem in L^2 space is written as a generalized eigenvalue problem for self-adjoint operators in a constrained L^2 space. While the degeneracy of the zero eigenvalue is not so important in the context of the spectral stability, it becomes the key element in analysis of the nonlinear stability of localized modes.

For simplicity of presentation, we consider the generalized Gross–Pitaevskii equation with a localized potential,

$$\mathrm{i}\psi_t = -\psi_{xx} + V(x)\psi - f(|\psi|^2)\psi, \qquad (4.4.1)$$

where $\psi(x,t) : \mathbb{R} \times \mathbb{R}_+ \to \mathbb{C}$ is the wave function, $V(x) : \mathbb{R} \to \mathbb{R}$ is a bounded potential with an exponential decay to zero as $|x| \to \infty$, and $f(|\psi|^2) : \mathbb{R}_+ \to \mathbb{R}$ is a C^1 function such that $f(0) = 0$ and $f'(u) > 0$ for all $u \in \mathbb{R}_+$. The only role of $V(x)$ is to break the translational invariance of the nonlinear Schrödinger equation and to reduce the algebraic multiplicity of the zero eigenvalue of the spectral stability problem. Some arguments on the construction of localized modes and eigenvectors of the spectral stability problem can be simplified if $V(x)$ is even on $\mathbb{R}$.

A similar analysis can be developed for other versions of the nonlinear Schrödinger equation and the Korteweg–de Vries equations. Accounts of these works can be found in Tao [202] and Angulo Pava [11].

4.4.1 Linearized stability of localized modes

Let us assume the existence of the localized mode of the Gross–Pitaevskii equation (4.4.1),

$$\psi(x,t) = \phi(x)e^{-i\omega t}, \tag{4.4.2}$$

where $\phi(x) : \mathbb{R} \to \mathbb{R}$ and $\omega \in \mathbb{R}$ are found from the stationary generalized Gross–Pitaevskii equation,

$$-\phi''(x) + V(x)\phi(x) - f(\phi^2)\phi(x) = \omega\phi(x), \quad x \in \mathbb{R}. \tag{4.4.3}$$

We assume that the map $\mathbb{R} \ni \omega \mapsto \phi \in H^2(\mathbb{R})$ is C^1. If $V(-x) = V(x)$ for all $x \in \mathbb{R}$, we also assume that

$$\phi(-x) = \phi(x), \quad x \in \mathbb{R}.$$

Thanks to the exponential decay of $V(x)$ to zero as $|x| \to \infty$, if $\omega < 0$, then the localized mode ϕ enjoys the exponential decay to zero such that

$$\exists C > 0 : \quad |\phi(x)| \leq Ce^{-\sqrt{|\omega|}|x|}, \quad x \in \mathbb{R}.$$

Applying the standard linearization

$$\psi(x,t) = [\phi(x) + z(x,t)]\, e^{-i\omega t},$$

we obtain the linearized Gross–Pitaevskii equation,

$$iz_t = -z_{xx} + Vz - \omega z - f(\phi^2)z - \phi^2 f'(\phi^2)(z + \bar{z}). \tag{4.4.4}$$

Separation of variables

$$z(x,t) = [u(x) + iw(x)]\, e^{\lambda t}$$

results in the spectral stability problem

$$L_+u = -\lambda w, \quad L_-w = \lambda u, \tag{4.4.5}$$

associated with the Schrödinger operators,

$$L_+ = -\partial_x^2 + V - \omega - f(\phi^2) - 2\phi^2 f'(\phi^2),$$
$$L_- = -\partial_x^2 + V - \omega - f(\phi^2).$$

Since

$$L_-\phi = 0, \quad L_+\partial_\omega\phi = \phi, \tag{4.4.6}$$

the spectral stability problem (4.4.5) admits a non-trivial null space. Assuming

$$\mathrm{Ker}(L_+) = \varnothing \quad \text{and} \quad \langle \phi, \partial_\omega\phi \rangle_{L^2} = \frac{1}{2}\frac{d}{d\omega}\|\phi\|_{L^2}^2 \neq 0,$$

which are standard assumptions in Sections 4.2 and 4.3, the zero eigenvalue has algebraic multiplicity one and geometric multiplicity two.

We say that $(u, w) \in L^2(\mathbb{R}) \times L^2(\mathbb{R})$ is symplectically orthogonal to the eigenvectors of the generalized kernel of the spectral stability problem (4.4.5) if

$$\langle u, \phi \rangle_{L^2} = 0, \quad \langle w, \partial_\omega \phi \rangle_{L^2} = 0. \tag{4.4.7}$$

Under this condition, $(u, w) \in L^2(\mathbb{R}) \times L^2(\mathbb{R})$ belong to the invariant subspace of the linearized problem (4.4.5), which is symplectically orthogonal to its two-dimensional null space. Note that u belongs to the same constrained subspace $L_c^2 \subset L^2$ as in Sections 4.2 and 4.3, where the spectral stability problem (4.4.5) was reduced to a generalized eigenvalue problem. In this section, however, we shall work with the full system (4.4.5).

Besides the zero eigenvalue, the spectral problem (4.4.5) may have other isolated eigenvalues and the continuous spectrum with possible embedded eigenvalues. To single out the main ingredient of this section, we shall assume that *no isolated and embedded eigenvalues of the spectral stability problem (4.4.5) exist except for the double zero eigenvalue.* This assumption is not critical since symplectically orthogonal complements of invariant subspaces for isolated and embedded eigenvalues can be constructed to reduce $L^2(\mathbb{R})$ to an invariant subspace for the continuous spectrum of the spectral stability problem as in the work of Cuccagna *et al.* [43].

Under the assumption of no nonzero eigenvalues, we present the scattering theory for the eigenfunctions of the spectral stability problem (4.4.5) associated with the continuous spectrum. We shall prove the symplectic orthogonality and completeness of these eigenfunctions in L^2 following the works of Buslaev & Perelman [25], Pelinovsky [153], Cuccagna [41], and Krieger & Schlag [122].

For convenience of working with purely imaginary λ, we set

$$\lambda = i\Omega, \quad u = U, \quad \text{and} \quad w = iW,$$

obtaining then $L_+U = \Omega W$, $L_-W = \Omega U$, or explicitly

$$\begin{cases} -U''(x) + cU(x) + V_+(x)U(x) = \Omega W(x), \\ -W''(x) + cW(x) + V_-(x)W(x) = \Omega U(x), \end{cases} \tag{4.4.8}$$

where $c = -\omega > 0$ and

$$V_+ := V - f(\phi^2) - 2\phi^2 f'(\phi^2), \quad V_- := V - f(\phi^2).$$

By the Weyl Theorem (Appendix B.15), the continuous spectrum of system (4.4.8) coincides with the purely continuous spectrum of the limiting system

$$-U_0''(x) + cU_0(x) = \Omega W_0(x), \quad -W_0''(x) + cW_0(x) = \Omega U_0(x). \tag{4.4.9}$$

Using the Fourier transform, we can see that the only solutions of system (4.4.9) in $L^\infty(\mathbb{R})$ exist for $\Omega \in (-\infty, -c] \cup [c, \infty)$. Let us focus on the positive branch $[c, \infty)$ and define

$$k = \sqrt{\Omega - c}, \quad p = \sqrt{\Omega + c}.$$

It is clear that c is a branch point of the two-sheet Riemann surface for solutions (U, W) of system (4.4.8) on the complex plane of Ω. If $\arg(\Omega - c) = 0$, then $k \in \mathbb{R}_+$, whereas if $\arg(\Omega - c) = 2\pi$, then $k \in \mathbb{R}_-$. In both cases, we can choose $p \in \mathbb{R}_+$.

We shall define two particular solutions $\mathbf{u} = (U, W)$ of system (4.4.8) for $\Omega = c + k^2$ with $k \in \mathbb{R}$ according to the limiting behavior

$$\lim_{x \to -\infty} \mathbf{u}_1(x, k)e^{-px} = \mathbf{e}_-, \qquad \lim_{x \to -\infty} \mathbf{u}_2(x, k)e^{-ikx} = \mathbf{e}_+, \qquad (4.4.10)$$

where $\mathbf{e}_\pm = (1, \pm 1)$. These two solutions exist thanks to the exponential decay of $V_\pm(x)$ to zero as $|x| \to \infty$. If $\Omega \in [c, \infty)$, the first solution $\mathbf{u}_1(x, k)$ decays exponentially to zero as $x \to -\infty$, while the second solution $\mathbf{u}_2(x, k)$ remains bounded as $x \to -\infty$.

Interpreting the terms with the potentials $V_\pm(x)$ as source terms, we rewrite the linear system (4.4.8) for $\mathbf{u}_1(x)$ in the equivalent integral form

$$\mathbf{u}_1(x, k) = e^{px}\mathbf{e}_- - \frac{1}{2} \int_{-\infty}^x \begin{bmatrix} g_1 + g_2 & g_1 - g_2 \\ g_1 - g_2 & g_1 + g_2 \end{bmatrix} (y - x) \begin{bmatrix} V_+(y) & 0 \\ 0 & V_-(y) \end{bmatrix} \mathbf{u}_1(y, k)\,dy,$$

where

$$g_1(x) = \frac{\sin(kx)}{k}, \qquad g_2(x) = \frac{\sinh(px)}{p}.$$

Since $V_\pm(x)$ decay to zero exponentially as $x \to -\infty$, there exists a unique solution for $\mathbf{u}_1(x, k)$ for all $x \in \mathbb{R}$ and all $k \in \mathbb{R}$ in the class of functions decaying like e^{px} as $x \to -\infty$. When the solution $\mathbf{u}_1(x, k)$ is extended to the opposite limit $x \to +\infty$, it may grow exponentially to infinity according to the limiting behavior

$$\lim_{x \to +\infty} \mathbf{u}_1(x, k)e^{-px} = A(k)\mathbf{e}_-, \qquad (4.4.11)$$

where

$$A(k) = 1 - \frac{1}{4\kappa} \int_{\mathbb{R}} \left(V_+(y)U_1(y, k) - V_-(y)W_1(y, k) \right) e^{-py}\,dy$$

and $\mathbf{u}_1 = (U_1, W_1)$.

The integral equation for a bounded solution $\mathbf{u}_2(x, k)$ cannot be found from the linear system (4.4.8) in the same way as the solution $\mathbf{u}_1(x, k)$ because the integral operator with the kernel $g_2(x)$ is not well-defined on the class of bounded functions. To derive the equivalent integral equation, we need to follow the steps of a reduction procedure described by Buslaev & Perelman [25].

Let us first rewrite the system (4.4.8) in vector variable

$$\boldsymbol{\varphi} = \left(\frac{U + W}{2}, \frac{U - W}{2} \right)$$

as follows:

$$\left(-\partial_x^2 + c \right) \boldsymbol{\varphi} + \hat{V}\boldsymbol{\varphi} = \Omega \sigma_3 \boldsymbol{\varphi}, \qquad (4.4.12)$$

where

$$\hat{V} = \begin{bmatrix} V - f(\phi^2) - \phi^2 f'(\phi^2) & -\phi^2 f'(\phi^2) \\ -\phi^2 f'(\phi^2) & V - f(\phi^2) - \phi^2 f'(\phi^2) \end{bmatrix} = \begin{bmatrix} V_1 & V_2 \\ V_2 & V_1 \end{bmatrix}. \qquad (4.4.13)$$

The solution $\mathbf{u}_1(x,k)$ transforms to the solution $\varphi_1(x,k)$ that satisfies the integral equation

$$\varphi_1(x,k) = e^{px}\mathbf{e}_2 - \int_{-\infty}^{x} \begin{bmatrix} g_1(y-x) & 0 \\ 0 & g_2(y-x) \end{bmatrix} \hat{V}(y)\varphi_1(y,k)dy,$$

where $\{\mathbf{e}_1,\mathbf{e}_2\}$ is a standard basis in $\mathbb{C}^2$. Let us look for the bounded solution $\varphi_2(x,k)$ in the form

$$\varphi_2(x,k) = z_0(x,k)\varphi_1(x,k) + z_1(x,k)\mathbf{e}_1, \quad \partial_x z_0(x,k) = \frac{z_2(x,k)}{\varphi_{12}(x,k)} \quad (4.4.14)$$

where $z_1(x,k)$ and $z_2(x,k)$ are new variables and $\varphi_{12}(x,k) \neq 0$ at least for large negative x.

Substitution of (4.4.14) into system (4.4.12) leads to the system of equations

$$-\left(\partial_x^2 + k^2\right)z_1 + \left(V_1 - \frac{\varphi_{11}}{\varphi_{12}}V_2\right)z_1 - 2\left(\frac{\partial_x\varphi_{11}}{\varphi_{12}} - \frac{\varphi_{11}\partial_x\varphi_{12}}{\varphi_{12}^2}\right)z_2 = 0,$$

$$-\left(\partial_x + p\right)z_2 + V_2 z_1 + \left(p - \frac{\partial_x\varphi_{12}}{\varphi_{12}}\right)z_2 = 0.$$

The Green function can be constructed for large negative values of x because the potential-free equation for z_2 has only a growing solution as $x \to -\infty$ and the new potential terms decay exponentially as $x \to -\infty$ thanks to the representation

$$\tilde{V} = \begin{bmatrix} V_1 - \frac{\varphi_{11}}{\varphi_{12}}V_2 & -2\left(\frac{\partial_x\varphi_{11}}{\varphi_{12}} - \frac{\varphi_{11}\partial_x\varphi_{12}}{\varphi_{12}^2}\right) \\ V_2 & p - \frac{\partial_x\varphi_{12}}{\varphi_{12}} \end{bmatrix} \equiv \begin{bmatrix} V_{11} & V_{12} \\ V_{21} & V_{22} \end{bmatrix}.$$

Incorporating the boundary condition for $\mathbf{u}_2(x,k)$ (and thus for $\varphi_2(x,k)$), we rewrite the differential system for $z_1(x,k)$ and $z_2(x,k)$ in the equivalent integral form

$$z_1(x,k) = e^{ikx} - \int_{-\infty}^{x} \frac{\sin k(y-x)}{k}\left(V_{11}(y,k)z_1(y,k) + V_{12}(y,k)z_2(y,k)\right)dy,$$

$$z_2(x,k) = \int_{-\infty}^{x} e^{p(y-x)}\left(V_{21}(y,k)z_1(y,k) + V_{22}(y,k)z_2(y,k)\right)dy.$$

It follows from the system of integral equations that $z_2(x,k)$ decays exponentially to zero as $x \to -\infty$ with the exponential rate of the potential $V_{21} = V_2$, while $z_1(x,k)$ remains bounded as $x \to -\infty$. Integrating the first-order equation for $z_0(x,k)$, we obtain the bounded solution $\varphi_2(x,k)$ in the final form

$$\varphi_2(x,k) = z_1(x,k)\mathbf{e}_1 + \left(\int_{x_0}^{x} \frac{z_2(y,k)dy}{\varphi_{12}(y,k)}\right)\varphi_1(x,k),$$

where x_0 is at our disposal. Note that the term $\varphi_1(x,k)$ multiplied by the integral term decays exponentially to zero as $x \to -\infty$ with the exponential rate of the potential V_2, while the integral term may grow as $x \to -\infty$. Therefore, $x_0 \neq -\infty$.

If $A(k) \neq 0$ in the asymptotic behavior (4.4.11), then $\tilde{V}(x,k)$ decays to zero exponentially fast as $x \to +\infty$, so that $z_1(x,k)$ remains bounded and $z_2(x,k)$

decays exponentially to zero as $x \to +\infty$. In this case, we can pick $x_0 = +\infty$. When the integral representation for $z_1(x, k)$ is extended to $x \to +\infty$, we obtain the limiting behavior

$$z_1(x, k) \to a(k)e^{ikx} + b(k)e^{-ikx} \quad \text{as} \quad x \to +\infty,$$

where the exponentially decaying term is not written and

$$a(k) = 1 - \frac{1}{2ik} \int_{\mathbb{R}} e^{-iky} \left(V_{11}(y, k)z_1(y, k) + V_{12}(y, k)z_2(y, k) \right) dy,$$

$$b(k) = \frac{1}{2ik} \int_{\mathbb{R}} e^{iky} \left(V_{11}(y, k)z_1(y, k) + V_{12}(y, k)z_2(y, k) \right) dy.$$

Coefficients $a(k)$ and $b(k)$ are referred to as *scattering data* and the construction of eigenfunctions of the continuous spectrum resembles the construction of inverse scattering for Lax operators (Section 1.4.1). The following limiting behavior of the solution $\mathbf{u}_2(x, k)$ follows from the construction above for any $A(k) \neq 0$:

$$\mathbf{u}_2(x, k) \to \left[a(k)e^{ikx} + b(k)e^{-ikx} \right] \mathbf{e}_+ \quad \text{as} \quad x \to +\infty. \tag{4.4.15}$$

Combining the previous computations, we have proved the following lemma.

Lemma 4.21 *There exist bounded solutions $\mathbf{u}_1(x, k)e^{-px}$ and $\mathbf{u}_2(x, k)e^{-ikx}$ for any $x \in \mathbb{R}$ and $k \in \mathbb{R}\backslash\{0\}$ of the linear system (4.4.8) that satisfy the boundary conditions (4.4.10) and (4.4.11). Moreover, if $A(k) \neq 0$, then $\mathbf{u}_2(x, k)$ also satisfies the asymptotic behavior (4.4.15).*

Since the system of equations (4.4.8) is real-valued for real Ω, we obtain immediately that

$$\mathbf{u}_1(x, k) = \bar{\mathbf{u}}_1(x, -k), \quad \mathbf{u}_2(x, k) = \bar{\mathbf{u}}_2(x, -k), \quad x \in \mathbb{R}, \quad k \in \mathbb{R},$$

so that

$$a(k) = \bar{a}(-k), \quad b(k) = \bar{b}(-k), \quad k \in \mathbb{R}. \tag{4.4.16}$$

We need two more solutions of system (4.4.8), which would exhibit similar behavior as $x \to +\infty$. If we adopt the assumption that $V(x)$ is even on $\mathbb{R}$ (merely for a simplification of algebra), these two solutions are obtained by the inversion $x \to -x$,

$$\mathbf{w}_1(x, k) = \mathbf{u}_1(-x, k), \quad \mathbf{w}_2(x, k) = \mathbf{u}_2(-x, k), \quad x \in \mathbb{R}, \quad k \in \mathbb{R}. \tag{4.4.17}$$

Exercise 4.32 Let $\mathbf{u}_1 = (U_1, W_1)$ and $\mathbf{u}_2 = (U_2, W_2)$ be any two solutions of system (4.4.8) for Ω_1 and Ω_2 respectively. Let $\mathcal{W}[\mathbf{u}_1, \mathbf{u}_2]$ be the Wronskian between the two solutions in the form

$$\mathcal{W}[\mathbf{u}_1, \mathbf{u}_2] = U_1(x)U_2'(x) - U_1'(x)U_2(x) + W_1(x)W_2'(x) - W_1'(x)W_2(x). \tag{4.4.18}$$

Show that

$$\frac{d}{dx}\mathcal{W}[\mathbf{u}_1, \mathbf{u}_2] = (\Omega_1 - \Omega_2)\left(U_1(x)W_2(x) + W_1(x)U_2(x) \right), \quad x \in \mathbb{R}, \tag{4.4.19}$$

so that $\mathcal{W}[\mathbf{u}_1, \mathbf{u}_2]$ does not depend on x if $\Omega_1 = \Omega_2$.

Lemma 4.22 *Let* $\mathbf{u}_1(x, k)$ *and* $\mathbf{u}_2(x, k)$ *be solutions of (4.4.8), which satisfy the boundary conditions (4.4.10), (4.4.11) with* $A(k) \neq 0$, *and (4.4.15). Then, for any* $k \in \mathbb{R} \backslash \{0\}$, *we have*

$$|a(k)|^2 - |b(k)|^2 = 1, \quad \bar{b}(k) = -b(k). \qquad (4.4.20)$$

Proof The first scattering relation follows from the direct computations of the Wronskian $\mathcal{W}[\mathbf{u}_2, \bar{\mathbf{u}}_2]$ for $x \to \pm\infty$:

$$\mathcal{W}[\mathbf{u}_2, \bar{\mathbf{u}}_2] = -4ik = -4ik(|a(k)|^2 - |b(k)|^2), \quad k \in \mathbb{R}.$$

The second scattering relation follows from similar computations:

$$\mathcal{W}[\mathbf{u}_2, \mathbf{w}_2] = -4ika(k), \quad W(\mathbf{u}_2, \bar{\mathbf{w}}_2) = -4ik\bar{b}(k) = 4ikb(k), \quad k \in \mathbb{R}.$$

The value $k = 0$ is to be excluded from consideration. $\qquad \square$

We shall now identify important cases when this generic construction of bounded solutions of system (4.4.8) may break.

Definition 4.9 We say that $\Omega \in [c, \infty)$ is an *embedded eigenvalue* if $A(k) = 0$ and there exists $B(k) \neq 0$ such that $\mathbf{u}_1(x, k)$ satisfies

$$\lim_{x \to +\infty} \mathbf{u}_1(x, k)e^{px} = B(k)\mathbf{e}_-.$$

We say that $\Omega \in [c, \infty)$ is a *resonance* if $A(k) \neq 0$ and there exists $b(k) \neq 0$ such that $\mathbf{u}_2(x, k)$ satisfies

$$\lim_{x \to +\infty} \mathbf{u}_2(x, k)e^{ikx} = b(k)\mathbf{e}_+.$$

Lemma 4.23 *No resonances may occur for* $\Omega \in (c, \infty)$.

Proof If $\Omega \in (c, \infty)$, that is $k \in \mathbb{R} \backslash \{0\}$, the resonance condition is equivalent to $a(k) = 0$ in the limiting behavior (4.4.15). However, the condition $a(k) = 0$ contradicts the constraint (4.4.20) for any $k \in \mathbb{R} \backslash \{0\}$. Therefore, resonance may occur only for $k = 0$ ($\Omega = c$). $\qquad \square$

Remark 4.9 Resonances may lead to bifurcations of exponentially decaying solutions of the linearized system if a perturbation to the potential terms is applied. Resonances with $k = 0$ are referred to as the edge bifurcations [107, 42], for which new isolated eigenvalues bifurcate from the end point of the continuous spectrum. Note that $a(k)$ and $b(k)$ are singular as $k \to 0$ if $k = 0$ is not a resonance and they are bounded as $k \to 0$ if $k = 0$ is a resonance.

To define the symplectic orthogonality of the bounded eigenfunctions, we use the symplectic inner product defined by

$$\mathcal{J}[\mathbf{u}_1, \mathbf{u}_2] := \int_{\mathbb{R}} \left(U_1(x)\bar{W}_2(x) + W_1(x)\bar{U}_2(x) \right) dx,$$

for any two solutions $\mathbf{u}_1 = (U_1, W_1)$ and $\mathbf{u}_2 = (U_2, W_2)$ of system (4.4.8). Although the eigenfunctions $\mathbf{u}_2(x, k)$ are not orthogonal because of the quadratic dependence of $\Omega = c + k^2$ versus k, one can correct the definition of the wave functions to regain the standard orthogonality relations of the eigenfunctions. The following lemma gives the relevant calculation.

Lemma 4.24 *Let $\mathbf{u}_2(x,k)$ and $\mathbf{w}_2(x,k)$ be solutions of system (4.4.8) for $\Omega = c + k^2$, which satisfy (4.4.10), (4.4.15), and (4.4.17). The wave functions*

$$\mathbf{v}(x,k) = \frac{1}{\sqrt{4\pi a(k)}} \begin{cases} \mathbf{u}_2(x,k), & k > 0, \\ \mathbf{w}_2(x,-k), & k < 0, \end{cases} \tag{4.4.21}$$

are orthogonal such that

$$\mathcal{J}[\mathbf{v}(\cdot,k), \mathbf{v}(\cdot,k')] = \delta(k - k'), \tag{4.4.22}$$

where $\delta(k)$ is Dirac's delta function in the distribution sense.

Proof For any solutions $\mathbf{u}(x,k)$ of system (4.4.8) with $\Omega = c + k^2$, the Wronskian identity (4.4.19) implies that

$$\mathcal{J}[\mathbf{u}(\cdot,k), \mathbf{u}(\cdot,k')] = \frac{\mathcal{W}[\mathbf{u}(\cdot,k), \bar{\mathbf{u}}(\cdot,k')]}{k^2 - (k')^2} \bigg|_{x \to -\infty}^{x \to +\infty}. \tag{4.4.23}$$

Using the standard identity for the Dirac delta function,

$$\lim_{x \to \pm\infty} \frac{e^{ikx}}{i\pi k} = \pm\delta(k), \quad k \in \mathbb{R},$$

we can see that the main contribution of the right-hand side in (4.4.23) occurs at $k' = k$ and $k' = -k$. Both contributions are to be considered separately because of the piecewise definition (4.4.21).

Without loss of generality, we take $k > 0$ and $\mathbf{u}(x,k) = \mathbf{u}_2(x,k)$. For $k' = k > 0$, we take $\mathbf{u}(x,k') = \mathbf{u}_2(x,k')$ and for $k' = -k < 0$, we take $\mathbf{u}(x,k') = \mathbf{w}_2(x,-k')$. Performing tedious but straightforward computations with the use of boundary values (4.4.10) and (4.4.15), we obtain

$$\begin{aligned} \mathcal{J}[\mathbf{u}(\cdot,k), \mathbf{u}(\cdot,k')] &= 2\pi \left[1 + |a(k)|^2 + |b(k)|^2\right] \delta(k - k') + 2\pi \left[\bar{b}(k) + b(k)\right] \delta(k + k') \\ &= 4\pi |a(k)|^2 \delta(k - k'), \end{aligned}$$

where the last equality is obtained from identities (4.4.16) and (4.4.20). The orthogonality condition (4.4.22) follows from the normalization of $\mathbf{v}(x,k)$ in the definition (4.4.21). $\qquad\square$

The branch $(-\infty, -c]$ of the continuous spectrum can be obtained from the symmetry of system (4.4.8). If $\mathbf{u} = (U, W)$ solves (4.4.8) for $\Omega > 0$, then $\sigma_3\mathbf{u} = (U, -W)$ solves (4.4.8) for $-\Omega < 0$.

With the help of Lemma 4.24, one can uniquely decompose any element in the invariant subspace of L^2 associated to the continuous spectrum of the linear problem (4.4.8) by the generalized Fourier integrals over the wave functions $\mathbf{v}(x,k)$ and $\sigma_3\mathbf{v}(x,k)$. This decomposition is given by

$$\mathbf{u}(x) = \int_{\mathbb{R}} \hat{a}_+(k)\mathbf{u}_2(x,k)dk + \int_{\mathbb{R}} \hat{a}_-(k)\sigma_3\mathbf{u}_2(x,k)dk, \tag{4.4.24}$$

where the symplectic orthogonality gives the coefficients of the decomposition,

$$\hat{a}_+(k) = \mathcal{J}[\mathbf{u}, \mathbf{v}(\cdot,k)], \quad \hat{a}_-(k) = \mathcal{J}[\mathbf{u}, \sigma_3\mathbf{v}(\cdot,k)], \quad k \in \mathbb{R}. \tag{4.4.25}$$

This representation formula implies the completeness relation for the wave functions $\mathbf{v}(x,k)$ and $\sigma_3\mathbf{v}(x,k)$,

$$\frac{1}{4\pi}\int_{\mathbb{R}}[\mathbf{v}(x,k)\otimes\bar{\mathbf{v}}(y,k)+\sigma_3\mathbf{v}(x,k)\otimes\sigma_3\bar{\mathbf{v}}_2(y,k)]\,dk = I\delta(x-y),\quad x,y\in\mathbb{R},$$

$$(4.4.26)$$

where I is the 2×2 identity matrix and

$$\mathbf{u}=\begin{bmatrix} u_1 \\ u_2 \end{bmatrix},\quad \mathbf{w}=\begin{bmatrix} w_1 \\ w_2 \end{bmatrix}:\quad \mathbf{u}\otimes\mathbf{w}:=\begin{bmatrix} u_1w_2 & u_1w_1 \\ u_2w_2 & u_2w_1 \end{bmatrix}.$$

The completeness relation (4.4.26) follows from analysis of the resolvent of the linearized operator, which has a jump across the continuous spectrum at $(-\infty,-c]\cup[c,\infty)$.

Let us illustrate the orthogonal decomposition (4.4.24) for the explicit example of the cubic focusing NLS equation,

$$i\psi_t + \psi_{xx} + 2|\psi|^2\psi = 0.$$

Therefore, we take $V(x)\equiv 0$ and $f(|\psi|^2) = 2|\psi|^2$, and we recall that $\phi(x) = \sqrt{|\omega|}\,\text{sech}(\sqrt{|\omega|}x)$. Eigenfunctions of the linear system (4.4.8) were explicitly written by Kaup [109], who proved the completeness relation (4.4.26) by explicit computations.

Since the linear system (4.4.12) admits $\hat{V}(x)$ in the explicit form

$$\hat{V}(x) = 2|\omega|\text{sech}^2(\sqrt{|\omega|}x)\begin{bmatrix} 2 & 1 \\ 1 & 2 \end{bmatrix},$$

we can set $\omega = -1$ by a scaling transformation. The eigenfunctions $\mathbf{u}_1(x,k)$ and $\mathbf{u}_2(x,k)$ of the continuous spectrum $[1,\infty)$ of the linear system (4.4.8) with $c=1$ are available in the closed form [109]:

$$\mathbf{u}_1(x;k) = e^{px}\left[\left(1-\frac{2p\exp(x)}{(p+1)^2\cosh(x)}\right)\begin{pmatrix} 1 \\ -1 \end{pmatrix} - \frac{2}{(p+1)^2\cosh^2(x)}\begin{pmatrix} 1 \\ 0 \end{pmatrix}\right]$$

and

$$\mathbf{u}_2(x;k) = e^{ikx}\left[\left(1+\frac{2ik\exp(x)}{(k-i)^2\cosh(x)}\right)\begin{pmatrix} 1 \\ 1 \end{pmatrix} + \frac{2}{(k-i)^2\cosh^2(x)}\begin{pmatrix} 1 \\ 0 \end{pmatrix}\right].$$

Looking at the limits $x\to+\infty$, we obtain

$$A(k) = \left(\frac{p-1}{p+1}\right)^2,\quad a(k) = \left(\frac{k+i}{k-i}\right)^2,\quad b(k) = 0,\quad k\in\mathbb{R}.$$

Recall that $p = \sqrt{1+\Omega}$, $k = \sqrt{\Omega-1}$, and $\Omega\in[1,\infty)$. Therefore, $A(k)\neq 0$ for all $k\in\mathbb{R}$ and no embedded eigenvalues exist. On the other hand, $k=0$ ($\Omega=1$) is the resonance according to Definition 4.9.

Exercise 4.33 Confirm the orthogonality and completeness relations (4.4.26) for the eigenfunctions $\mathbf{v}(x,k)$ and $\sigma_3\mathbf{v}(x,k)$ by explicit integration.

Combining the contributions from the discrete and continuous spectrum of the spectral problem (4.4.5), a general solution of the linearized Gross–Pitaevskii equation (4.4.4) can be written in the form $z(x,t) = u(x,t) + iw(x,t)$ with

$$
\begin{bmatrix} u(x,l) \\ w(x,t) \end{bmatrix} = \int_{\mathbb{R}} \hat{a}_+(k) e^{i\sqrt{c+k^2}\,t} \mathbf{v}(x,k)\,dk + \int_{\mathbb{R}} \hat{a}_-(k) e^{-i\sqrt{c+k^2}\,t} \sigma_3 \mathbf{v}(x,k)\,dk
$$

$$
+ \sum_j \hat{a}_j e^{\lambda_j t} \left(\mathbf{u}_j^{(1)}(x) + t\mathbf{u}_j^{(2)}(x) + \cdots + \frac{1}{(m_j-1)!} t^{m_j-1} \mathbf{u}_j^{(m_j)}(x) \right),
$$

where the summation $\sum_j$ includes each isolated and embedded eigenvalue λ_j of the spectral problem (4.4.5) with the account of their algebraic multiplicity m_j and the set of generalized eigenvectors $\{ \mathbf{u}_j^{(1)}, \mathbf{u}_j^{(2)}, ..., \mathbf{u}_j^{(m_j)} \}$.

Exercise 4.34   Write explicitly the decomposition in terms of eigenvectors (4.4.6) for the zero eigenvalue of the spectral problem (4.4.5) and show that the linear growth is absent if $(\mathbf{u}, \mathbf{w})$ satisfies the symplectic orthogonality conditions (4.4.7).

Thanks to the completeness of the eigenvectors of the spectral problem (4.4.5), the Cauchy problem for the linearized Gross–Pitaevskii equation (4.4.4) is well-posed and there exists a global unique solution

$$
z(t) \in C(\mathbb{R}, H^2(\mathbb{R})) \cap C^1(\mathbb{R}, L^2(\mathbb{R})).
$$

This implies that the series of eigenfunctions converges absolutely for any $t \in \mathbb{R}$, that is, the eigenvectors provide a Schauder basis in $L^2(\mathbb{R})$. The following theorem summarizes the linearized stability of the localized mode ϕ in the sense of Definition 4.2.

Theorem 4.14   *Consider the linearized Gross–Pitaevskii equation (4.4.4) with $V(x)$ and $\phi(x)$ decaying exponentially to zero as $|x| \to \infty$. Then, the following hold:*

(i) *If there exists at least one eigenvalue with $\mathrm{Re}(\lambda) > 0$ in the spectral problem (4.4.5), the localized mode ϕ is linearly unstable and the perturbation $z(t)$ grows exponentially as $t \to +\infty$.*

(ii) *If there exists a pair of multiple purely imaginary eigenvalues or the zero eigenvalue of algebraic multiplicity higher than two in the spectral problem (4.4.5), the localized mode ϕ is linearly unstable and the perturbation $z(t)$ grows algebraically as $t \to +\infty$.*

(iii) *If the spectrum of the spectral problem (4.4.5) has only semi-simple purely imaginary eigenvalues (isolated or embedded), the zero eigenvalue of algebraic multiplicity two and no other eigenvalues, the localized mode ϕ is linearly stable and the perturbation $z(t)$ remains bounded as $t \to +\infty$.*

Exercise 4.35   Assume that the zero eigenvalue of the spectral problem (4.4.5) has algebraic multiplicity four and geometric multiplicity one (under the condition that $\frac{d}{d\omega}\|\phi\|_{L^2}^2 = 0$). Prove that the localized mode ϕ is linearly unstable and the perturbation $z(t) = u(t) + iw(t)$ grows algebraically as $t \to +\infty$ even if (u, w) satisfy the constraints (4.4.7).

Exercise 4.36 Assume that the spectral problem (4.4.5) has a pair of purely imaginary eigenvalues of algebraic multiplicity two and geometric multiplicity one. Prove that the localized mode ϕ is linearly unstable and the perturbation $z(t)$ grows algebraically as $t \to +\infty$.

The key to the linearized analysis is the completeness property of the bounded eigenfunctions of the linearized Gross–Pitaevskii equation. We note that for a general non-self-adjoint operator, the eigenfunctions may only form a dense set in space L^2 but not a basis in L^2 [36].

The completeness property of the eigenfunctions can also be studied by other techniques. In the space of three dimensions, the wave operator formalism shows isomorphism of the linearized problem with exponentially decaying potentials and the free linearized problem without potentials [43]. Unfortunately, the wave operator methods do not work in the space of two spatial dimensions.

For problems with a purely discrete spectrum, e.g. arising in the linearization of periodic wave solutions, the completeness property of eigenfunctions can be established with the spectral theory of non-self-adjoint operators [85].

4.4.2 Orbital stability of localized modes

Let us assume again the existence of the localized mode (4.4.2) in the stationary Gross–Pitaevskii equation (4.4.3). The localized mode ϕ is a critical point of the energy functional $E_\omega(u) = H(u) - \omega Q(u)$ in function space $H^1(\mathbb{R})$ (Section 3.1), where $H(u)$ and $Q(u)$ are conserved energy and power of the Gross–Pitaevskii equation (4.4.1),

$$H(u) = \int_{\mathbb{R}} \left(|\partial_x u|^2 + V|u|^2 - \int_0^{|u|^2} f(s)ds \right) dx, \quad Q(u) = \int_{\mathbb{R}} |u|^2 dx.$$

Let $D(\omega) = E_\omega(\phi)$ and assume that the map $\mathbb{R} \ni \omega \mapsto D$ is C^2. Since ϕ is real-valued, we have

$$D'(\omega) = 2\langle \nabla_{\bar{u}} E(\phi) - \omega \nabla_{\bar{u}} Q(\phi), \partial_\omega \phi \rangle_{L^2} - Q(\phi) = -\|\phi\|_{L^2}^2 \qquad (4.4.27)$$

and

$$D''(\omega) = -\frac{d}{d\omega} \|\phi\|_{L^2}^2. \qquad (4.4.28)$$

Assume that, for this localized mode ϕ, the linearized operator L_+ has only one negative eigenvalue and no zero eigenvalues (that is, $n(L_+) = 1$ and $z(L_+) = 0$) and that the linearized operator L_- has no negative eigenvalues and a simple zero eigenvalue (that is, $n(L_-) = 0$ and $z(L_-) = 1$). Theorem 4.8 holds under these assumptions and states that the localized mode ϕ is spectrally stable if $D''(\omega) > 0$ and is spectrally unstable if $D''(\omega) < 0$.

We will now show that the variational structure of the Gross–Pitaevskii equation (4.4.1) can be used to prove orbital stability from spectral stability avoiding the complexity of the linearized stability analysis. This theory is the oldest in the stability analysis of localized modes and dates back to the first papers by Shatah & Strauss [189] and Weinstein [211], summarized in the two papers of Grillakis *et al.* [74, 75]. We will only review the stability part of the orbital stability/instability theorems. The proof of the stability theorem is similar to the proof of the first Lyapunov Theorem on stability of a center point in a dynamical system with a sign-definite energy.

Theorem 4.15 *Assume that the stationary Gross–Pitaevskii equation (4.4.3) admits an exponentially decaying solution $\phi_0 \in H^1(\mathbb{R})$ for a fixed $\omega_0 < 0$. Assume that the map $\mathbb{R} \ni \omega \mapsto \phi \in H^2(\mathbb{R})$ is C^1 near ω_0. Furthermore, assume for this ω_0 that*

$$n(L_+) = 1, \quad z(L_+) = 0, \quad n(L_-) = 0, \quad z(L_-) = 1.$$

The solution ϕ_0 is orbitally stable in energy space $H^1(\mathbb{R})$ in the sense of Definition 4.3 if $D''(\omega_0) > 0$.

Proof Let us expand the energy functional $E_\omega(\psi)$ using the decomposition

$$\psi = \phi_0 + u + iw,$$

where $(u, w) \in H^1(\mathbb{R}) \times H^1(\mathbb{R})$ are real. Then, we obtain

$$E_{\omega_0}(\psi) = D(\omega_0) + \langle L_+ u, u \rangle_{L^2} + \langle L_- w, w \rangle_{L^2} + \tilde{E}_{\omega_0}(u, w), \tag{4.4.29}$$

where the remainder term $\tilde{E}_\omega(u, w)$ contains cubic and higher-order terms in $\|u\|_{H^1}$ and $\|w\|_{H^1}$.

Let us consider the constrained energy space,

$$H_c^1 = \{u \in H^1(\mathbb{R}) : \quad \langle u, \phi_0 \rangle_{L^2} = 0\} \tag{4.4.30}$$

and use the orthogonal projection operator $P_c : L^2 \to L_c^2 \subset L^2$. Because

$$\langle L_+^{-1} \phi_0, \phi_0 \rangle_{L^2} = \langle \partial_\omega \phi_0, \phi_0 \rangle_{L^2} = -\frac{1}{2} D''(\omega_0),$$

operator $P_c L_+ P_c$ has no negative or zero eigenvalues if $D''(\omega_0) > 0$ (Theorem 4.1). Therefore, the quadratic form associated with $P_c L_+ P_c$ is coercive and

$$\exists C > 0 : \quad \forall u \in H_c^1 : \quad \langle L_+ u, u \rangle_{L^2} \geq C \|u\|_{H^1}^2. \tag{4.4.31}$$

On the other hand, ϕ_0 is an eigenvector of L_- for the simple zero eigenvalue, hence, operator $P_c L_- P_c$ also has no negative or zero eigenvalues. We infer again that the quadratic form associated with $P_c L_- P_c$ is also coercive and

$$\exists C > 0 : \quad \forall w \in H_c^1 : \quad \langle L_- w, w \rangle_{L^2} \geq C \|w\|_{H^1}^2. \tag{4.4.32}$$

Let $T(\theta)\phi_0 := e^{-i\theta}\phi_0$ be the orbit of the localized mode ϕ_0 in $H^1(\mathbb{R})$ for a fixed $\epsilon > 0$. Let $\Phi_\epsilon \subset H^1(\mathbb{R})$ be an open ϵ-neighborhood of the orbit defined by

$$\Phi_\epsilon = \{\psi \in H^1(\mathbb{R}) : \quad \inf_{\theta \in \mathbb{R}} \|\psi - e^{-i\theta}\phi_0\|_{H^1} < \epsilon\}. \tag{4.4.33}$$

Let us consider the decomposition

$$\forall \psi \in \Phi_\epsilon : \quad \psi = (\phi(\omega) + u + iw) e^{-i\theta}, \tag{4.4.34}$$

where $\phi(\omega)$ is the map $\mathbb{R} \ni \omega \mapsto \phi \in H^2(\mathbb{R})$ near $\omega = \omega_0$ and the parameters $(\omega, \theta) \in \mathbb{R}^2$ are defined by the constraints $(u, w) \in H_c^1 \times H_c^1$. The constraints can be rewritten as a scalar complex-valued equation,

$$F(\theta, \omega) := \langle e^{i\theta}\psi, \phi_0 \rangle_{L^2} - \langle \phi(\omega), \phi_0 \rangle_{L^2} = 0. \qquad (4.4.35)$$

For any small $\epsilon > 0$ and any $\psi \in \Phi_\epsilon$, let θ_0 be the argument of $\inf_{\theta \in \mathbb{R}} \|\psi - e^{-i\theta}\phi_0\|_{H^1}$. Therefore, if $\psi = e^{-i\theta_0}\phi_0 + \tilde{\psi}$, then $\|\tilde{\psi}\|_{H^1} < \epsilon$ and

$$F(\theta_0, \omega_0) = \langle e^{i\theta_0}\tilde{\psi}, \phi_0 \rangle_{L^2} \quad \Rightarrow \quad \exists C > 0 : \quad |F(\theta_0, \omega_0)| \le C\epsilon.$$

On the other hand, derivatives of $F(\theta, \omega)$ at (θ_0, ω_0) can be computed in the form

$$\partial_\theta F(\theta_0, \omega_0) = i\|\phi_0\|_{L^2}^2 + i\langle e^{i\theta_0}\tilde{\psi}, \phi_0 \rangle_{L^2},$$

$$\partial_\omega F(\theta_0, \omega_0) = -\langle \partial_\omega \phi_0, \phi_0 \rangle_{L^2} = \frac{1}{2}D''(\omega_0).$$

Since both derivatives are nonzero for small $\epsilon > 0$, the Implicit Function Theorem states that there is a unique solution of equation (4.4.35) for (θ, ω) near (θ_0, ω_0) for small $\epsilon > 0$ such that

$$\exists C > 0 : \quad \forall \psi \in \Phi_\epsilon : \quad |\omega - \omega_0| \le C\epsilon, \quad |\theta - \theta_0| \le C\epsilon. \qquad (4.4.36)$$

Expanding the energy function $E_\omega(\psi)$ using the decomposition (4.4.34), we obtain

$$E_\omega(\psi) = D(\omega) + \langle L_+ u, u \rangle_{L^2} + \langle L_- w, w \rangle_{L^2} + \tilde{E}_\omega(u, w),$$

where operators $L_\pm$ are now computed at ϕ and ω. Thanks to computation (4.4.27), we know that

$$E_\omega(\phi(\omega)) - E_\omega(\phi_0) = D(\omega) - D(\omega_0) + (\omega - \omega_0)Q(\phi_0)$$
$$= \frac{1}{2}D''(\omega_0)(\omega - \omega_0)^2 + \mathcal{O}(\omega - \omega_0)^3.$$

On the other hand, the quadratic forms involving L_+ and L_- can be computed at $\phi = \phi_0$ and $\omega = \omega_0$ with the truncation error of the order

$$\mathcal{O}(|\omega - \omega_0|(\|u\|_{H^1}^2 + \|w\|_{H^1}^2)).$$

As a result, we obtain the expansion

$$E_\omega(\psi) - E_\omega(\phi_0) = \frac{1}{2}D''(\omega_0)(\omega - \omega_0)^2 + \langle L_+ u, u \rangle_{L^2} + \langle L_- w, w \rangle_{L^2}$$
$$+ \mathcal{O}\left((\omega - \omega_0)^3 + (\omega - \omega_0)(\|u\|_{H^1}^2 + \|w\|_{H^1}^2) + (\|u\|_{H^1} + \|w\|_{H^1})^3\right).$$

We note that for all $a \in \mathbb{R}$ and all $u \in H_c^1$,

$$\langle L_+(a\partial_\omega \phi_0 + u), (a\partial_\omega \phi_0 + u) \rangle_{L^2} = -\frac{1}{2}a^2 D''(\omega_0) + \langle L_+ u, u \rangle_{L^2}. \qquad (4.4.37)$$

Thanks to bounds (4.4.31) and (4.4.32), equality (4.4.37), and the positivity of $D''(\omega_0)$, we obtain the lower bound

$$\langle L_+ u, u \rangle_{L^2} + \langle L_- w, w \rangle_{L^2} \ge C\left(\|(\omega - \omega_0)\partial_\omega \phi_0 + u\|_{H^1}^2 + \|w\|_{H^1}^2\right)$$
$$= C\left(\|\phi(\omega) - \phi_0 + u\|_{H^1}^2 + \|w\|_{H^1}^2\right) + \mathcal{O}((\omega - \omega_0)^3, (\omega - \omega_0)^2\|u\|_{H^1})$$
$$= C\|e^{i\theta}\psi - \phi_0\|_{H^1}^2 + \mathcal{O}((\omega - \omega_0)^3, (\omega - \omega_0)^2\|u\|_{H^1}).$$

As a result, for sufficiently small $\epsilon > 0$, there is $C > 0$ such that for all $\psi \in \Phi_\epsilon$, we have

$$E_\omega(\psi) - E_\omega(\phi_0) \geq \frac{1}{2}D''(\omega_0)(\omega - \omega_0)^2 + C \inf_{\theta \in \mathbb{R}} \|\psi - e^{-i\theta}\phi_0\|_{H^1}^2. \quad (4.4.38)$$

If ψ is a time-dependent solution of the Gross–Pitaevskii equation (4.4.1), then $E(\psi)$ and $Q(\psi)$ are constant in time. The left-hand side of the bound (4.4.38) can be related to the conserved quantities by

$$E_\omega(\psi) - E_\omega(\phi_0) = E(\psi) - E(\phi_0) + \omega_0(Q(\psi) - Q(\phi_0)) + (\omega - \omega_0)(Q(\psi) - Q(\phi_0)).$$

The linear term in $(\omega - \omega_0)$ can be combined with the positive quadratic term on the right-hand side of the bound (4.4.38) to obtain

$$C \inf_{\theta \in \mathbb{R}} \|\psi - e^{-i\theta}\phi_0\|_{H^1}^2 \leq E(\psi) - E(\phi_0) + \omega_0(Q(\psi) - Q(\phi_0)) + \frac{[Q(\psi) - Q(\phi_0)]^2}{2D''(\omega_0)},$$

and

$$\frac{1}{2}D''(\omega_0)\left(\omega - \omega_0 - \frac{Q(\psi) - Q(\phi_0)}{D''(\omega_0)}\right)^2$$
$$\leq E(\psi) - E(\phi_0) + \omega_0(Q(\psi) - Q(\phi_0)) + \frac{[Q(\psi) - Q(\phi_0)]^2}{2D''(\omega_0)}.$$

Let $\psi|_{t=0} = \psi_0$ be the initial data for the Gross–Pitaevskii equation (4.4.1). Then, $E(\psi) = E(\psi_0)$ and $Q(\psi) = Q(\psi_0)$ for all $t \geq 0$. For any $\epsilon > 0$, there is $\delta > 0$ such that for any $\psi_0 \in \Phi_\delta$, we have $\psi \in \Phi_\epsilon$ for all $t \geq 0$ under the condition

$$\exists C > 0: \quad C\epsilon^2 \leq E(\psi_0) - E(\phi_0) + \omega_0(Q(\psi_0) - Q(\phi_0)) + \frac{[Q(\psi_0) - Q(\phi_0)]^2}{2D''(\omega_0)},$$

where the right-hand side goes to zero as $\delta \to 0$. This is precisely the orbital stability of the localized mode ϕ_0 in the sense of Definition 4.3. $\qquad \square$

Under the same assumptions, the instability theorem from [74, 75] states that the localized solution ϕ_0 for a fixed $\omega_0 < 0$ is unstable if $D''(\omega_0) < 0$. Since the spectral stability problem has a real positive eigenvalue under the same condition $D''(\omega_0) < 0$, the instability theorem is not surprising. It simply confirms that the spectral and linearized instabilities imply orbital instability. The proof of the instability theorem is similar to the proof of the second Lyapunov Theorem on instability of a saddle point in a dynamical system with a sign-indefinite energy function.

Exercise 4.37 Prove orbital stability of the fundamental localized mode of the discrete nonlinear Schrödinger equation,

$$i\dot{\psi}_n + \epsilon(\psi_{n+1} - 2\psi_n + \psi_{n-1}) + |\psi_n|^2\psi_n = 0, \quad n \in \mathbb{Z},$$

which converges as $\epsilon \to 0$ to the solution $\psi_n(t) = e^{it}\delta_{n,0}$ supported at the node $n = 0$.

4.4.3 Asymptotic stability of localized modes

Let us assume again the existence of the localized mode (4.4.2) in the stationary Gross–Pitaevskii equation (4.4.3). Assume further that the spectral stability problem (4.4.5) has no nonzero eigenvalues and no resonances at the end points of the continuous spectrum in the sense of Definition 4.9. We also assume that the condition

$$D''(\omega) = -\frac{d}{d\omega} \|\phi\|_{L^2}^2 > 0$$

is satisfied so that the localized mode ϕ is stable spectrally, linearly, and orbitally.

To simplify computations, we substitute $f(|\phi|^2) = |\psi|^{2p}$ into the Gross–Pitaevskii equation (4.4.1),

$$i\psi_t = -\psi_{xx} + V(x)\psi - |\psi|^{2p}\psi, \qquad (4.4.39)$$

where $p > 0$. Under the above restrictions, we shall prove the asymptotic stability of the localized mode ϕ in the sense of Definition 4.4.

Asymptotic stability of solitary waves in nonlinear Schrödinger equations has been considered in many papers. The latest development includes analysis of the one-dimensional NLS equation by Buslaev & Sulem [27] for $p \geq 4$ and by Mizumachi [141] and Cuccagna [41] for $p \geq 2$.

Using a fixed bounded potential $V(x)$, Mizumachi [141] proved asymptotic stability of small bound states bifurcating from the lowest eigenvalue of the Schrödinger operator $L = -\partial_x^2 + V(x)$ (Section 3.2.1). He needed only the spectral theory of the self-adjoint operator since spectral projections and small nonlinear terms were controlled in the corresponding norm. Pioneering works along the same lines are attributed to Soffer & Weinstein [193, 194, 195], Pillet & Wayne [169], and Yau & Tsai [218, 219, 220].

Compared to this approach, Cuccagna [41] proved the asymptotic stability of nonlinear symmetric bound states in the energy space of the nonlinear Schrödinger equation with $V(x) \equiv 0$. He invoked the spectral theory of non-self-adjoint operators arising in the linearization of the nonlinear Schrödinger equation, following earlier works of Buslaev & Perelman [25, 26], Buslaev & Sulem [27], and Gang & Sigal [62, 63]. If additional isolated eigenvalues are present in the spectral stability problem, these eigenvalues can be studied with the Fermi Golden Rule by the normal form transformations [27, 41, 62, 63].

Let us decompose the solution ψ to the Gross–Pitaevskii equation (4.4.39) into a sum of the localized mode ϕ with time-varying parameters (ω, θ) and the dispersive remainder z using the substitution

$$\psi(x,t) = e^{-i\theta(t)} \left(\phi(x; \omega(t)) + z(x,t) \right), \qquad (4.4.40)$$

where parameters $(\omega, \theta)(t) : \mathbb{R}_+ \to \mathbb{R}^2$ represent a two-dimensional orbit of the stationary solutions $e^{-i(\omega t + \theta)}\phi$ (their time evolution will be specified later) and $z(x,t) : \mathbb{R} \times \mathbb{R}_+ \to \mathbb{C}$ is a solution of the time evolution equation

$$
\begin{aligned}
iz_t = &-z_{xx} + Vz - \omega z - \phi^{2p}z - p\phi^{2p}(z + \bar{z}) \\
&+ (\omega - \dot{\theta})(\phi + z) - i\dot{\omega}\partial_\omega\phi - N(z),
\end{aligned}
\qquad (4.4.41)
$$

where

$$N(z) = |\phi + z|^{2p}(\phi + z) - \phi^{2p+1} - \phi^{2p}z - p\phi^{2p}(z + \bar{z}).$$

The linearized evolution is characterized by the linearized Gross–Pitaevskii equation (4.4.4), which transforms to the non-self-adjoint eigenvalue problem (4.4.5) in variables $u = \mathrm{Re}(z)$ and $w = \mathrm{Im}(z)$.

To determine the time evolution of varying parameters (ω, θ) in the evolution equation (4.4.41), we shall add conditions (4.4.7) that tell us that z is symplectically orthogonal to the two-dimensional null space of the linearized problem (4.4.5), or equivalently,

$$\langle \mathrm{Re}(z), \phi \rangle_{L^2} = \langle \mathrm{Im}(z), \partial_\omega \phi \rangle_{L^2} = 0. \tag{4.4.42}$$

Note that the symplectic orthogonality (4.4.42) is different from the constraints on $\mathrm{Im}(z) \equiv w \in L_c^2$ used in the proof of the orbital stability theorem (Theorem 4.15). We shall prove that the symplectic orthogonality conditions (4.4.42) define a unique decomposition (4.4.40) if ψ belongs to a neighborhood of the localized mode ϕ_0 for a fixed $\omega_0 < 0$. (This neighborhood was denoted by Φ_ϵ in Section 4.4.2.)

Lemma 4.25 *Fix $\omega_0 < 0$ and denote $\phi(x; \omega_0)$ by ϕ_0. There exists $\delta > 0$ such that any $\psi \in H^2(\mathbb{R})$ satisfying*

$$\|\psi - \phi_0\|_{H^2} \leq \delta \tag{4.4.43}$$

can be uniquely decomposed by (4.4.40) and (4.4.42). Moreover, there exists $C > 0$ such that

$$|\omega - \omega_0| \leq C\delta, \quad |\theta| \leq C\delta, \quad \|z\|_{H^2} \leq C\delta \tag{4.4.44}$$

and the map $H^2(\mathbb{R}) \ni \psi \mapsto (\omega, \theta, z) \in \mathbb{R} \times \mathbb{R} \times H^2(\mathbb{R})$ is a C^1 diffeomorphism.

Proof Let us rewrite the decomposition (4.4.40) in the form

$$z = e^{i\theta}(\psi - \phi_0) + (e^{i\theta}\phi_0 - \phi). \tag{4.4.45}$$

First, we show that the constraints (4.4.42) give unique values of (ω, θ) satisfying bounds (4.4.44) provided that bound (4.4.43) holds. To do so, we rewrite (4.4.42) and (4.4.45) as a fixed-point equation $\mathbf{F}(\omega, \theta) = \mathbf{0}$, where $\mathbf{F}(\omega, \theta) : \mathbb{R}^2 \to \mathbb{R}^2$ is given by

$$\mathbf{F}(\omega, \theta) = \mathbf{F}_1(\omega, \theta) + \mathbf{F}_2(\omega, \theta),$$

with

$$\mathbf{F}_1(\omega, \theta) = \begin{bmatrix} \langle \phi_0 \cos\theta - \phi, \phi \rangle_{L^2} \\ \langle \phi_0 \sin\theta, \partial_\omega \phi \rangle_{L^2} \end{bmatrix}, \quad \mathbf{F}_2(\omega, \theta) = \begin{bmatrix} \langle \mathrm{Re}(\psi - \phi_0)e^{i\theta}, \phi \rangle_{L^2} \\ \langle \mathrm{Im}(\psi - \phi_0)e^{i\theta}, \partial_\omega \phi \rangle_{L^2} \end{bmatrix}.$$

We note that the map $\mathbb{R}^2 \ni (\omega, \theta) \mapsto \mathbf{F}_1 \in \mathbb{R}^2$ is C^1 such that $\mathbf{F}_1(\omega_0, 0) = \mathbf{0}$ and

$$D\mathbf{F}_1(\omega_0, 0) = \begin{bmatrix} \langle -\partial_\omega \phi_0, \phi_0 \rangle_{L^2} & 0 \\ 0 & \langle \phi_0, \partial_\omega \phi_0 \rangle_{L^2} \end{bmatrix}.$$

On the other hand, there is $C > 0$ such that for all $(\omega, \theta) \in \mathbb{R}^2$, we have

$$\|\mathbf{F}_2(\omega, \theta)\| \leq C\delta,$$

thanks to the bound (4.4.43). Since $\langle \partial_\omega \phi_0, \phi_0 \rangle_{L^2} < 0$, $D\mathbf{F}_1(\omega_0, 0)$ is invertible. By the Implicit Function Theorem, there exists a unique root of $\mathbf{F}(\omega, \theta) = \mathbf{0}$ near $(\omega_0, 0)$ for any ψ satisfying (4.4.43) such that the first two bounds (4.4.44) are satisfied and the map $H^2(\mathbb{R}) \ni \psi \mapsto (\omega, \theta) \in \mathbb{R}^2$ is a C^1 diffeomorphism thanks to the fact that $\mathbf{F}$ is linear in ψ. Finally, a unique z and the third bound (4.4.44) follow from the representation (4.4.45) and the triangle inequality. $\square$

Recall that there exists a $T > 0$ and a local solution

$$\psi(t) \in C([0, T], H^2(\mathbb{R})) \cap C^1([0, T], L^2(\mathbb{R}))$$

to the Gross–Pitaevskii equation (4.4.39) (Section 1.3.1). Assuming that

$$(\omega, \theta) \in C^1([0, T], \mathbb{R}^2),$$

we define the time evolution of (ω, θ) from the symplectic projections (4.4.42) of the evolution equation (4.4.41). The resulting system is written in the matrix–vector form,

$$\mathbf{A}(\omega, z) \left[\begin{array}{c} \dot\omega \\ \dot\theta - \omega \end{array} \right] = \mathbf{g}(\omega, z), \tag{4.4.46}$$

where

$$\mathbf{A}(\omega, z) = \left[\begin{array}{cc} \langle \partial_\omega \phi, \phi \rangle - \langle \mathrm{Re}(z), \partial_\omega \phi \rangle_{L^2} & \langle \mathrm{Im}(z), \phi \rangle_{L^2} \\ \langle \mathrm{Im}(z), \partial_\omega^2 \phi \rangle_{L^2} & \langle (\phi + \mathrm{Re}(z)), \partial_\omega \phi \rangle_{L^2} \end{array} \right]$$

and

$$\mathbf{g}(\omega, z) = - \left[\begin{array}{c} \langle \mathrm{Im}(N(z)), \phi \rangle_{L^2} \\ \langle \mathrm{Re}(N(z)), \partial_\omega \phi \rangle_{L^2} \end{array} \right].$$

Using an elementary property for power functions, for all $a, b \in \mathbb{C}$ there is $C > 0$ such that

$$\left| |a+b|^{2p}(a+b) - |a|^{2p}a - (1+p)|a|^{2p}b - p|a|^{2p-2}a^2\bar{b} \right| \leq C(|a|^{2p-1}|b|^2 + |b|^{2p+1}).$$

As a result, there is $C > 0$ such that the vector fields of the evolution equations (4.4.41) and (4.4.46) are bounded by

$$|N(z)| \leq C \left(|\phi^{2p-1}z^2| + |z|^{2p+1} \right), \tag{4.4.47}$$

$$\|\mathbf{g}(\omega, z)\| \leq C \left(\|\phi^{2p}z^2\|_{L^1} + \|\phi z^{2p+1}\|_{L^1} \right). \tag{4.4.48}$$

Similar to the proof of Lemma 4.25, it follows that if z is small in L^2, then $\mathbf{A}(\omega, z)$ is invertible and there is $C > 0$ such that solutions of system (4.4.46) enjoy the estimate

$$|\dot\omega| + |\dot\theta - \omega| \leq C \left(\|\phi^{2p}z^2\|_{L^1} + \|\phi z^{2p+1}\|_{L^1} \right). \tag{4.4.49}$$

The bound (4.4.49) shows that if $\delta > 0$ is small and $\sup_{t \in [0,T]} \|z\|_{H^2} \leq C\delta$ for some $C > 0$, then

$$|\omega(t) - \omega(0)| \leq C(T)\delta^2, \quad \left| \theta(t) - \int_0^t \omega(t')dt' \right| \leq C(T)\delta^2, \quad t \in [0, T],$$

for some $C(T) = \mathcal{O}(T)$ as $T \to \infty$. These bounds are smaller than bounds (4.4.44) of Lemma 4.25 if T is finite. They become comparable with bounds (4.4.44) if $T = \mathcal{O}(\delta^{-1})$ as $\delta \to 0$. Our main task is to extend these bounds globally with $C(T) = \mathcal{O}(1)$ as $T = \infty$.

By Theorem 4.15, the localized mode $\phi_0 \in H^1(\mathbb{R})$ is orbitally stable, hence a trajectory of the Gross–Pitaevskii equation (4.4.39) originating from a point in a local neighborhood of the localized mode ϕ_0 remains in a local neighborhood of the orbit $e^{-i\theta}\phi_0$ for all $t \in \mathbb{R}_+$. To prove the main result on asymptotic stability, we need to show that the trajectory approaches the orbit $e^{-i\theta}\phi_\infty$ as $t \to \infty$, where ϕ_ω corresponds to ϕ for the value $\omega = \omega_\infty$ in a local neighborhood of the value $\omega = \omega_0$. The following theorem gives asymptotic stability of the localized mode ϕ_0.

Theorem 4.16 *Assume that $V(x)$ decays exponentially to zero as $x \to \pm\infty$ and supports no end-point resonances at the continuous spectrum. Fix $p \geq 2$, $\omega_0 < 0$, and small $\delta > 0$ such that $\theta(0) = 0$, $\omega(0) = \omega_0$, and*

$$\|u_0 - \phi_0\|_{H^2} \leq \delta.$$

There exist $\omega_\infty < 0$ near $\omega_0 < 0$, $(\omega, \theta) \in C^1(\mathbb{R}_+, \mathbb{R}^2)$, and

$$z(t) = e^{i\theta(t)}\psi(t) - \phi(\cdot; \omega(t)) \in C(\mathbb{R}_+, H^2(\mathbb{R})) \cap C^1(\mathbb{R}_+, L^2(\mathbb{R})) \cap L^4(\mathbb{R}_+, L^\infty(\mathbb{R}))$$

such that $\psi(t)$ solves the Gross–Pitaevskii equation (4.4.39) and

$$\lim_{t\to\infty} \omega(t) = \omega_\infty, \qquad \lim_{t\to\infty} \|\psi(\cdot, t) - e^{-i\theta(t)}\phi(\cdot; \omega(t))\|_{L^\infty} = 0.$$

In order to prove Theorem 4.16, we need dispersive decay estimates for the linearized Gross–Pitaevskii equation (4.4.4). Pointwise decay estimates were obtained by Buslaev & Perelman [25, 26] and the Strichartz estimates were considered by Mizumachi [141] and Cuccagna [41].

Let $P_+ : L^2 \to L^2$ be the symplectically orthogonal projection to the invariant subspace associated to the continuous spectrum of the linear problem (4.4.8) on $[c, \infty)$. From Section 4.4.1, we have

$$\forall \mathbf{u} \in L^2(\mathbb{R}, \mathbb{C}^2): \quad (e^{-iHt}P_+\mathbf{u})(x) := \int_{\mathbb{R}} e^{-i(c+k^2)t}\hat{a}_+(k)\mathbf{v}(x, k)dk,$$

where $\hat{a}_+(k) = \mathcal{J}[\mathbf{u}, \mathbf{v}(\cdot, k)]$ and

$$H = \begin{bmatrix} 0 & L_- \\ L_+ & 0 \end{bmatrix}.$$

We need the following definition to set up the dispersive decay estimates.

Definition 4.10 We say that (r, s) is a *Strichartz pair* for the Gross–Pitaevskii equation if $2 \leq r, s \leq \infty$ and

$$\frac{4}{r} + \frac{2}{s} \leq 1.$$

In particular, $(r, s) = (4, \infty)$ and $(r, s) = (\infty, 2)$ are end-point Strichartz pairs.

The linearized evolution $e^{-iHt}P_+\mathbf{u}$ satisfies the pointwise dispersive decay estimates [26], which are translated to the time-averaged dispersive decay estimates. Let us define the function space $L_t^p L_x^q$ for some $p, q \geq 2$ by

$$\|f\|_{L_t^p L_x^q} = \left(\int_0^T \|f(\cdot, t)\|_{L_x^q}^p dt \right)^{1/p}, \quad \|f\|_{L_x^q L_t^p} = \left(\int_{\mathbb{R}} \|f(x, \cdot)\|_{L_t^p}^q dx \right)^{1/q},$$

for any $T > 0$. The time-averaged dispersive decay estimates (or simply, *Strichartz estimates*) are given by the following lemma [141, 41].

Lemma 4.26 *Assume that $V(x)$ decays exponentially to zero as $x \to \pm\infty$ and supports no end-point resonances at the continuous spectrum in the sense of Definition 4.9. There exists a constant $C > 0$ such that*

$$\left\| e^{-iHt} P_+ \mathbf{f} \right\|_{L_t^4 L_x^\infty \cap L_t^\infty L_x^2} \leq C \|\mathbf{f}\|_{L_x^2}, \tag{4.4.50}$$

$$\left\| \int_0^t e^{-iH(t-s)} P_+ \mathbf{g}(s) ds \right\|_{L_t^4 L_x^\infty \cap L_t^\infty L_x^2} \leq C \|\mathbf{g}\|_{L_t^{4/3} L_x^1 + L_t^1 L_x^2}. \tag{4.4.51}$$

To control the evolution of the varying parameters (ω, θ), additional time-averaged estimates are needed in one dimension, because the time decay provided by the end-point Strichartz estimates is not sufficient to guarantee the integrability of $\dot\omega(t)$ and $\dot\theta(t) - \omega(t)$. Unless $\dot\omega \in L_t^1$ and $\dot\theta - \omega \in L_t^1$, the arguments on the decay of various norms of z satisfying the time evolution problem (4.4.41) cannot be closed. The additional time-averaged estimates were obtained by Mizumachi [141] and Cuccagna [41].

Lemma 4.27 *Under the same assumptions on V, there exists a constant $C > 0$ such that*

$$\left\| \langle x \rangle^{-3/2} e^{-iHt} P_+ \mathbf{f} \right\|_{L_x^\infty L_t^2} \leq C \|\mathbf{f}\|_{L_x^2}, \tag{4.4.52}$$

$$\left\| \langle x \rangle^{-3/2} \int_0^t e^{-iH(t-s)} P_+ \mathbf{g}(s) ds \right\|_{L_x^\infty L_t^2} \leq C \|\mathbf{g}\|_{L_t^1 L_x^2}, \tag{4.4.53}$$

$$\left\| \partial_x e^{-iHt} P_+ \mathbf{f} \right\|_{L_x^\infty L_t^2} \leq C \|\mathbf{f}\|_{H_x^{1/2}}, \tag{4.4.54}$$

$$\left\| \langle x \rangle^{-1} \int_0^t e^{-iH(t-s)} P_+ \mathbf{g}(s) ds \right\|_{L_x^\infty L_t^2} \leq C \|\langle x \rangle \mathbf{g}\|_{L_x^1 L_t^2}, \tag{4.4.55}$$

$$\left\| \partial_x \int_0^t e^{-iH(t-s)} P_+ \mathbf{g}(s) ds \right\|_{L_x^\infty L_t^2} \leq C \|\mathbf{g}\|_{L_x^1 L_t^2}, \tag{4.4.56}$$

$$\left\| \int_0^t e^{-iH(t-s)} P_+ \mathbf{g}(s) ds \right\|_{L_t^4 L_x^\infty \cap L_t^\infty L_x^2} \leq C \|\langle x \rangle^5 \mathbf{g}\|_{L_t^2 L_x^2}, \tag{4.4.57}$$

where $\langle x \rangle = (1 + x^2)^{1/2}$.

Remark 4.10 Bound (4.4.53) is added to the list of useful bounds thanks to the recent work [112] in the context of the discrete nonlinear Schrödinger equations.

Using dispersive decay estimates of Lemmas 4.26 and 4.27, we can now prove Theorem 4.16.

Proof of Theorem 4.16 Let $\mathbf{z} = (z, \bar{z})$ and write the evolution problem for $\mathbf{z}$ as

$$i\sigma_3 \dot{\mathbf{z}} = \hat{H}_\omega \mathbf{z} - (\dot{\theta} - \omega)\mathbf{z} + \mathbf{g}_1 + \mathbf{g}_2 + \mathbf{g}_3, \qquad (4.4.58)$$

where the new operator $\hat{H}_\omega = -\partial_x^2 - \omega + \hat{V}_\omega(x)$ contains the potential $\hat{V}_\omega(x)$ defined by (4.4.13) (Section 4.4.1) and

$$\mathbf{g}_1 = -(\dot{\theta} - \omega)\phi \begin{bmatrix} 1 \\ 1 \end{bmatrix}, \quad \mathbf{g}_2 = -i\dot{\omega}\partial_\omega\phi \begin{bmatrix} 1 \\ -1 \end{bmatrix}, \quad \mathbf{g}_3 = -\begin{bmatrix} N(z) \\ \bar{N}(z) \end{bmatrix}.$$

Since ω is time-dependent, the operator $\hat{H}_\omega$ is time-dependent too. However, by Theorem 4.15, we know that $\omega(t)$ remains locally close to the initial value $\omega(0) = \omega_0$ for all $t \geq 0$. Therefore, we represent

$$\hat{H}_\omega = \hat{H}_0 - (\omega - \omega_0) + \hat{V}_\omega(x) - \hat{V}_{\omega_0}(x),$$

where $\hat{H}_0$ is obtained from $\hat{H}$ after ω is replaced by ω_0 and, for any ω near ω_0 and any $r \geq 0$, there is $C_r > 0$ such that

$$\|\langle x \rangle^r (\hat{V}_\omega - \hat{V}_{\omega_0})\|_{L^\infty \cap L^1} \leq C_r |\omega - \omega_0|. \qquad (4.4.59)$$

Unlike the term $\dot{\theta} - \omega$, the other term $(\omega - \omega_0)$ is not in L^1 and it has no spatial decay in x. It introduces rotations of the phase of $z(t)$. Let us consider the linear evolution problem

$$\begin{cases} i\sigma_3 \dot{\mathbf{z}}(t) = \hat{H}_0 \mathbf{z} - (\omega - \omega_0)\mathbf{z}, \\ \mathbf{z}(0) = \mathbf{z}_0, \end{cases} \qquad (4.4.60)$$

for $\mathbf{z}_0 = (z_0, \bar{z}_0)$ and $z_0 \in H^2(\mathbb{R})$. If we denote

$$\hat{H}(t) := e^{-i\sigma_3 \int_0^t [\omega(t') - \omega_0] dt'} \hat{H}_0 e^{i\sigma_3 \int_0^t [\omega(t') - \omega_0] dt'},$$

then the solution of the linear evolution problem (4.4.60) is given by

$$\mathbf{z}(t) = e^{i\sigma_3 \int_0^t [\omega(t') - \omega_0] dt'} e^{-i\sigma_3 \int_0^t \hat{H}(t') dt'} \mathbf{z}_0 \equiv e^{-i\sigma_3 \hat{H} t} \mathbf{z}_0,$$

where the last notation is introduced to simplify the representation. Thanks to the proximity of $\omega(t)$ to ω_0 for all $t \geq 0$, we can still use the dispersive decay estimates of Lemmas 4.26 and 4.27 for the operator $\sigma_3 \hat{H}$ acting on the invariant subspace of $L^2(\mathbb{R})$ associated to the continuous spectrum of $\sigma_3 \hat{H}_\omega$ for any $t \geq 0$. Note that the modulation equations (4.4.46) guarantee that the right-hand side of the time evolution equation (4.4.58) belongs to the continuous spectrum of operator $\sigma_3 \hat{H}_\omega$ for any $t \geq 0$.

Let $P_c(\omega)$ be the orthogonal projection operator to the continuous spectrum of $\sigma_3 \hat{H}_\omega$ for any $t \geq 0$. By Duhamel's principle, equation (4.4.58) can be written in the integral form

$$\mathbf{z}(t) = e^{-i\sigma_3 \hat{H} t} \mathbf{z}_0 - i \int_0^t e^{-i\sigma_3 \hat{H}(t-s)} P_c(\omega)\sigma_3 \left(\mathbf{f}_1(s) + \mathbf{f}_2(s)\right) ds$$

$$- i \int_0^t e^{-i\sigma_3 \hat{H}(t-s)} P_c(\omega)\sigma_3 \left(\mathbf{g}_1(s) + \mathbf{g}_2(s) + \mathbf{g}_3(s)\right) ds, \qquad (4.4.61)$$

where $\mathbf{f}_1 = (\hat{V}_\omega - \hat{V}_{\omega_0})\mathbf{z}$ and $\mathbf{f}_2 = -(\dot{\theta} - \omega)\mathbf{z}$. We introduce the norms

$$M_1 = \|z\|_{L_t^4 L_x^\infty}, \quad M_2 = \|z\|_{L_t^\infty H_x^1}, \quad M_3 = \|\langle x \rangle^{-3/2} z\|_{L_x^\infty L_t^2},$$

$$M_4 = \|\omega - \omega_0\|_{L_t^\infty}, \quad M_5 = \left\| \theta - \int_0^t \omega(t')dt' \right\|_{L_t^\infty},$$

where the integration is performed on an interval $[0, T]$ for any $T \in (0, \infty)$. Our goal is to show that $\dot{\omega}, \dot{\theta} - \omega \in L_t^1$ and there exists T-independent constant $C > 0$ such that

$$M_1 + M_2 \leq C \left(\|z_0\|_{H_x^1} + M_1^2 + M_3^2 \right), \tag{4.4.62}$$

$$M_3 \leq C \left(\|z_0\|_{H_x^1} + M_2 M_3 \right), \tag{4.4.63}$$

$$M_4 + M_5 \leq C M_3^2, \tag{4.4.64}$$

where only quadratic terms are shown on the right-hand sides.

The estimates (4.4.62)–(4.4.64) allow us to conclude, by elementary continuation arguments, that

$$M_1 + M_2 + M_3 \leq C \|z_0\|_{H_x^1} \leq C\delta$$

and

$$\|\omega - \omega_0\|_{L_t^\infty} \leq C\delta^2, \quad \left\| \theta - \int_0^t \omega(t')dt' \right\|_{L_t^\infty} \leq C\delta^2,$$

for any $T \in (0, \infty)$. In particular, we have $z(t) \in L^4([0, T], L^\infty(\mathbb{R}))$.

Let us consider now the continuity of the parameterization $t \to [\omega(t), \theta(t), z(\cdot, t)]$. By Theorem 1.2 (Section 1.3.1), there exist $T > 0$ and a solution of the Gross–Pitaevskii equation (4.4.39) such that

$$\psi(t) \in C([0, T], H^2(\mathbb{R})) \cap C^1([0, T], L^2(\mathbb{R})).$$

Hence, by the decomposition (4.4.40) we infer that $z(t) \in L^\infty([0, T], H^2(\mathbb{R}))$. Once that is established, we have by (4.4.46) that $\dot{\omega}$ and $\dot{\theta} - \omega$ are locally bounded functions, hence $\omega(t)$ and $\theta(t)$ are continuous on $[0, T]$. A reexamination of the decomposition (4.4.40) yields that $z(t)$ is continuous on $[0, T]$ as well, which results in $\omega(t), \theta(t) \in C^1([0, T])$ from system (4.4.46). Then, the decomposition (4.4.40) implies that

$$z(t) \in C([0, T], H^2(\mathbb{R})) \cap C^1([0, T], L^2(\mathbb{R})).$$

Theorem 4.16 holds for $T = \infty$. In particular, since $\dot{\omega} \in L_t^1$ and $\|\omega - \omega_0\|_{L_t^\infty} \leq C\delta^2$, there exists $\omega_\infty := \lim_{t \to \infty} \omega(t)$ in a local neighborhood of point $\omega_0 < 0$. In addition, since $z(t) \in L^4(\mathbb{R}_+, L^\infty(\mathbb{R}))$ and $z(t) \in C(\mathbb{R}_+, L^\infty(\mathbb{R}))$, then

$$\lim_{t \to \infty} \|\psi(t) - e^{-i\theta(t)}\phi(\cdot; \omega(t))\|_{L^\infty} = \lim_{t \to \infty} \|z(t)\|_{L^\infty} = 0.$$

It remains to prove the bounds (4.4.62)–(4.4.64). By the bound (4.4.49) and Sobolev's embedding of $H^1(\mathbb{R})$ to $L^\infty(\mathbb{R})$, for any $p > \frac{1}{2}$, we have

$$\int_0^T |\dot\omega| dt \leq C \left(\|\langle x \rangle^3 \phi^{2p-1}\|_{L_x^1 L_t^\infty} \|\langle x \rangle^{-3/2} z\|_{L_x^\infty L_t^2}^2 \right.$$
$$\left. + \|\langle x \rangle^3 \phi\|_{L_x^1 L_t^\infty} \|z\|_{L_x^\infty L_t^\infty}^{2p-1} \|\langle x \rangle^{-3/2} z\|_{L_x^\infty L_t^2}^2 \right)$$
$$\leq C \left(M_3^2 + M_2^{2p-1} M_3^2 \right),$$

where we have used the fact that $\phi(x)$ decays exponentially to zero as $|x| \to \infty$ for any $\omega < 0$. We shall only keep the quadratic terms in $M_{1,2,3,4,5}$ on the right-hand sides of the bounds. As a result, we obtain

$$M_4 \leq \int_0^T |\dot\omega| dt \leq C M_3^2,$$

and, similarly,

$$M_5 \leq \int_0^T |\dot\theta - \omega| dt \leq C M_3^2,$$

which give the bound (4.4.64). Note that using a similar computation, we also obtain

$$\|\dot\omega\|_{L_t^\infty} + \|\dot\theta - \omega\|_{L_t^\infty} \leq C \left(\|\phi^{2p-1}\|_{L_t^\infty L_x^1} \|z\|_{L_t^\infty L_x^\infty}^2 + \|\phi\|_{L_t^\infty L_x^1} \|z\|_{L_t^\infty L_x^\infty}^{2p+1} \right)$$
$$\leq C M_2^2.$$

To estimate M_3, we use the bound (4.4.52) for the first term of the integral equation (4.4.61),

$$\|\langle x \rangle^{-3/2} e^{-i\sigma_3 \hat{H} t} \mathbf{z}_0\|_{L_x^\infty L_t^2} \leq C \|z_0\|_{L_x^2}.$$

Since $\dot\omega, \dot\theta - \omega \in L_t^1 \cap L_t^\infty$, we treat the terms of the integral equation (4.4.61) with $\mathbf{g}_1$ and $\mathbf{g}_2$ similarly. Using the bound (4.4.55), we obtain

$$\|\langle x \rangle^{-3/2} \int_0^t e^{-i\sigma_3 \hat{H}(t-s)} P_c(\omega) \sigma_3 \mathbf{g}_2(s) ds\|_{L_x^\infty L_t^2} \leq C \|\langle x \rangle \mathbf{g}_2\|_{L_x^1 L_t^2}$$
$$\leq C \|\dot\omega\|_{L_t^2} \|\langle x \rangle \partial_\omega \phi\|_{L_x^1 L_t^\infty} \leq C \|\dot\omega\|_{L_t^\infty}^{1/2} \|\dot\omega\|_{L_t^1}^{1/2} \leq C M_2 M_3.$$

For $\mathbf{g}_3$, we use the bounds (4.4.47), (4.4.53), and (4.4.55) to obtain

$$\|\langle x \rangle^{-3/2} \int_0^t e^{-i\sigma_3 \hat{H}(t-s)} P_c(\omega) \sigma_3 \mathbf{g}_3(s) ds\|_{L_x^\infty L_t^2}$$
$$\leq C (\|\langle x \rangle \phi^{2p-1} z^2\|_{L_x^1 L_t^2} + \|z^{2p+1}\|_{L_t^1 L_x^2})$$
$$\leq C \left(\|\langle x \rangle^{-3/2} z\|_{L_x^\infty L_t^2} \|z\|_{L_t^\infty L_x^\infty} \|\langle x \rangle^{5/2} \phi^{2p-1}\|_{L_x^\infty L_t^1} + \|z\|_{L_t^{2p+1} L_x^{2(2p+1)}}^{2p+1} \right)$$
$$\leq C \left(M_2 M_3 + \|z\|_{L_t^{2p+1} L_x^{2(2p+1)}}^{2p+1} \right).$$

To deal with the last term, we note that, if $p \geq 2$, then $(r,s) = ((2p+1), 2(2p+1))$ is the Strichartz pair in the sense of Definition 4.10. In particular, they satisfy

$$\frac{4}{2p+1} + \frac{2}{2(2p+1)} \leq 1.$$

As a result, we obtain

$$\|z\|_{L_t^{2p+1} L_x^{2(2p+1)}} \leq C \left(\|z\|_{L_t^4 L_x^\infty} + \|z\|_{L_t^\infty L_x^2} \right) = C(M_1 + M_2).$$

For the linear terms, we use the bound (4.4.59) and the previous estimates to obtain

$$\left\| \langle x \rangle^{-3/2} \int_0^t e^{-i\sigma_3 \hat{H}(t-s)} P_c(\omega) \sigma_3 \left(\mathbf{f}_1(s) + \mathbf{f}_2(s) \right) ds \right\|_{L_x^\infty L_t^2}$$
$$\leq C \left(\|\dot\theta - \omega\|_{L_t^1} \|z\|_{L_t^\infty L_x^2} + \|\langle x \rangle^{5/2} (\hat{V}_\omega - \hat{V}_{\omega_0})\|_{L_x^1 L_t^\infty} \|\langle x \rangle^{-3/2} z\|_{L_x^\infty L_t^2} \right)$$
$$\leq C(M_2 + M_3) M_3^2.$$

Combining all these bounds together, we obtain the bound (4.4.63), where only quadratic terms are written on the right-hand side.

To estimate M_1 and M_2, we use the bound (4.4.50) for the first term of the integral equation (4.4.61),

$$\|e^{-i\sigma_3 \hat{H} t} \mathbf{z}_0\|_{L_t^4 L_x^\infty \cap L_t^\infty L_x^2} \leq C\|z_0\|_{L_x^2}.$$

For the nonlinear terms involving $\mathbf{g}_{1,2}$, we use the bound (4.4.51) and obtain

$$\left\| \int_0^t e^{-i\sigma_3 \hat{H}(t-s)} P_c(\omega) \sigma_3 \mathbf{g}_2(s) ds \right\|_{L_t^4 L_x^\infty \cap L_t^\infty L_x^2}$$
$$\leq C\|\mathbf{g}_2\|_{L_t^1 L_x^2} \leq C\|\dot\omega\|_{L_t^1} \|\partial_\omega \phi\|_{L_x^\infty L_x^2} \leq C M_3^2.$$

For the nonlinear term involving $\mathbf{g}_3$, we use bounds (4.4.51) and (4.4.57) to obtain

$$\left\| \int_0^t e^{-i\sigma_3 \hat{H}(t-s)} P_c(\omega) \sigma_3 \mathbf{g}_3(s) ds \right\|_{L_t^4 L_x^\infty \cap L_t^\infty L_x^2}$$
$$\leq C\|\langle x \rangle^5 \phi^{2p-1} z^2\|_{L_t^2 L_x^2} + C\|z^{2p+1}\|_{L_t^1 L_x^2}.$$

The last two terms are estimated by

$$\|\langle x \rangle^5 \phi^{2p-1} z^2\|_{L_t^2 L_x^2} \leq \|z\|_{L_t^4 L_x^\infty}^2 \|\langle x \rangle^5 \phi^{2p-1}\|_{L_t^\infty L_x^2} \leq C M_1^2$$

and

$$\|z^{2p+1}\|_{L_t^1 L_x^2} \leq C\|z\|_{L_t^{2p+1} L_x^{2(2p+1)}}^{2p+1} \leq C(M_1 + M_2)^{2p+1},$$

where $p \geq 2$ is used again. For the linear terms, we use the bound (4.4.59) and the previous estimates to obtain

$$\left\| \int_0^t e^{-i\sigma_3 \hat{H}(t-s)} P_c(\omega) \sigma_3 \left(\mathbf{f}_1(s) + \mathbf{f}_2(s) \right) ds \right\|_{L_t^4 L_x^\infty \cap L_t^\infty L_x^2}$$
$$\leq C \left(\|\dot\theta - \omega\|_{L_t^1} \|z\|_{L_t^\infty L_x^2} + \|\langle x \rangle^{5+3/2} (\hat{V}_\omega - \hat{V}_{\omega_0})\|_{L_x^2 L_t^\infty} \|\langle x \rangle^{-3/2} z\|_{L_x^\infty L_t^2} \right)$$
$$\leq C(M_2 + M_3) M_3^2.$$

We also need estimates on the time evolution of $\|\partial_x z\|_{L_t^\infty L_x^2}$, which are obtained with the use of the bounds (4.4.54) and (4.4.56) [41, 141]. Combining all elements together, this construction completes the proof of the bound (4.4.62) and hence the proof of Theorem 4.16. $\qquad\square$

Recent works by Cuccagna & Tarulli [44] and Kevrekidis *et al.* [112] extended the proof of the asymptotic stability to the localized modes of the discrete nonlinear Schrödinger equation with power nonlinearity. The asymptotic stability of the stable localized modes in the DNLS equation holds for the powers $p \geq 3$ because of the slower dispersive decay estimates.

5

Traveling localized modes in lattices

Once we accept our limits, we go beyond them.
– Albert Einstein.

There is a point where in the mystery of existence contradictions meet; where movement is not all movement and stillness is not all stillness; where the idea and the form, the within and the without, are united; where infinite becomes finite, yet not.
– Rabindranath Tagore.

If we try titles like "mobility of breathers" or "moving solitons in lattices" on the Internet, the outcome will include a good hundred physics publications in the last twenty years. It is then surprising to hear from the author of this book that no traveling localized modes in lattices generally exist, except for some non-generic configurations of the discrete nonlinear Schrödinger equation in the space of one dimension. This chapter is written to elaborate this provocative point in relevant mathematical details.

It is true that traveling localized modes in nonlinear evolution equations with constant coefficients such as the nonlinear Schrödinger and nonlinear Dirac equations can often be found from stationary localized modes by means of the Lorentz transformation (Section 1.2.1). The periodic potentials break, however, the continuous translational invariance of the nonlinear evolution equations and destroy the existence of the Lorentz transformation.

To illustrate this point, we consider the cubic nonlinear Schrödinger equation,

$$iu_t + u_{xx} + |u|^2u = 0, \quad x \in \mathbb{R}, \quad t \in \mathbb{R},$$

which has the family of traveling solitons,

$$u(x,t) = \sqrt{2|\omega|} \operatorname{sech}\left[\sqrt{|\omega|}(x - 2ct - s)\right] e^{ic(x-ct)-i\omega t+i\theta},$$

with arbitrary parameters $\omega < 0$ and $(c, s, \theta) \in \mathbb{R}^3$. However, the Gross–Pitaevskii equation,

$$iu_t = -u_{xx} + V(x)u - |u|^2u, \quad x \in \mathbb{R}, \quad t \in \mathbb{R},$$

with a nonzero bounded periodic potential $V(x)$, only has stationary localized modes (Section 3.1),

$$u(x,t) = \phi(x)e^{-i\omega t+i\theta},$$

with arbitrary parameters $\omega < \omega_0$ and $\theta \in \mathbb{R}$, where $\omega_0 := \inf \sigma(L)$ and $L =$

$-\partial_x^2 + V(x)$. Moreover, these localized modes are centered at a countable discrete set of positions $\{s_n\}_{n \in \mathbb{Z}}$ on $\mathbb{R}$. No traveling localized modes with a nonzero wave speed generally exist in the Gross–Pitaevskii equation since the time and space variables are not separable for traveling solutions.

If this result is so negative and there is no hope of finding traveling localized modes in periodic potentials, what "moving breathers" are physicists talking about? At a rough interpretation, they are talking about nonlinear evolution equations with specially fabricated periodic potentials where localized modes propagate "easier" than in other potentials. For instance, some authors are happy with observations of moving localized modes with weakly decreasing amplitudes on a length of finitely many periods.

The language above is not acceptable for a book in applied mathematics. The author (and hopefully, the readers) will not be satisfied to rely upon visual observations of numerical or laboratory experiments. Therefore, we shall look for methods of dynamical system theory to analyze the existence of traveling localized modes in periodic potentials or lattices. Only if we can prove the existence of traveling and localized solutions in the original nonlinear evolution equations, can we conclude that the traveling localized modes exist.

From a qualitative point of view, it is intuitively clear why steady propagation of localized modes is impossible in periodic potentials. Stationary localized modes do not exist for all positions $s \in \mathbb{R}$ and, therefore, when we look for a traveling solution in the form $u(x,t) = \phi(x - ct)e^{-i\omega t}$, the localized mode $\phi(x)$ is assumed to be translated along non-existing stationary solutions. Each "translation" induces the dispersive radiation that detaches from a localized mode and results in its (slow or fast) decay. Even if we cook up an integrable model with the periodic potential where stationary modes may exist for all positions $s \in \mathbb{R}$ (Section 3.3.1), propagation of the localized mode still induces the dispersive radiation.

The qualitative picture above can be formalized, and the simplest formalism arises in the framework of the discrete nonlinear Schrödinger (DNLS) equation (justified in Section 2.4). Although some details of the dynamics of the radiation-induced decay of localized modes in periodic potentials are not present in the DNLS equation, the existence of stationary localized modes and non-existence of traveling localized modes are captured well by the reduced model. Therefore, in the main part of this chapter, we shall deal with the discrete counterpart of the Gross–Pitaevskii equation,

$$i\dot{u}_n + \frac{u_{n+1} - 2u_n + u_{n-1}}{h^2} + f(u_{n-1}, u_n, u_{n+1}) = 0, \quad n \in \mathbb{Z}, \quad t \in \mathbb{R},$$

where h is the parameter for lattice spacing, and $f(u_{n-1}, u_n, u_{n+1}) : \mathbb{C}^3 \to \mathbb{C}$ is a non-analytic nonlinear function. This equation is termed as the *generalized* DNLS equation.

On the one hand, the generalized DNLS equation is a spatial discretization of the nonlinear Schrödinger equation, where the second-order partial derivative is replaced with the second-order central difference on the grid $\{x_n = nh\}_{n \in \mathbb{Z}}$ and the nonlinearity $f(u_{n-1}, u_n, u_{n+1})$ incorporates the effects of on-site and neighbor-site couplings. On the other hand, this model is important on its own for modeling of various physical problems although the justification of this model in the semi-classical limit is shown to be impossible in the Gross–Pitaevskii equation with periodic linear and nonlinear coefficients [18].

The nonlinear function $f(u_{n-1}, u_n, u_{n+1})$ may take various forms that include the Salerno model [177],

$$f = 2(1-\alpha)|u_n|^2 u_n + \alpha|u_n|^2(u_{n+1} + u_{n-1}), \quad \alpha \in \mathbb{R},$$

which interpolates between the cubic DNLS equation at $\alpha = 0$ and the integrable Ablowitz–Ladik (AL) lattice at $\alpha = 1$. Generally, we can look for a class of nonlinear functions $f(u_{n-1}, u_n, u_{n+1})$ satisfying the following properties:

(P1) (gauge covariance) $f(e^{i\alpha}v, e^{i\alpha}u, e^{i\alpha}w) = e^{i\alpha}f(v, u, w), \alpha \in \mathbb{R},$
(P2) (symmetry) $f(v, u, w) = f(w, u, v),$
(P3) (reversibility) $\overline{f(v, u, w)} = f(\bar{v}, \bar{u}, \bar{w}).$

These properties originate from applications of the generalized DNLS equation to the modeling of the envelope of modulated nonlinear dispersive waves in a non-dissipative isotropic system. If, in addition, $f(u_{n-1}, u_n, u_{n+1})$ is a homogeneous cubic polynomial in variables (u_{n-1}, u_n, u_{n+1}), then the nonlinear function is uniquely represented by the polynomial

$$\begin{aligned}
f = {} & \alpha_1|u_n|^2 u_n + \alpha_2|u_n|^2(u_{n+1} + u_{n-1}) + \alpha_3 u_n^2(\bar{u}_{n+1} + \bar{u}_{n-1}) \\
& + \alpha_4(|u_{n+1}|^2 + |u_{n-1}|^2)u_n + \alpha_5(\bar{u}_{n+1}u_{n-1} + u_{n+1}\bar{u}_{n-1})u_n \\
& + \alpha_6(u_{n+1}^2 + u_{n-1}^2)\bar{u}_n + \alpha_7 u_{n+1}u_{n-1}\bar{u}_n + \alpha_8(|u_{n+1}|^2 u_{n+1} + |u_{n-1}|^2 u_{n-1}) \\
& + \alpha_9(u_{n+1}^2\bar{u}_{n-1} + u_{n-1}^2\bar{u}_{n+1}) + \alpha_{10}(|u_{n+1}|^2 u_{n-1} + |u_{n-1}|^2 u_{n+1}),
\end{aligned}$$

where $(\alpha_1, ..., \alpha_{10}) \in \mathbb{R}^{10}$ are arbitrary parameters. Stationary solutions of the generalized DNLS equation are defined by

$$u_n(t) = \phi_n e^{-i\omega t + i\theta},$$

with two parameters $(\omega, \theta) \in \mathbb{R}^2$, whereas traveling solutions are defined by

$$u_n(t) = \phi(hn - 2ct - s)e^{-i\omega t + i\theta},$$

with four parameters $(\omega, c, \theta, s) \in \mathbb{R}^4$. Since $n \in \mathbb{Z}$ is a discrete variable, the existence of traveling solutions imposes a constraint on the dynamics of amplitudes $\{u_n(t)\}_{n \in \mathbb{Z}}$, in particular,

$$u_{n+1}(t) = u_n(t - \tau)e^{-i\alpha}, \quad n \in \mathbb{Z}, \quad t \in \mathbb{R},$$

where $\tau = h/(2c)$ and $\alpha = \omega h/(2c)$. In both cases of stationary and traveling solutions, we will only be interested in the existence of a fundamental (single-pulse) localized mode $\{\phi_n\}_{n \in \mathbb{Z}}$ and $\phi(z)$ on $\mathbb{R}$, which corresponds to the sech solitons of the continuous NLS equation.

Although the separation of variables seems to be straightforward both for stationary and traveling solutions, we will see that there are serious obstacles that prevent the existence of traveling localized modes in the generalized DNLS equation. These obstacles are related to the previous discussion that traveling localized modes in periodic potentials induce radiation and decay.

Besides the homogeneous cubic polynomials for $f(u_{n-1}, u_n, u_{n+1})$, other non-linear functions were also considered recently, including the DNLS equation with a saturable nonlinearity [139] and the cubic–quintic DNLS equation [29]. These DNLS equations share the same properties with the generalized DNLS equation in the sense that the traveling localized modes may only occur as a result of the codimension-one bifurcation.

Before drilling into details, we should add a disclaimer that traveling localized modes in lattices are known in many other models, where no obstacles on the existence of these solutions arise. For instance, traveling localized modes exist generically in the Fermi–Pasta–Ulam lattices, which can be regarded as the spatial discretization of the Korteweg–de Vries and Boussinesq equations [60]. Similarly, traveling periodic waves can be found with both variational [147] and topological [56] methods in a large number of lattice models including the DNLS equations. Still the same problem of non-existence of traveling kinks or traveling time-periodic space-localized breathers are common for the Klein–Gordon lattices [95, 96, 97, 146].

5.1 Differential advance–delay operators

We shall first analyze the linear differential advance–delay operators. Properties of an associated linear operator are again the keys to understanding the nonlinear analysis of localized modes in the nonlinear differential advance–delay equations.

5.1.1 Asymptotically hyperbolic operators

Consider an abstract differential advance–delay operator,

$$(LU)(Z) = -2c\frac{dU}{dZ} + V_0(Z)U(Z)$$
$$+ (1 + V_+(Z))U(Z+1) + (1 + V_-(Z))U(Z-1),$$

and assume that $V_0(Z), V_\pm(Z) : \mathbb{R} \to \mathbb{R}$ are bounded functions that decay exponentially to zero as $|Z| \to \infty$. Operator L maps continuously $H^1(\mathbb{R})$ to $L^2(\mathbb{R})$ for any $c \neq 0$. Related to the inner product in $L^2(\mathbb{R})$, the adjoint operator L^* is written in the form

$$(L^*U)(Z) = 2c\frac{dU}{dZ} + V_0(Z)U(Z)$$
$$+ (1 + V_+(Z-1))U(Z-1) + (1 + V_-(Z+1))U(Z+1).$$

Let L_0 denote the constant-coefficient part of L,

$$(L_0 U)(Z) = -2c\frac{dU}{dZ} + U(Z+1) + U(Z-1).$$

Using the Fourier transform, we obtain for any $c \in \mathbb{R}$ that the spectrum of L_0 is purely continuous and located at

$$\sigma(L_0) = \{\lambda \in \mathbb{C} : \quad \lambda = \lambda_0(k) := -2ick + 2\cos(k), \quad k \in \mathbb{R}\}. \quad (5.1.1)$$

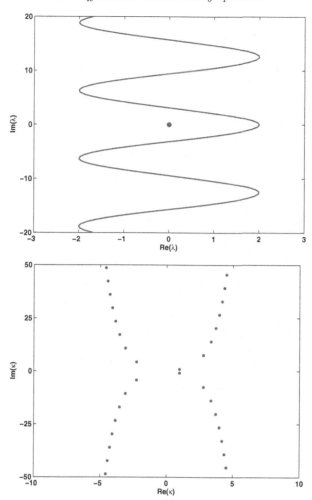

Figure 5.1 Top: the spectrum of L_0 for $c = 1$. The dot shows the position of $\lambda = 0$. Bottom: roots κ of $D(\kappa; c) = 0$ for $c = 1$.

Figure 5.1 (top) shows the spectrum of L_0. If $c \neq 0$, the location of $\sigma(L_0)$ is bounded away from $\lambda = 0$. Solutions of the homogeneous equation $L_0 U = 0$ are given by a linear superposition of $U(Z) = e^{\kappa Z}$, where κ solves the characteristic equation in the form

$$D(\kappa; c) := -2c\kappa + 2\cosh(\kappa) = 0. \tag{5.1.2}$$

Because $D(\kappa; c)$ is analytic in κ, roots κ of $D(\kappa; c) = 0$ are all isolated and of finite multiplicities. Figure 5.1 (bottom) shows the location of roots κ of $D(\kappa; c) = 0$. There are infinitely many roots which diverge to infinity in the complex κ plane.

Because the location of $\sigma(L_0)$ is bounded away from $\lambda = 0$ for $c \neq 0$, roots κ of $D(\kappa; c) = 0$ are bounded away from the imaginary axis $\mathrm{Re}(\kappa) = 0$. We say that L_0 is a *hyperbolic* operator and L is an *asymptotically hyperbolic* operator if the roots κ of $D(\kappa; c) = 0$ are bounded away from $\mathrm{Re}(\kappa) = 0$. Under this condition, the Fredholm Alternative Theorem for differential advance–delay operators is proved by Mallet-Paret [138] (Theorems A, B, and C).

Theorem 5.1 *Assume that $c \neq 0$ and L is an asymptotically hyperbolic operator with the exponential decay of $V_0(Z)$ and $V_\pm(Z)$ to zero as $|Z| \to \infty$. Operator $L : H^1(\mathbb{R}) \to L^2(\mathbb{R})$ is a Fredholm operator with property*

$$\dim\left(\mathrm{Ker}(L^*)\right) = \mathrm{codim}\left(\mathrm{Ran}(L)\right), \quad \dim\left(\mathrm{Ker}(L)\right) = \mathrm{codim}\left(\mathrm{Ran}(L^*)\right).$$

The Fredholm index is

$$\mathrm{ind}(L) := \dim\left(\mathrm{Ker}(L)\right) - \dim\left(\mathrm{Ker}(L^*)\right) = -\mathrm{ind}(L^*).$$

If $L \to L_0$ as $Z \to \pm\infty$, then $\mathrm{ind}(L) = 0$.

The spectrum of the Fredholm operator L with $c \neq 0$ can be divided into the two disjoint sets: isolated eigenvalues and the continuous spectrum with embedded eigenvalues. Thanks to the exponential decay of $V_0(Z)$ and $V_\pm(Z)$ to zero as $Z \to \pm\infty$, the Weyl Theorem (Appendix B.15) implies that the continuous spectrum of L coincides with $\sigma(L_0)$ shown in the top panel of Figure 5.1. Regarding the isolated and embedded eigenvalues, the following lemma shows that if there is one eigenvalue, then there are infinitely many eigenvalues of L.

Lemma 5.1 *Assume that there exists an eigenvalue–eigenvector pair*

$$(\lambda_0, u_0) \in \mathbb{C} \times L^2(\mathbb{R})$$

of $Lu = \lambda u$. For any $c \neq 0$, there is an infinite set of eigenvalue–eigenvector pairs of $Lu = \lambda u$,

$$\lambda_m = \lambda_0 + 4\pi m c i, \quad u_m(Z) = u_0(Z) e^{-2\pi m i Z}, \quad m \in \mathbb{Z}.$$

Proof The proof follows by direct substitution with the use of $e^{2\pi n i} = 1$ for all $n \in \mathbb{Z}$. $\qquad\qquad\qquad\qquad\qquad\qquad\qquad\qquad\qquad\qquad\qquad\qquad\qquad\qquad\qquad\square$

Lemma 5.1 shows that the differential advance–delay operators with exponentially decaying potentials are more complicated than differential operators (Section 4.1). In particular, the limit $c \to 0$ is singular because every (isolated or embedded) eigenvalue of the advance–delay operator has an infinite multiplicity. The advance–delay operator L with $c = 0$ is no longer a Fredholm operator.

5.1.2 Non-hyperbolic operators

Consider an abstract differential advance–delay operator

$$(LU)(Z) = -2c\frac{dU}{dZ} + V_0(Z)U(Z)$$
$$+ (1 + V_+(Z))\,U(Z+1) - (1 + V_-(Z))\,U(Z-1),$$

and assume again that $V_0(Z), V_\pm(Z) : \mathbb{R} \to \mathbb{R}$ are bounded functions that decay exponentially to zero as $|Z| \to \infty$. The adjoint operator L^* is now

$$L^*U = 2c\frac{dU}{dZ} + V_0(Z)U(Z)$$
$$+ (1 + V_+(Z-1))U(Z-1) - (1 + V_-(Z+1))U(Z+1).$$

Let L_0 denote the constant-coefficient part of L,

$$L_0U = -2c\frac{dU}{dZ} + U(Z+1) - U(Z-1).$$

Using the Fourier transform again, we obtain for any $c \in \mathbb{R}$ that the spectrum of L_0 is purely continuous and located at

$$\sigma(L_0) = \{\lambda \in \mathbb{C} : \quad \lambda = \lambda_0(k) := 2\mathrm{i}\left[\sin(k) - ck\right], \quad k \in \mathbb{R}\}. \tag{5.1.3}$$

If $c \neq 0$, the spectrum of L_0 covers $\mathrm{i}\mathbb{R}$. If $|c| > 1$, then $\lambda_0(k) : \mathbb{R} \to \mathrm{i}\mathbb{R}$ is one-to-one and onto.

Thanks to the exponential decay of $V_0(Z)$ and $V_\pm(Z)$ to zero as $Z \to \pm\infty$, the Weyl Theorem (Appendix B.15) implies that the continuous spectrum of L coincides with $\sigma(L_0)$. There exist typically embedded eigenvalues of L, because the translational symmetry of the differential advance–delay equation implies that L has zero eigenvalue. By Lemma 5.1, if $\lambda = 0$ is an eigenvalue with an eigenvector $u_0(Z)$, then $\lambda = 4\pi cm\mathrm{i}$ is another eigenvalue with an eigenvector $u_0(Z)e^{-2\pi m\mathrm{i}Z}$ for any $m \in \mathbb{Z}$. Therefore, there are infinitely many embedded eigenvalues of L if $0 \in \sigma(L)$.

Exercise 5.1 Show that if $\lambda = 0$ is a double eigenvalue with an eigenfunction $u_0(Z)$ and a generalized eigenfunction $u_1(Z)$ in the sense of

$$Lu_0 = 0, \quad Lu_1 = u_0,$$

then $\lambda = 4\pi cm\mathrm{i}$ is also a double eigenvalue for any $m \in \mathbb{Z}$.

To separate embedded eigenvalues of L and the continuous spectrum L, we use the technique of exponential weighted spaces,

$$L_\mu^2(\mathbb{R}) = \left\{U \in L_{\mathrm{loc}}^2(\mathbb{R}) : e^{\mu Z}U(Z) \in L^2(\mathbb{R})\right\}, \quad \mu > 0. \tag{5.1.4}$$

The adjoint space is $L_{-\mu}^2(\mathbb{R})$. Similarly, we introduce the weighted H^1 spaces, $H_\mu^1(\mathbb{R})$ and $H_{-\mu}^1(\mathbb{R})$.

Under the transformation (5.1.4), operator L in $L_\mu^2(\mathbb{R})$ is equivalent to operator $L_\mu = e^{\mu Z}Le^{-\mu Z}$ in $L^2(\mathbb{R})$, whereas L^* in $L_{-\mu}^2(\mathbb{R})$ is equivalent to $L_{-\mu}^* = e^{-\mu Z}L^*e^{\mu Z}$ in $L^2(\mathbb{R})$.

Lemma 5.2 *Fix $c > 1$. There exists $\mu_0 > 0$ such that the continuous spectrum of L_μ and $L_{-\mu}^*$ for $0 < \mu < \mu_0$ is located along a curve contained within the strip*

$$D = \{\lambda \in \mathbb{C} : \quad \lambda_- < \mathrm{Re}(\lambda) < \lambda_+\},$$

for some $0 < \lambda_- < \lambda_+$.

Proof Thanks to the exponential decay of $V_0(Z)$ and $V_\pm(Z)$, there exists $\mu_1 > 0$ such that the potential terms in operator L_μ for $|\mu| < \mu_1$ decay exponentially to zero as $|Z| \to \infty$. As a result, the continuous spectrum $\sigma_c(L_\mu)$ coincides with that of $e^{\mu Z} L_0 e^{-\mu Z}$. Therefore, it is located at

$$\sigma_c(L_\mu) = \{\lambda \in \mathbb{C} : \lambda = \lambda_\mu(k), \ k \in \mathbb{R}\}, \tag{5.1.5}$$

where

$$\lambda_\mu(k) = \lambda_0(k + i\mu) = 2\left[\mu c - \sinh(\mu)\cos(k)\right] + 2i\left[\cosh(\mu)\sin(k) - kc\right].$$

In particular, if $c > 1$, then $\mathrm{Re}\,\lambda_\mu(k) > 0$ for $0 < \mu < \mu_2$ and $\frac{d}{dk}\mathrm{Im}\,\lambda_\mu(k) < 0$ if $|\mu| < \mu_3$, where μ_2 is the root of $\sinh(\mu)/\mu = c$ and $\mu_3 = \cosh^{-1}(c)$. Therefore, if $0 < \mu < \mu_0 := \min\{\mu_1, \mu_2, \mu_3\}$, the continuous spectrum of L_μ is a one-to-one map from $k \in \mathbb{R}$ to $\lambda \in \mathbb{C}$, where λ oscillates in the strip D, where

$$\lambda_\pm = 2\mu\left[c \pm \frac{\sinh(\mu)}{\mu}\right] > 0.$$

The continuous spectrum of $L^*_{-\mu}$ is located along the curve determined by

$$\lambda = -\lambda_0(k - i\mu) = \lambda_\mu(-k) = \bar{\lambda}_\mu(k), \quad k \in \mathbb{R}.$$

This curve on the λ plane is the same as $\lambda = \lambda_\mu(k)$ but it is traversed in the reverse direction as k increases. $\qquad\square$

Using Lemma 5.2, we introduce eigenfunctions of the continuous spectrum of L and add an assumption that, besides the sequence of double eigenvalues $\{4\pi mci\}_{m\in\mathbb{Z}}$, no other eigenvalues exist near the imaginary axis of λ.

Definition 5.1 Fix $c > 1$ and $\mu \in (0, \mu_0)$. Let $U_\mu(Z; k)$ and $W_\mu(Z; k)$ be *eigenfunctions of the continuous spectrum of L_μ and $L^*_{-\mu}$,*

$$LU_\mu(Z; k) = \lambda_\mu(k)U_\mu(Z; k), \quad L^*W_\mu(Z; k) = \bar{\lambda}_\mu(k)W_\mu(Z; k), \tag{5.1.6}$$

subject to the boundary conditions

$$\lim_{Z \to \infty} e^{-ikZ + \mu Z}U_\mu(Z; k) = 1, \quad \lim_{Z \to \infty} e^{-ikZ - \mu Z}W_\mu(Z; k) = 1. \tag{5.1.7}$$

Let $\{a_\mu(k), b_\mu(k)\}$ be the *scattering coefficients* defined by the boundary conditions

$$\lim_{Z \to -\infty} e^{-ikZ + \mu Z}U_\mu(Z; k) = a_\mu(k), \quad \lim_{Z \to -\infty} e^{-ikZ - \mu Z}W_\mu(Z; k) = b_\mu(k), \tag{5.1.8}$$

assuming that the limits exist.

Assumption 5.1 *There exist constants $\Lambda_- < 0 < \Lambda_+$ such that no eigenvalues of L with eigenfunctions in $H^1_\mu(\mathbb{R})$ for any $|\mu| < \mu_0$ exist for any $\lambda \in (\Lambda_-, \Lambda_+)$, except for the set of double eigenvalues $\{4\pi mci\}_{m\in\mathbb{Z}}$. Furthermore, assume that there exists $u_0, u_1 \in H^1_\mu(\mathbb{R})$ for any $|\mu| < \mu_0$ such that*

$$Lu_0 = 0, \quad Lu_1 = u_0,$$

and there exists no $u_2 \in H^1_\mu(\mathbb{R})$ such that $Lu_2 = u_1$.

The following lemma describes properties of the eigenfunctions of the continuous spectrum of L under Assumption 5.1.

Lemma 5.3 *Under Assumption 5.1, the set of eigenfunctions $U_\mu(Z;k)$ and $W_\mu(Z;k)$ in Definition 5.1 satisfies the orthogonality relation*

$$\int_{\mathbb{R}} U_\mu(Z;k)\bar{W}_\mu(Z;k')dZ = \frac{4\pi c}{c - \cosh(\mu)\cos(k)}\delta(k' - k), \quad k,k' \in \mathbb{R}. \qquad (5.1.9)$$

In addition, $\bar{b}_\mu(k) = 1/a_\mu(k)$ for all $k \in \mathbb{R}$.

Proof Let us consider the homogeneous equations (5.1.6) for $k \in \mathbb{R}$. Integrating

$$\bar{W}_\mu(Z;k')LU_\mu(Z;k) - U_\mu(Z;k)L^*\bar{W}_\mu(Z;k') \qquad (5.1.10)$$

for all $Z \in [-L, L]$ and extending the limit $L \to \infty$, we obtain that

$$\int_{\mathbb{R}} U_\mu(Z;k)\bar{W}_\mu(Z;k')dZ$$
$$= -2c\left(\lim_{L\to\infty}\frac{\bar{W}_\mu(L;k')U_\mu(L;k)}{\lambda_\mu(k) - \lambda_\mu(k')} - \lim_{L\to-\infty}\frac{\bar{W}_\mu(L;k')U_\mu(L;k)}{\lambda_\mu(k) - \lambda_\mu(k')}\right).$$

Using the asymptotic representations (5.1.7) and (5.1.8) for the eigenfunctions $U_\mu(Z;k)$ and $\bar{W}_\mu(Z;k')$ as $|Z| \to \infty$ and the property of the Dirac delta function,

$$\lim_{L\to\pm\infty}\frac{e^{i(k-k')L}}{i(k - k')} = \pm\pi\delta(k - k'), \qquad (5.1.11)$$

we obtain the orthogonality relation

$$\int_{\mathbb{R}} U_\mu(Z;k)\bar{W}_\mu(Z;k')dZ = \frac{2\pi c\left[1 + a_\mu(k)\bar{b}_\mu(k)\right]}{c - \cosh(\mu)\,\cos(k)}, \quad k,k' \in \mathbb{R}. \qquad (5.1.12)$$

Let us now consider equation (5.1.10) with $k = k' \in \mathbb{R}$. Using the same integration on $Z \in [-L, L]$ and extending the limit $L \to \infty$, we obtain that

$$2c\left[a_\mu(k)\bar{b}_\mu(k) - 1\right] = 0, \quad k \in \mathbb{R}.$$

Therefore, $\bar{b}_\mu(k) = 1/a_\mu(k)$ and the orthogonality relation (5.1.9) follows from (5.1.12). $\qquad\square$

Lemma 5.4 *Under Assumption 5.1, there exist $w_0, w_1 \in H^1_{-\mu}(\mathbb{R})$ for $\mu \in (0, \mu_0)$ such that*

$$L^*w_0 = 0, \quad L^*w_1 = w_0.$$

Proof Since the differential operator $L_\mu = e^{\mu Z}Le^{-\mu Z}$ is a Fredholm operator of zero index for $\mu \in (0, \mu_0)$, Theorem 5.1 states that the adjoint operator $L^*_{-\mu} = e^{-\mu Z}L^*e^{\mu Z}$ has a one-dimensional geometric kernel and a two-dimensional generalized kernel for the same value of $\mu \in (0, \mu_0)$. $\qquad\square$

Assumption 5.2 *Eigenfunctions $\{U_\mu(Z;k), W_\mu(Z;k)\}$ and scattering coefficients $\{a_\mu(k), b_\mu(k)\}$ in Definition 5.1 are bounded for all $k \in \mathbb{R}$ in the limit $\mu \downarrow 0$.*

The following theorem gives the main result for analysis of bifurcations of traveling localized modes.

Theorem 5.2 *Let $F \in L^2_\mu(\mathbb{R})$ with $|\mu| < \mu_0$ for some $\mu_0 > 0$. Under Assumptions 5.1 and 5.2, there exists a solution $U \in H^1_\mu(\mathbb{R})$ with $\mu \in (0, \mu_0)$ of the linear inhomogeneous equation $LU = F$ if and only if*

$$\int_{\mathbb{R}} w_0(Z)F(Z)dZ = 0. \tag{5.1.13}$$

Moreover, $U \in H^1(\mathbb{R})$ if and only if F satisfies both (5.1.13) and

$$\int_{\mathbb{R}} W_0(Z; 0)F(Z)dZ = 0. \tag{5.1.14}$$

Proof The first constraint (5.1.13) follows by the Fredholm Alternative Theorem (Appendix B.4), since the zero eigenvalue of L_μ is isolated from the continuous spectrum of L_μ if $\mu \in (0, \mu_0)$. Under Assumption 5.1, the spectrum of L_μ for $\mu \in (0, \mu_0)$ consists of the continuous spectrum, the set of double eigenvalues $\{4\pi mci\}_{m \in \mathbb{Z}}$, and the other isolated eigenvalues outside the strip $\Lambda_- < \lambda < \Lambda_+$.

Assume that the condition (5.1.13) is satisfied. Thanks to Lemma 5.3, we represent the solution $U \in H^1_\mu(\mathbb{R})$ of $LU = F$ for $\mu \in (0, \mu_0)$ by the generalized Fourier transform

$$U(Z) = \int_{\mathbb{R}} B_\mu(k)U_\mu(Z; k)dk + A_0 u_0(Z) + A_1 u_1(Z) + \sum_{\lambda_j \in \sigma_d(L_\mu)\backslash\{0\}} A_j u_j(Z),$$

$$\tag{5.1.15}$$

where $A_0 \in \mathbb{R}$ is arbitrary,

$$B_\mu(k) = \frac{c - \cosh(\mu)\,\cos(k)}{4\pi c \lambda_\mu(k)}\,\langle W_\mu(\cdot; k), F\rangle_{L^2}, \qquad A_1 = \frac{\langle w_1, F\rangle_{L^2}}{\langle w_1, u_0\rangle_{L^2}},$$

and A_j for $\lambda_j \in \sigma_d(L_\mu)\backslash\{0\}$ are projections to the eigenfunctions for nonzero eigenvalues of L_μ. We note that the location of these eigenvalues λ_j is not affected by the weight parameter μ for small $0 < \mu < \mu_0$ thanks to the fast decay of eigenfunctions.

Under Assumption 5.2, the functions $B_\mu(k)\lambda_\mu(k)$ and $U_\mu(Z; k)$ in the representation (5.1.15) are bounded in $k \in \mathbb{R}$ as $\mu \downarrow 0$. The integrand of (5.1.15) for $\mu = 0$ has only one singularity at $k = 0$ from the simple zero of $\lambda_0(k)$ at $k = 0$ and this singularity is a simple pole. The integral can be split into two parts:

$$\lim_{\mu \downarrow 0} \int_{\mathbb{R}} B_\mu(k)U_\mu(Z; k)dk = \pi i \, \mathrm{Res}\,[B_0(k)U_0(Z; k), k = 0]$$

$$+ \lim_{\epsilon \downarrow 0} \left(\int_{-\infty}^{-\epsilon} + \int_{\epsilon}^{\infty}\right) B_0(k)U_0(Z; k)dk$$

$$= -\frac{1}{8c}\langle W_0(\cdot; 0), F\rangle_{L^2} U_0(Z; 0) + \mathrm{p.v.}\int_{\mathbb{R}} B_0(k)U_0(Z; k)dk,$$

where p.v. denotes the principal value of singular integrals with a simple pole.

Thanks to the linear growth of $\lambda_0(k)$ in k as $|k| \to \infty$, the second term is in $H^1(\mathbb{R})$ if $F \in L^2(\mathbb{R})$. Since the first term is bounded but non-decaying and all other eigenfunctions $u_0(Z)$, $u_1(Z)$, and $u_j(Z)$ are in $H^1(\mathbb{R})$, it is clear that $U \in H^1(\mathbb{R})$ if and only if $\langle W_0(\cdot; 0), F\rangle_{L^2} = 0$, which yields the condition (5.1.14). $\qquad\square$

5.2 Integrable discretizations for stationary localized modes

Direct substitution of

$$u_n(t) = \phi_n e^{-i\omega t + i\theta}, \qquad n \in \mathbb{Z}, \quad t \in \mathbb{R}, \tag{5.2.1}$$

into the generalized DNLS equation,

$$i\dot{u}_n + \frac{u_{n+1} - 2u_n + u_{n-1}}{h^2} + f(u_{n-1}, u_n, u_{n+1}) = 0, \qquad n \in \mathbb{Z}, \quad t \in \mathbb{R}, \tag{5.2.2}$$

results in the second-order difference equation,

$$\frac{\phi_{n+1} - 2\phi_n + \phi_{n-1}}{h^2} + \omega\phi_n + f(\phi_{n-1}, \phi_n, \phi_{n+1}) = 0, \quad n \in \mathbb{Z}. \tag{5.2.3}$$

When $f = \phi_n^{2p+1}$, two fundamental solutions of the second-order difference equation (5.2.3) exist for $p > 0$ (Section 3.3.3): one localized mode is symmetric about a selected lattice node and the other one is symmetric about a midpoint between two adjacent nodes. The first solution is referred to as the *on-site* or *site-symmetric* soliton and the other one is referred to as the *inter-site* or *bond-symmetric* soliton.

We shall find the conditions on $f(u_{n-1}, u_n, u_{n+1})$ under which the two localized modes of the second-order difference equation (5.2.3) can be interpolated into a continuous family of localized modes,

$$\phi_n = \phi(hn - s), \quad n \in \mathbb{Z}, \tag{5.2.4}$$

where $s \in \mathbb{R}$ is an arbitrary parameter and $\phi(z) : \mathbb{R} \to \mathbb{R}$ is a continuous single-humped function with the exponential decay to zero at infinity as $|z| \to \infty$. If $\phi(z)$ is an even function, the on-site and inter-site localized modes correspond to the values $s = 0$ and $s = \frac{h}{2}$ of the continuous family (5.2.4). This solution is often referred to as the *translationally invariant* family of localized modes. The existence of the continuous family (5.2.4) is non-generic for nonlinear lattices because the latter have no continuous translational invariance. In the situations when the DNLS equation (5.2.2) admits a conserved Hamilton function, the existence of translationally invariant localized modes implies vanishing of the *Peierls–Nabarro* barrier, which is defined as a difference between the energies of the on-site and inter-site solitons.

Exercise 5.2 Show that the stationary localized modes of the cubic DNLS equation,

$$\frac{\phi_{n+1} - 2\phi_n + \phi_{n-1}}{h^2} + \omega\phi_n + |\phi_n|^2 \phi_n = 0, \quad n \in \mathbb{Z},$$

can be obtained by the first variation of the energy,

$$E_\omega(u) = \sum_{n \in \mathbb{Z}} \left(\frac{|u_{n+1} - u_n|^2}{h^2} - \frac{1}{2}|u_n|^4 - \omega|u_n|^2 \right).$$

Prove that if the translationally invariant family (5.2.4) of localized modes exists, then $E_\omega(\phi)$ is independent of $s \in \mathbb{R}$, so that

$$\Delta E := E_\omega(\phi(hn)) - E_\omega(\phi(hn - h/2)) = 0.$$

Remark 5.1 The converse statement to Exercise 5.2 is not true. That is, the zero Peierls–Nabarro barrier $\Delta E = 0$ does not imply the existence of the translationally invariant family (5.2.4) of localized modes.

To find the nonlinear function $f(u_{n-1}, u_n, u_{n+1})$ supporting the existence of translationally invariant localized modes, we shall look for conditions which give integrability of the second-order difference equation (5.2.3) and the conservation in $n \in \mathbb{Z}$ of the first-order invariant

$$E = \frac{1}{h^2}|\phi_{n+1} - \phi_n|^2 + \frac{1}{2}\omega\left(\phi_n\bar{\phi}_{n+1} + \bar{\phi}_n\phi_{n+1}\right) + g(\phi_n, \phi_{n+1}), \quad n \in \mathbb{Z}, \quad (5.2.5)$$

where E is constant and $g(\phi_n, \phi_{n+1}) : \mathbb{C}^2 \to \mathbb{R}$ is a non-analytic function with the properties

(S1) (gauge invariance) $g(e^{i\alpha}u, e^{i\alpha}w) = g(u, w),\ \alpha \in \mathbb{R}$,
(S2) (symmetry) $g(u, w) = g(w, u)$,
(S3) (reversibility) $\overline{g(u, w)} = g(\bar{u}, \bar{w})$.

The most general homogeneous quartic polynomial $g(\phi_n, \phi_{n+1})$ that satisfies properties (S1)–(S3) is

$$g = \tilde{\alpha}_1(|\phi_n|^2 + |\phi_{n+1}|^2)(\bar{\phi}_{n+1}\phi_n + \phi_{n+1}\bar{\phi}_n) + \tilde{\alpha}_2|\phi_n|^2|\phi_{n+1}|^2$$
$$+ \tilde{\alpha}_3(\phi_n^2\bar{\phi}_{n+1}^2 + \bar{\phi}_n^2\phi_{n+1}^2) + \tilde{\alpha}_4(|\phi_n|^4 + |\phi_{n+1}|^4),$$

where $(\tilde{\alpha}_1, ..., \tilde{\alpha}_4) \in \mathbb{R}^4$ are arbitrary parameters.

The idea behind the search for the conserved quantity E in the form (5.2.5) comes from the continuous limit $h \to 0$ when the second-order difference equation (5.2.3) with a cubic polynomial $f(\phi_{n-1}, \phi_n, \phi_{n+1})$ can be reduced to the second-order differential equation

$$\phi''(x) + \omega\phi(x) + 2|\phi(x)|^2\phi(x) = 0, \quad x \in \mathbb{R}, \quad (5.2.6)$$

that admits the first integral

$$E = |\phi'(x)|^2 + \omega|\phi(x)|^2 + |\phi(x)|^4, \quad x \in \mathbb{R}. \quad (5.2.7)$$

If $\phi(x)$ is a localized mode, then $E = 0$ and

$$\phi(x) = \sqrt{|\omega|}\,\mathrm{sech}(\sqrt{|\omega|}(x - s)), \quad \omega < 0, \quad s \in \mathbb{R}. \quad (5.2.8)$$

Reductions of the second-order difference equations to the first-order invariants are old in the classical field theory and have been exploited for the discrete Klein–Gordon equation by Speight [196]. This idea was later developed by Kevrekidis [111], Dmitriev *et al.* [48, 49], Barashenkov *et al.* [14, 15], and Pelinovsky [154].

For simplification of this presentation, we shall consider only real-valued solutions $\{\phi_n\}_{n\in\mathbb{Z}}$ of the difference equations (5.2.3) and (5.2.5). Therefore, we write the homogeneous polynomials for $f(\phi_{n-1}, \phi_n, \phi_{n+1})$ and $g(\phi_n, \phi_{n+1})$ as

$$f = \beta_1\phi_n^3 + \beta_2\phi_n^2(\phi_{n+1} + \phi_{n-1}) + \beta_3\phi_n(\phi_{n+1}^2 + \phi_{n-1}^2) + \beta_4\phi_{n+1}\phi_{n-1}\phi_n$$
$$+ \beta_5(\phi_{n+1}^3 + \phi_{n-1}^3) + \beta_6\phi_{n-1}\phi_{n+1}(\phi_{n+1} + \phi_{n-1}), \quad (5.2.9)$$

and

$$g = \tilde{\beta}_1 \phi_n^2 \phi_{n+1}^2 + \tilde{\beta}_2 (\phi_n^2 + \phi_{n+1}^2) \phi_n \phi_{n+1} + \tilde{\beta}_3 (\phi_n^4 + \phi_{n+1}^4), \qquad (5.2.10)$$

where $(\beta_1, ..., \beta_6) \in \mathbb{R}^6$ and $(\tilde{\beta}_1, \tilde{\beta}_2, \tilde{\beta}_3) \in \mathbb{R}^3$ are arbitrary parameters.

Subtracting (5.2.5) from the same equation with $n+1$ replaced by n and assuming that $\{\phi_n\}_{n \in \mathbb{Z}} \in \mathbb{R}^{\mathbb{Z}}$ is real-valued, we obtain the second-order difference equation (5.2.3) under the constraint

$$g(\phi_n, \phi_{n+1}) - g(\phi_{n-1}, \phi_n) = (\phi_{n+1} - \phi_{n-1}) f(\phi_{n-1}, \phi_n, \phi_{n+1}). \qquad (5.2.11)$$

If $f(\phi_{n-1}, \phi_n, \phi_{n+1})$ and $g(\phi_n, \phi_{n+1})$ are polynomials in the form (5.2.9) and (5.2.10), the constraint (5.2.11) gives the following constraints on parameters of the nonlinear function:

$$\beta_1 = \beta_3 = \beta_4, \quad \beta_5 = \beta_6. \qquad (5.2.12)$$

Therefore, three parameters $\beta_2 = \tilde{\beta}_1$, $\beta_4 = \tilde{\beta}_2$, and $\beta_6 = \tilde{\beta}_3$ are still arbitrary.

We shall now prove that the existence of the first-order invariant (5.2.5) is sufficient for the existence of the translationally invariant family (5.2.4) of localized modes for small values of $h > 0$.

Note that if the sequence $\{\phi_n\}_{n \in \mathbb{Z}}$ decays to zero as $|n| \to \infty$, then $E = 0$. Expressing $(\phi_{n+1} - \phi_n)^2$ from the first-order invariant (5.2.5), we obtain the initial-value problem for the sequence $\{\phi_n\}_{n \in \mathbb{Z}}$ in the implicit form

$$\begin{cases} (\phi_{n+1} - \phi_n)^2 = h^2 Q(\phi_n, \phi_{n+1}), & n \in \mathbb{Z}, \\ \phi_0 = \varphi, \end{cases} \qquad (5.2.13)$$

where $\varphi \in \mathbb{R}$ is the initial data, iterations in both positive and negative directions of $n \in \mathbb{Z}$ can be considered, and

$$Q(\phi_n, \phi_{n+1}) = -\omega \phi_n \phi_{n+1} - g(\phi_n, \phi_{n+1}).$$

Let $g(\phi_n, \phi_{n+1})$ be given by the polynomial (5.2.10). We shall assume the following constraint on parameters of $g(\phi_n, \phi_{n+1})$,

$$\tilde{\beta}_1 + 2\tilde{\beta}_2 + 2\tilde{\beta}_3 = 1,$$

which ensures that the discrete first-order invariant (5.2.5) recovers the continuous first integral (5.2.7) in the continuous limit $h \to 0$ with $g(\phi, \phi) = \phi^4$. Localized modes (5.2.8) exist in the continuous limit only for $\omega < 0$, which is also assumed in what follows.

To prove the existence of the translationally invariant localized modes, we prove the existence of two families of solutions of the initial-value problem (5.2.13), one monotonically decreasing for $n > 0$ and the other monotonically increasing for $n < 0$. Both families depend continuously on the initial data φ.

Lemma 5.5 *Fix $\omega < 0$. There exists $h_0 > 0$ and $L > \sqrt{|\omega|}$ such that for any $h \in (0, h_0)$, the algebraic (quartic) equation*

$$(y - x)^2 = h^2 Q(x, y) \qquad (5.2.14)$$

defines a convex, simply connected, closed curve inside the domain $[0, L] \times [0, L]$. *The curve is symmetric about the diagonal* $y = x$ *and has two intersections with the diagonal at* $(x, y) = (0, 0)$ *and* $(x, y) = (\sqrt{|\omega|}, \sqrt{|\omega|})$.

Proof Symmetry of the curve about the diagonal $y = x$ follows from the fact that $Q(x, y) = Q(y, x)$. Intersections with the diagonal $y = x$ follows from the continuity condition that gives $Q(x, x) = x^2(|\omega| - x^2)$. To prove convexity, we consider the behavior of the curve for small h in two regions of the (x, y) plane: in the quadrant $[0, x_0] \times [0, x_0]$ for $x_0 = \sqrt{|\omega|} - r_0 h^2$ and in the disk centered at $(x, y) = (\sqrt{|\omega|}, \sqrt{|\omega|})$ with radius $r_0 h^2$, where

$$r_0 > \left(\frac{\sqrt{|\omega|}}{2} \right)^3$$

is an arbitrary h-independent number.

Let $y = x + z$ and consider two branches of the algebraic equation (5.2.14) in the implicit form

$$F_\pm(x, z, h) = z \mp h\sqrt{Q(x, x + z)} = 0. \tag{5.2.15}$$

For any $x \geq 0$, we have

$$F_\pm(x, 0, h) = \mp hx\sqrt{|\omega| - x^2}, \qquad \partial_z F_\pm(x, 0, h) = 1 \mp \frac{h(|\omega| - 2x^2)}{2\sqrt{|\omega| - x^2}}.$$

It is clear that $F_\pm(x, 0, 0) = 0$ and $\partial_z F_\pm(x, 0, 0) = 1$, while $F_\pm(x, 0, h)$ is continuously differentiable in x and h and $\partial_z F_\pm(x, 0, h)$ is uniformly bounded in x and h on $[0, x_0] \times [0, h_0]$, for $x_0 = \sqrt{|\omega|} - r_0 h^2$, where $h_0 > 0$ is small and $r_0 > 0$ is h-independent. By the Implicit Function Theorem, the implicit equations (5.2.15) define unique roots $z = \pm h S_\pm(x, h)$ in the domain $[0, x_0] \times [0, h_0]$, where $h S_\pm(x, h)$ are positive, continuously differentiable functions in x and h. Therefore, the algebraic equation (5.2.14) defines two strictly increasing curves located above and below the diagonal $y = x$ in the square $[0, x_0] \times [0, x_0]$. In the limit $h_0 \to 0$, the two branches converge to the diagonal $y = x$ for $x \in [0, \sqrt{|\omega|}]$.

Derivatives of the algebraic equation (5.2.14) in x are defined for any branch of the curve by

$$y' \left[2(y - x) - h^2(|\omega|x - \partial_y g(x, y)) \right] = 2(y - x) + h^2(|\omega|y - \partial_x g(x, y)),$$

$$y'' \left[2(y - x) - h^2(|\omega|x - \partial_y g(x, y)) \right] = -2(y' - 1)^2 + h^2(2|\omega|y' - \tilde{g}),$$

where $\tilde{g} = \partial^2_{xx} g(x, y) - 2y' \partial^2_{xy} g(x, y) - (y')^2 \partial^2_{yy} g(x, y)$. Therefore, $y' = -1$ at $(x, y) = (\sqrt{|\omega|}, \sqrt{|\omega|})$ for any $h > 0$ and there exists a small neighborhood of the point $(x, y) = (\sqrt{|\omega|}, \sqrt{|\omega|})$ where the upper branch of the curve is strictly decreasing.

Let B_h^+ be the upper semi-disk centered at $(x, y) = (\sqrt{|\omega|}, \sqrt{|\omega|})$ with a radius $r_0 h^2$, where $r_0 > (\sqrt{|\omega|}/2)^3$ is h-independent. There exists an h-independent constant $C > 0$ such that

$$2(y - x) - h^2(|\omega|x - \partial_y g(x, y)) \geq Ch^2, \qquad (x, y) \in B_h^+.$$

From the second derivative, there is $C_1 > 0$ such that $y'' \leq -C_1/h^2$ for any $(x,y) \in B_h^+$. Using the rescaled variables in B_h^+,

$$x = \sqrt{|\omega|} - Xh^2, \quad y = \sqrt{|\omega|} + Yh^2,$$

we find that the curvature of the curve in new variables is bounded by $Y''(X) \leq -C_1$ and therefore the first derivatives $Y'(X)$ and $y'(x)$ may only change by a finite number in $(x,y) \in B_h$. Therefore, there exists $0 < C_2 < \infty$ such that $y'(x) = C_2$ for some $x \leq x_0 = \sqrt{|\omega|} - r_0 h^2$. By the first part of the proof, the upper branch of the curve is monotonically increasing in the square $[0, x_0] \times [0, x_0]$. By the second part of the proof, the curve $y(x)$ has a single maximum for $x_0 \leq x \leq \sqrt{|\omega|}$. Thus, the curve defined by the quartic equation (5.2.14) is convex for $y \geq x \geq 0$ (and for $0 \leq y \leq x$ by symmetry). $\qquad \square$

Let ψ_0 be the maximal value of y on the curve defined by equation (5.2.14) and φ_0 be the corresponding value of x. By Lemma 5.5, we have

$$\varphi_0 < \sqrt{|\omega|} < \psi_0.$$

Lemma 5.6 *Fix $\omega < 0$ and small $h > 0$. The initial-value problem (5.2.13) admits a unique monotonically decreasing sequence $\{\phi_n\}_{n \geq 0}$ for any $\varphi \in (0, \psi_0)$ that converges to zero from above as $n \to \infty$. The sequence $\{\phi_n\}_{n \geq 0}$ is continuous with respect to φ and h.*

Proof By Lemma 5.5, there exists a unique lower branch of the curve defined by the implicit equation (5.2.14) below the diagonal $y = x$ in the form $y - x = -hS_-(x, h)$, where $hS_-(x, h) > 0$ is continuously differentiable with respect to x and h in the domain $[0, \sqrt{|\omega|}] \times [0, h_0]$. Let $\{\phi_n\}_{n \geq 0}$ satisfy the initial-value problem

$$\begin{cases} \phi_{n+1} = \phi_n - hS_-(\phi_n, h), & n \in \mathbb{N}, \\ \phi_0 = \varphi, \end{cases} \qquad (5.2.16)$$

for any $\varphi \in (0, \psi_0)$. It is clear that $\{\phi_n\}_{n \geq 0}$ is monotonically decreasing as long as ϕ_n remains positive. We shall prove that $\{\phi_n\}_{n \geq 0}$ converges to zero from above, where 0 is the root of $S_-(\varphi, h) = 0$. Since $g(x, y)$ is a quartic polynomial, there exists a constant $C > 0$ that depends on ω and is independent of h, such that

$$\forall |\phi_n| \leq \sqrt{|\omega|} : \quad |\phi_{n+1} - \phi_n| \leq Ch\phi_n.$$

If h is sufficiently small, then $Ch < 1$ and $0 < \phi_{n+1} < \phi_n$. Therefore, the sequence $\{\phi_n\}_{n \geq 0}$ is bounded from below by 0. By the Weierstrass Theorem, the monotonically decreasing and bounded-from-below sequence $\{\phi_n\}_{n \geq 0}$ converges as $n \to \infty$ to the fixed point $\phi = 0$. Continuity of the sequence $\{\phi_n\}_{n \geq 0}$ in h and φ follows from the continuity of $hS_-(\phi, h)$ in φ and h. $\qquad \square$

Lemma 5.7 *Fix $\omega < 0$ and small $h > 0$. The initial-value problem (5.2.13) admits a unique monotonically increasing sequence $\{\phi_n\}_{n \leq 0}$ for any $\varphi \in (0, \psi_0)$ that converges to zero from above as $n \to -\infty$. The sequence $\{\phi_n\}_{n \leq 0}$ is continuous with respect to φ and h.*

Proof By Lemma 5.5, there exists a unique upper branch of the curve defined by (5.2.14) above the diagonal $y = x$ in the form $y - x = hS_+(x, h)$, where $hS_+(x, h) > 0$ is continuously differentiable with respect to x and h in the domain $[0, \sqrt{|\omega|}] \times [0, h_0]$. Let $\{\phi_n\}_{n \leq 0}$ satisfy the initial-value problem

$$\begin{cases} \phi_{n+1} = \phi_n + hS_+(\phi_n, h), & (-n) \in \mathbb{N}, \\ \phi_0 = \varphi, \end{cases} \tag{5.2.17}$$

for any $\varphi \in (0, \sqrt{|\omega|})$. The proof that the monotonically increasing sequence $\{\phi_n\}_{n \leq 0}$ converges to zero from above as $n \to -\infty$ repeats the proof of Lemma 5.6. $\square$

By Lemmas 5.6 and 5.7, for fixed $\omega < 0$ and small $h > 0$, the initial-value problem (5.2.13) admits the following solutions:

(i) For any given $\varphi \in (0, \psi_0]$, there exists a symmetric single-humped sequence $\{\phi_n\}_{n \in \mathbb{Z}}$ with maximum at $n = 0$, such that $\phi_n = \phi_{-n}$. Let S_{on} denote the countable set of values of $\{\phi_n\}_{n \in \mathbb{Z}}$ corresponding to $\varphi = \psi_0$.
(ii) For $\varphi = \sqrt{\omega}$, there exists a symmetric two-site single-humped sequence. Let S_{inter} denote the countable set of values of $\{\phi_n\}_{n \in \mathbb{Z}}$ corresponding to $\varphi = \sqrt{|\omega|}$.
(iii) For any $\varphi \in (0, \psi_0) \backslash \{S_{\mathrm{on}}, S_{\mathrm{inter}}\}$, there exists a unique non-symmetric single-humped sequence $\{\phi_n\}_{n \in \mathbb{Z}}$ with $\phi_k \neq \phi_m$ for all $k \neq m$.

Figures 5.2 and 5.3 show the construction of the symmetric sequences (i) and (ii) from solutions of the quartic equation (5.2.14) for

$$Q(x, y) = |\omega| xy - \frac{1}{2}(x^4 + y^4).$$

Let us consider the single-humped solution (iii) for $\phi_0 \in (0, \psi_0) \backslash \{S_{\mathrm{on}}, S_{\mathrm{inter}}\}$ and complete it by solutions (i) for S_{on} and (ii) for S_{inter}. By Lemmas 5.6 and 5.7, the single-humped sequence is continuous in $\varphi \in (0, \psi_0]$.

We claim that the sequence $\{\phi_n\}_{n \in \mathbb{Z}}$ converges pointwise to the sequence $\{\phi_s(hn - s))\}_{n \in \mathbb{Z}}$ for some $s \in \mathbb{R}$, where

$$\phi_s(x) = \sqrt{|\omega|} \operatorname{sech}\left(\sqrt{|\omega|}x\right).$$

Recall that $\phi_s(x)$ solves the second-order differential equation (5.2.6), which is related to the second-order difference equation (5.2.3) in the continuous limit $h \to 0$.

By the translational invariance of the solution $\{\phi_n\}_{n \in \mathbb{Z}}$, we can place the maximum at $n = 0$ if $\varphi = \psi_0$, which is equivalent to the choice $s = 0$ (S_{on}). Using the power series expansions in h^2 for solutions of the difference equation (5.2.3), we obtain the following theorem that justifies the existence of the translationally invariant family (5.2.4) of localized modes.

Theorem 5.3 *Fix $\omega < 0$ and small $h > 0$. There exists a continuous family of single-humped localized modes $\{\phi_n\}_{n \in \mathbb{Z}}$ of the initial-value problem (5.2.13) for any $\varphi \in (0, \psi_0]$. Moreover, there exists $C > 0$ and $s \in \mathbb{R}$ such that*

$$\max_{n \in \mathbb{Z}} |\phi_n - \sqrt{|\omega|} \operatorname{sech}(\sqrt{|\omega|}(nh - s))| \leq Ch^2. \tag{5.2.18}$$

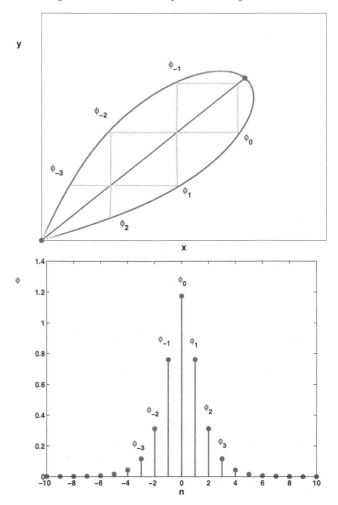

Figure 5.2 Top: the symmetric single-humped sequence (i). Bottom: the corresponding solution $\{\phi_n\}_{n\in\mathbb{Z}}$ for $\varphi = \psi_0$.

Remark 5.2 The initial-value problem (5.2.13) admits other families of single-humped localized modes in addition to the translational invariant family (5.2.4).

Exercise 5.3 Show for the AL lattice with $f = |\phi_n|^2(\phi_{n+1} + \phi_{n-1})$ that the second-order difference equation (5.2.3) admits the exact solution,

$$\phi_n = \psi_0 \operatorname{sech}[\kappa(hn - s)], \quad \psi_0 = \frac{1}{h}\sinh(\kappa h), \quad \omega = \frac{4}{h^2}\sinh^2\left(\frac{\kappa h}{2}\right),$$

where $(\kappa, s) \in \mathbb{R}^2$ are arbitrary parameters.

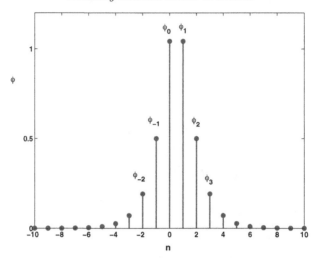

Figure 5.3 The symmetric two-site single-humped sequence (ii) for $\varphi = \sqrt{\omega}$.

Substituting constraints (5.2.12) into the nonlinear function (5.2.9), we obtain

$$f = \beta_2\phi_n^2(\phi_{n+1} + \phi_{n-1}) + \beta_4\phi_n\left(\phi_n^2 + \phi_{n+1}^2 + \phi_{n-1}^2 + \phi_{n+1}\phi_{n-1}\right)$$
$$+ \beta_6(\phi_{n+1} + \phi_{n-1})(\phi_{n+1}^2 + \phi_{n-1}^2), \tag{5.2.19}$$

where $(\beta_2, \beta_4, \beta_6) \in \mathbb{R}^3$ are arbitrary parameters. The second-order difference equation (5.2.3) with nonlinearity (5.2.19) admits a translationally invariant family (5.2.4) of localized modes. Although our analysis was developed for real-valued solutions of the second-order difference equation (5.2.3), computations can also be developed for complex-valued solutions.

Exercise 5.4 Using the first-order invariant (5.2.5) for complex-valued solutions of the second-order difference equation (5.2.3), obtain a more general nonlinear function for the generalized DNLS equation (5.2.2),

$$f = \alpha_2|u_n|^2(u_{n+1} + u_{n-1}) + \alpha_3 u_n^2(\bar{u}_{n+1} + \bar{u}_{n-1})$$
$$+ \alpha_4\left[(|u_n|^2 + |u_{n+1}|^2 + |u_{n-1}|^2)u_n + u_{n+1}u_{n-1}\bar{u}_n\right],$$
$$+ \alpha_6\left[(|u_n|^2 + u_{n+1}\bar{u}_{n-1} + \bar{u}_{n+1}u_{n-1})u_n + (u_{n+1}^2 + u_{n-1}^2 - u_{n+1}u_{n-1})\bar{u}_n\right]$$
$$+ \alpha_8(|u_{n+1}|^2 + |u_{n-1}|^2)(u_{n+1} + u_{n-1})$$
$$+ \alpha_9\left(u_{n+1}^2\bar{u}_{n-1} + u_{n-1}^2\bar{u}_{n+1} - |u_{n+1}|^2u_{n-1} - |u_{n-1}|^2u_{n+1}\right), \tag{5.2.20}$$

where $(\alpha_2, \alpha_3, \alpha_4, \alpha_6, \alpha_8, \alpha_9) \in \mathbb{R}^6$ are arbitrary parameters.

It can be checked by direct differentiation with respect to the time variable t that the generalized DNLS equation (5.2.2) with the nonlinear function $f(u_{n-1}, u_n, u_{n+1})$ in (5.2.20) has the conserved quantity,

$$P = i \sum_{n \in \mathbb{Z}} (\bar{u}_{n+1} u_n - u_{n+1} \bar{u}_n).$$

This conserved quantity can be thought to be a discretization of the momentum,

$$P = i \int_{\mathbb{R}} (\bar{u} u_x - u \bar{u}_x) \, dx$$

of the continuous NLS equation $iu_t + u_{xx} + 2|u|^2 u = 0$ (Section 1.2.1).

Although discrete equations have no continuous translational invariance, the generalized DNLS equation (5.2.2) with the special nonlinear function $f(u_{n-1}, u_n, u_{n+1})$ in (5.2.20) inherits the momentum conservation of the continuous NLS equation. This was the original idea of Kevrekidis [111] (without a proof) to identify all exceptional nonlinear functions $f(u_{n-1}, u_n, u_{n+1})$. It was proved by Pelinovsky [154] that all such DNLS equations with homogeneous cubic polynomials $f(u_{n-1}, u_n, u_{n+1})$ admit the first-order invariant (5.2.5) for solutions of the second-order difference equation (5.2.3).

Exercise 5.5 Consider the generalized DNLS equation (5.2.2) with a general nonlinear function $f(u_{n-1}, u_n, u_{n+1})$ and find the constraints on $f(u_{n-1}, u_n, u_{n+1})$ from the two independent conditions: (1) P is constant in t and (2) the second-order difference equation (5.2.3) possesses the first-order invariant (5.2.5). Can you show that the two conditions are equivalent?

Exercise 5.6 Find all possible homogeneous cubic polynomials $f(u_{n-1}, u_n, u_{n+1})$ in the generalized DNLS equation (5.2.2) that conserve the modified momentum

$$\tilde{P} = i \sum_{n \in \mathbb{Z}} (\bar{u}_{n+2} u_n - u_{n+2} \bar{u}_n),$$

in time $t \in \mathbb{R}$. Can you find the first-order invariant for the second-order difference equation (5.2.3) with this special nonlinear function?

As the opposite to the result on the conservation of P, it can also be checked by direct substitution that the generalized DNLS equation (5.2.2) with the nonlinear function $f(u_{n-1}, u_n, u_{n+1})$ in (5.2.20) does not admit the standard symplectic Hamiltonian structure,

$$i \frac{du_n}{dt} = \frac{\partial H}{\partial \bar{u}_n}, \quad H = \sum_{n \in \mathbb{Z}} \left[\frac{|u_{n+1} - u_n|^2}{h^2} - F(u_n, u_{n+1}) \right], \tag{5.2.21}$$

where $F(u_n, u_{n+1}) : \mathbb{C}^2 \to \mathbb{R}$ satisfies properties (S1)–(S3). This negative result can be a disappointment for physicists working with the Peierls–Nabarro potential.

Nevertheless, this result does not exclude the possibility of a more complicated symplectic Hamiltonian structure for the DNLS equation (5.2.2). For instance, the integrable AL lattice with $f = |u_n|^2(u_{n+1} + u_{n-1})$ is known to possess a non-standard symplectic Hamiltonian structure,

$$\mathrm{i}\frac{du_n}{dt} = -(1 + h^2|u_n|^2)\frac{\partial H}{\partial \bar{u}_n},$$

with the Hamiltonian function

$$H = \frac{1}{h^2}\sum_{n\in\mathbb{Z}}(\bar{u}_n u_{n+1} + u_n \bar{u}_{n+1}) - \frac{2}{h^4}\sum_{n\in\mathbb{Z}}\log(1 + h^2|u_n|^2).$$

The following exercises are based on the technical computations of Barashenkov *et al.* [14]. They can be used to generate additional classes of special nonlinear functions $f(u_{n-1}, u_n, u_{n+1})$ in the generalized DNLS equation (5.2.2) with translationally invariant solutions. Because the first-order invariant in these examples is linear with respect to $(\phi_{n+1} - \phi_n)$, the relevant translationally invariant solutions are not localized modes but monotonically increasing (or decreasing) *kinks* that approach the nonzero boundary conditions at infinity.

Exercise 5.7 Consider the second-order difference equation (5.2.3) for real-valued solutions $\{\phi_n\}_{n\in\mathbb{Z}}$ and find the constraints on the nonlinear function (5.2.9) from the existence of another conserved quantity,

$$\frac{1}{h}(\phi_{n+1} - \phi_n) = \tilde{g}(\phi_n, \phi_{n+1}), \quad n \in \mathbb{Z}, \tag{5.2.22}$$

where

$$\tilde{g}(\phi_n, \phi_{n+1}) = \gamma_0 + \gamma_1(\phi_n^2 + \phi_{n+1}^2) + \gamma_2\phi_n\phi_{n+1},$$

and $(\gamma_0, \gamma_1, \gamma_2) \in \mathbb{R}^3$ are arbitrary parameters. Show that if $\{\phi_n\}_{n\in\mathbb{Z}}$ is a solution of the first-order difference equation (5.2.22), then

$$\mathcal{I} := (\phi_n - \phi_{n-1})\tilde{g}(\phi_{n+1}, \phi_n) - (\phi_{n+1} - \phi_n)\tilde{g}(\phi_n, \phi_{n-1}) = 0, \quad n \in \mathbb{Z}.$$

Therefore, $\mathcal{I}$ multiplied by an arbitrary parameter can be added to the corresponding nonlinear function $f(\phi_{n-1}, \phi_n, \phi_{n+1})$.

Exercise 5.8 Show that if $\{\phi_n\}_{n\in\mathbb{Z}}$ is a solution of the first-order difference equation

$$\frac{1}{h}(\phi_{n+1} - \phi_n) = -1 + \phi_n\phi_{n+1}, \quad n \in \mathbb{Z}, \tag{5.2.23}$$

then

$$\tilde{\mathcal{I}} := \left(1 + \frac{h^2}{4}\right)\phi_n(\phi_{n+1} + \phi_{n-1}) - 2\phi_{n+1}\phi_{n-1} - \frac{h^2}{2} = 0, \quad n \in \mathbb{Z}.$$

Again, $\tilde{\mathcal{I}}$ multiplied by an arbitrary parameter can be added to the corresponding nonlinear function $f(\phi_{n-1}, \phi_n, \phi_{n+1})$. Find the explicit solution of the first-order difference equation (5.2.23) in the form $\phi_n = \tanh(bhn)$, $n \in \mathbb{Z}$, for some $b > 0$.

5.3 Melnikov integral for traveling localized modes

Although the generalized DNLS equation (5.2.2) may have translationally invariant family (5.2.4) of localized modes, this fact alone does not imply the existence of traveling localized modes with nonzero wave speed because traveling solutions satisfy a more complicated equation compared with the second-order difference equation (5.2.3). Direct substitution of

$$u_n(t) = \phi(hn - 2ct - s)e^{-i\omega t + i\theta}, \quad n \in \mathbb{Z}, \quad t \in \mathbb{R}, \tag{5.3.1}$$

into the generalized DNLS equation (5.2.2) with the nonlinear function f satisfying properties (P1)–(P3) at the start of this chapter results in the differential advance–delay equation,

$$2\mathrm{i}c\frac{d\phi}{dz} = \frac{\phi(z+h) - 2\phi(z) + \phi(z-h)}{h^2}$$
$$+ \omega\phi(z) + f(\phi(z-h), \phi(z), \phi(z+h)), \quad z \in \mathbb{R}. \tag{5.3.2}$$

We are looking for a single-humped solution $\phi \in H^1(\mathbb{R})$ of the differential advance–delay equation (5.3.2). To simplify the formalism, we shall assume that $f(u_{n-1}, u_n, u_{n+1})$ is a homogeneous cubic polynomial.

Besides parameters of the nonlinear function $f(u_{n-1}, u_n, u_{n+1})$ and the lattice spacing h, the differential advance–delay equation (5.3.2) has two "internal" parameters ω and c. It is convenient to replace (ω, c) by new parameters (κ, β) according to the parameterization

$$\omega = -\frac{2}{h}\beta c - \frac{2}{h^2}\left(\cos(\beta)\cosh(\kappa) - 1\right), \quad c = \frac{1}{h\kappa}\sin(\beta)\sinh(\kappa), \tag{5.3.3}$$

and to transform the variables $(z, \phi(z))$ to new variables $(Z, \Phi(Z))$, where

$$\phi(z) = \frac{1}{h}\Phi(Z)e^{\mathrm{i}\beta Z}, \quad Z = \frac{z}{h}.$$

The new function $\Phi(Z)$ satisfies the differential advance–delay equation

$$2\mathrm{i}\sin(\beta)\frac{\sinh(\kappa)}{\kappa}\Phi'(Z)$$
$$= \Phi(Z+1)e^{\mathrm{i}\beta} + \Phi(Z-1)e^{-\mathrm{i}\beta} - 2\cos(\beta)\cosh(\kappa)\,\Phi(Z)$$
$$+ f(\Phi(Z-1)e^{-\mathrm{i}\beta}, \Phi(Z), \Phi(Z+1)e^{\mathrm{i}\beta}), \quad Z \in \mathbb{R}, \tag{5.3.4}$$

where the lattice spacing h has been scaled out.

For some special nonlinear functions $f(u_{n-1}, u_n, u_{n+1})$, there exist exact solutions of the differential advance–delay equation (5.3.4). By direct differentiation, one can check that there exists a two-parameter family of traveling localized modes,

$$\Phi(Z) = \sinh(\kappa)\mathrm{sech}(\kappa Z), \quad \kappa > 0, \quad \beta \in [0, 2\pi], \tag{5.3.5}$$

for the integrable AL lattice with $f = |u_n|^2(u_{n+1} + u_{n-1})$. Similarly, there exists a one-parameter family of traveling localized modes,

$$\Phi(Z) = \frac{\sinh(\kappa)}{\sqrt{\alpha_2 - \alpha_3}}\mathrm{sech}\,(\kappa Z), \quad \kappa > 0, \quad \beta = \frac{\pi}{2}, \tag{5.3.6}$$

in the non-integrable generalized DNLS equation with

$$f = \alpha_2 |u_n|^2 (u_{n+1} + u_{n-1}) + \alpha_3 u_n^2 (\bar{u}_{n+1} + \bar{u}_{n-1}), \quad \alpha_2 > \alpha_3 \neq 0. \tag{5.3.7}$$

We shall consider two statements about solutions of the differential advance–delay equation (5.3.4), which seem to contradict each other at first glance. On the one hand, we show that no localized single-pulse localized modes exist in the general case. On the other hand, we prove that, if a solution exists along a curve in the parameter plane (κ, β), then these solutions generally persist in a neighborhood of the curve with respect to perturbations in the nonlinear function $f(u_{n-1}, u_n, u_{n+1})$. Using the language of dynamical system theory, we say that the existence of traveling solutions in the differential advance–delay equation (5.3.4) is a *codimension-one bifurcation*. Hence one-parameter existence curves in the two-parameter plane (κ, β) are not generic but, if they exist for one model, they are structurally stable with respect to the parameter continuation in this model.

Since the exact solution (5.3.6) is constructed for $\beta = \frac{\pi}{2}$, we are particularly interested in the localized modes of the differential advance–delay equation (5.3.4) near $\kappa > 0$ and $\beta = \frac{\pi}{2}$, although the analysis is expected to hold also for $\beta \neq \frac{\pi}{2}$. In what follows, κ is a fixed positive number. This analysis is based on the work of Pelinovsky *et al.* [157].

Let us rewrite the differential advance–delay equation (5.3.4) in a convenient form, which is suitable for separation of the real and imaginary parts in the function $\Phi(Z)$. To this end, we obtain

$$\cos(\beta) \left(\Phi_+ + \Phi_- - 2\cosh(\kappa)\, \Phi \right) + \mathrm{i} \sin(\beta) \left(\Phi_+ - \Phi_- - 2\frac{\sinh(\kappa)}{\kappa} \frac{d\Phi}{dZ} \right)$$
$$+ f_r + \mathrm{i} f_i = 0, \tag{5.3.8}$$

where $\Phi_\pm = \Phi(Z \pm 1)$ and $f_r + \mathrm{i} f_i = f(\Phi_- e^{-\mathrm{i}\beta}, \Phi, \Phi_+ e^{\mathrm{i}\beta})$. In particular, if

$$f = \alpha_2 |u_n|^2 (u_{n+1} + u_{n-1}) + \alpha_3 u_n^2 (\bar{u}_{n+1} + \bar{u}_{n-1})$$
$$+ \alpha_8 (|u_{n+1}|^2 + |u_{n-1}|^2)(u_{n+1} + u_{n-1}),$$

then the differential advance–delay equation (5.3.4) with $\beta = \frac{\pi}{2}$ and $\Phi(Z) : \mathbb{R} \to \mathbb{R}$ reduces to the scalar equation

$$2\frac{\sinh(\kappa)}{\kappa} \frac{d\Phi}{dZ} = \left[1 + (\alpha_2 - \alpha_3)\Phi^2 + \alpha_8 (\Phi_+^2 + \Phi_-^2) \right] (\Phi_+ - \Phi_-), \quad Z \in \mathbb{R}. \tag{5.3.9}$$

We note that the function (5.3.6) is an exact solution of equation (5.3.9) for $\alpha_8 = 0$ and $\alpha_2 > \alpha_3$. Therefore, we ask if the exact solution (5.3.6) is structurally stable in the scalar equation (5.3.9) with respect to $\alpha_8 \neq 0$ and in the complex-valued equation (5.3.8) for a general homogeneous cubic nonlinear function $f(u_{n-1}, u_n, u_{n+1})$.

Exercise 5.9 Consider the generalized DNLS equation with the nonlinear function,

$$f = \alpha_1 |u_n|^2 u_n + \alpha_4 (|u_{n+1}|^2 + |u_{n-1}|^2) u_n + \alpha_6 (u_{n+1}^2 + u_{n-1}^2) \bar{u}_n$$
$$+ \alpha_8 \left(2|u_n|^2 (u_{n+1} + u_{n-1}) + u_n^2 (\bar{u}_{n+1} + \bar{u}_{n-1}) + |u_{n+1}|^2 u_{n+1} + |u_{n-1}|^2 u_{n-1} \right),$$

which is known to admit the standard symplectic Hamiltonian structure (5.2.21). Find the constraints on $(\alpha_1, \alpha_4, \alpha_6, \alpha_8) \in \mathbb{R}^4$, under which the differential advance–delay equation (5.3.4) with $\beta = \frac{\pi}{2}$ can be reduced to the scalar equation (5.3.9) for $\Phi(Z) : \mathbb{R} \to \mathbb{R}$.

To explain why $\beta = \frac{\pi}{2}$ is so special in the existence of localized modes of the differential advance–delay equation (5.3.4), we shall consider the following linear equation,

$$\cos(\beta)\left(\Phi_+ + \Phi_- - 2\cosh(\kappa)\,\Phi\right) + i\sin(\beta)\left(\Phi_+ - \Phi_- - 2\frac{\sinh(\kappa)}{\kappa}\frac{d\Phi}{dZ}\right) = 0.$$
(5.3.10)

Applying the Laplace transform to the linear equation (5.3.10), we infer that all linear eigenmodes are given by $e^{\lambda Z}$, where λ is a root of the characteristic equation

$$D(\lambda; \kappa, \beta) := \cos(\beta)\left[\cosh(\lambda) - \cosh(\kappa)\right] + i\sin(\beta)\left[\sinh(\lambda) - \frac{\sinh(\kappa)}{\kappa}\lambda\right] = 0.$$
(5.3.11)

Roots of $D(\lambda; \kappa, \beta) = 0$ with $\text{Re}(\lambda) > 0$ and $\text{Re}(\lambda) < 0$ correspond to the exponentially growing and decaying eigenmodes that provide the spatial decay of the solution $\Phi(Z)$, while the roots with $\text{Re}(\lambda) = 0$ correspond to bounded eigenmodes that result in the oscillatory behavior of the solution $\Phi(Z)$. For any $\beta \in [0, 2\pi]$, there always exists a pair of real roots $\lambda = \pm\kappa$, which are responsible for localization of the single-humped solution $\Phi(Z)$. However, for any $\beta \neq \{0, \pi\}$, there exist also roots with $\text{Re}(\lambda) = 0$ which may destroy the localization of $\Phi(Z)$.

If $\beta = \frac{\pi}{2}$, the only root of $D(\lambda; \kappa, \beta) = 0$ with $\text{Re}(\lambda) = 0$ is at $\lambda = 0$. Near the line $\beta = \frac{\pi}{2}$ and $\kappa \in \mathbb{R}_+$, there exists a single root $\lambda = ik$ with $k \in \mathbb{R}$. When β is reduced to zero or increased to π, the number of roots of $D(\lambda; \kappa, \beta) = 0$ with $\text{Re}(\lambda) = 0$ increases to infinity. The limits $\beta \to 0$ and $\beta \to \pi$ are singular: no roots exist for $\beta = 0$ and $\beta = \pi$, but many roots with large values of $\text{Im}(\lambda)$ exist for small values of $\beta \neq 0$ and $\beta \neq \pi$. This behavior is illustrated in Figure 5.4, which shows the behavior of $D(ik; \kappa, \beta)$ versus k for $\kappa = 1$ and different values of β.

Note that the values of $\beta = 0$ and $\beta = \pi$ correspond to the stationary solutions of the advance–delay equation (5.3.2) with $c = 0$. Existence of the continuous family of single-pulse localized modes was considered in Section 5.2 using the second-order difference equation (5.2.3). However, persistence of the stationary solutions for $c \neq 0$ is a singular perturbation problem.

Exercise 5.10 Consider the differential advance–delay equation for traveling localized modes of the discrete Klein–Gordon equation,

$$c^2 \phi''(z) = \frac{\phi(z+h) - 2\phi(z) + \phi(z-h)}{h^2} - \phi(z) + f(\phi(z-h), \phi(z), \phi(z+h)),$$

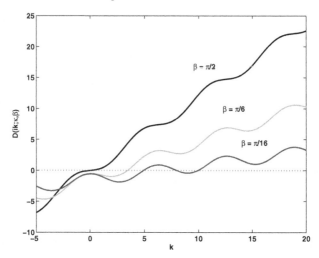

Figure 5.4 The behavior of $D(ik; \kappa, \beta)$ versus k for $\kappa = 1$ and different values of β.

where c is the wave speed and f is a homogeneous cubic nonlinear function in its variables. Find the dispersion relation in the linearization around the zero solution and show that for small $h > 0$ and $c \in (0, 1)$, there exist finitely many purely imaginary roots but the number of such roots becomes infinite as c approaches zero.

To consider persistence of the exact solution (5.3.6) in the scalar equation (5.3.9) with $\alpha_2 > \alpha_3$ and $\alpha_8 \in \mathbb{R}$, we replace $\Phi(Z)$ by $\Phi(Z) + U(Z)$, where $\Phi(Z)$ is the exact solution (5.3.6) and $U(Z)$ satisfies the differential advance–delay equation

$$L_+ U = N(U) + F(\Phi + U), \tag{5.3.12}$$

with

$$L_+ U = -2\frac{\sinh(\kappa)}{\kappa}\frac{dU}{dZ} + (1 + (\alpha_2 - \alpha_3)\Phi^2)(U_+ - U_-) + 2(\alpha_2 - \alpha_3)\Phi(\Phi_+ - \Phi_-)U,$$

$$N(U) = -(\alpha_2 - \alpha_3)\left(U^2(\Phi_+ - \Phi_-) + 2\Phi U(U_+ - U_-) + U^2(U_+ - U_-)\right),$$

$$F(\Phi) = -\alpha_8(\Phi_+^2 + \Phi_-^2)(\Phi_+ - \Phi_-).$$

Operator L_+ can also be written in the explicit form

$$L_+ U = -2\frac{\sinh(\kappa)}{\kappa}\frac{dU}{dZ} + \left(1 + \sinh^2(\kappa)\,\text{sech}^2(\kappa Z)\right)(U(Z+1) - U(Z-1))$$
$$- 4\sinh^3(\kappa)\,\tanh(\kappa Z)\,\text{sech}(\kappa Z + \kappa)\,\text{sech}(\kappa Z - \kappa)U(Z),$$

which shows that it is independent of parameters $(\alpha_2, \alpha_3) \in \mathbb{R}^2$.

The linearized operator L_+ belongs to the class of non-hyperbolic differential advance–delay operators (Section 5.1.2) with

$$c = \frac{\sinh(\kappa)}{\kappa} > 1.$$

Thanks to the continuous translational invariance of the differential advance–delay equation (5.3.9), L_+ has a two-dimensional generalized kernel spanned by eigenvectors $\{u_0, u_1\}$,

$$u_0 = \frac{\partial \Phi}{\partial Z}, \quad u_1 = \frac{\kappa^2}{2(\kappa \cosh(\kappa) - \sinh(\kappa))} \frac{\partial \Phi}{\partial \kappa}, \tag{5.3.13}$$

so that $L_+ u_0 = 0$ and $L_+ u_1 = u_0$.

Exercise 5.11 Show that the eigenfunction $U_0(Z; k)$ of the continuous spectrum of L_+ in Definition 5.1 is given by

$$U_0(Z; k) = e^{ikZ} \frac{1 - \cos(k)\cosh(\kappa) + i\sin(k)\sinh(\kappa)\tanh(\kappa Z) + \sinh^2(\kappa)\text{sech}^2(\kappa Z)}{1 - \cos(k)\cosh(\kappa) + i\sin(k)\sinh(\kappa)}, \tag{5.3.14}$$

which correspond to the spectral parameter,

$$\lambda = \lambda_0(k) := 2i\left[\sin(k) - k\frac{\sinh(\kappa)}{\kappa}\right], \quad k \in \mathbb{R}.$$

From this explicit expression, show that

$$a_0(k) = \frac{1 - \cos(k)\ \cosh(\kappa) - i\sin(k)\ \sinh(\kappa)}{1 - \cos(k)\ \cosh(\kappa) + i\sin(k)\ \sinh(\kappa)} = \frac{1 - \cosh(\kappa + ik)}{1 - \cosh(\kappa - ik)}, \tag{5.3.15}$$

where the scattering coefficient $a_0(k)$ is also defined in Definition 5.1.

According to the explicit expression (5.3.14), the eigenfunction $U_0(Z; k)$ is analytically extended in the strip $-\kappa < \text{Im}(k) < \kappa$ and the eigenfunctions in the weighted space $H_\mu^1(\mathbb{R})$ are given by $U_\mu(Z; k) = U_0(Z; k + i\mu)$ for any $\mu \in (-\kappa, \kappa)$. Similarly, $a_\mu(k) = a_0(k + i\mu)$. No eigenvalues of L_+ exist in this strip except for the double zero eigenvalue and the associated sequence of eigenvalues at $\{4\pi cmi\}_{m\in\mathbb{Z}}$. Since the two eigenvectors in the generalized kernel of L_+ decay exponentially to zero with the decay rate $\kappa > 0$, we have $u_0, u_1 \in H_\mu^1(\mathbb{R})$ for $\mu \in (-\kappa, \kappa)$. Furthermore, the eigenfunction $U_\mu(Z; k)$ and the scattering coefficient $a_\mu(k)$ are bounded and nonzero for any $k \in \mathbb{R}$ and $\mu \in (-\kappa, \kappa)$. Therefore, Assumptions 5.1 and 5.2 (Section 5.1.2) are satisfied for the given operator L_+ and the result of Theorem 5.2 applies.

To solve uniquely equation (5.3.12) for small values of α_8, we apply the Implicit Function Theorem (Appendix B.7), for which we need the following two technical results.

Lemma 5.8 *The maps*

$$N(U), F(\Phi + U) : H_{\text{ev}}^1(\mathbb{R}) \to H_{\text{odd}}^1(\mathbb{R})$$

are C^1 with respect to U and there exist constants $C_1, C_2, C_3 > 0$ such that

$$\|N(U)\|_{H^1} \leq C_1 \|U\|_{H^1}^2 + C_2 \|U\|_{H^1}^3, \qquad \|F(\Phi + U)\|_{H^1} \leq C_3 \|\Phi + U\|_{H^1}^3.$$

Proof The assertions follow by the Banach algebra of $H^1(\mathbb{R})$ with respect to point-wise multiplication (Appendix B.1) and the symmetry, since $\Phi \in H^1_{\mathrm{ev}}(\mathbb{R})$ and $\Phi_+ - \Phi_- \in H^1_{\mathrm{odd}}(\mathbb{R})$.   □

Lemma 5.9   *The operator*

$$L_+ : H^1_{\mathrm{ev}}(\mathbb{R}) \to L^2_{\mathrm{odd}}(\mathbb{R})$$

is continuously invertible and

$$\exists C > 0: \quad \forall F \in L^2_{\mathrm{odd}}(\mathbb{R}): \quad \|L_+^{-1} F\|_{H^1} \leq C \|F\|_{L^2}.$$

Proof Since $\Phi \in H^1(\mathbb{R})$, operator L_+ maps continuously $H^1(\mathbb{R})$ to $L^2(\mathbb{R})$. The change of parity follows from the symmetry of potentials in L_+.

To prove that there exists a unique solution $U \in H^1_{\mathrm{ev}}(\mathbb{R})$ of the linear inhomogeneous equation $L_+ U = F \in L^2_{\mathrm{odd}}(\mathbb{R})$, we apply Theorem 5.2, since operator L_+ satisfies Assumptions 5.1 and 5.2. By the explicit expressions (5.3.13), (5.3.14), and (5.3.15), $U_0(Z;0)$ is even on $\mathbb{R}$ and $u_0(Z)$ is odd on $\mathbb{R}$. Similarly, $W_0(Z;0)$ and $w_0(Z)$ are even on $\mathbb{R}$. As a result, the conditions of Theorem 5.2 are satisfied, because

$$\langle w_0, F \rangle_{L^2} = \langle W_0, F \rangle_{L^2} = 0.$$

Uniqueness follows from the fact that $u_0 \notin H^1_{\mathrm{ev}}(\mathbb{R})$.   □

As a result of Lemmas 5.8 and 5.9, the Jacobian operator $L_+ = D_U N(U)$ is invertible in a local neighborhood of the point $U = 0 \in H^1_{\mathrm{ev}}(\mathbb{R})$ and the nonlinear vector field $N(U) + F(\Phi + U)$ is Lipschitz continuous in $H^1(\mathbb{R})$. The Implicit Function Theorem gives the following result.

Theorem 5.4   *There exist $\epsilon > 0$ and $C > 0$ such that for any $|\alpha_8| < \epsilon$, there exists a unique solution $U \in H^1_{\mathrm{ev}}(\mathbb{R})$ of the scalar equation (5.3.12) satisfying*

$$\|U\|_{H^1} \leq C|\alpha_8|.$$

In other words, Theorem 5.4 states that there exists a unique continuation of the exact solution (5.3.6) in the scalar equation (5.3.9) for fixed $\alpha_2 > \alpha_3$ and $\kappa > 0$ with respect to small $\alpha_8 \neq 0$.

We now consider the system of differential advance–delay equations (5.3.8) for real and imaginary parts of $\Phi(Z)$. We represent the nonlinear function $f(u_{n-1}, u_n, u_{n+1})$ in the form

$$f = \alpha_2 |u_n|^2 (u_{n+1} + u_{n-1}) + \alpha_3 u_n^2 (\bar{u}_{n+1} + \bar{u}_{n-1}) + \epsilon \tilde{f}(u_{n-1}, u_n, u_{n+1}), \quad (5.3.16)$$

where ϵ is a small parameter. We replace $\Phi(Z)$ by $\Phi(Z) + U(Z) + iV(Z)$, where $\Phi(Z)$ is the exact solution (5.3.6) and $U(Z)$ and $V(Z)$ satisfy the system of differential advance–delay equations

$$L_+ U = N_+(U, V) + \mu M_+(\Phi + U, V) + \epsilon F_+(\Phi + U, V; \mu), \quad (5.3.17)$$

$$L_- V = N_-(U, V) + \mu M_-(\Phi + U, V) + \epsilon F_-(\Phi + U, V; \mu). \quad (5.3.18)$$

Here L_+ is the same as in the scalar equation (5.3.12) and L_- is given by

$$L_- V = -2\frac{\sinh(\kappa)}{\kappa}\frac{dV}{dZ} + \left(1 + (\alpha_2 + \alpha_3)\Phi^2\right)(V_+ - V_-) - 2\alpha_3\Phi(\Phi_+ - \Phi_-)V,$$

the vector fields $N_\pm(U, V)$ contain the quadratic and cubic terms in (U, V) from the expansion of the nonlinear function (5.3.16) for $\epsilon = 0$, $\mu = \cot(\beta)$ is a small parameter, $M_\pm(\Phi, V)$ is given by

$$\begin{cases} M_+(\Phi, V) = 2\cosh(\kappa)V - V_+ - V_- - \operatorname{Im} M(\Phi, V), \\ M_-(\Phi, V) = \Phi_+ + \Phi_- - 2\cosh(\kappa)\Phi + \operatorname{Re} M(\Phi, V), \end{cases} \qquad (5.3.19)$$

with

$$\begin{aligned} M(\Phi, V) &= \alpha_2|\Phi + iV|^2(\Phi_+ + \Phi_- + i(V_+ + V_-)) \\ &\quad + \alpha_3(\Phi + iV)^2(\Phi_+ + \Phi_- - i(V_+ + V_-)), \end{aligned}$$

and $\epsilon F_\pm(\Phi, V; \mu)$ contain the remainder terms from the expansion of the nonlinear function (5.3.16) in ϵ. If $f(u_{n-1}, u_n, u_{n+1})$ is a homogeneous cubic nonlinear function in its variables, then $F_\pm(\Phi, V; \mu)$ is a cubic polynomial in (Φ, V).

Exercise 5.12 Using the Banach algebra property of $H^1(\mathbb{R})$ and the symmetry of $\Phi(Z)$, show that if $f(u_{n-1}, u_n, u_{n+1})$ is a homogeneous cubic nonlinear function (5.2.20), then

$$\begin{cases} N_+(U, V), M_+(U, V), F_+(U, V) : H^1_{\mathrm{ev}}(\mathbb{R}) \times H^1_{\mathrm{odd}}(\mathbb{R}) \to H^1_{\mathrm{odd}}(\mathbb{R}), \\ N_-(U, V), M_-(U, V), F_-(U, V) : H^1_{\mathrm{ev}}(\mathbb{R}) \times H^1_{\mathrm{odd}}(\mathbb{R}) \to H^1_{\mathrm{ev}}(\mathbb{R}). \end{cases}$$

Moreover, show that

$$\begin{cases} \epsilon F_+(\Phi, 0; \mu) = -\alpha_8(\Phi_+^2 + \Phi_-^2)(\Phi_+ - \Phi_-) + \mathcal{O}(\mu), \\ \epsilon F_-(\Phi, 0; \mu) = (\alpha_4 + \alpha_6)\Phi^3 + (\alpha_4 - \alpha_6)\Phi(\Phi_+^2 + \Phi_-^2) \\ \qquad\qquad + (\alpha_4 - 3\alpha_6)\Phi\Phi_+\Phi_- + \mathcal{O}(\mu). \end{cases}$$

By Lemma 5.9, there exists a C^1 map

$$H^1_{\mathrm{odd}}(\mathbb{R}) \times \mathbb{R} \times \mathbb{R} \ni (V, \epsilon, \mu) \mapsto U \in H^1_{\mathrm{ev}}(\mathbb{R}),$$

such that U satisfies equation (5.3.17) for sufficiently small ϵ and μ and

$$\exists C > 0 : \quad \|U\|_{H^1} \le C\left(|\epsilon| + |\mu|\|V\|_{H^1} + \|V\|_{H^1}^2\right),$$

where the special form of $M_+(\Phi, V)$ in (5.3.19) is used.

When U is substituted into equation (5.3.18), we obtain a closed equation on V, where operator L_- is not invertible on the space of even functions because solutions of the homogeneous equation $L_-^* V = 0$ may not be odd on $\mathbb{R}$.

Using the exact solution (5.3.6) for $\Phi(Z)$, we can write L_- in the explicit form

$$\begin{aligned} L_- V &= -2\frac{\sinh(\kappa)}{\kappa}\frac{dV}{dZ} + \left(1 + (1 + 2\nu)\sinh^2(\kappa)\operatorname{sech}^2(\kappa Z)\right)(V(Z+1) - V(Z-1)) \\ &\quad + 4\nu\sinh^3(\kappa)\,\tanh(\kappa Z)\,\operatorname{sech}(\kappa Z + \kappa)\operatorname{sech}(\kappa Z - \kappa)V(Z), \end{aligned}$$

where $\nu = \alpha_3/(\alpha_2 - \alpha_3)$. The integrable AL lattice corresponds to the case $\alpha_3 = 0$ ($\nu = 0$).

For any $\nu \in \mathbb{R}$, operator L_- has a geometric kernel $\mathrm{Ker}(L_-) = \mathrm{span}\{\Phi\}$ thanks to the gauge invariance of the differential advance–delay equation (5.3.8). If $\nu = 0$, operator L_- has a two-dimensional generalized kernel spanned by $\{u_0, u_1\}$,

$$u_0 = \Phi(Z), \quad u_1 = \frac{\kappa}{2(\kappa \cosh(\kappa) - \sinh(\kappa))} Z\Phi(Z), \qquad (5.3.20)$$

so that $L_- u_0 = 0$ and $L_- u_1 = u_0$. Similarly to Exercise 5.11, the eigenfunction $U_0(Z; k)$ of the continuous spectrum of L_- in the integrable case $\nu = 0$ can be found in the explicit form

$$U_0(Z; k) = e^{ikZ} \frac{1 - \cos(k)\,\cosh(\kappa) + \mathrm{i}\sin(k)\,\sinh(\kappa)\,\tanh(\kappa Z)}{1 - \cos(k)\,\cosh(\kappa) + \mathrm{i}\sin(k)\,\sinh(\kappa)},$$

and the scattering coefficient $a_0(k)$ has the same expression (5.3.15). By the same arguments, operator L_- satisfies Assumptions 5.1 and 5.2 (Section 5.1.2) in the integrable case $\nu = 0$. However $\nu \neq 0$ destroys the integrability of the AL lattice and also destroys the two-dimensional generalized kernel of L_- in $H^1(\mathbb{R})$. The following exercise is solved in Appendix B of Pelinovsky *et al.* [157].

Exercise 5.13 Use perturbation series expansions in powers of ν and show that the double zero eigenvalue of L_- splits generally into two simple eigenvalues, one of which is zero.

Because of the splitting of the zero eigenvalue for small $\nu \neq 0$, different assumptions are needed for the non-hyperbolic differential advance–delay operator L_- with $\nu \neq 0$ instead of Assumptions 5.1 and 5.2. As in Section 5.1, we use an abstract notation L instead of the particular L_- with $c = \sinh(\kappa)/\kappa > 1$.

Assumption 5.3 *There exist constants $\Lambda_- < 0 < \Lambda_+$ such that no eigenvalues of L with eigenvectors in $H^1_\mu(\mathbb{R})$ for any $\mu \in (-\mu_0, \mu_0)$ exist for any $\lambda \in (\Lambda_-, \Lambda_+)$, except for the set of simple eigenvalues $\{4\pi c m \mathrm{i}\}_{m \in \mathbb{Z}}$. Furthermore, assume that there exist $u_0 \in H^1_\mu(\mathbb{R})$ for any $|\mu| < \mu_0$ and $w_0 \in H^1_{-\mu}(\mathbb{R})$ for any $\mu \in (0, \mu_0)$ such that*

$$L u_0 = 0, \quad L^* w_0 = 0,$$

and that there exists no $u_1 \in H^1_\mu(\mathbb{R})$ such that $L u_1 = u_0$.

Remark 5.3 Assumption 5.3 implies that $\langle w_0, u_0 \rangle_{L^2} \neq 0$, which is only possible if $w_0(Z)$ has a component of the same parity as $u_0(Z)$. The operator L can satisfy the assertion that $w_0 \notin H^1(\mathbb{R})$ only if the solution of $L w_0 = 0$ has two bounded non-decaying functions (even and odd), a linear combination of which would generate a function $w_0 \in H^1_{-\mu}(\mathbb{R})$ for $0 < \mu < \mu_0$.

Figures 5.5 and 5.6 show numerical approximations of the spectrum of L_- for $\kappa = 1$ and for $\nu = 0$ and $\nu = 0.2$ respectively. The numerical method is based on the sixth-order finite-difference approximation of the derivative operator and the truncation of the computational domain for Z on $[-L, L]$ with $L = 10$ and step size $h = 0.1$. The number of grid points is odd so that the number of eigenvalues in the truncated matrix problem is also odd. For $\nu = 0$ (Figure 5.5), the smallest

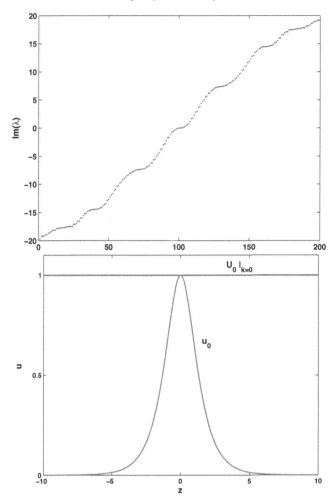

Figure 5.5 Eigenvalues of the operator L_- for $\nu = 0$ (top) and the normalized eigenvectors for the three smallest eigenvalues (bottom). Reproduced from [157].

eigenvalue with $|\lambda| = 2.9201 \times 10^{-15}$ corresponds to the bounded eigenfunction, while the next two eigenvalues with $|\lambda| = 8.5718 \times 10^{-5}$ correspond to the decaying eigenfunctions. This picture corresponds to Assumption 5.1. For $\nu = 0.2$ (Figure 5.6), the smallest eigenvalue with $|\lambda| = 1.6511 \times 10^{-13}$ corresponds to the decaying eigenfunction, while the next two eigenvalues with $|\lambda| = 0.0390$ correspond to the bounded oscillatory complex-valued eigenfunctions. This picture corresponds to Assumption 5.3.

Assumption 5.4 *Eigenfunction $W_\mu(Z; k)$ and the scattering coefficient $b_\mu(k)$ in Definition 5.1 are uniformly bounded for all $k \in \mathbb{R}$ in the limit $\mu \downarrow 0$. The only singularity of the eigenfunction $U_\mu(Z; k)$ and the scattering coefficient $a_\mu(k)$ in the*

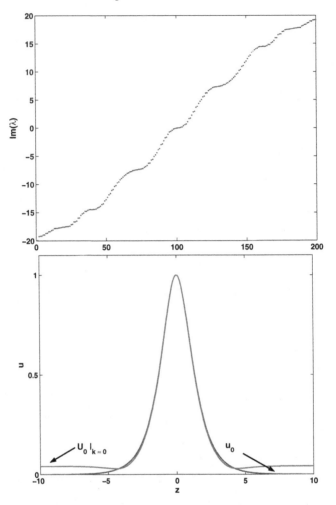

Figure 5.6 Eigenvalues of the operator L_- for $\nu = 0.2$ (top) and the normalized eigenvectors for the three smallest eigenvalues (bottom). Reproduced from [157].

limit $\mu \downarrow 0$ *is a simple pole at* $k = 0$. *Moreover, there exist* $C_1, C_2, C_3 > 0$ *such that*

$$\lim_{k \to 0} W_0(Z; k) = C_1 w_0(Z) \in H^1_{-\mu}(\mathbb{R}), \quad \mu \in (0, \mu_0), \tag{5.3.21}$$

$$\lim_{k \to 0} k U_0(Z; k) = C_2 u_0(Z) \in H^1_{\mu}(\mathbb{R}), \quad \mu \in (-\mu_0, \mu_0), \tag{5.3.22}$$

$$\lim_{k \to 0} k a_0(k) = C_3. \tag{5.3.23}$$

In spite of the difference in the construction of solutions of the homogeneous equations $LU = 0$ and $L^*W = 0$, the following theorem is analogous to Theorem 5.2.

Theorem 5.5 *Let $F \in L^2_\mu(\mathbb{R})$ with $|\mu| < \mu_0$ for some $\mu_0 > 0$. Under Assumptions 5.3 and 5.4, there exists a solution $U \in H^1_\mu(\mathbb{R})$ with $\mu \in (0, \mu_0)$ of the linear inhomogeneous equation $LU = F$ if and only if*

$$\int_{\mathbb{R}} W_0(Z; 0) F(Z) dZ = 0. \qquad (5.3.24)$$

Proof Thanks to the limit (5.3.21), (5.3.24) is equivalent to $\langle w_0, F \rangle_{L^2} = 0$. Under this condition, we represent the solution $U \in H^1_\mu(\mathbb{R})$ of $LU = F$ for $\mu \in (0, \mu_0)$ by the generalized Fourier transform

$$U(Z) = \int_{\mathbb{R}} B_\mu(k) U_\mu(Z; k) dk + A_0 u_0(Z) + \sum_{\lambda_j \in \sigma_d(L_\mu) \backslash \{0\}} A_j u_j(Z), \qquad (5.3.25)$$

where $A_0 \in \mathbb{R}$ is arbitrary,

$$B_\mu(k) = \frac{c - \cosh(\mu) \, \cos(k)}{4\pi c \lambda_\mu(k)} \, \langle W_\mu(\cdot; k), F \rangle_{L^2},$$

and A_j for $\lambda_j \in \sigma_d(L_\mu) \backslash \{0\}$ are projections to the eigenvectors for nonzero eigenvalues of L_μ.

Under Assumptions 5.3 and 5.4, the integral in the representation (5.3.25) has a double pole at $k = 0$ as $\mu \downarrow 0$ if $\hat{F}_0(0) \neq 0$. Condition (5.3.24) gives $\hat{F}_0(0) = 0$, after which the integral can be split into two parts as in the proof of Theorem 5.2. The residue term produces now the function $\lim_{k \to 0} k U_0(Z; k)$, which is proportional to $u_0 \in H^1(\mathbb{R})$ thanks to limit (5.3.22). The principal value integral produces a function in $H^1(\mathbb{R})$ if $F \in L^2(\mathbb{R})$. Therefore, $U \in H^1(\mathbb{R})$ if and only if $\hat{F}_0(0) = 0$. □

Coming back to the operator L_- in both cases $\nu = 0$ and $\nu \neq 0$, Theorems 5.2 and 5.5 imply the following lemma.

Lemma 5.10 *For $\nu = 0$ and $\nu \neq 0$, operator L_- is a continuous map*

$$L_- : H^1_{\text{odd}}(\mathbb{R}) \to L^2_{\text{ev}}(\mathbb{R}).$$

There exists a unique solution $U \in H^1_{\text{odd}}(\mathbb{R})$ of the linear inhomogeneous equation $L_- U = F$ if $F \in L^2_{\text{ev}}(\mathbb{R})$ and $\langle W_0, F \rangle_{L^2} = 0$, where $W_0(Z) := W_0(Z; 0)$.

Proof In the integrable case $\nu = 0$, the zero eigenvalue is double, but the explicit expressions for eigenvectors imply that $w_0(Z)$ is odd on $\mathbb{R}$, whereas $u_0(Z)$ is even on $\mathbb{R}$. Therefore, $\langle w_0, F \rangle_{L^2} = 0$ is satisfied trivially, whereas (5.1.14) gives $\langle W_0, F \rangle_{L^2} = 0$.

In the non-integrable case $\nu \neq 0$, the zero eigenvalue is simple and $\langle W_0, F \rangle_{L^2} = 0$ coincides with condition (5.3.24) for existence of $U \in H^1_{\text{odd}}(\mathbb{R})$. □

By Lemma 5.10, operator L_- in the second equation (5.3.18) can be inverted under the condition that $(U, V) \in H^1_{\text{ev}}(\mathbb{R}) \times H^1_{\text{odd}}(\mathbb{R})$ satisfies the constraint $\Delta(\epsilon, \mu) = 0$, where

$$\Delta(\epsilon, \mu) := \langle W_0, N_-(U, V) + \mu M_-(\Phi + U, V) + \epsilon F_-(\Phi + U, V; \mu) \rangle_{L^2}. \qquad (5.3.26)$$

The function $\Delta(\epsilon, \mu)$ is referred to as the *Melnikov integral*. It determines persistence of localized modes of the system (5.3.17)–(5.3.18) for $\epsilon \neq 0$. Since the Jacobian operator of the system (5.3.17)–(5.3.18) is continuously invertible in a local neighborhood of the point $(U, V) = (0, 0) \in H^1_{ev}(\mathbb{R}) \times H^1_{odd}(\mathbb{R})$, $\epsilon = 0 \in \mathbb{R}$, and $\mu = 0 \in \mathbb{R}$ provided that $\Delta(\epsilon, \mu) = 0$, the Implicit Function Theorem gives the following result.

Theorem 5.6 *Under Assumptions 5.1 and 5.2 for L_+ and L_- with $\nu = 0$ and Assumptions 5.3 and 5.4 for L_- with $\nu \neq 0$, there exists a unique solution $(U, V) \in H^1_{ev}(\mathbb{R}) \times H^1_{odd}(\mathbb{R})$ of the system (5.3.17)–(5.3.18) for sufficiently small ϵ and μ such that*

$$\exists C_1, C_2 > 0 : \quad \|U\|_{H^1} \leq C_1(|\epsilon| + \mu^2), \quad \|V\|_{H^1} \leq C_2(|\epsilon| + |\mu|),$$

provided that $\Delta(\epsilon, \mu) = 0$. Moreover, the map $\mathbb{R}^2 \ni (\epsilon, \mu) \mapsto \Delta \in \mathbb{R}$ is C^1 near $\epsilon = \mu = 0$.

Corollary 5.1 *Under the assumptions of Theorem 5.6, the following hold:*

(i) *There exists a unique continuation of the exact solution (5.3.6) with respect to parameter ϵ if $\partial_\mu \Delta(0, 0) \neq 0$.*

(ii) *No continuation of the exact solution (5.3.6) with respect to parameter ϵ exists if $\Delta(0, \mu) = 0$ and $\partial_\epsilon \Delta(0, 0) \neq 0$.*

Proof The result of the corollary follows from the Implicit Function Theorem for the root of $\Delta(\epsilon, \mu) = 0$, where $\Delta(0, 0) = 0$ and $\Delta(\epsilon, \mu)$ is C^1 near $\epsilon = \mu = 0$. $\square$

Remark 5.4 Recall that

$$\mu = \cot(\beta) = -\left(\beta - \frac{\pi}{2}\right) + \mathcal{O}\left(\beta - \frac{\pi}{2}\right)^3,$$

which implies that the result of Corollary 5.1(i) implies persistence of the solution curve on the plane (κ, β) near $\beta = \frac{\pi}{2}$ for a fixed $\kappa > 0$.

We now show that the one-parameter family (5.3.6) of traveling localized modes of the non-integrable DNLS equation with the nonlinear function (5.3.7) persists with respect to parameter continuation if $\alpha_3 \neq 0$. On the other hand, the two-parameter family (5.3.5) of traveling localized modes of the integrable AL lattice does not persist generally with respect to parameter continuation. Hence traveling localized modes of the integrable AL lattice are, in this sense, less structurally stable than traveling localized modes of the non-integrable generalized DNLS equation.

We apply Corollary 5.1(i) to the exact solution (5.3.6) of the non-integrable generalized DNLS equation with the nonlinear function (5.3.7) for fixed $\alpha_2 > \alpha_3 \neq 0$. Using (5.3.26), we obtain

$$\partial_\mu \Delta(0, 0) = \langle W_0, M_-(\Phi, 0) \rangle_{L^2}$$
$$= \langle W_0, (1 + (\alpha_2 + \alpha_3)\Phi^2)(\Phi_+ + \Phi_-) - 2\cosh(\kappa)\Phi \rangle_{L^2}$$
$$= 2\alpha_3 \langle W_0, \Phi^2(\Phi_+ + \Phi_-) \rangle_{L^2} \neq 0,$$

where the special form of $M_-(\Phi, V)$ in (5.3.19) is used. By Corollary 5.1(i), there exists a unique continuation of the solution (5.3.6) with respect to perturbations in the nonlinear function $f(u_{n-1}, u_n, u_{n+1})$ near $\beta = \frac{\pi}{2}$.

We note that $\partial_\mu \Delta(0, 0) = 0$ if $\alpha_3 = 0$, because the exact solution (5.3.6) is a member of the two-parameter family of the exact solutions (5.3.5) in the integrable AL lattice. Moreover, in this case, we actually have $\Delta(0, \mu) = 0$ for any $\mu \in \mathbb{R}$. If, in addition, $\partial_\epsilon \Delta(0, 0) \neq 0$, then the exact solution (5.3.6) cannot be continued with respect to (ϵ, μ) by Corollary 5.1(ii). To illustrate this situation, let us consider the Salerno model with the nonlinear function,

$$f = 2(1 - \alpha) |u_n|^2 u_n + \alpha |u_n|^2 (u_{n+1} + u_{n-1}), \quad \alpha \in \mathbb{R}. \qquad (5.3.27)$$

The differential advance–delay equation (5.3.4) with the nonlinear function (5.3.27) can be rewritten in the compact form

$$(1 + |\Phi|^2)(\Phi_+ - \Phi_-) - 2 \frac{\sinh(\kappa)}{\kappa} \Phi'(Z)$$
$$= i\mu \left[(1 + |\Phi|^2)(\Phi_+ + \Phi_-) - 2 \cosh(\kappa)\Phi \right] + i\epsilon |\Phi|^2 \Phi,$$

where

$$\epsilon = \frac{2(1 - \alpha)}{\alpha \sin(\beta)}, \quad \mu = \cot(\beta),$$

and the amplitude of $\Phi(Z)$ is rescaled by the factor $\sqrt{\alpha}$ for $\alpha \neq 0$. Since $\Phi(Z) = \sinh(\kappa) \operatorname{sech}(\kappa Z)$ is a solution of the advance–delay equation,

$$(1 + |\Phi|^2)(\Phi_+ + \Phi_-) - 2 \cosh(\kappa)\Phi = 0,$$

it is clear that $\Delta(0, \mu) = 0$. On the other hand, we have for any $\kappa > 0$

$$\partial_\epsilon \Delta(0, 0) = \langle W_0, \Phi^3 \rangle_{L^2} \neq 0.$$

Therefore, the two-parameter family of exact solutions (5.3.5) terminates near $\beta = \frac{\pi}{2}$ in the Salerno model with the nonlinear function (5.3.27) for $\alpha \neq 1$.

To illustrate the persistence of traveling localized modes, we consider the differential advance–delay equation (5.3.4) with the nonlinear function (5.2.20) under the constraints

$$\alpha_4 = \alpha_6, \quad \alpha_9 = 0, \quad \alpha_2 + \alpha_3 + 4\alpha_6 + 2\alpha_8 = 1,$$

where $(\alpha_3, \alpha_6, \alpha_8) \in \mathbb{R}^3$ are arbitrary parameters and the last condition is the normalization of $f(u, u, u) = 2|u|^2 u$.

Figure 5.7 illustrates persistence of the solution curve in the (κ, β) plane near $\beta = \frac{\pi}{2}$ for fixed values $(\alpha_3, \alpha_8) = (-1, 1)$ and different values of

$$\alpha_6 \in \{0.5, 0, -0.5, -1, -1.5, -2\}.$$

All the solution curves can be seen to intersect the point $(\kappa, \beta) = (0, \frac{\pi}{2})$.

Figure 5.8 shows that the family of traveling localized modes undergo a fold bifurcation for positive values of α_6 as κ is increased. At the fold bifurcation, two solution branches merge, one corresponding to a single-humped solution and the other one becoming a double-humped solution as the parameter moves away from

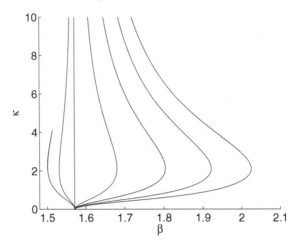

Figure 5.7 Persistence of traveling localized modes for $(\alpha_3, \alpha_8) = (-1, 1)$. Different curves correspond to different values of $\alpha_6 = 0.5, 0.25, 0, -0.5, -1, -1.5, -2$ from left to right. Reproduced from [157].

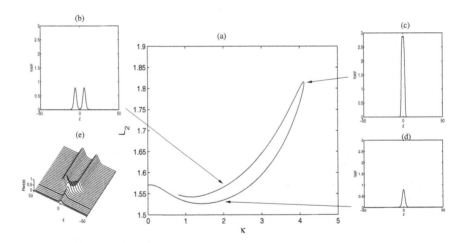

Figure 5.8 (a) Fold bifurcation at which single-humped and double-humped solutions coalesce for $(\alpha_3, \alpha_6, \alpha_8) = (-1, 0.5, 1)$. Solution profiles are shown in panels (b)–(d). Panel (e) shows the fold bifurcation for fixed $\kappa = 1$ as α_6 varies. The fold bifurcation only occurs when α_6 is positive. Reproduced from [157].

the fold point. The insets to the figure show the solution profiles along the solution curves for $\alpha_6 = 0.5$. The amplitudes of both single-humped and double-humped solutions grow with increasing values of κ up to the maximum amplitude at $\kappa \approx 4$.

The other outcome of the previous analysis is that the family of traveling localized modes (5.3.5) of the integrable AL lattice does not persist near $\beta = \frac{\pi}{2}$ in the Salerno model with the nonlinear function (5.3.27) near $\alpha = 1$. Figure 5.9(a) shows the tail

(a)

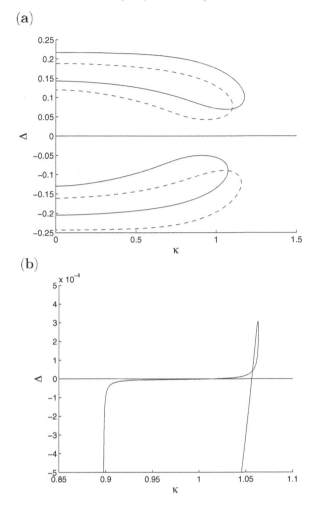

(b)

Figure 5.9 (a) Tail amplitude of the Salerno model as a function of κ for $\beta = \frac{13\pi}{20}$ (solid curve) and $\beta = \frac{7\pi}{20}$ (dashed curve). The curves with $\Delta > 0$ correspond to $\alpha = 1.1$ and the curves with $\Delta < 0$ correspond to $\alpha = 0.9$. (b) Existence of zeros of the tail amplitude for $\alpha = 0.7$ and $\beta = \frac{7\pi}{8}$.

amplitude Δ of the numerical solution versus κ for different values of α and β. In agreement with the above conclusion, the tail amplitude remains nonzero in a neighborhood of the point $(\alpha, \beta) = (1, \frac{\pi}{2})$. However, Figure 5.9(b) shows that zeros of the tail amplitude Δ can appear far from this point: two solutions with zero tail amplitudes are found for two values of κ when $\alpha = 0.7$ and $\beta = \frac{7\pi}{8}$. This result indicates that in addition to the known solutions of the AL lattice at $\alpha = 1$, other traveling localized modes of the Salerno model can exist away from the values

$\beta = \frac{\pi}{2}$ and $\alpha = 1$. This numerical observation is elaborated further in Section 5.5 using the Stokes constant computations.

5.4 Normal forms for traveling localized modes

If no explicit analytical solution is available for a given differential advance–delay equation, no persistence problem can be formulated and we are left wondering if the given equation admits any traveling localized modes. The only analytical method that can be employed to attack the existence problem in the general case is the local bifurcation theory from a linear limit, where the traveling localized mode has a small amplitude and it is slowly varying over the lattice sites. Formal asymptotic multi-scale expansion methods (Section 1.1) are helpful for predictions of the nontrivial bifurcations and we will explain how to make the arguments rigorous using the formulation of the differential advance–delay equation as the spatial dynamical system. This formalism originates from the work of Iooss & Kirchgassner [94] and Iooss [92] in the context of Fermi–Pasta–Ulam lattices.

Let us consider again the differential advance–delay equation (5.3.4),

$$2\mathrm{i}\sin(\beta)\,\frac{\sinh(\kappa)}{\kappa}\Phi'(Z)$$

$$= \Phi(Z+1)e^{\mathrm{i}\beta} + \Phi(Z-1)e^{-\mathrm{i}\beta} - 2\cos(\beta)\cosh(\kappa)\,\Phi(Z)$$

$$+ f(\Phi(Z-1)e^{-\mathrm{i}\beta},\Phi(Z),\Phi(Z+1)e^{\mathrm{i}\beta}),\quad Z \in \mathbb{R}, \qquad (5.4.1)$$

where $\kappa > 0$ and $\beta \in [0, 2\pi]$. The characteristic equation (5.3.11) for the linearized equation is rewritten as

$$D(\lambda;\kappa,\beta) := \cos(\beta)\left[\cosh(\lambda) - \cosh(\kappa)\right] + \mathrm{i}\sin(\beta)\left[\sinh(\lambda) - \frac{\sinh(\kappa)}{\kappa}\lambda\right] = 0.$$
$$(5.4.2)$$

Roots of this equation determine solutions $\Phi(Z) \sim e^{\lambda Z}$ of the linearized equation and the set of roots always includes the pair of real roots $\lambda = \pm\kappa$. Moreover, these are the only real-valued roots of the characteristic equation $D(\lambda;\kappa,\beta) = 0$. Other complex-valued roots exist (Section 5.3). The following lemma shows that all but finitely many roots of $D(\lambda;\kappa,\beta) = 0$ are bounded away from the imaginary axis $\mathrm{Re}(\lambda) = 0$.

Lemma 5.11 *Fix $\beta \in (0,\pi)$ and $\kappa > 0$. All but finitely many roots of the characteristic equation (5.4.2) are bounded away from the line $\mathrm{Re}(\lambda) = 0$. Moreover, for small $\kappa > 0$, there exists $p_0 = p_0(\beta) > 0$ such that $|\mathrm{Re}(\lambda)| \geq p_0$ for all roots with $\mathrm{Re}(\lambda) \neq 0$.*

Proof Since $D(\lambda;\kappa,\beta)$ is analytic in $\lambda \in \mathbb{C}$, roots cannot accumulate to a finite point in $\mathbb{C}$. Therefore, it suffices to consider a sequence of roots $\{\lambda_n\}_{n \in \mathbb{N}}$ such that $\lim_{n \to \infty} |\lambda_n| = \infty$ and to prove that the sequence of roots is bounded away from the line $\mathrm{Re}(\lambda) = 0$. Let $\lambda_n = p_n + ik_n$ with $p_n, k_n \in \mathbb{R}$ and rewrite $D(\lambda;\kappa,\beta) = 0$

in the vector form

$$\begin{cases} \cos(\beta)\left[\cosh(p_n)\cos(k_n) - \cosh(\kappa)\right] + \sin(\beta)\left[\dfrac{\sinh(\kappa)}{\kappa}k_n - \cosh(p_n)\sin(k_n)\right] = 0, \\ \cos(\beta)\sinh(p_n)\sin(k_n) + \sin(\beta)\left[\sinh(p_n)\cos(k_n) - \dfrac{\sinh(\kappa)}{\kappa}p_n\right] = 0. \end{cases}$$

It follows from the system that, for a fixed $\beta \in (0,\pi)$ and all $\kappa > 0$,

$$|k_n| \leq \frac{\kappa}{\sinh(\kappa)}\left[\cosh(p_n) + \cot(\beta)(\cosh(p_n) + \cosh(\kappa))\right]. \qquad (5.4.3)$$

By contradiction, if the sequence accumulates to the imaginary axis, then

$$\lim_{n \to \infty} p_n = 0$$

and, thanks to the bound (5.4.3), for small $\kappa > 0$, there exists $k_0 = k_0(\beta) < \infty$ such that $\lim_{n\to\infty}|k_n| \leq k_0$. This fact contradicts the analyticity of $D(\lambda;\kappa,\beta)$ in $\lambda \in \mathbb{C}$. Therefore, the sequence $\{\lambda_n\}_{n\in\mathbb{N}}$ remains bounded away from the line $\mathrm{Re}(\lambda) = 0$.

On the other hand, we have

$$\cosh(p_n) = \frac{\cos(\beta)\cosh(\kappa) - \sin(\beta)\sinh(x)k_n/x}{\cos(\beta + k_n)}. \qquad (5.4.4)$$

For small $\kappa > 0$, there exists $p_0 = p_0(\beta) > 0$ such that $|\mathrm{Re}(\lambda)| \geq p_0$ for all roots with $\mathrm{Re}(\lambda) \neq 0$. $\qquad\square$

Thanks to Lemma 5.11, only finitely many purely imaginary roots exist and they are determined by the roots of the real function

$$D(ik;\kappa,\beta) = \cos(\beta)\left[\cos(k) - \cosh(\kappa)\right] + \sin(\beta)\left[\frac{\sinh(\kappa)}{\kappa}k - \sin(k)\right] = 0. \quad (5.4.5)$$

The behavior of $D(ik;\kappa,\beta)$ is shown on Figure 5.2 (Section 5.3).

It follows from Lemma 5.11 that, among the roots with $\mathrm{Re}(\lambda) \neq 0$, the roots $\lambda = \pm\kappa$ are closest to the line $\mathrm{Re}(\lambda) = 0$ for sufficiently small $\kappa > 0$. If a localized mode exists in the differential advance–delay equation (5.4.1), then $\Phi(Z)$ decays exponentially to zero exactly as $e^{-\kappa|Z|}$ for small $\kappa > 0$. Therefore, the bifurcation curve corresponding to the localized mode that is slowly varying over Z on $\mathbb{R}$ is determined by the line $\kappa = 0$. The characteristic equation in this limit becomes

$$D(\lambda;0,\beta) = \cos(\beta)(\cosh(\lambda) - 1) + i\sin(\beta)(\sinh(\lambda) - \lambda) = 0. \qquad (5.4.6)$$

Exercise 5.14 Show that the set of roots of $D(\lambda;0,\beta) = 0$ with $\mathrm{Re}(\lambda) = 0$ consists of a zero root of multiplicity three for $\beta = \frac{\pi}{2}$ and of a zero root of multiplicity two and a simple nonzero root for small $|\beta - \frac{\pi}{2}|$. Show that the number of roots of $D(\lambda;0,\beta) = 0$ with $\mathrm{Re}(\lambda) = 0$ is always odd and it increases without any bounds as $\beta \to 0$ or $\beta \to \pi$.

We shall study the local bifurcation of traveling localized modes separately for two different cases $\beta = \frac{\pi}{2}$ and $\beta \neq \frac{\pi}{2}$. The first case results in the third-order differential equation, which is a normal form for the local bifurcation of traveling localized modes. Existence of localized modes in the normal form gives a necessary condition for the existence of localized modes in the differential advance–delay equation (5.4.1). The second case results in the beyond-all-orders asymptotic analysis with another necessary condition for the existence of localized modes (Section 5.5).

Let us derive the normal form for the local bifurcation of localized modes at $\beta = \frac{\pi}{2}$ using formal asymptotic multi-scale expansions (Section 1.1). Let κ be a small positive parameter and use the scaling transformation

$$\Phi(Z) = \kappa^\gamma \varphi(\zeta), \quad \zeta = \kappa Z,$$

for solutions of the differential advance–delay equation (5.4.1), where $\gamma > 0$ is a parameter to be determined. To expand the linear terms, we assume that $\Phi(Z) \in C^\infty(\mathbb{R})$ and use the Taylor series

$$\begin{cases} \varphi(\zeta + \kappa) - \varphi(\zeta - \kappa) - 2\sinh(\kappa)\varphi'(\zeta) = \frac{1}{3}\kappa^3 \left[\varphi'''(\zeta) - \varphi'(\zeta)\right] + \mathcal{O}(\kappa^5), \\ \varphi(\zeta + \kappa) + \varphi(\zeta - \kappa) - 2\cosh(\kappa)\varphi(\zeta) = \kappa^2 \left[\varphi''(\zeta) - \varphi(\zeta)\right] + \mathcal{O}(\kappa^4). \end{cases}$$

The balance between two linear terms occurs near $\kappa = 0$ and $\beta = \frac{\pi}{2}$ if the two small parameters $\mu = \cot(\beta)$ and κ are related by $\mu = \Omega\kappa$, where Ω is κ-independent. Under this scaling, the differential advance–delay equation (5.4.1) can be rewritten at the leading order of κ as

$$\frac{\mathrm{i}}{3} \left[\varphi'''(\zeta) - \varphi'(\zeta)\right] + \Omega \left[\varphi''(\zeta) - \varphi(\zeta)\right]$$
$$+ \kappa^{2\gamma-3} f(-\mathrm{i}\varphi(\zeta - \kappa), \varphi(\zeta), \mathrm{i}\varphi(\zeta + \kappa)) + \mathcal{O}(\kappa^2, \kappa^{2\gamma-2}) = 0, \quad (5.4.7)$$

where we have assumed again that $f(u_{n-1}, u_n, u_{n+1})$ is a homogeneous cubic nonlinear function in its variables.

In the case of the cubic DNLS equation with $f = |u_n|^2 u_n$, the leading-order equation (5.4.7) can be truncated after the choice $\gamma = \frac{3}{2}$ at the normal form

$$\frac{\mathrm{i}}{3} \left[\varphi'''(\zeta) - \varphi'(\zeta)\right] + \Omega \left(\varphi''(\zeta) - \varphi(\zeta)\right) + |\varphi(\zeta)|^2 \varphi(\zeta) = 0, \quad \zeta \in \mathbb{R}. \quad (5.4.8)$$

Equation (5.4.8) for the cubic DNLS equation will be justified in this section.

Exercise 5.15 Consider the nonlinearity function $f(u_{n-1}, u_n, u_{n+1})$ in the explicit form (5.2.20) and compute the Taylor series expansion for the nonlinear terms

$$f(-\mathrm{i}\varphi(\zeta - \kappa), \varphi(\zeta), \mathrm{i}\varphi(\zeta + \kappa)) = 4(\alpha_4 - \alpha_6)|\varphi|^2 \varphi$$
$$+ 2\mathrm{i}\kappa \left(\alpha_2 + 2\alpha_8 - 2\alpha_9\right) |\varphi(\zeta)|^2 \varphi'(\zeta) - 2\mathrm{i}\kappa \left(\alpha_3 - 2\alpha_9\right) \varphi^2(\zeta)\bar{\varphi}'(\zeta) + \mathcal{O}(\kappa^2).$$

Choose $\alpha_4 = \alpha_6$ and $\gamma = 1$, and show that the leading-order equation (5.4.7) can be truncated at a different normal form

$$\frac{\mathrm{i}}{3} \left[\varphi'''(\zeta) - \varphi'(\zeta)\right] + \Omega \left[\varphi''(\zeta) - \varphi(\zeta)\right] + 2\mathrm{i} \left(\alpha_2 + 2\alpha_8 - 2\alpha_9\right) |\varphi(\zeta)|^2 \varphi'(\zeta)$$
$$- 2\mathrm{i} \left(\alpha_3 - 2\alpha_9\right) \varphi^2(\zeta)\bar{\varphi}'(\zeta) = 0, \quad \zeta \in \mathbb{R}. \quad (5.4.9)$$

The difference between the normal form equations (5.4.8) and (5.4.9) implies different outcomes in the local bifurcation of traveling localized modes near the point $(\kappa, \beta) = (0, \frac{\pi}{2})$. The normal form equation (5.4.8) admits no single-humped localized modes for any $\Omega \in \mathbb{R}$ [78, 214]. On the other hand, the normal form equation (5.4.9) admits a reduction to a real-valued equation for $\Omega = 0$,

$$\varphi'''(\zeta) - \varphi'(\zeta) + 6\alpha\varphi^2(\zeta)\varphi'(\zeta) = 0, \quad \zeta \in \mathbb{R},$$

where $\alpha = \alpha_2 - \alpha_3 + 2\alpha_8$. There exists a single-humped localized mode of this equation for any $\alpha > 0$,

$$\varphi(\zeta) = \alpha^{-1/2} \operatorname{sech}(\zeta).$$

This localized mode persists as the exact solution of the differential advance–delay equation (5.4.1) for any $\kappa > 0$ and $\beta = \frac{\pi}{2}$ if $\alpha_2 > \alpha_3$ and $\alpha_4 = \alpha_6 = \alpha_8 = \alpha_9 = 0$. Moreover, it is structurally stable with respect to continuation in parameters $\alpha_4, \alpha_6, \alpha_8, \alpha_9 \in \mathbb{R}$ in the (κ, β) plane near $\beta = \frac{\pi}{2}$ if $\alpha_3 \neq 0$ (Section 5.3). Thus, the normal form equation (5.4.9) gives the necessary condition for the existence of localized modes in the differential advance–delay equation (5.4.1). In particular, the existence curve of localized modes in the (κ, β) plane intersects the point $(0, \frac{\pi}{2})$ if the normal form equation (5.4.9) has the corresponding localized mode $\varphi(\zeta)$.

We shall now focus on the justification of the normal form equation (5.4.8). Our presentation follows the works of Pelinovsky & Rothos [158] and Iooss & Pelinovsky [95]. To proceed with the center manifold reductions of the spatial dynamical system, we rewrite the differential advance–delay equation (5.4.1) as an infinite-dimensional evolution equation.

Let $P \in [-1, 1]$ be a new independent variable and $\mathbf{U} = \mathbf{U}(Z, P) = (U_1, U_2, U_3, U_4)$ be defined by

$$U_1 = \Phi(Z), \quad U_2 = \Phi(Z + P), \quad U_3 = \bar{\Phi}(Z), \quad U_4 = \bar{\Phi}(Z + P), \qquad (5.4.10)$$

so that $U_2(Z, 0) = U_1(Z)$ and $U_4(Z, 0) = U_3(Z)$. Let $D^\pm$ be the difference operators defined by $D^\pm \mathbf{U}(Z, P) = \mathbf{U}(Z, \pm 1)$. Let $\mathcal{D}$ and $\mathcal{H}$ be the Banach spaces for the vector $\mathbf{U}(Z, P)$:

$$\mathcal{D} = \left\{ \mathbf{U} \in \mathbb{C}^4 : U_1 = \bar{U}_3 \in C_b^1(\mathbb{R}), \ U_2 = \bar{U}_4 \in C_b^1(\mathbb{R}, [-1, 1]), \ U_1(Z) = U_2(Z, 0) \right\},$$

$$\mathcal{H} = \left\{ \mathbf{U} \in \mathbb{C}^4, \ U_1 = \bar{U}_3 \in C_b^0(\mathbb{R}), \ U_2 = \bar{U}_4 \in C_b^0(\mathbb{R}, [-1, 1]) \right\},$$

with the usual supremum norms. The differential advance–delay equation (5.4.1) is written as the infinite-dimensional evolution equation,

$$2ic\mathcal{J}\frac{d\mathbf{U}}{dZ} = \mathcal{L}_{\kappa,\beta}\mathbf{U} + \mathcal{M}(\mathbf{U}), \quad c = \sin(\beta)\frac{\sinh(\kappa)}{\kappa}, \qquad (5.4.11)$$

where $\mathcal{J}$ is the diagonal matrix of $\{1, 1, -1, -1\}$, $\mathcal{L}_{\kappa,\beta}$ is the linear operator,

$$\mathcal{L}_{\kappa,\beta} = \begin{bmatrix} -2\cos(\beta)\cosh(\kappa) & e^{i\beta}D^+ + e^{-i\beta}D^- & 0 & 0 \\ 0 & 2ic\,\partial/\partial P & 0 & 0 \\ 0 & 0 & -2\cos(\beta)\cosh(\kappa) & e^{-i\beta}D^+ + e^{i\beta}D^- \\ 0 & 0 & 0 & -2ic\,\partial/\partial P \end{bmatrix},$$

and $\mathcal{M}(\mathbf{U})$ is the nonlinear operator,

$$\mathcal{M}(\mathbf{U}) = \begin{bmatrix} U_1^2 U_3 \\ 0 \\ U_3^2 U_1 \\ 0 \end{bmatrix}.$$

The linear operator $\mathcal{L}_{\kappa,\beta}$ maps $\mathcal{D}$ into $\mathcal{H}$ continuously and it has a compact resolvent in $\mathcal{H}$. The nonlinear term $\mathcal{M}(\mathbf{U})$ is analytic in an open neighborhood of $\mathbf{U} = \mathbf{0} \in \mathcal{D}$, maps $\mathcal{D}$ into $\mathcal{D}$ continuously and

$$\exists C > 0: \quad \|\mathcal{M}(\mathbf{U})\|_{\mathcal{D}} \leq C \|\mathbf{U}\|_{\mathcal{D}}^3.$$

The spectrum of $\mathcal{L}_{\kappa,\beta}$ consists of an infinite set of isolated eigenvalues of finite multiplicities. Using the Laplace transform, linear eigenmodes of the evolution equation,

$$2\mathrm{i}c\mathcal{J}\frac{d\mathbf{U}}{dZ} = \mathcal{L}_{\kappa,\beta}\mathbf{U}$$

are found from the vector solution

$$\mathbf{U}(Z,P) = \left(u_1, u_1 e^{\lambda P}, u_3, u_3 e^{\lambda P}\right) e^{\lambda Z},$$

where λ is defined by the roots of the characteristic equations

$$D(\lambda; \kappa, \beta) = 0 \quad \text{if} \quad u_1 \neq 0, \quad \text{and} \quad \bar{D}(\lambda; \kappa, \beta) = 0 \quad \text{if} \quad u_3 \neq 0.$$

Here $D(\lambda; \kappa, \beta) = 0$ is, of course, the same characteristic equation (5.4.2). If λ is the root of $D(\lambda; \kappa, \beta) = 0$, then $\bar{\lambda}$ is the root of $\bar{D}(\lambda; \kappa, \beta) = 0$. By Lemma 5.11, all but a finite number of roots are isolated away from the line $\mathrm{Re}(\lambda) = 0$. Finitely many eigenvalues λ on $\mathrm{Re}(\lambda) = 0$ determine the center manifold of the spatial dynamical system, which is useful for a reduction of the infinite-dimensional evolution equation (5.4.11) to the finite-dimensional normal form equation (5.4.8). Many examples of center manifold reductions in spatial dynamical systems are discussed in the text of Iooss & Adelmeyer [93].

We are particularly interested in the bifurcation point $(\kappa, \beta) = (0, \frac{\pi}{2})$, when

$$D_0(\lambda) := D\left(\lambda; 0, \frac{\pi}{2}\right) = \mathrm{i}(\sinh(\lambda) - \lambda) = \frac{\mathrm{i}}{6}\lambda^3 \left(1 + \frac{\lambda^2}{20} + \mathcal{O}(\lambda^4)\right).$$

The only root of $D_0(\lambda)$ near the line $\mathrm{Re}(\lambda) = 0$ is the triple zero, so that the zero eigenvalue has the algebraic multiplicity *six* and the geometric multiplicity *two*. The two eigenvectors of the geometric kernel of $\mathcal{L}_0 := \mathcal{L}_{0,\pi/2}$ are

$$\mathbf{U}_0 = (1,1,0,0), \quad \mathbf{W}_0 = (0,0,1,1),$$

whereas the four generalized eigenvectors satisfy

$$\mathcal{L}_0 \mathbf{U}_k = 2\mathrm{i}\mathcal{J}\mathbf{U}_{k-1}, \quad \mathcal{L}_0 \mathbf{W}_k = 2\mathrm{i}\mathcal{J}\mathbf{W}_{k-1}, \quad k = 1,2,$$

or in the explicit form

$$\mathbf{U}_1 = (0,P,0,0), \quad \mathbf{W}_1 = (0,0,0,P), \quad \mathbf{U}_2 = \frac{1}{2}\left(0,P^2,0,0\right), \quad \mathbf{W}_2 = \frac{1}{2}\left(0,0,0,P^2\right).$$

Fix a small $\kappa > 0$ and a finite $\Omega \in \mathbb{R}$ in $\cot(\beta) = \Omega\kappa$. We rewrite the evolution equation (5.4.11) in the equivalent form,

$$2\mathrm{i}\,\mathcal{J}\frac{d\mathbf{U}}{dZ} = \mathcal{L}_0\mathbf{U} + \Omega\kappa\mathcal{L}_1\mathbf{U} + \kappa^2\mathcal{L}_2\mathbf{U} + \frac{\kappa}{\sinh(\kappa)}\sqrt{1+\Omega^2\kappa^2}\mathcal{M}(\mathbf{U}), \qquad (5.4.12)$$

where

$$\mathcal{L}_0 = \begin{bmatrix} 0 & \mathrm{i}(D^+ - D^-) & 0 & 0 \\ 0 & 2\mathrm{i}\,\partial/\partial P & 0 & 0 \\ 0 & 0 & 0 & -\mathrm{i}(D^+ - D^-) \\ 0 & 0 & 0 & -2\mathrm{i}\,\partial/\partial P \end{bmatrix},$$

$$\mathcal{L}_1 = \frac{\kappa}{\sinh(\kappa)}\begin{bmatrix} -2\cosh(\kappa) & (D^+ + D^-) & 0 & 0 \\ 0 & 0 & 0 & 0 \\ 0 & 0 & -2\cosh(\kappa) & (D^+ + D^-) \\ 0 & 0 & 0 & 0 \end{bmatrix},$$

and

$$\mathcal{L}_2 = \frac{1}{\kappa^2}\left(\frac{\kappa}{\sinh(\kappa)} - 1\right)\begin{bmatrix} 0 & \mathrm{i}(D^+ - D^-) & 0 & 0 \\ 0 & 0 & 0 & 0 \\ 0 & 0 & 0 & -\mathrm{i}(D^+ - D^-) \\ 0 & 0 & 0 & 0 \end{bmatrix}.$$

The initial-value problem for the time evolution problem (5.4.12) is ill-posed because the eigenvalues of $\mathcal{L}_0$ diverge both to the left and right halves of the complex λ plane. Instead of dealing with an initial-value problem, we are interested in bounded solutions on the entire Z-axis. These bounded solutions can be constructed with the decomposition of the solution to the finite-dimensional subspace related to the zero eigenvalue and to the infinite-dimensional subspace related to all other nonzero complex eigenvalues of $\mathcal{L}_0$. After the decomposition, we project the time evolution problem (5.4.12) to the corresponding subspaces and then truncate the resulting system of equations to obtain the third-order differential equation (5.4.8). This technique relies on the solution of the resolvent equation

$$(2\mathrm{i}\lambda\mathcal{J} - \mathcal{L}_0)\mathbf{U} = \mathbf{F}, \quad \mathbf{U} \in \mathcal{D}, \quad \mathbf{F} \in \mathcal{H}, \quad \lambda \in \mathbb{C}. \qquad (5.4.13)$$

If λ is not a root of $D_0(\lambda) = 0$ or $\bar{D}_0(\lambda) = 0$, the explicit solution of the resolvent equation (5.4.13) is obtained in the form

$$U_1 = -\frac{1}{2D_0(\lambda)}\left[F_1 - \frac{1}{2}\int_0^1 F_2(P)e^{\lambda(1-P)}dP - \frac{1}{2}\int_{-1}^0 F_2(P)e^{-\lambda(1+P)}dP\right],$$

$$U_2 = U_1 e^{\lambda P} - \frac{1}{2\mathrm{i}}\int_0^P F_2(P')e^{\lambda(P-P')}dP',$$

$$U_3 = \frac{1}{2\bar{D}_0(\lambda)}\left[F_3 - \frac{1}{2}\int_0^1 F_4(P)e^{\lambda(1-P)}dP - \frac{1}{2}\int_{-1}^0 F_4(P)e^{-\lambda(1+P)}dP\right],$$

$$U_4 = U_3 e^{\lambda P} + \frac{1}{2\mathrm{i}}\int_0^P F_4(P')e^{\lambda(P-P')}dP'.$$

The solution $\mathbf{U}$ has a triple pole at $\lambda = 0$. Let us decompose the solution of the time evolution problem (5.4.12) into two parts,

$$\mathbf{U}(Z) = \mathbf{U}_c(Z) + \mathbf{U}_h(Z),$$

where $\mathbf{U}_c$ is the projection to the six-dimensional subspace of the zero eigenvalue,

$$\mathbf{U}_c(Z) = A(Z)\mathbf{U}_0 + B(Z)\mathbf{U}_1 + C(Z)\mathbf{U}_2 + \bar{A}(Z)\mathbf{W}_0 + \bar{B}(Z)\mathbf{W}_1 + \bar{C}(Z)\mathbf{W}_2,$$

and $\mathbf{U}_h$ is the projection to the complementary invariant subspace of operator $\mathcal{L}_0$. We notice in particular that

$$U_{c1} = A, \quad U_{c2} = A + PB + \frac{1}{2}P^2C.$$

When the decomposition is substituted into the time evolution equation (5.4.12), we obtain

$$2\mathrm{i}\mathcal{J}\frac{d\mathbf{U}_h}{dZ} - \mathcal{L}_0\mathbf{U}_h = \mathbf{F}_h(A, B, C, \mathbf{U}_h), \tag{5.4.14}$$

where

$$\mathbf{F}_h = -2\mathrm{i}\mathcal{J}\frac{d\mathbf{U}_c}{dZ} + \mathcal{L}_0\mathbf{U}_c + \Omega\kappa\mathcal{L}_1(\mathbf{U}_c + \mathbf{U}_h) + \kappa^2\mathcal{L}_2(\mathbf{U}_c + \mathbf{U}_h)$$
$$+ \frac{\kappa}{\sinh(\kappa)}\sqrt{1 + \Omega^2\kappa^2}\mathcal{M}(\mathbf{U}_c + \mathbf{U}_h).$$

By the Fredholm Alternative Theorem (Appendix B.4), there exists a solution for $\mathbf{U}_h$ if and only if $\mathbf{F}_h$ is orthogonal to the subspace of the adjoint linear operator associated to the zero eigenvalue. Equivalently, one can consider the solution of the resolvent equation (5.4.13) in the Laplace transform form $\hat{\mathbf{U}}_h(\lambda) = (2\mathrm{i}\lambda\mathcal{J} - \mathcal{L}_0)^{-1}\hat{\mathbf{F}}_h$ and remove the pole singularities from the function $\hat{\mathbf{U}}_h(\lambda)$ near $\lambda = 0$.

To make this procedure algorithmic, we expand the explicit solution of the resolvent equation (5.4.13) into the Laurent series at $\lambda = 0$:

$$\mathbf{U} = \frac{a_{-3}\mathbf{U}_0 + b_{-3}\mathbf{W}_0}{\lambda^3} + \frac{a_{-2}\mathbf{U}_0 + a_{-3}\mathbf{U}_1 + b_{-2}\mathbf{W}_0 + b_{-3}\mathbf{W}_1}{\lambda^2}$$
$$+ \frac{(a_{-1} - \frac{1}{20}a_{-3})\mathbf{U}_0 + a_{-2}\mathbf{U}_1 + a_{-3}\mathbf{U}_2}{\lambda}$$
$$+ \frac{(b_{-1} - \frac{1}{20}b_{-3})\mathbf{W}_0 + b_{-2}\mathbf{W}_1 + b_{-3}\mathbf{W}_2}{\lambda} + \mathbf{R_U},$$

where $\mathbf{R_U}$ is analytic in the neighborhood of $\lambda = 0$ and the projection operators are given explicitly by

$$a_{-3} = 3\mathrm{i}\left(F_1 - \frac{1}{2}\int_{-1}^1 F_2(P)dP\right),$$

$$a_{-2} = -\frac{3\mathrm{i}}{2}\left(\int_0^1 (1-P)F_2(P)dP - \int_{-1}^0 (1+P)F_2(P)dP\right),$$

$$a_{-1} = -\frac{3\mathrm{i}}{4}\left(\int_0^1 (1-P)^2 F_2(P)dP + \int_{-1}^0 (1+P)^2 F_2(P)dP\right),$$

with similar expressions for b_{-3}, b_{-2}, b_{-1} in terms of F_3 and F_4.

Since a_{-3}, a_{-2}, a_{-1} do not depend on λ, we can use $\mathbf{F}_h$ instead of its Laplace transform $\hat{\mathbf{F}}_h$. Substituting $\mathbf{F}_h$ into the integral expressions for a_{-3}, a_{-2}, a_{-1}, we obtain

$$a_{-3} = -C'(Z) + 3\mathrm{i}\Omega\kappa C + 3\mathrm{i}|A|^2 A + R_{-3}(A,B,C,\mathbf{U}_h) = 0, \qquad (5.4.15)$$

$$a_{-2} = -B'(Z) + C + R_{-2}(A,B,C,\mathbf{U}_h) = 0, \qquad (5.4.16)$$

$$a_{-1} = -A'(Z) + B - \frac{1}{20}C'(Z) + R_{-1}(A,B,C,\mathbf{U}_h) = 0, \qquad (5.4.17)$$

where R_{-3}, R_{-2}, R_{-1} are remainder terms.

Using the scaling transformation

$$A = \kappa^{3/2}\phi_1(\zeta), \quad B = \kappa^{5/2}\phi_2(\zeta), \quad C = \kappa^{7/2}\phi_3(\zeta), \quad \mathbf{U}_h = \kappa^{5/2}\mathbf{V}_h, \quad \zeta = \kappa Z,$$

we reduce the system of differential equations (5.4.15)–(5.4.17) and obtain

$$\frac{d}{d\zeta}\begin{bmatrix}\phi_1\\\phi_2\\\phi_3\end{bmatrix} = \begin{bmatrix}0 & 1 & 0\\0 & 0 & 1\\0 & 0 & 0\end{bmatrix}\begin{bmatrix}\phi_1\\\phi_2\\\phi_3\end{bmatrix} + 3\mathrm{i}\begin{bmatrix}0\\0\\1\end{bmatrix}(\Omega\phi_3 + |\phi_1|^2\phi_1) + \kappa\mathbf{R}_{\phi}(\boldsymbol{\phi},\mathbf{V}_h),$$

$$(5.4.18)$$

where $\boldsymbol{\phi} = (\phi_1,\phi_2,\phi_3)$ depends on $\zeta = \kappa Z$ and $\mathbf{R}_{\phi}$ is a remainder term. This equation is complemented with the residual equation (5.4.13), which now becomes

$$2\mathrm{i}\mathcal{J}\frac{d\mathbf{V}_h}{dZ} - \mathcal{L}_0\mathbf{V}_h = \kappa\mathbf{R}_{\mathbf{V}}(\boldsymbol{\phi},\mathbf{V}_h), \qquad (5.4.19)$$

where $\mathbf{R}_{\mathbf{V}}$ is a remainder term.

The formal truncation of system (5.4.18) and (5.4.19) at $\kappa = 0$ recovers the scalar third-order equation (5.4.8) for $\varphi(\zeta) = \phi_1(\zeta)$, rewritten again as

$$\frac{\mathrm{i}}{3}[\varphi'''(\zeta) - \varphi'(\zeta)] + \Omega[\varphi''(\zeta) - \varphi(\zeta)] + |\varphi(\zeta)|^2\varphi(\zeta) = 0, \quad \zeta \in \mathbb{R}. \qquad (5.4.20)$$

The justification of the center manifold reduction is described by the following theorem.

Theorem 5.7 *Fix $M > 0$ and let $\kappa > 0$ be small enough. For any given $\boldsymbol{\phi} \in C_b^0(\mathbb{R},\mathbb{C}^3)$ such that $\|\boldsymbol{\phi}\|_{L^\infty} \leq M$, there exists a unique solution $\mathbf{V}_h \in C_b^0(\mathbb{R},\mathcal{D})$ of system (5.4.19) such that*

$$\exists K > 0: \quad \|\mathbf{V}_h\|_{\mathcal{D}} \leq K\kappa.$$

The proof of existence of center manifolds for the class of differential advance–delay equations is developed by Iooss & Kirchgässner [94] using the Green function technique and by Iooss & Pelinovsky [95] using the Implicit Function Theorem.

By Theorem 5.7, bounded solutions of the system (5.4.18) generate bounded solutions in the full system (5.4.12). System (5.4.18) can be interpreted as a perturbation of the normal form equation (5.4.20). Therefore, one needs first to construct bounded solutions of the normal form equation (5.4.20) and then to develop the persistence analysis of these bounded solutions in the system (5.4.18). In particular, we are interested in localized modes of the normal form equation (5.4.20).

It is clear that the linear part of equation (5.4.20) has three fundamental solutions $\varphi(\zeta) \sim e^{\pm\zeta}$ and $\varphi(\zeta) \sim e^{i3\Omega\zeta}$, where the first two solutions correspond to the one-dimensional stable and unstable manifolds whereas the last solution corresponds to the one-dimensional center manifold. Existence of localized solutions in the systems with a one-dimensional center manifold is a bifurcation of codimension one and hence it is non-generic. Using the beyond-all-orders asymptotic method (Section 5.5), Grimshaw [78] showed that no single-humped localized modes exist in the normal form equation (5.4.20) for any value of $\Omega \in \mathbb{R}$. This negative result "kills" the problem of persistence of the single-humped localized modes in the system (5.4.18).

On the other hand, Calvo & Akylas [28] proved analytically that double-humped and multi-humped localized modes of the normal form equation (5.4.20) exist for special values of parameter Ω. Double-humped localized modes were approximated numerically by Yang & Akylas [214]. Different double-humped localized modes are characterized by the different distances between the two individual humps. Although it is expected that double-humped localized modes persist in a neighborhood of these curves within system (5.4.18) and thus in the full dynamical system (5.4.12), persistence analysis of these solutions is a delicate open problem of analysis.

Exercise 5.16   Consider traveling localized modes in the discrete Klein–Gordon equation,

$$c^2 \phi''(z) = \frac{\phi(z+h) - 2\phi(z) + \phi(z-h)}{h^2} - \phi(z) + \phi^3(z)$$

and show with the formal Taylor expansions that the scaling

$$c^2 = 1 + \epsilon\gamma, \quad h^2 = \epsilon^2 \tau, \quad \zeta = \frac{z}{\sqrt{\epsilon}},$$

with two parameters (γ, τ), results in the normal form equation

$$\frac{\tau}{12}\phi''''(\zeta) - \gamma\phi''(\zeta) - \phi(\zeta) + \phi^3(\zeta) = 0, \quad \zeta \in \mathbb{R}.$$

Exercise 5.17   Consider the discrete Klein–Gordon equation,

$$\ddot{u}_n \frac{1 - u_{n-1}u_{n+1}}{1 - u_n^2} = \frac{u_{n+1} - 2u_n + u_{n-1}}{h^2} + \frac{1}{2}u_n(1 - u_{n+1}u_{n-1}),$$

and prove that it admits an exact traveling kink solution for any $c \in (-1, 1)$,

$$u_n(t) = \phi(n - ct) = \tanh\left(\frac{n - ct}{\sqrt{1 - c^2}}\right).$$

Derive the normal form equation similarly to Exercise 5.16 and show that the normal form inherits the same exact solution for arbitrary values of (γ, τ).

5.5 Stokes constants for traveling localized modes

In Section 5.4, local bifurcation of traveling localized modes was considered in the differential advance–delay equation (5.4.1) near a special point $(\kappa, \beta) = (0, \frac{\pi}{2})$. Near this point, non-existence of single-humped localized modes can be checked at the algebraic order of the perturbation series using truncation of the normal form equation. Here we shall look at other values of $\beta \neq \frac{\pi}{2}$ along the bifurcation curve $\kappa = 0$. Although the center manifold reduction looks relatively simple at algebraic orders of the perturbation theory for $\beta \neq \frac{\pi}{2}$, non-existence of single-humped localized modes can only be established if beyond-all-orders terms are captured in the asymptotic expansion.

Asymptotic analysis of the beyond-all-orders perturbation theory originated from the work of Kruskal & Segur [123] and has been employed by many authors in the context of differential equations. It was extended by Pomeau et al. [172] to allow the computation of the splitting constants from the Borel summation of series rather than from the numerical solution of differential equations. Rigorous justification of this method for differential and difference equations was developed by Tovbis and his coworkers [203, 204, 205, 206].

We shall describe the beyond-all-orders method on a formal (already complicated) level of arguments. Our presentation follows Melvin et al. [140], who considered the Salerno model with the nonlinearity function,

$$f(u_{n-1}, u_n, u_{n+1}) = 2(1 - \alpha) |u_n|^2 u_n + \alpha |u_n|^2 (u_{n+1} + u_{n-1}), \quad \alpha \in \mathbb{R}. \quad (5.5.1)$$

Oxtoby & Barashenkov [145] developed a similar application of the beyond-all-orders method for the saturable DNLS equation.

The integrable AL lattice, which arises from the Salerno model with the nonlinear function (5.5.1) for $\alpha = 1$, has a two-parameter family of exact traveling localized modes. However, these solutions do not persist for $\alpha \neq 1$ away from the integrable limit (Section 5.3). Numerical results suggest, however, that the Salerno model can still support traveling localized modes for some $\alpha \neq 1$ (Figure 5.9). The beyond-all-orders method is the tool to capture non-trivial bifurcations of the solution curves in the parameter plane (κ, β) that intersects the bifurcation curve $\kappa = 0$.

We start with the differential advance–delay equation (5.4.1) for the Salerno model (5.5.1) rewritten in variables $\Phi(Z) = \kappa \varphi(\zeta)$ and $\zeta = \kappa Z$:

$$\cos(\beta) [\varphi(\zeta + \kappa) + \varphi(\zeta - \kappa) - 2\cosh(\kappa)\varphi(\zeta)]$$
$$+ \, i\sin(\beta) [\varphi(\zeta + \kappa) - \varphi(\zeta - \kappa) - 2\sinh(\kappa)\varphi'(\zeta)]$$
$$+ \kappa^2 f(\varphi(\zeta - \kappa)e^{-i\beta}, \varphi(\zeta), \varphi(\zeta + \kappa)e^{i\beta}) = 0, \quad (5.5.2)$$

where the nonlinear function is

$$f = 2(1 - \alpha)\varphi(\zeta)|\varphi(\zeta)|^2 + \alpha \cos(\beta)(\varphi(\zeta + \kappa) + \varphi(\zeta - \kappa))|\varphi(\zeta)|^2$$
$$+ \, i\sin(\beta)(\varphi(\zeta + \kappa) - \varphi(\zeta - \kappa))|\varphi(\zeta)|^2.$$

The beyond-all-orders method consists of three steps.

- A regular asymptotic solution is derived in the continuous limit $\kappa \to 0$, when the differential advance–delay operator is formally replaced by the series of differential operators of increasing orders. Although this solution is correct to all orders of the asymptotic expansion, it does not capture the bounded oscillatory tails as these terms are exponentially small in κ.
- To capture the oscillatory tails, we continue analytically the solution into the complex plane where the asymptotic expansion blows up. The Stokes constants measure the residue coefficients of the singular part of the beyond-all-orders terms corresponding to the splitting between stable and unstable manifolds.
- We compute the Stokes constant in the region of the parameter space where there is only one eigenvalue of the linearized equation on the imaginary axis besides the zero eigenvalue. Zeros of the Stokes constant indicate bifurcations of traveling localized modes in the differential advance–delay equation (5.5.2).

This analytic scheme is complemented by numerical approximations of the Stokes constant. Location of bifurcation curves captured by the beyond-all-orders method give good hints for numerical searches of the localized solutions of the differential advance–delay equation (5.5.2).

We seek a regular asymptotic expansion of equation (5.5.2) in powers of κ,

$$\varphi(\zeta) = \sum_{n=0}^{\infty} (i\kappa)^n \varphi_n(\zeta), \tag{5.5.3}$$

where $\varphi_n(\zeta)$ are real-valued functions. The advance and delay terms are expanded in power series by

$$\varphi(\zeta + \kappa) - \varphi(\zeta - \kappa) - 2\sinh(\kappa)\varphi'(\zeta) = \frac{1}{3}\kappa^3 \left[\varphi'''(\zeta) - \varphi'(\zeta)\right] + \mathcal{O}(\kappa^5),$$

$$\varphi(\zeta + \kappa) + \varphi(\zeta - \kappa) - 2\cosh(\kappa)\varphi(\zeta) = \kappa^2 \left[\varphi''(\zeta) - \varphi(\zeta)\right] + \mathcal{O}(\kappa^4),$$

while the nonlinear function f is expanded as

$$f = 2\left(1 + \alpha(\cos(\beta) - 1)\right)|\varphi(\zeta)|^2 \varphi(\zeta) + 2i\kappa\alpha\sin(\beta)|\varphi(\zeta)|^2 \varphi'(\zeta)$$
$$+ \kappa^2 \alpha\cos(\beta)|\varphi(\zeta)|^2 \varphi''(\zeta) + \mathcal{O}(\kappa^3).$$

Substituting all expansions into (5.5.2) gives to leading order, $\mathcal{O}(\kappa^2)$,

$$\cos(\beta)\left(\varphi_0''(\zeta) - \varphi_0(\zeta)\right) + 2\left(1 + \alpha(\cos(\beta) - 1)\right)\varphi_0^3(\zeta) = 0,$$

which admits the localized mode

$$\varphi_0(\zeta) = S\,\text{sech}\,(\zeta), \quad S = \frac{\sqrt{\cos(\beta)}}{\sqrt{1 + \alpha(\cos(\beta) - 1)}}. \tag{5.5.4}$$

The region of parameter space for which $S \in \mathbb{R}$ is defined by

$$(1 - \cos(\beta))\alpha < 1, \quad \beta \in \left[0, \tfrac{\pi}{2}\right] \cup \left[\tfrac{3\pi}{2}, 2\pi\right],$$
$$(1 - \cos(\beta))\alpha > 1, \quad \beta \in \left[\tfrac{\pi}{2}, \tfrac{3\pi}{2}\right].$$

At the next order, $\mathcal{O}(\kappa^3)$, we obtain the linear inhomogeneous equation

$$L_-\varphi_1 = \frac{2\sin(\beta)S^3(1 - \alpha)}{\cos^2(\beta)}\,\text{sech}^3(\zeta)\tanh(\zeta),$$

where $L_- = -\partial_\zeta^2 + 1 - 2\,\mathrm{sech}^2(\zeta)$. Since $L_- \varphi_0 = 0$ and $\varphi_0(\zeta)$ is even on $\mathbb{R}$, a unique odd solution exists for $\varphi_1(\zeta)$, in fact, in the explicit form

$$\varphi_1(\zeta) = \frac{\sin(\beta) S^3 (1-\alpha)}{2\cos^2(\beta)}\,\mathrm{sech}(\zeta)\tanh(\zeta). \tag{5.5.5}$$

To compute the solution to higher orders, $\mathcal{O}(\kappa^n)$ for $n \geq 4$, we have to solve a system of inhomogeneous linear equations separately in odd and even orders of n:

$$L_+ \varphi_{2k} = g_{2k}(\varphi_0, \varphi_1, ..., \varphi_{2k-1}), \quad L_- \varphi_{2k+1} = g_{2k+1}(\varphi_0, \varphi_1, ..., \varphi_{2k}), \tag{5.5.6}$$

where $L_+ = -\partial_\zeta^2 + 1 - 6\,\mathrm{sech}^2(\zeta)$, $k \in \mathbb{N}$, and g_n contains linear and cubic powers of $(\varphi_0, \varphi_1, ..., \varphi_{n-1})$ and their derivatives up to the $(n+2)$th order.

Thanks to the parity of the first two terms, we check inductively that $g_{2k}(\zeta)$ is even on $\mathbb{R}$ and $g_{2k+1}(\zeta)$ is odd on $\mathbb{R}$. Since $L_+ \varphi_0'(\zeta) = 0$ and $L_- \varphi_0(\zeta) = 0$, where $\varphi_0'(\zeta)$ is odd on $\mathbb{R}$ and $\varphi_0(\zeta)$ is even on $\mathbb{R}$, respectively, there exist unique solutions of the linear inhomogeneous equations (5.5.6) in the space of even functions for $\varphi_{2k}(\zeta)$ and odd functions for $\varphi_{2k+1}(\zeta)$. Therefore, the asymptotic expansion (5.5.3) can be computed up to any order of κ.

To go beyond all orders of the asymptotic expansion in powers of κ, we rescale dependent and independent variables near the singularity of the regular asymptotic expansion in a complex plane. The leading-order term (5.5.4) of the regular expansion (5.5.3) has its first singularity at $\zeta = \frac{i\pi}{2}$. Therefore, we take

$$\Psi(Z) = \kappa \varphi(\zeta), \quad \Theta(Z) = \kappa \overline{\varphi}(\zeta), \quad \zeta = \kappa Z + \frac{i\pi}{2}. \tag{5.5.7}$$

Note that because $\zeta \in \mathbb{C}$, it is no longer true that $\overline{\Theta(Z)} = \Psi(Z)$. The change of coordinates leads to a new system of equations, which we only write at the leading order, $\mathcal{O}(\kappa^0)$,

$$\cos(\beta)\left[\Psi_0(Z+1) + \Psi_0(Z-1) - 2\Psi_0(Z)\right]$$
$$+\,i\sin(\beta)\left[\Psi_0(Z+1) - \Psi_0(Z-1) - 2\Psi_0'(Z)\right] + F(\Psi_0, \Theta_0) = 0, \tag{5.5.8}$$

$$\cos(\beta)\left[\Theta_0(Z+1) + \Theta_0(Z-1) - 2\Theta_0(Z)\right]$$
$$-\,i\sin(\beta)\left[\Theta_0(Z+1) - \Theta_0(Z-1) - 2\Theta_0'(Z)\right] + \bar{F}(\Theta_0, \Psi_0) = 0, \tag{5.5.9}$$

where

$$F(\Psi, \Theta) = 2(1-\alpha)\Psi^2(Z)\Theta(Z) + \alpha\cos(\beta)\Psi(Z)\Theta(Z)\left(\Psi(Z+1) + \Psi(Z-1)\right)$$
$$+\,i\alpha\sin(\beta)\Psi(Z)\Theta(Z)\left(\Psi(Z+1) - \Psi(Z-1)\right).$$

The zero index indicates the leading order of the rescaled asymptotic expansion in powers of κ,

$$\Psi(Z) = \Psi_0(Z) + \sum_{n=1}^{\infty} \kappa^{2n}\Psi_n(Z), \quad \Theta(Z) = \Theta_0(Z) + \sum_{n=1}^{\infty} \kappa^{2n}\Theta_n(Z). \tag{5.5.10}$$

By substituting $\Theta_0(Z) = e^{-pZ}$ into the second equation (5.5.9) linearized around the zero solution, we recover the characteristic equation (5.4.6),

$$D_0(p; \beta) := \cos(\beta)\left(\cosh(p) - 1\right) + i\sin(\beta)\left(\sinh(p) - p\right) = 0. \tag{5.5.11}$$

By substituting $\Psi_0(Z) = e^{-pZ}$ into the first equation (5.5.8) linearized around the zero solution, we obtain the conjugate characteristic equation $\bar{D}_0(p;\beta) = 0$.

If $\beta \neq \frac{\pi}{2}$, $D_0(p;\beta) = 0$ has a double root $p = 0$ and a varying number of imaginary roots, $p \in i\mathbb{R}$, depending upon the value of β. Let $p = ik_0$ be the smallest root of $D_0(p;\beta)$ on the imaginary axis; moreover k_0 is the only nonzero root on the imaginary axis for β near $\beta = \frac{\pi}{2}$. A symmetric root $p = -ik_0$ exists for $\bar{D}_0(p;\beta) = 0$.

The system of differential advance–delay equations (5.5.8)–(5.5.9) can be reformulated as a system of integral equations using the Laplace transforms,

$$\Psi_0(Z) = \int_\gamma V_0(p)e^{-pZ}dp, \quad \Theta_0(Z) = \int_\gamma W_0(p)e^{-pZ}dp, \qquad (5.5.12)$$

which gives

$$\bar{D}_0(p;\beta)V_0 + V_0 * W_0 * [((1-\alpha) + \alpha\cos(\beta)\cosh(p) - i\alpha\sin(\beta)\sinh(p))\,V_0] = 0,$$
$$D_0(p;\beta)W_0 + W_0 * V_0 * [((1-\alpha) + \alpha\cos(\beta)\cosh(p) + i\alpha\sin(\beta)\sinh(p))\,W_0] = 0,$$

where the convolution operator is defined by

$$(V * W)(p) = \int_0^p V(p - p_1)W(p_1)dp_1,$$

and the integration is performed along a curve γ in the complex p plane.

Exercise 5.18 Show that the symmetry of integral equations implies the reduction

$$W_0(p) = V_0(-p), \quad p \in \mathbb{C}.$$

Let us now consider various solutions of the system of differential advance–delay equations (5.5.8)–(5.5.9).

First, we define two particular solutions $\{\Psi_0^s(Z), \Theta_0^s(Z)\}$ and $\{\Psi_0^u(Z), \Theta_0^u(Z)\}$ of this system, which lie on the stable and unstable manifolds respectively,

$$\lim_{\mathrm{Re}(Z)\to+\infty} \Psi_0^s(Z) = \lim_{\mathrm{Re}(Z)\to+\infty} \Theta_0^s(Z) = 0,$$
$$\lim_{\mathrm{Re}(Z)\to-\infty} \Psi_0^u(Z) = \lim_{\mathrm{Re}(Z)\to-\infty} \Theta_0^u(Z) = 0.$$

If a localized mode exists, then the two particular solutions coincide. The Laplace transform (5.5.12) generates the solution $\{\Psi_0^s(Z), \Theta_0^s(Z)\}$ when the contour of integration $\gamma = \gamma_s$ lies in the first quadrant of the complex p plane and produces the solution $\{\Psi_0^u(Z), \Theta_0^u(Z)\}$ if $\gamma = \gamma_u$ lies in the second quadrant. If there are no singularities between the two integration contours γ_s and γ_u and $\{V_0(p), W_0(p)\}$ are uniformly bounded as $|p| \to \infty$ in the upper half of the complex p plane, then the contours could be continuously deformed into each other, implying that the solution generated by each contour is the same, i.e.

$$\Psi_0^s(Z) = \Psi_0^u(Z), \quad \Theta_0^s(Z) = \Theta_0^u(Z),$$

therefore a localized mode would exist. However, as mentioned previously, there is at least one resonance $p = i|k_0|$ on the positive imaginary axis. A deformation of the integration contours γ_s and γ_u leads to a residue contribution around the singularity at $p = i|k_0|$. Apart from the double root at $p = 0$, $p = i|k_0|$ is the only root on the positive imaginary axis for β near $\beta = \frac{\pi}{2}$ and all other roots are bounded away from $\mathrm{Re}(p) = 0$ by Lemma 5.11. It is hence possible to define γ_s and γ_u to lie above these points.

Second, we are interested in the existence of the inverse power series solutions

$$\Psi_0(Z) = \sum_{n=1}^{\infty} \frac{a_n}{Z^n}, \quad \Theta_0(Z) = \sum_{n=1}^{\infty} \frac{b_n}{Z^n}. \tag{5.5.13}$$

Substituting (5.5.13) into system (5.5.8)–(5.5.9) gives at leading order, $\mathcal{O}(Z^{-3})$,

$$a_1 S^2 + a_1^2 b_1 = 0, \quad b_1 S^2 + b_1^2 a_1 = 0,$$

which has a symmetric solution $a_1 = b_1 = -\mathrm{i}S$. This solution corresponds to the leading-order term in the expansion

$$\varphi_0(\zeta) = S \,\mathrm{sech}(\zeta) = \frac{S}{\mathrm{i}\kappa Z + \mathcal{O}(\kappa^3 Z^3)} = \frac{-\mathrm{i}S}{\kappa Z}\left(1 + \mathcal{O}(\kappa^2 Z^2)\right) \quad \text{as} \quad \kappa \to 0.$$

The inverse power series in Z for $\Psi_0(Z)$ and $\Theta_0(Z)$ become the power series expansions in p for $V_0(p)$ and $W_0(p)$,

$$V_0(p) = \sum_{n=0}^{\infty} V_n p^n, \quad W_0(p) = \sum_{n=0}^{\infty} W_n p^n, \tag{5.5.14}$$

where $V_0 = W_0 = -\mathrm{i}S$. Since no singularities are present in the neighborhood of $p = 0$, the power series (5.5.14) converge near the origin but diverge as p approaches the singularity at $p = \mathrm{i}|k_0|$.

Exercise 5.19 Substituting (5.5.14) into the system of integral equations, find the first two terms of the power series in the form

$$n = 2: \quad V_0 = W_0 = -\mathrm{i}S,$$

$$n = 3: \quad V_1 = -W_1 = \frac{\sin(\beta)S^3(1 - \alpha)}{2\cos^2(\beta)}.$$

Prove that if $\alpha = 1$ (the integrable AL lattice), the power series (5.5.14) are truncated at the first term with $n = 0$.

To study the singular behavior of power series (5.5.14) near $p = \mathrm{i}|k_0|$, we now consider a linearization of system (5.5.8)–(5.5.9) about the inverse power series (5.5.13). To accommodate both cases $k_0 > 0$ and $k_0 < 0$, we write

$$\Psi_0(Z) = \frac{-\mathrm{i}S}{Z} + \Gamma\hat{\Psi}(Z)e^{-\mathrm{i}|k_0|Z} + \sum_{n=2}^{\infty} \frac{a_n}{Z^n},$$

$$\Theta_0(Z) = \frac{-\mathrm{i}S}{Z} + \Gamma\hat{\Theta}(Z)e^{-\mathrm{i}|k_0|Z} + \sum_{n=2}^{\infty} \frac{b_n}{Z^n},$$

where the exponential term

$$e^{-\mathrm{i}|k_0|Z} = e^{-\pi|k_0|/2\kappa}e^{-\mathrm{i}|k_0|\zeta/\kappa}$$

describes the beyond-all-orders effects due to the coefficient $e^{-\pi|k_0|/2\kappa}$ becoming exponentially small in the limit $\kappa \downarrow 0$. The coefficient Γ is referred to as the *Stokes constant* and it is a measure of the amplitude of the bounded oscillatory tail. A localized mode exists only if $\Gamma = 0$.

Substituting the expansion into system (5.5.8)–(5.5.9) and truncating the nonlinear terms to $\mathcal{O}(\zeta^{-3})$ gives the linearized system for $\hat{\Psi}$ and $\hat{\Theta}$,

$$\cos(\beta)\left[\hat{\Psi}(Z+1)e^{-i|k_0|} + \hat{\Psi}(Z-1)e^{i|k_0|} - 2\hat{\Psi}(Z)\right]\left(1 - \frac{\alpha S^2}{Z^2}\right)$$

$$+ i\sin(\beta)\left[\left(\hat{\Psi}(Z+1)e^{-i|k_0|} - \hat{\Psi}(Z-1)e^{i|k_0|}\right)\left(1 - \frac{\alpha S^2}{Z^2}\right) - 2\hat{\Psi}'(Z) + 2i|k_0|\hat{\Psi}(Z)\right]$$

$$- \frac{2\cos(\beta)}{Z^2}\left[2\hat{\Psi}(Z) + \hat{\Theta}(Z)\right] + \mathcal{O}(Z^{-3}) = 0,$$

$$\cos(\beta)\left[\hat{\Theta}(Z+1)e^{-i|k_0|} + \hat{\Theta}(Z-1)e^{i|k_0|} - 2\hat{\Theta}(Z)\right]\left(1 - \frac{\alpha S^2}{Z^2}\right)$$

$$- i\sin(\beta)\left[\left(\hat{\Theta}(Z+1)e^{-i|k_0|} - \hat{\Theta}(Z-1)e^{i|k_0|}\right)\left(1 - \frac{\alpha S^2}{Z^2}\right) - 2\hat{\Theta}'(Z) + 2i|k_0|\hat{\Theta}(Z)\right]$$

$$- \frac{2\cos(\beta)}{Z^2}\left[2\hat{\Theta}(Z) + \hat{\Psi}(Z)\right] + \mathcal{O}(Z^{-3}) = 0.$$

Using Laurent expansions with some integer exponent r, yet to be determined,

$$\hat{\Psi}(Z) = \rho_1 Z^r + \rho_2 Z^{r-1} + \rho_3 Z^{r-2} + \mathcal{O}(Z^{r-3}),$$

$$\hat{\Theta}(Z) = \eta_1 Z^r + \eta_2 Z^{r-1} + \eta_3 Z^{r-2} + \mathcal{O}(Z^{r-3}),$$

the linearized system can be solved at each successive power of Z. We hence obtain from the first equation

$$\mathcal{O}(Z^r): \quad \rho_1 D_0(-i|k_0|;\beta) = 0,$$

$$\mathcal{O}(Z^{r-1}): \quad \rho_2 D_0(-i|k_0|;\beta) + \rho_1 r D_0'(-i|k_0|;\beta) = 0,$$

$$\mathcal{O}(Z^{r-2}): \quad \rho_3 D_0(-i|k_0|;\beta) + \rho_2(r-1)D_0'(-i|k_0|;\beta) + \rho_1\frac{r(r-1)}{2}D_0''(-i|k_0|;\beta)$$

$$- \cos(\beta)\left(2\rho_1 + \eta_1\right) - \sin(\beta)\alpha S^2|k_0|\rho_1 = 0,$$

and from the second equation

$$\mathcal{O}(Z^r): \quad \eta_1 D_0(i|k_0|;\beta) = 0,$$

$$\mathcal{O}(Z^{r-1}): \quad \eta_2 D_0(i|k_0|;\beta) + \eta_1 r D_0'(i|k_0|;\beta) = 0,$$

$$\mathcal{O}(Z^{r-2}): \quad \eta_3 D_0(i|k_0|;\beta) + \eta_2(r-1)D_0'(i|k_0|;\beta) + \eta_1\frac{r(r-1)}{2}D_0''(i|k_0|;\beta)$$

$$- \cos(\beta)\left(2\eta_1 + \rho_1\right) + \sin(\beta)\alpha S^2|k_0|\eta_1 = 0,$$

where derivatives of $D_0(p;\beta)$ are taken with respect to p at $p = \pm i|k_0|$. If $\beta < \frac{\pi}{2}$, then $k_0 > 0$ and $p = ik_0$ is a root of $D_0(i|k_0|;\beta) = 0$ but not a root of $D_0(-i|k_0|;\beta) \neq 0$. To avoid the trivial solution, we normalize $\eta_1 = 1$. Since $D_0'(i|k_0|;\beta) \neq 0$, we find $r = 0$ and obtain unique values for the coefficients of the power series, e.g.

$$\rho_1 = 0, \quad \rho_2 = 0, \quad \rho_3 = \frac{\cos(\beta)}{D_0(-i|k_0|;\beta)}, \dots$$

and

$$\eta_1 = 1, \quad \eta_2 = \frac{-2\cos(\beta) + \sin(\beta)\alpha S^2 |k_0|}{D_0'(\mathrm{i}|k_0|;\beta)}, \cdots .$$

These computations give the first terms in the solution

$$\Psi_0(Z) = \frac{-\mathrm{i}S}{Z} + \Gamma\left[\frac{\cos(\beta)}{D_0(-\mathrm{i}|k_0|;\beta)}\frac{1}{Z^2} + \mathcal{O}\left(\frac{1}{Z^3}\right)\right]e^{-\mathrm{i}|k_0|Z} + \sum_{n=2}^{\infty}\frac{a_n}{Z^n},$$

$$\Theta_0(Z) = \frac{-\mathrm{i}S}{Z} + \Gamma\left[1 + \frac{-2\cos(\beta) + \sin(\beta)\alpha S^2 |k_0|}{D_0'(\mathrm{i}|k_0|;\beta)}\frac{1}{Z} + \mathcal{O}\left(\frac{1}{Z^2}\right)\right]e^{-\mathrm{i}|k_0|Z} + \sum_{n=2}^{\infty}\frac{b_n}{Z^n}.$$

If $\beta > \frac{\pi}{2}$, then $k_0 < 0$ and $p = -\mathrm{i}|k_0|$ is the root of $D_0(-\mathrm{i}|k_0|;\beta) = 0$ but not a root of $\bar{D}_0(\mathrm{i}|k_0|;\beta) \neq 0$. As a result, the role of components $\hat{\Psi}(Z)$ and $\hat{\Theta}(Z)$ is opposite to the one in the solution for $\beta < \frac{\pi}{2}$.

Exercise 5.20 Compute the expansion for $\hat{\Psi}(Z)$ and $\hat{\Theta}(Z)$ for the case $\beta > \frac{\pi}{2}$.

After the expansions for $\hat{\Psi}(Z)$ and $\hat{\Theta}(Z)$ are substituted into the Laplace transforms (5.5.12), the functions $V_0(p)$ and $W_0(p)$ are found in the form

$$V_0(p) = \mathcal{O}\left((p - \mathrm{i}|k_0|)\log(p - \mathrm{i}|k_0|)\right) - \mathrm{i}S + \sum_{n=2}^{\infty} a_n p^{n-1},$$

$$W_0(p) = \frac{\Gamma}{2\pi\mathrm{i}(p - \mathrm{i}|k_0|)} + \mathcal{O}(\log(p - \mathrm{i}|k_0|)) - \mathrm{i}S + \sum_{n=2}^{\infty} b_n p^{n-1}, \tag{5.5.15}$$

where the pole term arises from the term $e^{-\mathrm{i}|k_0|Z}$, the logarithmic terms arise from the terms $Z^{-m}e^{-\mathrm{i}|k_0|Z}$ with $m \geq 1$, and the power terms arise from the inverse power terms Z^{-m} with $m \geq 1$.

The pole singularity of the solution (5.5.15) is recovered from the power series solution (5.5.14) if the coefficients of the power series satisfy the matching condition

$$\sum_{n=0}^{\infty} W_n p^n \xrightarrow[p\to\mathrm{i}|k_0|]{} \frac{\Gamma}{2\pi}\frac{1}{|k_0| + \mathrm{i}p} = \frac{\Gamma}{2\pi|k_0|}\sum_{n=0}^{\infty}\frac{(-\mathrm{i}p)^n}{|k_0|^n}. \tag{5.5.16}$$

The Stokes constant is hence given by

$$K(\alpha,\beta) := \mathrm{i}\Gamma = 2\pi\mathrm{i}|k_0| \lim_{n\to\infty} (\mathrm{i}|k_0|)^n W_n. \tag{5.5.17}$$

Formula (5.5.17) is used for numerical computations of the Stokes constant. We shall only work with those values of β for which $p = \pm\mathrm{i}k_0$ are the only nonzero roots of the dispersion relations $D_0(p;\beta) = 0$ and $\bar{D}_0(p;\beta) = 0$ on the imaginary axis. For other values of β far from $\beta = \frac{\pi}{2}$, finding a simple zero of $K(\alpha;\beta)$ is not sufficient for the bifurcation of a localized mode of the differential advance–delay equation (5.5.2) as we would also have to compute the Stokes constants for other roots of the characteristic equations $D_0(p;\beta) = 0$ and $\bar{D}_0(p;\beta) = 0$ on the imaginary axis.

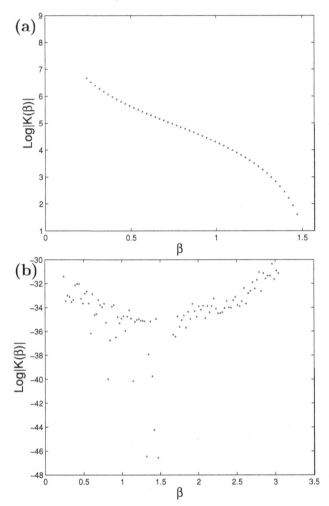

Figure 5.10 The Stokes constant $K(\alpha, \beta)$ versus β for (a) $\alpha = 0$ (the cubic DNLS equation) and (b) $\alpha = 1$ (the integrable AL lattice). There are no zeros of the Stokes constant for the cubic DNLS equation while the Stokes constant for the integrable AL lattice is always zero up to the numerical round off error. Reprinted from [140].

Exercise 5.21 Show that if $V(p)$ and $W(p)$ are given by the power series (5.5.14), then the convolution operator is given by

$$(V * W)(p) = \sum_{n_1=0}^{\infty} \sum_{n_2=0}^{\infty} \frac{n_1! n_2!}{(n_1 + n_2 + 1)!} V_{n_1} W_{n_2} p^{n_1+n_2+1}.$$

Let us first consider the two limiting cases: the cubic DNLS equation with $\alpha = 0$, where we expect $K(0, \beta) \neq 0$ for all $\beta \in (0, \pi)$ [145], and the integrable AL lattice with $\alpha = 1$, where $K(1, \beta) \equiv 0$ (Exercise 5.19). The results shown on Figure 5.10 agree with our predictions.

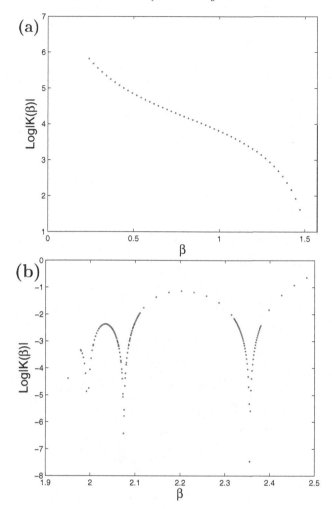

Figure 5.11 The Stokes constant $K(\alpha, \beta)$ versus β for the Salerno model with (a) $\alpha = 0.25$ and (b) $\alpha = 0.75$. Zeros of $K(\alpha, \beta)$ exist at $\beta \approx 1.98, 2.08, 2.36$ for $\alpha = 0.75$. Reprinted from [140].

Now we consider the intermediate values $\alpha \in (0, 1)$. If $\alpha < 0.5$ then $S \in \mathbb{R}$ in the domain $\beta \in \left(0, \frac{\pi}{2}\right) \cup \left(\frac{3\pi}{2}, 2\pi\right)$. Figure 5.11 (top) shows that no zeros of $K(\alpha, \beta)$ are found for $\alpha = 0.25$. If $\alpha > 0.5$, computations are extended to the domain $\beta \in \left(\frac{\pi}{2}, \frac{3\pi}{2}\right)$, where $S \in \mathbb{R}$. Figure 5.11 (bottom) shows a number of zeros of $K(\alpha, \beta)$ in the interval with $\beta > \frac{\pi}{2}$. The number of zeros of $K(\alpha, \beta)$ depends on the value of α and each zero moves towards the point $\beta = \frac{\pi}{2}$ as $\alpha \to 1$.

We would expect the solution curves for localized modes of the differential advance–delay equation (5.5.2) to approach the curves for the zeros of $K(\alpha, \beta)$ as κ is reduced towards zero. Numerically, it is easier to compute the solution curves for

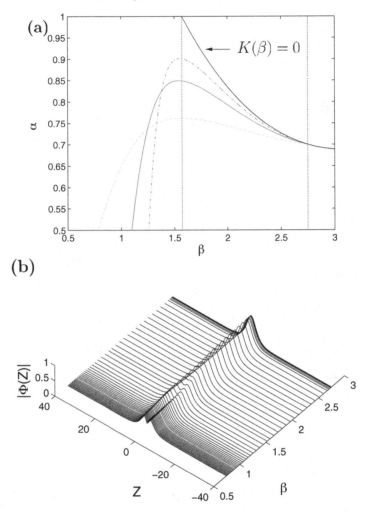

Figure 5.12 (a) Continuation of numerical solutions to the differential advance–delay equation (5.5.2) for varying α and β with $\kappa = 0.3$ (dash-dot line), $\kappa = 0.5$ (solid line) and $\kappa = 1$ (dashed line). As β is reduced, a fold bifurcation occurs and the single-humped localized mode splits into a double-humped one. Continuation of the first zero of the Stokes constant $K(\alpha, \beta)$ is shown as a solid black line. Dotted vertical lines indicate the special points $\beta = \frac{\pi}{2}$ and $\beta = \beta_1$ (more resonances occur for $\beta > \beta_1$). (b) Profiles $|\Phi(Z)|$ along the continuation branch with $\kappa = 0.5$ showing the splitting of single-humped localized modes into double-humped ones. Reprinted from [140].

a fixed nonzero value of κ in the (α, β) plane and reduce the value of κ. Such continuations for nonzero values of κ are shown in Figure 5.12(a). Three existence curves for $\kappa = 0.3, 0.5, 1$ are computed numerically and these curves approach the line $K(\alpha, \beta) = 0$ for $\beta > \frac{\pi}{2}$ as κ becomes smaller. All existence curves have a fold point at the maximum value of α for some value of $\beta < \frac{\pi}{2}$, which approaches the point

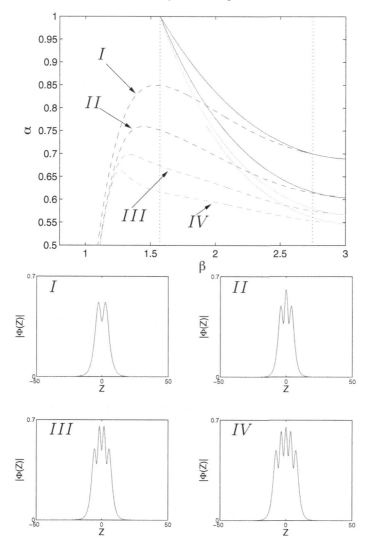

Figure 5.13 Top: continuation of zeros of the Stokes constant $K(\alpha, \beta)$ (solid lines) along with the corresponding branches of numerical solutions for $\kappa = 0.5$ (dashed lines). For $\beta < \frac{\pi}{2}$ the localized modes are multi-humped. The number of humps corresponding to each branch (I–IV) is shown in the other panels for $\alpha = 0.5$ and $\beta = 1.327$ (I), $\beta = 1.206$ (II), $\beta = 1.217$ (III), and $\beta = 1.175$ (IV). Reprinted from [140].

$(\alpha, \beta) = \left(1, \frac{\pi}{2}\right)$ as $\kappa \to 0$. For β less than this fold point the single-humped localized mode splits into a double-humped localized mode as shown in Figure 5.12(b). The existence of localized modes for $\beta > \frac{\pi}{2}$ agrees with Figure 5.9 (Section 5.3), where a localized mode was found numerically for $\alpha = 0.7$, $\beta = \frac{7\pi}{8}$, and some values of κ.

Figure 5.13 (top) illustrates the correspondence between subsequent zeros of $K(\alpha, \beta)$ shown by solid lines and the existence curves shown by dashed lines, for

$\kappa = 0.5$. As in the previous case, all the lines of $K(\alpha, \beta) = 0$ originate from the point $(\alpha, \beta) = \left(1, \frac{\pi}{2}\right)$ and the numerical approximations to solutions of the differential advance–delay equation (5.5.2) match up well with the lines of $K(\alpha, \beta) = 0$ for $\beta > \frac{\pi}{2}$. In contrast, for $\beta < \frac{\pi}{2}$ the branches experience a fold with respect to α and the corresponding single-humped localized modes become multi-humped localized modes as β is decreased. The number of humps on each branch is shown on the other panels of Figure 5.13. The number of humps increases as α is decreased for a fixed value of β.

Figures 5.12 and 5.13 illustrate the two analytical results in Sections 5.3 and 5.4. First, the branches of single-humped localized modes for $\beta > \frac{\pi}{2}$ terminate for values of $\alpha > 0.5$ away from the cubic DNLS equation with $\alpha = 0$, while the branches of multi-humped localized modes for $\beta < \frac{\pi}{2}$ extend to the cubic DNLS equation with $\alpha = 0$. On the other hand, both figures suggest that all existence curves for a fixed value of κ have a fold point for $\alpha < 1$ away from the integrable AL lattice with $\alpha = 1$, indicating that the solutions of the integrable AL lattice do not persist for $\alpha \neq 1$ and a fixed value of $\kappa > 0$.

5.6 Traveling localized modes in periodic potentials

All examples of the previous sections were devoted to traveling localized modes in the generalized DNLS equation. The DNLS equation is a reduction of the Gross–Pitaevskii equation in the limit of large-amplitude periodic potentials (Section 2.4). In particular, the spectrum of the linearized DNLS equation is a zoom of a single spectral band of the Schrödinger operator with a periodic potential. As a consequence, a traveling localized mode may have only one resonance with linear eigenmodes of the DNLS equation.

Let us now consider the existence of traveling localized modes in the Gross–Pitaevskii equation,

$$iu_t = -u_{xx} + V(x)u - |u|^2u, \qquad (5.6.1)$$

where $u(x, t) : \mathbb{R} \times \mathbb{R}_+ \to \mathbb{C}$ and $V(x) : \mathbb{R} \to \mathbb{R}$ is a bounded and 2π-periodic potential.

We will see that a traveling localized mode has infinitely many resonances with linear eigenmodes of the Schrödinger operator $L = -\partial_x^2 + V(x)$ that has infinitely many spectral bands. As a result, bifurcations of traveling localized modes in the Gross–Pitaevskii equation (5.6.1) have *codimension infinity*. In view of the delicacy in the search of traveling localized modes in the DNLS equation (Sections 5.3–5.5), the presence of infinitely many resonances rules out any realistic hope that the traveling localized modes would exist in the Gross–Pitaevskii equation (5.6.1). This may explain why no exact traveling localized modes were found in the Gross–Pitaevskii equation although such solutions were successfully constructed for the DNLS equation, e.g. by Flach & Kladko [57] and Khare *et al.* [114].

Although direct analysis of the Gross–Pitaevskii equation (5.6.1) is also possible, we shall simplify the formalism in the limit of the small-amplitude periodic potential

$V(x)$. In this limit, the nonlinear Dirac equations describe narrow band gaps between two adjacent spectral bands (Section 2.2) and the nonlinear Schrödinger equation describes a semi-infinite band gap below the lowest spectral band (Section 2.3). Our analysis follows the work of Pelinovsky & Schneider [163] where the nonlinear Dirac equations are considered; however, details are further simplified using the nonlinear Schrödinger equation for the semi-infinite band gap.

Let us represent the small-amplitude periodic potential $V(x)$ by the Fourier series

$$V(x) = \epsilon \sum_{m \in \mathbb{Z}} V_m e^{imx}, \qquad (5.6.2)$$

where $\epsilon > 0$ is a small parameter, $V_0 = 0$, and $V_m = V_{-m} = \bar{V}_m$ for all $m \in \mathbb{N}$. These constraints imply that $V(x)$ is a real-valued 2π-periodic function with zero mean and even symmetry on $\mathbb{R}$.

Recall that an asymptotic solution of the Gross–Pitaevskii equation (5.6.1) for small values of $\epsilon > 0$ can be represented in the form

$$u(x,t) = \epsilon^{1/2} a(\epsilon^{1/2} x, \epsilon t) + \mathcal{O}(\epsilon^{3/2}), \qquad (5.6.3)$$

where $a(X,T)$ satisfies the cubic NLS equation,

$$ia_T + a_{XX} + |a|^2 a = 0, \qquad (5.6.4)$$

in slow variables $X = \epsilon^{1/2} x$ and $T = \epsilon t$. There exists a traveling soliton of the NLS equation (5.6.4),

$$a(X,T) = A(X - 2\gamma T) \, e^{i\gamma(X - \gamma T) - i\Omega T}, \quad A(X) = \sqrt{2|\Omega|} \, \text{sech}(\sqrt{|\Omega|} X), \quad (5.6.5)$$

for any $\Omega < 0$ and $\gamma \in \mathbb{R}$. Note that

$$A(X) = \bar{A}(-X) = \bar{A}(X), \qquad (5.6.6)$$

because the trivial parameters of spatial translations and phase rotations are set to zero.

When $\gamma = 0$ in (5.6.5), the stationary localized mode of the NLS equation (5.6.4) persists as the stationary localized mode of the Gross–Pitaevskii equation (5.6.1) (Section 2.3.2). Here we shall study persistence of the traveling localized mode of the NLS equation (5.6.4) when $\gamma \neq 0$ in (5.6.5). We show that the corresponding traveling solution of the Gross–Pitaevskii equation (5.6.1) has a bounded oscillatory tail in the far-field profile, which is not accounted for in the NLS equation (5.6.4).

From a technical point of view, our analysis relies on the modification of spatial dynamics methods developed for the persistence of small-amplitude localized modulated pulses in the nonlinear Klein–Gordon and Maxwell equations by Groves & Schneider [80, 81, 82]. Similar to these works, we will show that a local center manifold of a spatial dynamical system for traveling localized modes of the Gross–Pitaevskii equation (5.6.1) spanned by bounded oscillatory modes destroys an exponential localization of the solutions along the directions of the slow stable and unstable manifolds. As a result, the spatial field of these solutions decays to small-amplitude oscillatory modes in the far-field regions. This is yet another indication that resonances with linear oscillation modes become obstacles in the bifurcation of traveling localized modes in lattices and periodic potentials.

Let us also mention that propagation of a moving solitary wave in the focusing Gross–Pitaevskii equation (5.6.1) with a periodic potential $V(x)$ of a large period is an old physical problem (see review by Sánchez & Bishop [178]). An effective particle equation is usually derived in this limit by a heuristic asymptotic expansion. The particle equation describes a steady propagation of the moving solitary wave. The radiation effects from the moving solitary wave in the far-field regions appear beyond all orders of the asymptotic expansion. This picture gives a good intuition on what to expect in the time evolution of a traveling localized mode in a periodic potential but it lacks rigor. The spatial dynamics methods replace this qualitative picture with rigorous analysis.

Since the space and time variables of the Gross–Pitaevskii equation (5.6.1) are not separated for a traveling localized mode, we shall look for traveling solutions in the form

$$u(x,t) = \psi(x,y)e^{ic(x-ct)-i\omega t}, \qquad y = x - 2ct,$$

where $(\omega, c) \in \mathbb{R}^2$ are parameters. Coordinates (x, y) are linearly independent if $c \neq 0$. For simplicity, we only consider the case $c > 0$. The envelope function $\psi(x, y)$ satisfies the partial differential equation

$$\left(\omega + 2ic\partial_x + \partial_x^2 + 2\partial_x\partial_y + \partial_y^2\right)\psi(x,y) = V(x)\psi(x,y) - |\psi(x,y)|^2\psi(x,y). \quad (5.6.7)$$

To accommodate the traveling localized mode according to the leading-order representations (5.6.3) and (5.6.5), we shall look for 2π-periodic functions $\psi(x, y)$ in variable x and bounded solutions $\psi(x, y)$ in variable y on $\mathbb{R}$. Therefore, we write

$$\psi(x,y) = \epsilon^{1/2} \sum_{m \in \mathbb{Z}} \psi_m(y)e^{imx}, \qquad (5.6.8)$$

where $\epsilon > 0$ is the same small parameter as in the potential $V(x)$.

The series representation (5.6.8) transforms the partial differential equation (5.6.7) to an infinite system of ordinary differential equations

$$\psi_m''(y) + 2im\psi_m'(y) + \left(\omega - m^2 - 2cm\right)\psi_m(y) = \epsilon \sum_{m_1 \in \mathbb{Z}} V_{m-m_1}\psi_{m_1}(y)$$

$$-\epsilon \sum_{m_1 \in \mathbb{Z}} \sum_{m_2 \in \mathbb{Z}} \psi_{m_1}(y)\bar{\psi}_{-m_2}(y)\psi_{m-m_1-m_2}(y), \quad m \in \mathbb{Z}. \quad (5.6.9)$$

The left-hand side of system (5.6.9) represents the linearized system for $\epsilon = 0$,

$$\psi_m''(y) + 2im\psi_m'(y) + \left(\omega - m^2 - 2cm\right)\psi_m(y) = 0. \quad (5.6.10)$$

Solutions of this system are given by the eigenmodes $\psi_m(y) \sim e^{\kappa y}$, where κ is determined by the roots of the quadratic equation,

$$\kappa^2 + 2im\kappa + \omega - m^2 - 2cm = 0, \qquad m \in \mathbb{Z}. \quad (5.6.11)$$

If $\omega = 0$ (bifurcation case), the two roots of the quadratic equations (5.6.11) are given by

$$\omega = 0: \quad \kappa = \kappa_m^\pm = -im \pm \sqrt{2cm}, \quad m \in \mathbb{Z}. \quad (5.6.12)$$

If $c > 0$ (for convenience) and $m > 0$, both roots are complex-valued with

$$\text{Re}(\kappa_m^{\pm}) = \pm\sqrt{2cm} \neq 0 \quad \text{and} \quad \text{Im}(\kappa_m^{\pm}) = -m. \qquad (5.6.13)$$

If $m \leq 0$, both roots κ are purely imaginary with

$$\kappa_m^{\pm} = ik_m^{\pm} \quad \text{and} \quad k_m^{\pm} = -m \pm \sqrt{2c|m|}. \qquad (5.6.14)$$

Let $E_m^{\pm} = \text{span}\{e^{\kappa_m^{\pm} y}\}$ be the subspace in a phase space of the linearized system (5.6.10). The following two lemmas give useful results about this system.

Lemma 5.12   *Let $\omega = 0$ and $c > 0$. The phase space of the linearized system (5.6.10) decomposes into a direct sum of infinite-dimensional subspaces $E^s \oplus E^u \oplus E^0 \oplus E^c$, where*

$$E^u = \bigoplus_{m>0} E_m^+, \quad E^s = \bigoplus_{m>0} E_m^-, \quad E^0 = E_0^+ \oplus E_0^-$$

and

$$E^c = \left(\bigoplus_{m<0} E_m^+\right) \oplus \left(\bigoplus_{m<0} E_m^-\right).$$

The double zero root $\kappa = 0$ is associated to the subspace E^0. The purely imaginary roots $\kappa \in i\mathbb{R}$ are semi-simple. All other roots $\kappa \in \mathbb{C}$ are simple.

Proof   Let $\kappa = \kappa_0$ be a double root of the quadratic equation (5.6.11). Then, $\kappa_0 = -im$ and this implies that if $\omega = 0$, then $m = 0$. Therefore, all nonzero roots for $m \neq 0$ are semi-simple. When $m > 0$, all roots are complex-valued and simple.   □

Remark 5.5   Subspaces E^s, E^u, and $E^0 \oplus E^c$ define stable, unstable, and center manifolds of the linearized system (5.6.10).

Lemma 5.13   *Consider a linear inhomogeneous equation*

$$\left(-\partial_y^2 - 2im\partial_y + m^2 + 2mc\right)\psi_m(y) = F_m(y), \quad m > 0,$$

where $F_m \in C_b^0(\mathbb{R})$. There exists a unique solution $\psi_m \in C_b^2(\mathbb{R})$, such that

$$\|\psi_m\|_{L^\infty} \leq \frac{1}{2cm}\|F_m\|_{L^\infty}.$$

Proof   Let $\psi_m(y) = e^{-imy}\varphi_m(y)$. The function $\varphi_m(y)$ solves the inhomogeneous Schrödinger equation

$$\left(\beta^2 - \partial_y^2\right)\varphi_m(y) = F_m(y)e^{imy}, \quad \text{where} \quad \beta = \sqrt{2cm}.$$

Since the solutions of the homogeneous equation are exponentially decaying and growing as $\varphi_m \sim e^{\pm\beta y}$, there exists a unique bounded solution of the inhomogeneous equation in the integral form

$$\varphi_m(y) = \frac{1}{2\beta}\int_{-\infty}^{\infty} e^{-\beta|y-y'|}F_m(y')e^{imy'}\,dy',$$

such that $\|\varphi_m\|_{L^\infty} \leq (1/\beta^2)\|F_m\|_{L^\infty}$.   □

Using Lemma 5.13 and the Implicit Function Theorem (Appendix B.7), one can solve for small $\epsilon > 0$ all equations of system (5.6.9) for $m > 0$ and parameterize solutions $\psi_m(y)$ for $m > 0$ in terms of bounded components $\psi_m(y)$ for $m \leq 0$. However, we do not perform this parameterization, since we intend to write system (5.6.9) as a Hamiltonian system with a local symplectic structure and use the formalism of near-identity transformations and normal forms. The following theorem formulates the main result of this section.

Theorem 5.8 *Let $c > 0$, $\omega = \epsilon\Omega$ with an ϵ-independent $\Omega < 0$. Fix $N \in \mathbb{N}$. Assume that $V \in H^s_{\mathrm{per}}(\mathbb{R})$ for a fixed $s > \frac{1}{2}$ and $V(-x) = V(x)$ for all $x \in \mathbb{R}$. For sufficiently small ϵ, there are ϵ-independent constants $L > 0$ and $C > 0$, such that the Gross–Pitaevskii equation (5.6.1) admits an infinite-dimensional, continuous family of traveling solutions in the form*

$$u(x,t) = \epsilon^{1/2}\psi(x,y)e^{ic(x-ct)-i\omega t}, \quad y = x - 2ct,$$

where the function $\psi(x,y)$ is a periodic function of x satisfying the reversibility constraint

$$\psi(x,y) = \bar\psi(x,-y),$$

and the bound

$$\left|\psi(x,y) - A_\epsilon(\epsilon^{1/2}y)\right| \leq C\epsilon^N, \quad x \in \mathbb{R}, \ y \in [-L/\epsilon^{N+1}, L/\epsilon^{N+1}].$$

Here $A_\epsilon(Y) = A(Y) + \mathcal{O}(\epsilon)$ for $Y = \epsilon^{1/2}y \in \mathbb{R}$ is an exponentially decaying solution as $|Y| \to \infty$ satisfying the reversibility constraint

$$A_\epsilon(Y) = \bar A_\epsilon(-Y),$$

with $A(Y) = \sqrt{2|\Omega|}\,\mathrm{sech}(\sqrt{|\Omega|}Y)$.

Remark 5.6 The solution $\psi(x,y)$ is a bounded non-decaying function on a large finite interval $[-L/\epsilon^{N+1}, L/\epsilon^{N+1}] \subset \mathbb{R}$ but we do not claim that the solution $\psi(x,y)$ can be extended globally on $\mathbb{R}$. The localized solution $A_\epsilon(Y)$ is defined up to the terms of $\mathcal{O}(\epsilon^N)$ and it satisfies an extended NLS equation. The bounded oscillatory tails are expected to be exponentially small in ϵ on an exponentially large scale of y-axis according to the analysis of Groves & Schneider [82].

The proof of Theorem 5.8 relies on the Hamiltonian formulation subject to reversibility constraints, the near-identity transformations, and the construction of a local center–stable manifold in the spatial dynamical system.

The system of second-order equations (5.6.9) can be written as the system of first-order equations which admits a symplectic Hamiltonian structure. Let $\omega = \epsilon\Omega$ and denote

$$\phi_m(y) = \psi'_m(y) + im\psi_m(y), \quad m \in \mathbb{Z}.$$

System (5.6.9) is equivalent to the first-order system

$$
\begin{cases}
\dfrac{d\psi_m}{dy} = \phi_m - im\psi_m, \\[2mm]
\dfrac{d\phi_m}{dy} = 2cm\psi_m - im\phi_m - \epsilon\Omega\psi_m \\[2mm]
\qquad + \epsilon \displaystyle\sum_{m_1\in\mathbb{Z}} V_{m-m_1}\psi_{m_1} - \epsilon \displaystyle\sum_{m_1\in\mathbb{Z}}\sum_{m_2\in\mathbb{Z}} \psi_{m_1}\bar{\psi}_{-m_2}\psi_{m-m_1-m_2}.
\end{cases}
\tag{5.6.15}
$$

Let $\boldsymbol{\psi}$ denote a vector consisting of elements of the set $\{\psi_m\}_{m\in\mathbb{Z}}$. The variables $\{\boldsymbol{\psi}, \boldsymbol{\phi}, \bar{\boldsymbol{\psi}}, \bar{\boldsymbol{\phi}}\}$ are canonical because system (5.6.15) is equivalent to the following system:

$$
\frac{d\psi_m}{dy} = \frac{\partial H}{\partial \bar{\phi}_m}, \quad \frac{d\phi_m}{dy} = -\frac{\partial H}{\partial \bar{\psi}_m}, \quad m \in \mathbb{Z},
\tag{5.6.16}
$$

where $H = H(\boldsymbol{\psi}, \boldsymbol{\phi}, \bar{\boldsymbol{\psi}}, \bar{\boldsymbol{\phi}})$ is the Hamiltonian function given by

$$
H = \sum_{m\in\mathbb{Z}} \left(|\phi_m|^2 - 2cm|\psi_m|^2 - im(\psi_m\bar{\phi}_m - \bar{\psi}_m\phi_m) + \epsilon\Omega|\psi_m|^2 \right)
$$

$$
- \epsilon \sum_{m\in\mathbb{Z}}\sum_{m_1\in\mathbb{Z}} V_{m-m_1}\psi_{m_1}\bar{\psi}_m + \frac{\epsilon}{2}\sum_{m\in\mathbb{Z}}\sum_{m_1\in\mathbb{Z}}\sum_{m_2\in\mathbb{Z}} \psi_{m_1}\bar{\psi}_{-m_2}\psi_{m-m_1-m_2}\bar{\psi}_m.
$$

Since $l_s^2(\mathbb{Z})$ is a Banach algebra with respect to convolution sums for any $s > \frac{1}{2}$ (Appendix B.1), the convolution sums in the nonlinear system (5.6.15) map $\boldsymbol{\psi} \in l_s^2(\mathbb{Z})$ to an element of $l_s^2(\mathbb{Z})$ if $\mathbf{V} \in l_s^2(\mathbb{Z})$ for a fixed $s > \frac{1}{2}$. The linear terms of system (5.6.15) map, however, an element of $D_s \subset X$ to an element of X, where

$$
D_s = \left\{ (\boldsymbol{\psi}, \boldsymbol{\phi}, \bar{\boldsymbol{\psi}}, \bar{\boldsymbol{\phi}}) \in l_{s+1}^2(\mathbb{Z}', \mathbb{C}^4) \right\}, \quad X = \left\{ (\boldsymbol{\psi}, \boldsymbol{\phi}, \bar{\boldsymbol{\psi}}, \bar{\boldsymbol{\phi}}) \in l_s^2(\mathbb{Z}', \mathbb{C}^4) \right\}, \quad s \geq 0.
$$

Combining the two facts together, we note that X with $s > \frac{1}{2}$ can be chosen as the phase space of the dynamical system (5.6.15).

If $V_m = V_{-m}$ for all $m \in \mathbb{Z}$ (thanks to the symmetry of $V(x)$ about $x = 0$), the Hamiltonian system (5.6.15) is reversible and its solutions are invariant under the transformation

$$
\boldsymbol{\psi}(y) \mapsto \bar{\boldsymbol{\psi}}(-y), \quad \boldsymbol{\phi}(y) \mapsto -\bar{\boldsymbol{\phi}}(-y), \quad y \in \mathbb{R}.
\tag{5.6.17}
$$

Reversible solutions satisfy the reduction

$$
\boldsymbol{\psi}(y) = \bar{\boldsymbol{\psi}}(-y), \quad \boldsymbol{\phi}(y) = -\bar{\boldsymbol{\phi}}(-y), \quad y \in \mathbb{R}.
\tag{5.6.18}
$$

If a local solution of system (5.6.15) is constructed on $y \in \mathbb{R}_+$ and it intersects at $y = 0$ with the reversibility hyperplane,

$$
\Sigma = \left\{ (\boldsymbol{\psi}, \boldsymbol{\phi}, \bar{\boldsymbol{\psi}}, \bar{\boldsymbol{\phi}}) \in D_s : \quad \mathrm{Im}(\boldsymbol{\psi}) = \mathbf{0}, \ \mathrm{Re}(\boldsymbol{\phi}) = \mathbf{0} \right\},
\tag{5.6.19}
$$

then the solution is extended to a global reversible solution in D_s for all $y \in \mathbb{R}$ using the reversibility transformation (5.6.17).

Exercise 5.22 Introduce normal coordinates,

$$
m < 0: \quad \psi_m = \frac{c_m^+(y) + c_m^-(y)}{\sqrt{2}\sqrt[4]{-2cm}}, \quad \phi_m = i\sqrt[4]{-2cm}\,\frac{c_m^+(y) - c_m^-(y)}{\sqrt{2}}
$$

and

$$m > 0 : \quad \psi_m = \frac{c_m^+(y) + c_m^-(y)}{\sqrt{2}\sqrt[4]{2cm}}, \quad \phi_m = \sqrt[4]{2cm}\frac{c_m^+(y) - c_m^-(y)}{\sqrt{2}}.$$

Show that the Hamiltonian function $H(\psi, \phi, \bar{\psi}, \bar{\phi})$ at $\epsilon = 0$ transforms to the form

$$H = |\phi_0|^2 + \sum_{m<0} \left(k_m^+ |c_m^+|^2 - k_m^- |c_m^-|^2 \right) + \sum_{m>0} \left(\kappa_m^- c_m^- \bar{c}_m^+ - \kappa_m^+ c_m^+ \bar{c}_m^- \right),$$

where $k_m^\pm$ and $\kappa_m^\pm$ are defined by (5.6.13) and (5.6.14).

The formal truncation of the Hamiltonian function H at the mode with $m = 0$ leads to the NLS equation. To make it rigorous, we need to use the near-identity transformations of the Hamiltonian function and to derive an extended NLS equation, where we can prove persistence of a homoclinic orbit under the reversibility constraint. To do so, let us consider the subspace of the phase space of the dynamical system (5.6.15):

$$S = \{\phi_m = \psi_m = 0, \quad m \in \mathbb{Z}\backslash\{0\}\}. \qquad (5.6.20)$$

The formal truncation of H on the subspace S gives

$$H|_S = |\phi_0|^2 + \epsilon\Omega|\psi_0|^2 + \frac{\epsilon}{2}|\psi_0|^4,$$

where we have set $V_0 = 0$ for convenience. The Hamiltonian function $H|_S$ generates the system of equations, which results in the stationary NLS equation,

$$\psi_0''(Y) + \Omega\psi_0(Y) + |\psi_0(Y)|^2\psi_0(Y) = 0, \quad Y = \epsilon^{1/2}y, \qquad (5.6.21)$$

where $\phi_0(Y) = \epsilon^{1/2}\psi_0'(Y)$. Equation (5.6.21) is nothing but a reduction of the NLS equation (5.6.4) for solutions $a(X, T) = \psi_0(X)e^{-i\Omega T}$ with $X = Y$ and $\psi_0(X) = \sqrt{2|\Omega|}\,\text{sech}(\sqrt{|\Omega|}Y)$ for $\Omega < 0$.

Note that although the second-order equation (5.6.21) is formulated in the four-dimensional phase space, it has translational and gauge symmetries. A reversible homoclinic orbit satisfying the constraints

$$\psi_0(Y) = \bar{\psi}_0(-Y) = \bar{\psi}_0(Y), \quad Y \in \mathbb{R},$$

is unique and is defined on the phase plane (ψ_0, ϕ_0) of a planar Hamiltonian system generated by $H|_S$.

To use the reduction of the dynamical system (5.6.15) on the subspace S and persistence of the reversible homoclinic orbit in a planar Hamiltonian system, we extend the second-order equation (5.6.21) with near-identity transformations and the normal form theory.

Lemma 5.14 *Fix $c > 0$ and $s > \frac{1}{2}$. For each $N \in \mathbb{N}$ and sufficiently small $\epsilon > 0$, there is a near-identity, analytic, symplectic change of coordinates in a neighborhood of the origin in X, such that the Hamiltonian function H transforms to the normal form up to the order of $\mathcal{O}(\epsilon^{N+1})$,*

$$\tilde{H} = |\phi_0|^2 + \sum_{m<0} \left(k_m^+ |c_m^+|^2 - k_m^- |c_m^-|^2 \right) + \sum_{m>0} \left(\kappa_m^- c_m^- \bar{c}_m^+ - \kappa_m^+ c_m^+ \bar{c}_m^- \right)$$

$$+ \epsilon H_S(\psi_0, \bar{\psi}_0) + \epsilon H_T(\psi_0, \bar{\psi}_0, \mathbf{c}^+, \mathbf{c}^-) + \epsilon^{N+1} H_R(\psi_0, \bar{\psi}_0, \mathbf{c}^+, \mathbf{c}^-),$$

where H_S is a polynomial of degree $2N + 2$ in $(\psi_0, \bar{\psi}_0)$, H_T is a polynomial of degree $2N$ in $(\psi_0, \bar{\psi}_0)$ and of degree 4 with no linear terms in all other variables $(\mathbf{c}^+, \mathbf{c}^-)$, and H_R is a polynomial of degree $8N + 4$ in $(\psi_0, \bar{\psi}_0)$ and of degree 4 in all other variables $(\mathbf{c}^+, \mathbf{c}^-)$. All components H_S, H_T and H_R depend on ϵ as follows: H_S and H_T are polynomials in ϵ of degree $N - 1$, and H_R is a polynomial in ϵ of degree $3N - 1$. The reversibility constraint (5.6.18) is preserved by the change of the variables.

Remark 5.7 Vectors $\mathbf{c}^\pm$ with components $\{c_m^\pm\}_{m \in \mathbb{Z} \setminus \{0\}}$ in Lemma 5.14 correspond to the normal coordinates in Exercise 5.22.

Proof The existence of a near-identity symplectic transformation follows from the fact that the non-resonance conditions

$$l\kappa_0 - \kappa_m^\pm \neq 0, \quad l \in \mathbb{Z}, \quad m \in \mathbb{Z}$$

are satisfied since $\kappa_0 = 0$ and all other eigenvalues are nonzero and semi-simple for $\epsilon = 0$. The transformation is analytic in a local neighborhood of the origin in X as the vector field of the dynamical system (5.6.15) is analytic (given by a cubic polynomial). The reversibility constraint (5.6.18) is preserved by the symplectic change of variables. The count of the degree of polynomials H_S, H_T and H_R follows from the fact that the vector field of the dynamical system (5.6.15) contains only linear and cubic terms in normal coordinates and the near-identity transformation of $(\mathbf{c}^+, \mathbf{c}^-)$ up to the order $\mathcal{O}(\epsilon^{N+1})$ involves a polynomial of degree N in ϵ and of degree $2N + 1$ in $(\psi_0, \bar{\psi}_0)$. $\qquad\square$

For each $N \in \mathbb{N}$, the subspace S defined by (5.6.20) is an invariant subspace of the dynamical system (5.6.15) if the Hamiltonian function $\tilde{H}$ is truncated at $H_R \equiv 0$. The dynamics on S is given by the Hamiltonian system

$$\psi_0''(Y) + \frac{\partial H_S}{\partial \bar{\psi}_0} = 0, \quad Y = \epsilon^{1/2} y. \tag{5.6.22}$$

If $N = 1$, the extended equation (5.6.22) transforms to the stationary NLS equation (5.6.21), where a unique reversible homoclinic orbit exists for $\Omega < 0$. We shall prove the persistence of the reversible homoclinic orbit in the extended equation (5.6.22).

Lemma 5.15 *For each $N \in \mathbb{N}$ and a sufficiently small ϵ, there exists a reversible homoclinic orbit of system (5.6.22) for $\Omega < 0$ such that*

$$\psi_0(Y) = \bar{\psi}_0(-Y) = \bar{\psi}(Y) \tag{5.6.23}$$

and

$$\exists \gamma > 0, \ \exists C > 0: \quad |\psi_0(Y)| \leq Ce^{-\gamma|Y|}, \quad Y \in \mathbb{R}. \tag{5.6.24}$$

Proof Hamiltonian system (5.6.22) is integrable thanks to the first invariant

$$E = \left| \frac{d\psi_0}{dY} \right|^2 + H_s(\psi_0, \bar{\psi}_0),$$

where $E = 0$ for localized modes. In addition, it has real coefficients and reduces
to the second-order equation if $\psi = \bar{\psi}_0$. If $\epsilon = 0$, the equilibrium state $\psi_0 = 0$ is a
saddle point and there exists a turning point with $\psi_0'(0) = 0$ for $\Omega < 0$. Therefore,
a unique reversible homoclinic orbit exists. By Lemma 5.14, $H_s(\psi_0, \bar{\psi}_0)$ is given
by a polynomial of degree $2N + 2$ in ψ_0 and of degree $N - 1$ in ϵ. Therefore, the
equilibrium state $\psi_0 = 0$ persists as a saddle point and the turning point with
$\psi_0'(0) = 0$ persists in ϵ.   □

We study solutions of system (5.6.15) after the normal form transformations of
Lemma 5.14. We construct a local solution for all $y \in [0, L/\epsilon^{N+1}]$ for some ϵ-
independent constant $L > 0$, which is close in the function space X for $s > \frac{1}{2}$ to the
homoclinic orbit of Lemma 5.15 by the distance $C\epsilon^N$, where $C > 0$ is ϵ-independent.
This solution represents an infinite-dimensional local center–stable manifold and it
is spanned by small oscillatory and small exponentially decaying solutions near an
exponentially decaying solution of Lemma 5.15. Parameters of the local center–
stable manifold are chosen to ensure that the manifold intersects at $y = 0$ with the
reversibility hyperplane Σ given by (5.6.19). This construction completes the proof
of Theorem 5.8.

Using Lemma 5.14 and the explicit representation of the Hamiltonian function
$\tilde{H}$ in new variables (ψ_0, ϕ_0) and $\mathbf{c} = (\mathbf{c}^+, \mathbf{c}^-)$, we rewrite the set of equations in the
separated form

$$\psi_0''(y) = \epsilon F_S(\psi_0) + \epsilon F_T(\psi_0, \mathbf{c}) + \epsilon^{N+1} F_R(\psi_0, \mathbf{c}), \qquad (5.6.25)$$

$$\mathbf{c}'(y) = \Lambda_\epsilon \mathbf{c} + \epsilon \mathbf{F}_T(\psi_0, \mathbf{c}) + \epsilon^{N+1} \mathbf{F}_R(\psi_0, \mathbf{c}), \qquad (5.6.26)$$

where Λ_ϵ denotes the matrix operator for the linear terms of the system and
$(F_S, F_T, F_R, \mathbf{F}_T, \mathbf{F}_R)$ denote the nonlinear (polynomial) terms.

Lemma 5.16   *Let $c > 0$ and $\Omega < 0$. For sufficiently small $\epsilon > 0$, the linearization
of system (5.6.25)–(5.6.26) is topologically equivalent to the one for $\epsilon = 0$, except
that the double zero eigenvalue splits into a symmetric pair of real eigenvalues.*

Proof   Thanks to the explicit form, $\Lambda_\epsilon = \Lambda_0 + \epsilon \Lambda_1(\epsilon)$, where Λ_0 is a diagonal
unbounded matrix operator which consists of $\{ik_m^+, -ik_m^-\}$ for all $m < 0$ and of
$\{\kappa_m^+, \kappa_m^-\}$ for all $m > 0$. The matrix operator $\epsilon \Lambda_1(\epsilon)$ is a small perturbation to Λ_0
if $\mathbf{V} \in l_s^2(\mathbb{Z})$ for a fixed $s > \frac{1}{2}$ and $\epsilon > 0$ is sufficiently small. By Lemma 5.12,
all eigenvalues of Λ_0 are semi-simple. Therefore, they are structurally stable with
respect to perturbations for sufficiently small $\epsilon \neq 0$. The double zero eigenvalue
splits according to the linearization of the extended system (5.6.22).   □

We have seen that the truncated system (5.6.25)–(5.6.26) with $F_R \equiv 0$ and
$\mathbf{F}_R \equiv \mathbf{0}$ admits an invariant reduction on S. By Lemma 5.15, the extended second-
order equation (5.6.22) has a reversible homoclinic orbit, which satisfies the decay
bound (5.6.24). This construction enables us to rewrite $\psi_0(y)$ as $\psi_0(\epsilon^{1/2} y) + \varphi(y)$,
where $\psi_0(Y)$ is the reversible homoclinic orbit of Lemma 5.15 with $Y = \epsilon^{1/2} y$ and
$\varphi(y)$ is a perturbation term. As a result, system (5.6.25)–(5.6.26) is rewritten in

the equivalent form

$$\varphi''(y) = \epsilon L_\epsilon \varphi + \epsilon G_T(\varphi, \mathbf{c}) + \epsilon^{N+1} G_R(\psi_0 + \varphi, \mathbf{c}), \tag{5.6.27}$$

$$\mathbf{c}'(y) = \Lambda_\epsilon \mathbf{c} + \epsilon \mathbf{F}_T(\psi_0 + \varphi, \mathbf{c}) + \epsilon^{N+1} \mathbf{F}_R(\psi_0 + \varphi, \mathbf{c}), \tag{5.6.28}$$

where $L_\epsilon = D_{\psi_0} F_S(\psi_0)$ is the Jacobian operator and (G_T, G_R) is the nonlinear vector field obtained from system (5.6.25). We note that the function G_T combines the nonlinear terms in φ from F_S and the nonlinear terms in $\psi_0 + \varphi$ from F_T.

Exercise 5.23 Let $N = 2$ and compute the coefficients of L_ϵ, Λ_ϵ, G_T, and $\mathbf{F}_T$ explicitly.

Lemma 5.17 *Consider the linear inhomogeneous equation*

$$\frac{d^2\varphi}{dy^2} - \epsilon L_\epsilon \varphi = F_0(y), \tag{5.6.29}$$

where $F_0 \in C_b^0(\mathbb{R})$ is real-valued. If $F_0(y) = F_0(-y)$ for all $y \in \mathbb{R}$, then there exists a one-parameter family of solutions $\varphi \in C_b^2(\mathbb{R})$ in the form $\varphi = \alpha \psi_0'(Y) + \tilde{\varphi}(y)$, where $\alpha \in \mathbb{R}$ and $\tilde{\varphi}$ satisfies

$$\|\tilde{\varphi}\|_{L^\infty} \le \frac{C}{\epsilon} \|F_0\|_{L^\infty} \tag{5.6.30}$$

for an ϵ-independent constant $C > 0$.

Proof Thanks to the translational invariance of the extended second-order equation (5.6.22), $\psi_0'(Y)$ satisfies the homogeneous equation

$$\frac{d^2\varphi}{dy^2} - \epsilon L_\epsilon \varphi = 0.$$

The zero eigenvalue of the self-adjoint operator $-\partial_y^2 + \epsilon L_\epsilon$ is isolated from the rest of the spectrum. By the Fredholm Alternative Theorem (Appendix B.4), a solution of the linear inhomogeneous equation (5.6.29) is bounded for all $y \in \mathbb{R}$ if and only if $\langle \psi_0', F_0 \rangle_{L^2} = 0$. If F_0 is even on $\mathbb{R}$, then the constraint is trivially met. Since the norm in $L^\infty(\mathbb{R})$ is invariant with respect to the transformation $y \mapsto Y = \epsilon^{1/2} y$, there exists a solution $\tilde{\varphi} \in C_b^2(\mathbb{R})$ of the inhomogeneous equation (5.6.29) satisfying bound (5.6.30). A general solution of this equation is a linear combination in the form $\alpha \psi_0'(Y) + \tilde{\varphi}(y)$, where $\alpha \in \mathbb{R}$. $\qquad \square$

Lemma 5.18 *Fix $s > \frac{1}{2}$. The nonlinear vector field of system (5.6.27)–(5.6.28) satisfies the bounds*

$$|G_R| \le N_R \left(|\psi_0 + \varphi| + \|\mathbf{c}\|_X \right), \quad \|\mathbf{F}_R\|_X \le M_R \left(|\psi_0 + \varphi| + \|\mathbf{c}\|_X \right),$$

$$|G_T| \le N_T \left(|\varphi|^2 + \|\mathbf{c}\|_X^2 \right), \quad \|\mathbf{F}_T\|_X \le M_T \left(|\psi_0 + \varphi| + \|\mathbf{c}\|_X \right) \|\mathbf{c}\|_X,$$

for some $N_R, M_R, N_T, M_T > 0$.

Proof System (5.6.27)–(5.6.28) is semi-linear and the vector field is a polynomial acting on $D_s \subset X$, where X is the Banach algebra for $s > \frac{1}{2}$. Characterization of G_T and $\mathbf{F}_T$ is based on the fact that the Hamiltonian function H_T is quadratic with respect to $\mathbf{c}$ by Lemma 5.14. $\qquad \square$

A local neighborhood of the zero point in the phase space X can be decomposed into the subspaces determined by the respective subspaces of the linearized system (Lemma 5.12),

$$X = X_0 \oplus X_c \oplus X_u \oplus X_s = X_0 \oplus X_0^\perp, \qquad (5.6.31)$$

where X_s and X_u are associated with the subspaces E^s and E^u, while X_0 and X_c are associated with the subspaces E^0 and E^c, respectively. The following theorem completes the construction of a local center–stable manifold.

Theorem 5.9   *Let* $\mathbf{a} \in X_c$, $\mathbf{b} \in X_s$ *and* $\alpha \in \mathbb{R}$ *be small such that*

$$\|\mathbf{a}\|_{X_c} \le C_a \epsilon^N, \quad \|\mathbf{b}\|_{X_s} \le C_b \epsilon^N, \quad |\alpha| \le C_\alpha \epsilon^N \qquad (5.6.32)$$

for some ϵ-independent constants $C_a, C_b, C_\alpha > 0$. Under the conditions of Lemma 5.16, there exists a family of local solutions $\psi_0(y; \mathbf{a}, \mathbf{b}, \alpha)$ and $\mathbf{c}(y; \mathbf{a}, \mathbf{b}, \alpha)$ of system (5.6.27)–(5.6.28) such that

$$\mathbf{c}_c(0) = \mathbf{a}, \quad \mathbf{c}_s = e^{y\Lambda_s}\mathbf{b} + \tilde{\mathbf{c}}_s(y), \quad \varphi = \alpha\psi_0'(Y) + \tilde\varphi(y), \qquad (5.6.33)$$

where $\tilde{\mathbf{c}}_s(y)$ and $\tilde\varphi(y)$ are uniquely defined and the family of local solutions satisfies the bound

$$\sup_{y\in[0,L/\epsilon^{N+1}]} |\varphi| \le C_h \epsilon^N, \quad \sup_{y\in[0,L/\epsilon^{N+1}]} \|\mathbf{c}(y)\|_X \le C\epsilon^N, \qquad (5.6.34)$$

for some ϵ-independent constants $L > 0$ and $C_h, C > 0$.

Proof   We modify system (5.6.27)–(5.6.28) by the following trick. We multiply the nonlinear vector field of system (5.6.28) by the cut-off function $\chi_{[0,y_0]}(y)$, such that

$$\mathbf{c}'(y) - \Lambda_\epsilon \mathbf{c} = \epsilon\chi_{[0,y_0]}(y)\mathbf{F}_T(\psi_0 + \varphi, \mathbf{c}) + \epsilon^{N+1}\chi_{[0,y_0]}(y)\mathbf{F}_R(\psi_0 + \varphi, \mathbf{c}), \quad (5.6.35)$$

where $\chi_{[0,y_0]}(y)$ is the characteristic function on $[0, y_0]$. Similarly, we multiply the nonlinear vector field of system (5.6.27) by the cut-off function $\chi_{[0,y_0]}(y)$ and add a symmetrically reflected vector field multiplied by the cut-off function $\chi_{[-y_0,0]}(y)$, such that

$$\varphi''(y) - \epsilon L_\epsilon \varphi = \epsilon\chi_{[0,y_0]}(y)G_T(\varphi, \mathbf{c}) + \epsilon^{N+1}\chi_{[0,y_0]}(y)G_R(\psi_0 + \varphi, \mathbf{c})$$
$$+ \epsilon\chi_{[-y_0,0]}(y)G_T(\varphi, \mathbf{c}) + \epsilon^{N+1}\chi_{[-y_0,0]}(y)G_R(\psi_0 + \varphi, \mathbf{c}). \quad (5.6.36)$$

We are looking for a global solution of system (5.6.35)–(5.6.36) in the space of bounded continuous functions $C_b^0(\mathbb{R})$. By uniqueness of solutions of semi-linear differential equations, this global solution coincides with a local solution of system (5.6.27)–(5.6.28) for all $y \in [0, y_0] \subset \mathbb{R}$.

Let $\mathbf{c}_s(y) = e^{y\Lambda_s}\mathbf{b} + \tilde{\mathbf{c}}_s(y)$ and look for a solution $\tilde{\mathbf{c}}_s(y)$ and $\mathbf{c}_u(y)$ of system (5.6.35) projected to X_s and X_u. By Lemmas 5.13, 5.18, and the Implicit

Function Theorem (Appendix B.7), there exists a unique map from $C_b^0(\mathbb{R}, X_h \oplus X_c)$ to $C_b^0(\mathbb{R}, X_u \oplus X_s)$ such that

$$\sup_{\forall y \in [0, y_0]} (\|\mathbf{c}_u(y)\|_{X_u} + \|\mathbf{c}_s(y)\|_{X_s})$$

$$\leq \|\mathbf{b}\|_{X_s} + \epsilon M_1 \sup_{\forall y \in [0, y_0]} ((1 + |\varphi(y)| + \|\mathbf{c}_c(y)\|_{X_c}) \|\mathbf{c}_c(y)\|_{X_c})$$

$$+ \epsilon^{N+1} M_2 \sup_{\forall y \in [0, y_0]} (1 + |\varphi(y)| + \|\mathbf{c}_c(y)\|_{X_c}), \qquad (5.6.37)$$

for some $M_1, M_2 > 0$. To use Lemma 5.13, we note that the linear part of the system for $\mathbf{c}_s$ and $\mathbf{c}_u$ is not affected by the near-identity transformations of Lemma 5.14, so that it can be converted back to the scalar second-order equation.

Let $\varphi(y) = \alpha \psi_0'(Y) + \tilde{\varphi}(y)$ and look for a solution $\tilde{\varphi}(y)$ of system (5.6.36). Since the modified vector field is even for all $y \in \mathbb{R}$, by Lemmas 5.17, 5.18, bound (5.6.37), and the Implicit Function Theorem again, there exists a unique map from $C_b^0(\mathbb{R}, X_c)$ to $C_b^0(\mathbb{R}, X_h)$ such that

$$\sup_{\forall y \in [0, y_0]} |\varphi(y)| \leq |\alpha| + M_3 \sup_{\forall y \in [0, y_0]} \|\mathbf{c}_c(y)\|_{X_c}^2$$

$$+ \epsilon^N M_4 \sup_{\forall y \in [0, y_0]} (1 + \|\mathbf{c}_c(y)\|_{X_c}), \qquad (5.6.38)$$

for some $M_3, M_4 > 0$.

Since the spectrum of Λ_c consists of pairs of semi-simple purely imaginary eigenvalues, the operator Λ_c generates a strongly continuous group $e^{y\Lambda_c}$ for all $y \in \mathbb{R}$ on X_c such that

$$\exists K > 0 : \quad \sup_{\forall y \in \mathbb{R}} \|e^{y\Lambda_c}\|_{X_c \to X_c} \leq K. \qquad (5.6.39)$$

By variation of constant formula, the solution of system (5.6.35) projected to X_c can be rewritten in the integral form

$$\mathbf{c}_c(y) = e^{y\Lambda_c} \mathbf{a} + \epsilon \int_0^y e^{(y-y')\Lambda_c} P_c \mathbf{F}_T(\psi_0(\epsilon^{1/2} y') + \varphi(y'), \mathbf{c}(y')) dy'$$

$$+ \epsilon^{N+1} \int_0^y e^{(y-y')\Lambda_c} P_c \mathbf{F}_R(\psi_0(\epsilon^{1/2} y') + \varphi(y'), \mathbf{c}(y')) dy', \qquad (5.6.40)$$

where $\mathbf{a} = \mathbf{c}_c(0)$ and P_c is the projection operator to X_c.

Using bound (5.6.24) for $\psi_0(\epsilon^{1/2} y)$ and bounds (5.6.37) and (5.6.38) on the components $\mathbf{c}_s$, $\mathbf{c}_u$, and $\mathbf{c}_h$, the integral equation (5.6.40) results in the bound

$$\sup_{\forall y \in [0, y_0]} \|\mathbf{c}_c(y)\|_{X_c} \leq K \Bigg(\|\mathbf{a}\|_{X_c} + \|\mathbf{b}\|_{X_s} + |\alpha|$$

$$+ \epsilon M_5 \int_0^{y_0} |\psi_0(\epsilon^{1/2} y)| \|\mathbf{c}_c(y)\|_{X_c} dy + \epsilon y_0 M_6 \sup_{\forall y \in [0, y_0]} \|\mathbf{c}_c(y)\|_{X_c}^2$$

$$+ \epsilon^{N+1} M_7 \int_0^{y_0} |\psi_0(\epsilon^{1/2} y)| dy + \epsilon^{N+1} y_0 M_8 \sup_{\forall y \in [0, y_0]} \|\mathbf{c}_c(y)\|_{X_c} \Bigg)$$

for some $M_5, M_6, M_7, M_8 > 0$. By the Gronwall inequality (Appendix B.6), we have thus obtained that

$$\sup_{\forall y \in [0,y_0]} \|\mathbf{c}_c(y)\|_{X_c} \leq K e^{\epsilon K M_5 \int_0^{y_0} |\psi_0(\epsilon^{1/2}y)| dy} \left(\|\mathbf{a}\|_{X_c} + \|\mathbf{b}\|_{X_s} + |\alpha| + \epsilon^N M_9 \right.$$

$$\left. + \epsilon y_0 M_6 \sup_{\forall y \in [0,y_0]} \|\mathbf{c}_c(y)\|_{X_c}^2 + \epsilon^{N+1} y_0 M_8 \sup_{\forall y \in [0,y_0]} \|\mathbf{c}_c(y)\|_{X_c} \right), \qquad (5.6.41)$$

for some $M_9 > 0$. It follows by the decay bound (5.6.24) that there is an ϵ-independent $C > 0$ such that

$$\epsilon \int_0^{y_0} |\psi_0(\epsilon^{1/2}y)| dy \leq C \epsilon^{1/2}.$$

If we let $y_0 = L/\epsilon^M$, then the bound is consistent for $M \leq N + 1$, where the value $M = N + 1$ gives the balance of all terms in the upper bound (5.6.41). If arbitrary vectors $\mathbf{a}$, $\mathbf{b}$ and α satisfy bound (5.6.32), then we have constructed a local solution $\mathbf{c}_c(y)$ which satisfies the bound

$$\exists C > 0 : \qquad \sup_{\forall y \in [0, L/\epsilon^{N+1}]} \|\mathbf{c}_c(y)\|_{X_c} \leq C \epsilon^N. \qquad (5.6.42)$$

Using bounds (5.6.32), (5.6.37), (5.6.38), and (5.6.42), we have proved the bound (5.6.34) for some ϵ-independent constants $C_h, C > 0$. $\qquad \square$

We can now complete the proof of Theorem 5.8.

Proof of Theorem 5.8 By Theorem 5.9, we have constructed an infinite-dimensional continuous family of local bounded solutions of system (5.6.27)–(5.6.28) for all $y \in [0, L/\epsilon^{N+1}]$ for some ϵ-independent constant $L > 0$. The solutions are close to the reversible localized solution of the extended second-order equation (5.6.22). It remains to extend the local solution to the symmetric interval $[-L/\epsilon^{N+1}, L/\epsilon^{N+1}]$ under the reversibility constraints (5.6.18). To do so, we shall consider the intersections of the local invariant manifold of system (5.6.27)–(5.6.28) with the reversibility hyperplane Σ defined by (5.6.19).

Since the initial data $\mathbf{c}_c(0) = \mathbf{a}$ in the local center–stable manifold of Theorem 5.9 are arbitrary, the components of $\mathbf{a}$ can be chosen to lie in the symmetric section Σ, so that

$$\text{Im}(\mathbf{a})_m^+ = 0, \quad \text{Im}(\mathbf{a})_m^- = 0, \quad m < 0.$$

This construction still leaves infinitely many arbitrary parameters for

$$\text{Re}(\mathbf{a})_m^+, \quad \text{Re}(\mathbf{a})_m^-, \quad m < 0,$$

to be chosen in the bound (5.6.32). The initial data $\varphi(0)$ and $\mathbf{c}_{s,u}(0)$ are not arbitrary since we have used the Implicit Function Theorem for the mappings (5.6.37) and (5.6.38). Therefore, we have to show that the components of $\mathbf{b}$ and α can be chosen uniquely so that the local center–stable manifold intersects at $y = 0$ with the symmetric section Σ.

There are as many arbitrary parameters $\mathbf{b}$ and α in the local center–stable manifold as there are remaining constraints in the set Σ. First, let us consider constraints in the set Σ for all $m > 0$, namely

$$\mathrm{Re}(c_m^+(0)) = \mathrm{Re}(c_m^-(0)), \quad \mathrm{Im}(c_m^+(0)) = -\mathrm{Im}(c_m^-(0)), \quad m > 0.$$

Let $\mathbf{c}_s = e^{y\Lambda_s}\mathbf{b} + \tilde{\mathbf{c}}_s(y)$ and rewrite the constraints for any $m > 0$,

$$\mathrm{Re}(b_m) + \mathrm{Re}(\tilde{\mathbf{c}}_s)_m(0) = \mathrm{Re}(\mathbf{c}_u)_m(0),$$
$$\mathrm{Im}(b_m) + \mathrm{Im}(\tilde{\mathbf{c}}_s)_m(0) = -\mathrm{Im}(\mathbf{c}_u)_m(0),$$

where $\tilde{\mathbf{c}}_s(0)$ and $\mathbf{c}_u(0)$ are of order $\mathcal{O}(\epsilon^N)$ and depend on $\mathbf{b}$ in higher orders of $\mathcal{O}(\epsilon^k)$ with $k > N$. By the Implicit Function Theorem (Appendix B.7), there exists a unique solution $\mathbf{b}$ that satisfies the bound (5.6.32).

Finally, let us consider constraints in the set Σ for components of $\psi_0(y)$, namely $\psi_0'(0) = 0$. Let $\psi_0(y) = \alpha\psi_0'(\epsilon^{1/2}y) + \varphi(y)$ and note that $\psi_0''(0) \neq 0$. Let $\mathbf{a}$ and $\mathbf{b}$ be chosen so that $\mathbf{c}(0)$ belongs to the set Σ. Due to a construction of the modified vector field in system (5.6.36), if $\mathbf{c}(y)$ lies in the domain D_r and $\alpha = 0$, then the global solution $\tilde{\mathbf{c}}_s(y)$ constructed in Theorem 5.9 intersects the set Σ at $y = 0$. Therefore, the choice $\alpha = 0$ satisfies the constraint $\psi_0'(0) = 0$.

We have thus constructed a family of reversible solutions on the symmetric interval at $[-L/\epsilon^{N+1}, L/\epsilon^{N+1}]$. When all the coordinate transformations used in our analysis are traced back to the original variable $\psi(x, y)$, we obtain the statement of Theorem 5.8. $\qquad\qquad\square$

Unfortunately, we are not able to exclude the polynomial growth of the oscillatory tails in the far-field regions. This limitation of traveling solutions to a finite spatial scale is related with the *finite-time* applicability of the nonlinear Schrödinger equation in the Cauchy problem associated with the Gross–Pitaevskii equation (5.6.1) (Section 2.3.2).

Exercise 5.24   Consider the nonlinear Klein–Gordon equation,

$$E_{tt} - E_{xx} + V(x)E + E^3 = 0$$

for a 2π-periodic $V(x)$ and use the traveling wave ansatz in the form

$$E(x, t) = \sum_{m \in \mathbb{Z}} \psi_m(y)e^{imx - i\omega t}, \quad y = x - ct.$$

Obtain the dispersion relation for $V(x) \equiv 0$:

$$(1 - c^2)\kappa^2 + 2i(m - c\omega)\kappa + \omega - m^2 = 0, \quad m \in \mathbb{Z},$$

and prove that there are infinitely many purely imaginary roots $\kappa \in i\mathbb{R}$ in the resonant case $\omega = n^2$ for a fixed $n \in \mathbb{N}$.

Exercise 5.25   Consider the discrete Gross–Pitaevskii equation,

$$i\dot{E}_n = -E_{n+1} - E_{n-1} + V_n E_n + |E_n|^2 E_n$$

for a 1-periodic $\{V_n\}_{n\in\mathbb{Z}}$ and use the traveling wave ansatz in the form

$$E_n(t) = \sum_{m\in\mathbb{Z}} \psi_m(y)e^{i2\pi mn - i\omega t}, \quad y = n - ct.$$

Obtain the dispersion relation for $V_n = 0$:

$$\omega - ic\kappa - 2\cosh(\kappa + 2\pi m) = 0, \quad m \in \mathbb{Z},$$

and prove that there are infinitely many purely imaginary roots $\kappa \in i\mathbb{R}$ in the resonant case $\omega = 2\cosh(2\pi n)$ for a fixed $n \in \mathbb{N}$.

Appendix A

Mathematical notation

Sets:

- $\mathbb{N} := \{1, 2, 3, ...\}$ is the set of natural numbers (without zero).
- $\mathbb{N}_0 := \{0, 1, 2, 3, ...\}$ is the set of natural numbers (counting zero).
- $\mathbb{Z}$ is the set of integers.
- $\mathbb{R}$ is the set of real numbers.
- $\mathbb{R}_+ = \{x \in \mathbb{R} : x > 0\}$ is the set of positive real numbers.
- $\mathbb{C}$ is the set of complex numbers.
- $\mathbb{C}^n$, $n \in \mathbb{N}$ is the vector space of n-dimensional column vectors.
- $\mathbb{M}^{n \times n}$, $n \in \mathbb{N}$ is the vector space of square $n \times n$ matrices.

Inner products:

- $\forall \mathbf{u}, \mathbf{v} \in \mathbb{C}^n$, $\langle \mathbf{u}, \mathbf{v} \rangle_{\mathbb{C}^n} := \sum_{j=1}^n u_j \bar{v}_j$.
- $\forall u, v \in L^2(\mathbb{R}, \mathbb{C})$, $\langle u, v \rangle_{L^2} := \int_{\mathbb{R}} u(x) \overline{v(x)} dx$.

If $\mathbf{u}, \mathbf{v} \in \mathbb{R}^n$ or $u, v \in L^2(\mathbb{R}, \mathbb{R})$, the complex conjugation sign is dropped.

Function spaces on the infinite line:

- $C_b^n(\mathbb{R})$, $n \in \mathbb{N}$ is the space of n-times continuously differentiable functions with bounded derivatives up to the nth order, equipped with the norm

$$\|u\|_{C_b^n} := \sup_{x \in \mathbb{R}} \sum_{k=0}^n |\partial_x^k u(x)|.$$

- $L_s^p(\mathbb{R})$, $p, s \in \mathbb{R}_+$ is the space of functions with weight $(1 + x^2)^{s/2}$, whose p power is integrable, equipped with the norm

$$\|u\|_{L_s^p} := \left(\int_{\mathbb{R}} (1 + x^2)^{ps/2} |u(x)|^p dx \right)^{1/p}.$$

- $H^n(\mathbb{R})$, $n \in \mathbb{N}$ is the Sobolev space of functions with square integrable derivatives up to the nth order, equipped with the norm

$$\|u\|_{H^n} := \left(\int_{\mathbb{R}} \sum_{k=0}^n |\partial_x^k u(x)|^2 dx \right)^{1/2}.$$

Even and odd restrictions of these spaces can be defined. For instance, the even and odd restrictions of $H^n(\mathbb{R})$, $n \in \mathbb{N}$ are given by

$$H^n_{ev}(\mathbb{R}) = \{u \in H^n(\mathbb{R}) : \ u(-x) = u(x)\},$$

$$H^n_{odd}(\mathbb{R}) = \{u \in H^n(\mathbb{R}) : \ u(-x) = -u(x)\}.$$

If a function $f(x) : \mathbb{R} \to \mathbb{R}$ is square integrable on any compact subset of $\mathbb{R}$, we say that $f \in L^2_{loc}(\mathbb{R})$.

Function spaces on a periodic domain:

- $C^n_{per}([0, 2\pi])$, $n \in \mathbb{N}$ is the space of 2π-periodic n-times continuously differentiable functions, equipped with the norm

$$\|u\|_{C^n_{per}} := \sup_{x \in [0,2\pi]} \sum_{k=0}^{n} |\partial_x^k u(x)|.$$

- $H^n_{per}([0, 2\pi])$, $n \in \mathbb{N}$ is the Sobolev space of 2π-periodic functions with square integrable derivatives up to the nth order, equipped with the norm

$$\|u\|_{H^n_{per}} := \left(\int_0^{2\pi} \sum_{k=0}^{n} |\partial_x^k u(x)|^2 dx \right)^{1/2}.$$

Sequence spaces: Infinite sequences $\{\phi_n\}_{n \in \mathbb{Z}}$ are represented by vectors $\boldsymbol{\phi}$.

- $l^p_s(\mathbb{Z})$, $p, s \in \mathbb{R}_+$ is the space of sequences with weight $(1+n^2)^{s/2}$, whose p power is summable, equipped with the norm

$$\|\mathbf{u}\|_{l^p_s} := \left(\sum_{n \in \mathbb{Z}} (1+n^2)^{ps/2} |u_n|^p \right)^{1/p}.$$

- $l^\infty_s(\mathbb{Z})$ is the space of bounded sequences with weight $(1+n^2)^{s/2}$, equipped with the norm

$$\|\mathbf{u}\|_{l^\infty_s} := \sup_{n \in \mathbb{Z}} (1+n^2)^{s/2} |u_n|.$$

Fourier transforms: For any $u \in L^2(\mathbb{R})$, a Fourier transform $\hat{u} \in L^2(\mathbb{R})$ is given by

$$\hat{u}(k) := \frac{1}{(2\pi)^{1/2}} \int_{\mathbb{R}} u(x) e^{-ikx} dx, \quad k \in \mathbb{R}.$$

The inverse Fourier transform is given by

$$u(x) := \frac{1}{(2\pi)^{1/2}} \int_{\mathbb{R}} \hat{u}(k) e^{ikx} dk, \quad x \in \mathbb{R}.$$

A product of two functions $u(x)v(x)$ is represented by the convolution integral of their Fourier transforms

$$(\hat{u} \star \hat{v})(k) := \frac{1}{2\pi} \int_{\mathbb{R}} \hat{u}(k') \hat{v}(k - k') dk'.$$

Fourier series: For any $u \in L^2_{\text{per}}([0, 2\pi])$, a Fourier series is given by

$$u(x) := \sum_{n \in \mathbb{Z}} \hat{u}_n e^{inx}, \quad x \in \mathbb{R}.$$

The Fourier coefficients $\hat{u} \in l^2(\mathbb{Z})$ are given by

$$\hat{u}_n := \frac{1}{2\pi} \int_0^{2\pi} u(x) e^{-inx} dx, \quad n \in \mathbb{Z}.$$

A product of two functions $u(x)v(x)$ is represented by the convolution sum of their Fourier coefficients

$$(\hat{u} \star \hat{v})_n := \sum_{n' \in \mathbb{Z}} \hat{u}_{n'} \hat{v}_{n-n'}.$$

Kronecker's symbol: This symbol is defined by

$$\delta_{n,n'} = \begin{cases} 1, & n' = n, \\ 0, & n' \neq n, \end{cases} \quad n, n' \in \mathbb{Z}.$$

Dirac's delta function: This function takes formally the following values

$$\delta(x) = \begin{cases} \infty, & x = 0, \\ 0, & x \neq 0. \end{cases}$$

Dirac's delta function is understood in the sense of distributions

$$\forall f \in C_b^0(\mathbb{R}) : \quad \int_{\mathbb{R}} f(y)\delta(x - y)dy = f(x).$$

Heaviside's step function: This function is piecewise constant with the values

$$\Theta(x) = \begin{cases} 1, & x > 0, \\ 0, & x < 0. \end{cases}$$

At the jump discontinuity, Heaviside's step function can be defined differently. We shall define it as $\Theta(0) = 1$.

Pauli matrices: These 2×2 Hermitian matrices are given by

$$\sigma_1 = \begin{bmatrix} 0 & 1 \\ 1 & 0 \end{bmatrix}, \quad \sigma_2 = \begin{bmatrix} 0 & -i \\ i & 0 \end{bmatrix}, \quad \sigma_3 = \begin{bmatrix} 1 & 0 \\ 0 & -1 \end{bmatrix}.$$

Balls in Banach space: Let X be a Banach space. An open ball of radius $\delta > 0$ centered at $0 \in X$ is denoted by $B_\delta(X)$. The closed ball B with its boundary is denoted by $\bar{B}_\delta(X)$.

Order of asymptotic approximation: If A and B are two quantities depending on a parameter ϵ in a set $\mathcal{E}$, the notation $A(\epsilon) = \mathcal{O}(B(\epsilon))$ as $\epsilon \to 0$ indicates that $A(\epsilon)/B(\epsilon)$ remains bounded as $\epsilon \to 0$. The notation $A(\epsilon) \sim B(\epsilon)$ indicates that there is $C \in \mathbb{R}\backslash\{0\}$ such that $A(\epsilon)/B(\epsilon) \to C$ as $\epsilon \to 0$.

Constants: Different constants are denoted by C, C', C'' if they are independent of a small parameter of the problem. We say that a bound holds uniformly in

$0 < \epsilon \ll 1$ if there exists $\epsilon_0 > 0$ such that the bound holds uniformly for every $\epsilon \in (0, \epsilon_0)$.

Characteristic function of a set $S \subset \mathbb{R}$:

$$\chi_S(x) = \begin{cases} 1 & \text{if} \quad x \in S, \\ 0 & \text{if} \quad x \notin S. \end{cases}$$

Partial derivatives: If $f(v, u, w)$ is a function of three variables, then $\partial_{1,2,3} f(v, u, w)$ denotes partial derivatives with respect v, u, and w, respectively.

Appendix B

Selected topics of applied analysis

B.1 Banach algebra

X is a Banach space if X is a complete metric space with the norm $\| \cdot \|_X$.

Definition B.1 We say that a Banach space X forms a Banach algebra with respect to multiplication if there is $C > 0$ such that

$$\forall u, v \in X : \quad \|uv\|_X \leq C\|u\|_X\|v\|_X.$$

Examples of Banach algebra with respect to multiplication include Sobolev spaces $H^s(\mathbb{R}^d)$ for $s > \frac{d}{2}$ and discrete spaces $l_s^2(\mathbb{Z}^d)$ for any $s \geq 0$.

Theorem B.1 *If $s > \frac{d}{2}$, then $H^s(\mathbb{R}^d)$ is a Banach algebra with respect to multiplication.*

Proof We note the triangle inequality

$$\forall k, k' \in \mathbb{R}^d : \quad (1+|k|^2)^{s/2} \leq 2^s \left((1+|k-k'|^2)^{s/2} + (1+|k'|^2)^{s/2} \right).$$

By the Plancherel Theorem for Fourier transforms, we have

$$\|uv\|_{H^s} = \|\widehat{uv}\|_{L_s^2} = \|\hat{u} \star \hat{v}\|_{L_s^2}$$

where $\star$ is the convolution operator between two Fourier transforms. Using the triangle inequality, we have

$$\left| (1+|k|^2)^{s/2} \int_{\mathbb{R}} \hat{u}(k')\hat{v}(k-k')dk' \right|$$
$$\leq 2^s \left| \int_{\mathbb{R}} \hat{u}(k')(1+|k-k'|^2)^{s/2}\hat{v}(k-k')dk' \right| + 2^s \left| \int_{\mathbb{R}} (1+|k'|^2)^{s/2}\hat{u}(k')\hat{v}(k-k')dk' \right|.$$

Using Young's inequality,

$$\|u \star v\|_{L^r} \leq \|u\|_{L^p}\|v\|_{L^q}, \quad \frac{1}{p} + \frac{1}{q} = 1 + \frac{1}{r}, \quad p, q, r \geq 1,$$

with $r = p = 2$ and $q = 1$, we obtain

$$\|\hat{u} \star \hat{v}\|_{L_s^2} \leq 2^s \left(\|\hat{u}\|_{L_s^2}\|\hat{v}\|_{L^1} + \|\hat{u}\|_{L^1}\|\hat{v}\|_{L_s^2} \right) \leq C_s\|\hat{u}\|_{L_s^2}\|\hat{v}\|_{L_s^2},$$

where the last inequality is obtained from the Cauchy–Schwarz inequality

$$\|v\|_{L^1} = \int_{\mathbb{R}^d} |\hat{v}(k)| dk \leq \left(\int_{\mathbb{R}^d} \frac{dk}{(1+|k|^2)^s} \right)^{1/2} \left(\int_{\mathbb{R}^d} (1+|k|^2)^s |\hat{v}(k)|^2 dk \right)^{1/2},$$

which makes sense if $s > \frac{d}{2}$. By Definition B.1, $H^s(\mathbb{R}^d)$ is a Banach algebra with respect to multiplication if $s > \frac{d}{2}$.   ⊔

Remark B.1   Since $H^s(\mathbb{R}^d)$, $s > \frac{d}{2}$ is a Banach algebra with respect to multiplication, $L_s^2(\mathbb{R}^d)$, $s > \frac{d}{2}$ is a Banach algebra with respect to convolution integrals.

Similarly, the space of periodic functions $H_{per}^s([0, 2\pi]^d)$ is a Banach algebra with respect to multiplication for any $s > \frac{d}{2}$ and the discrete space $l_s^2(\mathbb{Z}^d)$ is a Banach algebra with respect to convolution sums for any $s > \frac{d}{2}$.

Because of the embedding of $l_s^2(\mathbb{Z}^d)$ to $l_s^\infty(\mathbb{Z}^d)$ thanks to the bound

$$\|u\|_{l_s^\infty} \leq \|u\|_{l_s^2}, \quad s \geq 0,$$

the same space $l_s^2(\mathbb{Z}^d)$ is also a Banach algebra with respect to multiplication for any $s \geq 0$.

B.2   Banach Fixed-Point Theorem

Let X be a Banach space with norm $\|\cdot\|_X$ and $A : X \to X$ be a nonlinear operator. The Banach Fixed-Point Theorem gives the conditions for existence and uniqueness of solutions of the operator equation

$$u = A(u), \quad u \in X.$$

This theorem is sometimes called the Contraction Mapping Principle.

Definition B.2   Let M be a closed non-empty set in a Banach space X. The operator $A : M \to M$ is called a contraction if there is $q \in [0, 1)$ such that

$$\forall u, v \in M : \quad \|Au - Av\|_X \leq q\|u - v\|_X.$$

Theorem B.2   *Let M be a closed non-empty set in the Banach space X and let $A : M \to M$ be a contraction operator. There exists a unique fixed point of A in M, that is, there exists a unique $u \in M$ such that $A(u) = u$.*

Proof   Since X is a complete metric space, any sequence $\{u_n\}_{n \in \mathbb{N}} \in X$ converges if and only if it is a Cauchy sequence, that is, for each $\epsilon > 0$, there is $N(\epsilon) \geq 1$ such that

$$\forall n, m \geq N(\epsilon) : \quad \|u_n - u_m\|_X < \epsilon.$$

Since M is a closed set in X, if a sequence $\{u_n\}_{n \in \mathbb{N}} \in M$ converges to u as $n \to \infty$, then $u \in M$.

We construct the sequence $\{u_n\}_{n \in \mathbb{N}} \in M$ from the iteration equation

$$u_{n+1} = A(u_n), \quad n \in \mathbb{N},$$

where $u_1 \in M$ is arbitrary. First, we show that $\{u_n\}_{n\in\mathbb{N}}$ is a Cauchy sequence. Indeed, if A is a contraction operator, then

$$\|u_{n+1} - u_n\|_X = \|A(u_n) - A(u_{n-1})\|_X \leq q\|u_n - u_{n-1}\|_X$$
$$\leq q^2\|u_{n-1} - u_{n-2}\|_X \leq \cdots \leq q^{n-1}\|u_2 - u_1\|_X.$$

By the triangle inequality, we obtain for any $m \in \mathbb{N}$

$$\|u_{n+m} - u_n\|_X \leq \|u_{n+m} - u_{n+m-1}\|_X + \cdots + \|u_{n+1} - u_n\|_X$$
$$\leq (q^{n-1} + q^n + \cdots + q^{n+m-2})\|u_2 - u_1\|_X$$
$$\leq q^{n-1}(1-q)^{-1}\|u_2 - u_1\|_X.$$

Because $q \in [0,1)$, we have $q^n \to 0$ as $n \to \infty$. Therefore, $\{u_n\}_{n\in\mathbb{N}}$ is the Cauchy sequence and there is $u \in M$ such that

$$\|u_n - u\|_X \to 0 \quad \text{as} \quad n \to \infty.$$

On the other hand, $A(u) \in M$ and since

$$\|A(u_n) - A(u)\|_X \leq q\|u_n - u\|_X \to 0 \quad \text{as} \quad n \to \infty,$$

we also have

$$\|A(u_n) - A(u)\|_X \to 0 \quad \text{as} \quad n \to \infty.$$

It follows from the iterative equation $u_{n+1} = A(u_n)$ in the limit $n \to \infty$ that u is a solution of $u = A(u)$, that is, u is a fixed point of A in M.

It remains to show that the solution is unique in M. By contradiction, we assume existence of two solutions $u \in M$ and $v \in M$. Then

$$\|u - v\|_X = \|A(u) - A(v)\|_X \leq q\|u - v\|_X,$$

and since $q \in [0,1)$, we have $\|u - v\|_X = 0$, that is $u = v$. $\square$

We also mention two modifications of the Banach Fixed-Point Theorem.

The Brouwer Fixed-Point Theorem states that if M is a compact, convex, non-empty set in a finite-dimensional normed space and $A : M \to M$ is a continuous operator, then there exists at least one fixed point of A in M.

The Schauder Fixed-Point Theorem states that if M is a bounded, closed, convex, non-empty subset of a Banach space X and $A : M \to M$ is a compact operator, then there exists at least one fixed point of A in M.

B.3 Floquet Theorem

We will start with an abstract lemma, which is often referred to as the Spectral Mapping Theorem.

Lemma B.1 *Let $C \in \mathbb{R}^{n \times n}$ be an invertible matrix. There exists a matrix $B \in \mathbb{C}^{n \times n}$ such that $e^B = C$.*

Proof Let us convert C to the Jordan block-diagonal form J using an invertible matrix P such that $P^{-1}CP = J$ and $C = PJP^{-1}$. If we can prove that $J = e^K$ for some matrix K, then $B = PKP^{-1}$.

Since C is invertible, the Jordan form J has nonzero blocks. We are looking for K in the block-diagonal form, which resembles the form of J. For a simple block of J with entry $\lambda_0 \neq 0$, we have a simple block of K with entry $\log(\lambda_0)$. For an $m \times m$ block of J in the form

$$
J_m = \begin{bmatrix}
\lambda_0 & 1 & 0 & \cdots & 0 \\
0 & \lambda_0 & 1 & \cdots & 0 \\
\vdots & \vdots & \vdots & \cdots & \vdots \\
0 & 0 & 0 & \cdots & \lambda_0
\end{bmatrix}
$$

with $\lambda_0 \neq 0$, we can rewrite it in the form

$$
J_m = \lambda_0 I_m + N_m,
$$

where I_m is the $m \times m$ identity matrix and N_m is the $m \times m$ nilpotent matrix of order m. We note that $N_m^m = O_m$, where O_m is the $m \times m$ zero matrix.

Using the power series for a logarithmic function, we obtain the corresponding block for matrix K,

$$
\begin{aligned}
K_m &= \log(\lambda_0)I_m + \log(I_m + \lambda_0^{-1}N_m) \\
&= \log(\lambda_0)I_m + \frac{1}{\lambda_0}N_m - \frac{1}{2\lambda_0^2}N_m^2 + \dots + \frac{(-1)^m}{(m-1)\lambda_0^{m-1}}N_m^{m-1},
\end{aligned}
$$

so that $e^{K_m} = J_m$. $\square$

Remark B.2   The choice of B is not unique since $e^{B+2\pi i k I} = C$ for any $k \in \mathbb{Z}$.

Consider the fundamental matrix solution of the system with T-periodic coefficients

$$
\begin{cases}
\dot{\Phi}(t) = A(t)\Phi(t), & t \in \mathbb{R}, \\
\Phi(0) = I,
\end{cases}
$$

where $A(t), \Phi(t) \in \mathbb{R}^{n \times n}$, $A(t + T) = A(t)$ for all $t \in \mathbb{R}$, and I is an identity matrix. The Floquet Theorem specifies a particular form for the fundamental matrix solution $\Phi(t)$ due to periodicity of the coefficient matrix $A(t)$.

Theorem B.3   *Let $A(t + T) = A(t) \in \mathbb{R}^{n \times n}$ be a continuous function for all $t \in \mathbb{R}$. Then, $\Phi(t) \in \mathbb{R}^{n \times n}$ is a continuously differentiable function in t and*

$$
\Phi(t + T) = \Phi(t)\Phi(T) \quad \text{for all } t \in \mathbb{R}.
$$

Moreover, there exists a constant matrix $B \in \mathbb{C}^{n \times n}$ and an invertible T-periodic continuously differentiable matrix $Q(t) \in \mathbb{C}^{n \times n}$ such that

$$
e^{TB} = \Phi(T), \quad Q(0) = Q(T) = I, \quad \text{and} \quad \Phi(t) = Q(t)e^{tB} \quad \text{for all } t \in \mathbb{R}.
$$

Proof Existence and uniqueness theorems for linear differential equations with continuous function $A(t)$ guarantee existence of the unique solution $\Phi(t)$ for all $t \in \mathbb{R}$, which is continuously differentiable in t and invertible for all $t \in \mathbb{R}$.

Let $\Psi(t) = \Phi(t + T)$ and compute

$$\dot{\Psi}(t) = \dot{\Phi}(t + T) = A(t + T)\Phi(t + T) = A(t)\Psi(t)$$

with initial condition $\Psi(0) = \Phi(T)$. By uniqueness of solutions and the linear superposition principle, we infer that

$$\Phi(t + T) = \Psi(t) = \Phi(t)\Psi(0) = \Phi(t)\Phi(T), \quad t \in \mathbb{R}.$$

Since $\Phi(T)$ is invertible, Lemma B.1 implies that there exists a constant matrix B such that $\Phi(T) = e^{TB}$. Let $Q(t) = \Phi(t)e^{-tB}$. Then, $Q(t)$ is continuously differentiable in $t \in \mathbb{R}$ and

$$Q(t + T) = \Phi(t + T)e^{-(t+T)B} = \Phi(t)\Phi(T)e^{-TB}e^{-tB} = \Phi(t)e^{-tB} = Q(t), \quad t \in \mathbb{R}.$$

Hence $Q(t)$ is a T-periodic continuously differentiable matrix in t. Moreover, it is invertible with the inverse $Q^{-1}(t) = e^{tB}\Phi^{-1}(t)$. $\qquad\square$

Remark B.3 It follows from the Floquet Theorem that the phase flow

$$\phi_t(x_0) = \Phi(t)x_0, \quad x_0 \in \mathbb{R}^n$$

of the linear T-periodic system

$$\begin{cases} \dot{x}(t) = A(t)x(t), & t \in \mathbb{R}, \\ x(0) = x_0, \end{cases}$$

forms a discrete group of translations such that

$$\phi_{t+T}(x_0) = \phi_t(\phi_T(x_0)), \quad x_0 \in \mathbb{R}^n, \quad t \in \mathbb{R}.$$

Remark B.4 Another consequence of the Floquet Theorem is a homogenization of the linear T-periodic system. Thanks to the invertibility of $Q(t)$, a substitution $x(t) = Q(t)y(t)$ with $y \in \mathbb{R}^n$ results in the linear system with constant coefficients:

$$\begin{cases} \dot{x}(t) = A(t)x(t), \\ x(0) = x_0, \end{cases} \quad \Rightarrow \quad \begin{cases} \dot{y}(t) = By(t), \\ y(0) = x_0. \end{cases}$$

B.4 Fredholm Alternative Theorem

Let H be a Hilbert space with the inner product $\langle \cdot, \cdot \rangle_H$ and L be a densely defined linear operator on H with the domain $\mathrm{Dom}(L) \subseteq H$. The adjoint operator L^* with the domain $\mathrm{Dom}(L^*) \subseteq H$ is defined by

$$\forall u \in \mathrm{Dom}(L), \ v \in \mathrm{Dom}(L^*): \quad \langle Lu, v \rangle_H = \langle u, L^*v \rangle_H.$$

If $L^* = L$, the operator L is called self-adjoint.

Definition B.3 We say that the linear operator $L : \mathrm{Dom}(L) \subseteq H \to H$ is a Fredholm operator of index zero if

$$\dim(\mathrm{Ker}(L)) = \dim(\mathrm{Ker}(L^*)) < \infty$$

and $\mathrm{Ran}(L)$ is closed.

Remark B.5   If L is a self-adjoint operator in H and the nonzero spectrum of L is bounded away from zero, then L is a Fredholm operator of index zero.

The following lemma gives two equivalent orthogonal decompositions of the Hilbert space H relative to the Fredholm operator L.

Lemma B.2   *Let L be a Fredholm operator of index zero. Then,*

$$H = \mathrm{Ker}(L^*) \oplus \mathrm{Ran}(L), \quad H = \mathrm{Ker}(L) \oplus \mathrm{Ran}(L^*).$$

Proof   We prove only the first decomposition, as the second one follows by replacement of L by L^*.

Let $u \in \mathrm{Ran}(L)$ and $v \in \mathrm{Ker}(L^*)$. Then, there exists $f \in \mathrm{Dom}(L)$ such that $u = Lf$. As a result, we compute

$$\langle u, v \rangle_H = \langle Lf, v \rangle_H = \langle f, L^*v \rangle_H = 0,$$

that is $\mathrm{Ker}(L^*) \subseteq [\mathrm{Ran}(L)]^\perp$.

On the other hand, let $u = Lf \in \mathrm{Ran}(L)$ for some $f \in \mathrm{Dom}(L)$ and $v \in [\mathrm{Ran}(L)]^\perp$. We have for all $u \in \mathrm{Ran}(L)$,

$$0 = \langle u, v \rangle_H = \langle Lf, v \rangle_H = \langle f, L^*v \rangle_H.$$

Because $\mathrm{Dom}(L)$ is dense in H, then $[\mathrm{Dom}(L)]^\perp = \{0\}$, which means that $L^*v = 0$, that is $[\mathrm{Ran}(L)]^\perp \subseteq \mathrm{Ker}(L^*)$. Therefore, $\mathrm{Ker}(L^*) \equiv [\mathrm{Ran}(L)]^\perp$.   □

The Fredholm Alternative Theorem gives the necessary and sufficient condition for existence of solutions of the inhomogeneous equation $Lu = f$ for a given $f \in H$ if L is a Fredholm operator on H.

Theorem B.4   *Let $L : \mathrm{Dom}(L) \subseteq H \to H$ be a Fredholm operator of index zero. There exists a solution $u \in \mathrm{Dom}(L)$ of the inhomogeneous equation $Lu = f$ for a given $f \in H$ if and only if $\langle f, v_0 \rangle_H = 0$ for all $v_0 \in \mathrm{Ker}(L^*)$. Moreover, the solution is unique under the constraint $\langle u, w_0 \rangle_H = 0$ for all $w_0 \in \mathrm{Ker}(L)$.*

Proof   The necessary condition is proved with an orthogonal projection. If there exists a solution $u \in \mathrm{Dom}(L)$ of $Lu = f$, then

$$\forall v_0 \in \mathrm{Ker}(L^*): \quad \langle f, v_0 \rangle_H = \langle Lu, v_0 \rangle_H = \langle u, L^*v_0 \rangle_H = 0.$$

To prove the sufficient condition, if $\langle f, v_0 \rangle_H = 0$ for all $v_0 \in \mathrm{Ker}(L^*)$, the orthogonal decomposition of H in Lemma B.2 implies that $f \in \mathrm{Ran}(L)$, so that there is $u \in \mathrm{Dom}(L)$ such that $Lu = f$.

To prove uniqueness, we note that there exists an orthogonal basis

$$\mathrm{Ker}(L) = \mathrm{span}\{u_1, ..., u_n\}$$

such that $\langle u_i, u_j \rangle_H = \delta_{i,j}$ and $n = \mathrm{Dim}(\mathrm{Ker}(L))$. If u_0 is one particular solution of $Lu = f$, then

$$u = u_0 - \sum_{j=1}^{n} \langle u_0, u_j \rangle_H u_j$$

is also a solution of $Lu = f$ such that $\langle u, w_0 \rangle_H = 0$ for all $w_0 \in \mathrm{Ker}(L)$. The solution u is unique under the condition $\langle u, w_0 \rangle_H = 0$, because addition to u of any projection to $\mathrm{Ker}(L)$ will violate the condition. □

The Fredholm Alternative Theorem can be applied both to symmetric matrices in $H = \mathbb{R}^n$ and to linear differential operators in $H = L^2$.

B.5 Gagliardo–Nirenberg inequality

The Gagliardo–Nirenberg inequality says that for any fixed $1 \leq p \leq d/(d-2)$ (if $d \geq 3$) and for any $p \geq 1$ (if $d = 1$ or $d = 2$), there is a constant $C_{p,d} > 0$ such that

$$\forall u \in H^1(\mathbb{R}^d): \quad \|u\|_{L^{2p}}^{2p} \leq C_{p,d} \|\nabla u\|_{L^2}^{d(p-1)} \|u\|_{L^2}^{d+p(2-d)}.$$

In particular, $H^1(\mathbb{R}^d)$ is continuously embedded to $L^{2p}(\mathbb{R}^d)$ for any $p \in [1, d/(d-2)]$ if $d \geq 3$ and for any $p \geq 1$ if $d = 1$ or $d = 2$.

Let us prove the Gagliardo–Nirenberg inequality for $d = 1$. By the Sobolev Embedding Theorem (Appendix B.10), if $u \in H^1(\mathbb{R})$, then $u \in L^\infty(\mathbb{R})$ and $u(x) \to 0$ as $|x| \to \infty$. As a result, for any $p \geq 1$ we have

$$\|u\|_{L^{2p}}^{2p} \leq \|u^2\|_{L^\infty}^{p-1} \|u\|_{L^2}^2 \leq 2^{p-1} \|\nabla u\|_{L^2}^{p-1} \|u\|_{L^2}^{p+1},$$

where the last inequality follows from the Cauchy–Schwarz inequality applied to

$$u^2(x) = 2 \int_{-\infty}^x u(x) u'(x) dx.$$

Therefore, the Gagliardo–Nirenberg inequality holds for any $p \geq 1$ if $d = 1$.

B.6 Gronwall inequality

The Gronwall inequality can be formulated in both the integral and differential form. This inequality leads to the Comparison Principle for differential equations.

Lemma B.3 *Let $C \geq 0$, $k(t)$ be a given continuous non-negative function for all $t \geq 0$, and $y(t)$ be a continuous function satisfying the integral inequality*

$$0 \leq y(t) \leq C + \int_0^t k(t') y(t') dt', \quad t \geq 0.$$

Then, $y(t) \leq C e^{\int_0^t k(t') dt'}$ for all $t \geq 0$.

Proof Define a solution of the integral equation

$$Y(t) = C + \int_0^t k(t') Y(t') dt', \quad t \geq 0,$$

which is equivalent to the initial-value problem

$$\begin{cases} \dot{Y}(t) = k(t) Y(t), & t \geq 0, \\ Y(0) = C. \end{cases}$$

The unique solution is $Y(t) = Ce^{\int_0^t k(t')dt'}$ for all $t \geq 0$. It is clear that

$$y(0) \leq C = Y(0).$$

For any $t \geq 0$, we have

$$y(t) - \int_0^t k(t')y(t')dt' \leq C = Y(t) - \int_0^t k(t')Y(t')dt',$$

or equivalently,

$$\frac{d}{dt}\left(e^{-\int_0^t k(t')dt'}\int_0^t k(t')y(t')dt'\right) \leq \frac{d}{dt}\left(e^{-\int_0^t k(t')dt'}\int_0^t k(t')Y(t')dt'\right), \quad t \geq 0.$$

From the comparison of slopes, we have the comparison of functions

$$e^{-\int_0^t k(t')dt'}\int_0^t k(t')y(t')dt' \leq e^{-\int_0^t k(t')dt'}\int_0^t k(t')Y(t')dt', \quad t \geq 0,$$

which proves that

$$y(t) \leq Y(t) + \int_0^t k(t')\left(y(t') - Y(t')\right)dt' \leq Y(t), \quad t \geq 0.$$

Therefore, $y(t) \leq Ce^{\int_0^t k(t')dt'}$ for all $t \geq 0$. $\qquad\square$

In the differential form, if $k(t)$ is a continuous non-negative function for all $t \geq 0$ and $y(t)$ is a continuously differentiable non-negative function satisfying the differential inequality

$$\begin{cases} \dot{y}(t) \leq k(t)y(t), & t \geq 0, \\ y(0) \leq C, \end{cases}$$

for any given $C \geq 0$, then $y(t) \leq Ce^{\int_0^t k(t')dt'}$ for all $t \geq 0$.

B.7 Implicit Function Theorem

The Implicit Function Theorem in Banach spaces is used for the unique continuation of solutions of operator equations beyond the particular point where an exact solution of the equation is available. It is also applied to the situations when a bifurcation (sudden change in the number and stability of solution branches) occurs. The Implicit Function Theorem is derived from the Banach Fixed-Point Theorem (Appendix B.2).

Theorem B.5 *Let X, Y, and Z be Banach spaces and let $F(x,y) : X \times Y \to Z$ be a C^1 map on an open neighborhood of the point $(x_0, y_0) \in X \times Y$. Assume that*

$$F(x_0, y_0) = 0$$

and that

$$D_x F(x_0, y_0) : X \to Z \quad \text{is one-to-one and onto.}$$

There are $r > 0$ and $\delta > 0$ such that for each y with $\|y - y_0\|_Y \leq \delta$ there exists a unique solution $x \in X$ of the operator equation $F(x,y) = 0$ with $\|x - x_0\|_X \leq r$. Moreover, the map $Y \ni y \mapsto x(y) \in X$ is C^1 near $y = y_0$.

Proof Let $x = x_0 + \xi$, $y = y_0 + \eta$, and

$$f(\xi, \eta) := D_x F(x_0, y_0)\xi - F(x_0 + \xi, y_0 + \eta) : X \times Y \to Z.$$

We note that $f(0,0) = 0$, operator $D_\xi f(\xi, \eta) = D_x F(x_0, y_0) - D_x F(x_0 + \xi, y_0 + \eta)$ is continuous at $(\xi, \eta) = (0,0)$, and $D_\xi f(0,0) = 0$. Therefore, function $f(\xi, 0)$ is super-linear in ξ in the sense

$$\limsup_{\|\xi\|_X \to 0} \frac{\|f(\xi, 0)\|_Z}{\|\xi\|_X} = 0.$$

Because $D_x F(x_0, y_0) : X \to Z$ is one-to-one and onto, the inverse operator $[D_x F(x_0, y_0)]^{-1} : Z \to X$ exists with the bound

$$C_0 := \|[D_x F(x_0, y_0)]^{-1}\|_{Z \to X} < \infty.$$

For each fixed $\eta \in Y$, let us define the new function

$$A(\xi) := [D_x F(x_0, y_0)]^{-1} f(\xi, \eta) = \xi - [D_x F(x_0, y_0)]^{-1} F(x_0 + \xi, y_0 + \eta) : X \times Y \to X.$$

For each fixed $\eta \in Y$, there is a root $\xi \in X$ of $F(x_0 + \xi, y_0 + \eta) = 0$ if and only if $\xi \in X$ is a fixed point of operator A, that is, $A(\xi) = \xi$.

To apply the Banach Fixed-Point Theorem (Appendix B.2), we note from the above analysis that for each $\|\xi\|_X \leq r$ and $\|\eta\|_Y \leq \delta$ there is $C > 0$ such that

$$\|A(\xi)\|_X \leq C_0 \|f(\xi, \eta)\|_Z \leq C(\epsilon(r)r + \delta),$$

and

$$\|A(\xi_1) - A(\xi_2)\|_X \leq C_0 \|f(\xi_1, \eta) - f(\xi_2, \eta)\|_Z \leq C(\epsilon(r) + \delta)\|\xi_1 - \xi_2\|_X,$$

where $\epsilon(r) \to 0$ as $r \to 0$. For sufficiently small $r > 0$ and $\delta > 0$, there is $q \in (0,1)$ such that

$$\|A(\xi)\|_X \leq r, \qquad \|A(\xi_1) - A(\xi_2)\|_X \leq q\|\xi_1 - \xi_2\|_X.$$

By the Banach Fixed-Point Theorem (Theorem B.2), for each $\eta \in Y$ such that $\|\eta\|_Y \leq \delta$ there exists a unique fixed point $\xi \in X$ of $A(\xi)$ such that $\|\xi\|_X \leq r$.

Consider now the continuity of the map $\eta \mapsto \xi(\eta)$ for the fixed point of $A(\xi)$. There is $q \in (0,1)$ such that

$$\|\xi(\eta_1) - \xi(\eta_2)\|_X$$
$$\leq C_0 \|f(\xi(\eta_1), \eta_1) - f(\xi(\eta_2), \eta_2)\|_Z$$
$$\leq C_0 \left(\|f(\xi(\eta_1), \eta_1) - f(\xi(\eta_2), \eta_1)\|_Z + \|f(\xi(\eta_2), \eta_1) - f(\xi(\eta_2), \eta_2)\|_Z \right)$$
$$\leq q\|\xi(\eta_1) - \xi(\eta_2)\|_X + C_0 \|f(\xi(\eta_2), \eta_1) - f(\xi(\eta_2), \eta_2)\|_Z.$$

We have $\|f(\xi(\eta_2), \eta_1) - f(\xi(\eta_2), \eta_2)\|_Z \to 0$ as $\|\eta_1 - \eta_2\|_Y \to 0$. Since $q \in (0,1)$, we then have $\|\xi(\eta_1) - \xi(\eta_2)\|_X \to 0$ as $\|\eta_1 - \eta_2\|_Y \to 0$.

To prove that $\xi(\eta)$ is C^1, let us consider equation

$$F(x_0 + \xi(\eta + h), y_0 + \eta + h) - F(x_0 + \xi(\eta), y_0 + \eta) = 0$$

for a fixed $\eta \in Y$ and an arbitrary small $h \in Y$. Since F is C^1 and $D_x F(x_0, y_0)$ is one-to-one and onto, we know that $[D_x F(x_0 + \xi(\eta), y_0 + \eta)]^{-1}$ exists and is continuous for all $\eta \in Y$ such that $\|\eta\|_Y \leq \delta$. By the Taylor series expansion, we have

$$\xi(\eta + h) - \xi(\eta) = -[D_x F(x_0 + \xi(\eta), y_0 + \eta)]^{-1} \left(D_y F(x_0 + \xi(\eta), y_0 + \eta)h + R(h) \right),$$

where the remainder term $R(h)$ is super-linear with respect to h and $\xi(\eta+h) - \xi(\eta)$. Therefore, for all $\eta \in Y$ such that $\|\eta\|_Y \leq \delta$ there is $C > 0$ such that

$$\|\xi(\eta + h) - \xi(\eta)\|_X \leq C\|h\|_Y,$$

and the map $Y \ni \eta \mapsto \xi(\eta) \in X$ is C^1 for all $\eta \in Y$ such that $\|\eta\|_Y \leq \delta$. $\square$

Remark B.6   If $F(x,y) : X \times Y \to Z$ is a C^n map near $(x_0, y_0) \in X \times Y$ for any integer $1 \leq n \leq \infty$, then the map $y \mapsto x(y)$ is C^n near $y = y_0$. Moreover, if $F(x,y)$ is analytic near $(x_0, y_0) \in X \times Y$, then $x(y)$ is analytic near $y = y_0$.

B.8 Mountain Pass Theorem

Let H be a Hilbert space with the inner product $\langle \cdot, \cdot \rangle_H$ and the induced norm $\|\cdot\|_H$. We set up some definitions to ensure that a continuously differentiable functional $I(u) : H \to \mathbb{R}$ has a critical point in H.

Definition B.4   A continuous functional $I(u) : H \to \mathbb{R}$ is said to have the mountain pass geometry if:

(i) $I(0) = 0$;
(ii) there exist $a > 0$ and $r > 0$ such that $I(u) \geq a$ for any $u \in H$ with $\|u\|_H = r$;
(iii) there exists an element $v \in H$ such that $I(v) < 0$ and $\|v\|_H > r$.

Definition B.5   A continuously differentiable functional $I(u) : H \to \mathbb{R}$ is said to satisfy the Palais–Smale compactness condition if each sequence $\{u_k\}_{k \geq 1}$ in H such that

(a) $\{I(u_k)\}_{k \geq 1}$ is bounded,
(b) $\frac{d}{d\epsilon} I(u_k + \epsilon v)|_{\epsilon=0} \to 0$ as $k \to \infty$ for every $v \in H$,

has a convergent subsequence in H.

For continuous functions in compact domains, the mountain pass geometry of the function $I(u)$ implies existence of a critical point of $I(u)$. Because Hilbert spaces are not generally compact function spaces, the functional $I(u)$ must also satisfy the Palais–Smale compactness condition in addition to the mountain pass geometry. The Mountain Pass Theorem guarantees the existence of a critical point of $I(u)$ in H under these two conditions.

Theorem B.6   *Assume that a continuously differentiable functional $I(u) : H \to \mathbb{R}$ satisfies the mountain pass geometry and the Palais–Smale compactness condition. For the element $v \in H$ with $I(v) < 0$, we define*

$$\Gamma := \{\mathbf{g} \in C([0,1], H) : \quad \mathbf{g}(0) = 0, \quad \mathbf{g}(1) = v\}.$$

Then,

$$c = \inf_{\mathbf{g} \in \Gamma} \max_{0 \le t \le 1} I(\mathbf{g}(t))$$

is a critical value of $I(u)$ in H.

The proof of the Mountain Pass Theorem can be found in Evans [55, Section 8.5].

B.9 Noether Theorem

We formulate the Noether Theorem variationally, in the context of Lagrangian mechanics. Consider the Lagrangian in the form $L = L(u, u_t, u_x, x)$, where $u = u(x, t) : \mathbb{R} \times \mathbb{R} \to \mathbb{R}$ is a solution of the Euler–Lagrange equation,

$$\frac{\partial L}{\partial u} = \frac{\partial}{\partial x} \frac{\partial L}{\partial u_x} + \frac{\partial}{\partial t} \frac{\partial L}{\partial u_t}.$$

Denote the action functional by

$$S = \int_{\mathbb{R}} dt \int_{\mathbb{R}} dx L(u, u_t, u_x, x).$$

We assume that L is C^2 in its variables and that the solution exists for all $(x, t) \in \mathbb{R} \times \mathbb{R}$. The Noether Theorem relates the existence of the continuous symmetry of the action function S and the existence of the balance equation

$$\frac{\partial P}{\partial t} + \frac{\partial Q}{\partial x} = 0,$$

where P and Q depend on (u, u_t, u_x) and, possibly, (x, t). Integrating the balance equation in x and using suitable decay conditions at infinity, we obtain that $\int_{\mathbb{R}} P dx$ is constant for all $t \in \mathbb{R}$.

Theorem B.7 *Assume that S is invariant under a continuous infinitesimal transformation*

$$\begin{cases} \tilde{t} = t + \epsilon T(t, x, u), \\ \tilde{x} = x + \epsilon X(t, x, u), \\ \tilde{u} = u + \epsilon U(t, x, u), \end{cases}$$

where $\epsilon \in \mathbb{R}$ and (T, X, U) are some functions. Then the balance equation holds with

$$P = \frac{\partial L}{\partial u_t}(u_t T + u_x X - U) - LT,$$

$$Q = \frac{\partial L}{\partial u_x}(u_t T + u_x X - U) - LX.$$

Proof Thanks to the invariance of the action functional, we have the expansion

$$0 = \int_{\mathbb{R}} d\tilde{t} \int_{\mathbb{R}} d\tilde{x} L(\tilde{u}, \tilde{u}_{\tilde{t}}, \tilde{u}_{\tilde{x}}, \tilde{x}) - \int_{\mathbb{R}} dt \int_{\mathbb{R}} dx L(u, u_t, u_x, x),$$

$$= \int_{\mathbb{R}} dt \int_{\mathbb{R}} dx \left(L(\tilde{u}, \tilde{u}_{\tilde{t}}, \tilde{u}_{\tilde{x}}, x) - L(u, u_t, u_x, x) \right)$$

$$+ \epsilon \int_{\mathbb{R}} dt \int_{\mathbb{R}} dx \left(\frac{\partial LT}{\partial t} + \frac{\partial LX}{\partial x} \right) + \mathcal{O}(\epsilon^2).$$

On the other hand, the chain rule gives

$$\tilde{u}(\tilde{x}, \tilde{t}) = u(\tilde{x} - \epsilon X, \tilde{t} - \epsilon T) + \epsilon U(t, x, u)$$
$$= u(\tilde{x}, \tilde{t}) + \epsilon \left(U - Tu_t - Xu_x \right) + \mathcal{O}(\epsilon^2).$$

As a result, we obtain

$$\int_{\mathbb{R}} dt \int_{\mathbb{R}} dx \left(L(\tilde{u}, \tilde{u}_{\tilde{t}}, \tilde{u}_{\tilde{x}}, x) - L(u, u_t, u_x, x) \right) = \epsilon \int_{\mathbb{R}} dt \int_{\mathbb{R}} dx \frac{\partial L}{\partial u} (U - Tu_t - Xu_x)$$

$$+ \epsilon \int_{\mathbb{R}} dt \int_{\mathbb{R}} dx \left(\frac{\partial L}{\partial u_t} \frac{\partial}{\partial t} (U - Tu_t - Xu_x) + \frac{\partial L}{\partial u_x} \frac{\partial}{\partial x} (U - Tu_t - Xu_x) \right) + \mathcal{O}(\epsilon^2).$$

Using the Euler–Lagrange equations to express $\partial L/\partial u$ and equating the integrand of the term $\mathcal{O}(\epsilon)$ to zero, we obtain the balance equation with the stated expressions for P and Q. $\qquad\square$

The Noether Theorem can be applied to the Gross–Pitaevskii equation,

$$iu_t = -u_{xx} + V(x)u + \sigma|u|^2 u,$$

where $u(x, t) : \mathbb{R} \times \mathbb{R} \to \mathbb{C}$, $V(x) \in L^\infty(\mathbb{R})$, and $\sigma \in \{1, -1\}$. The Lagrangian for the Gross–Pitaevskii equation is extended to the complex-valued functions,

$$L(u, \bar{u}, u_t, \bar{u}_t, u_x, \bar{u}_x, x) = \frac{i}{2} \left(\bar{u}u_t - \bar{u}_t u \right) - |u_x|^2 - V(x)|u|^2 - \frac{\sigma}{2}|u|^4.$$

Existence of the time translation symmetry implies that $T = 1$, $X = U = 0$, which gives the conservation of energy (Hamiltonian),

$$\int_{\mathbb{R}} \left(u_t \frac{\partial L}{\partial u_t} + \bar{u}_t \frac{\partial L}{\partial \bar{u}_t} - L \right) dx = \int_{\mathbb{R}} \left(|u_x|^2 + V|u|^2 + \frac{\sigma}{2}|u|^4 \right) dx.$$

If $V(x) \equiv 0$ for all $x \in \mathbb{R}$, existence of the space translation symmetry implies that $T = 0$, $X = 1$, and $U = 0$, which gives the conservation of momentum,

$$\int_{\mathbb{R}} \left(u_x \frac{\partial L}{\partial u_t} + \bar{u}_x \frac{\partial L}{\partial \bar{u}_t} \right) dx = \frac{i}{2} \int_{\mathbb{R}} (\bar{u}u_x - u\bar{u}_x) dx.$$

Existence of the gauge translation symmetry implies that $T = X = 0$, $U = -iu$, which gives the conservation of power,

$$\int_{\mathbb{R}} \left(\frac{\partial L}{\partial u_t} (-iu) + \frac{\partial L}{\partial \bar{u}_t} (iu) \right) dx = \int_{\mathbb{R}} |u|^2 dx.$$

B.10 Sobolev Embedding Theorem

The Sobolev Embedding Theorem relates functions in Sobolev space $H^s(\mathbb{R}^d)$ with continuously differentiable functions in $C_b^r(\mathbb{R}^d)$, provided that $s > r + \frac{d}{2}$. This property is ultimately related with the fact that Sobolev space $H^s(\mathbb{R}^d)$ forms a Banach algebra with respect to pointwise multiplication for $s > \frac{d}{2}$ (Appendix B.1).

The Riemann–Lebesgue Lemma does the same job for functions with Fourier transforms in $L_s^1(\mathbb{R}^d)$, provided that $s \geq r$. This space is referred to as the Wiener space and is denoted by $W^s(\mathbb{R}^d)$. Note that the Wiener space $W^s(\mathbb{R}^d)$ forms a

Banach algebra with respect to pointwise multiplication for any $s \geq 0$ (Appendix B.16).

Not only are the functions in $H^s(\mathbb{R}^d)$ with $s > r + \frac{d}{2}$ or in $W^s(\mathbb{R}^d)$ with $s \geq r$ r-times continuously differentiable but also the functions and their derivatives up to the rth order decay to zero at infinity. We start with the Riemann–Lebesgue Lemma.

Lemma B.4 *Let $u(x)$ have the Fourier transform $\hat{u}(k)$ in function space $L_s^1(\mathbb{R}^d)$ for an integer $s \geq 0$. Then, $u \in C_b^r(\mathbb{R}^d)$ for any $r \leq s$ and $u(x)$, $\partial_x u(x)$, ..., $\partial_x^r u(x)$ decay to zero as $|x| \to \infty$.*

Proof Using the Fourier transform, we obtain for any $r \leq s$,

$$\|\partial_x^r u\|_{L^\infty} \leq \frac{1}{(2\pi)^{d/2}}\|\hat{u}\|_{L_r^1} \leq \frac{1}{(2\pi)^{d/2}}\|\hat{u}\|_{L_s^1}.$$

If $\hat{u} \in L_s^1(\mathbb{R}^d)$, then $\partial_x^r u \in C_b^0(\mathbb{R}^d)$, so that $u \in C_b^r(\mathbb{R}^d)$. The decay of $u(x)$ and its derivatives up to the rth order is proved in Evans [55]. □

We can now formulate the Sobolev Embedding Theorem.

Theorem B.8 *For any fixed integer $r \geq 0$ and any $s > \frac{d}{2} + r$, the Sobolev space $H^s(\mathbb{R}^d)$ is continuously embedded in the space of functions $C_b^r(\mathbb{R}^d)$. Moreover, if $u \in H^s(\mathbb{R}^d)$, then $u(x)$, $\partial_x u(x)$, ..., $\partial_x^r u(x)$ decay to zero as $|x| \to \infty$.*

Proof We need to show that there exists $C_{s,r} > 0$ such that

$$\forall u \in H^s(\mathbb{R}^d): \quad \|u\|_{C^r} \leq C_{s,r}\|u\|_{H^s}.$$

Recall that the Cauchy–Schwarz inequality gives

$$\|\hat{u}\|_{L^1} \leq \left(\int_{\mathbb{R}^d} \frac{dk}{(1+|k|^2)^s}\right)^{1/2}\|\hat{u}\|_{L_s^2}, \quad s > \frac{d}{2}.$$

As a result, for an integer $r \geq 0$ and any $s > \frac{d}{2} + r$, there is $C_{s,r} > 0$ such that

$$\|\partial_x^r u\|_{L^\infty} \leq \frac{1}{(2\pi)^{d/2}}\|\hat{u}\|_{L_r^1} \leq C_{s,r}\|\hat{u}\|_{L_s^2} \leq C_{s,r}\|u\|_{H^s}.$$

Decay of $u(x)$, $\partial_x u(x)$, ..., $\partial_x^r u(z)$ to zero as $|x| \to \infty$ follows from the Riemann–Lebesgue Lemma. □

It follows from Theorem B.8 that $H^s(\mathbb{R})$ is continuously embedded into $L^p(\mathbb{R}^d)$ for any $p \geq 2$ if $s > \frac{d}{2}$. It is more interesting to find embedding of $H^s(\mathbb{R}^d)$ into $L^p(\mathbb{R}^d)$ space for $0 < s < \frac{d}{2}$. This embedding exists for any $2 \leq p \leq 2d/(d-2s)$ [55].

B.11 Spectral Theorem

Let H be a Hilbert space with the inner product $\langle \cdot, \cdot \rangle_H$ and the induced norm $\|\cdot\|_H$. Let $L : \text{Dom}(L) \subseteq H \to H$ be a self-adjoint linear operator in the sense

$$\forall f, g \in \text{Dom}(L): \quad \langle Lf, g \rangle_H = \langle f, Lg \rangle_H.$$

The spectrum of L, denoted as $\sigma(L)$, is the set of all points $\lambda \in \mathbb{C}$ for which $L - \lambda I$ is not invertible on H.

If $\lambda \in \sigma(L)$ is such that $\mathrm{Ker}(L - \lambda I) \neq \{0\}$, then λ is an eigenvalue of A of geometric multiplicity $N_g := \mathrm{Dim}\,\mathrm{Ker}(L - \lambda I)$ and algebraic multiplicity $N_a := \mathrm{Dim}\bigcup_{n \in \mathbb{N}} \mathrm{Ker}(L - \lambda I)^n$.

- The discrete spectrum of L, denoted by $\sigma_d(L)$, is the set of all eigenvalues of L with finite (algebraic) multiplicity and which are isolated points of $\sigma(L)$.
- The essential spectrum of L, denoted by $\sigma_{\mathrm{ess}}(L)$, is the complement of $\sigma_d(L)$ in $\sigma(L)$.

The Spectral Theorem characterizes the spectrum of a self-adjoint operator L.

Theorem B.9 *Let L be a self-adjoint operator in Hilbert space H. Then, $\sigma(L) \subset \mathbb{R}$ and the eigenvectors of L are orthogonal to each other.*

Proof Let $\lambda = \lambda_r + \mathrm{i}\lambda_i \in \sigma(L)$ with $\lambda_r, \lambda_i \in \mathbb{R}$ and compute

$$\forall u \in \mathrm{Dom}(L): \quad \|(L - \lambda I)u\|_H^2 = \|(L - \lambda_r I)u\|_H^2 + \lambda_i^2\|u\|_H^2 \geq \lambda_i^2\|u\|_H^2.$$

If $\lambda_i \neq 0$, then $\lambda \neq \sigma(L)$ because $(L - \lambda I)$ is invertible. Therefore, $\lambda_i = 0$ and $\lambda \in \mathbb{R}$.

To prove orthogonality of eigenvectors of L, we first assume that the two eigenvectors $u_1, u_2 \in \mathrm{Dom}(L) \subseteq H$ correspond to two distinct eigenvalues $\lambda_1, \lambda_2 \in \mathbb{R}$ such that $\lambda_1 \neq \lambda_2$. Then, we obtain

$$\lambda_1\langle u_1, u_2\rangle_H = \langle Lu_1, u_2\rangle_H = \langle u_1, Lu_2\rangle_H = \lambda_2\langle u_1, u_2\rangle_H.$$

If $\lambda_1 \neq \lambda_2$, then $\langle u_1, u_2\rangle_H = 0$. If an eigenvalue λ_0 is multiple, we will show that $\bigcup_{n \in \mathbb{N}} \mathrm{Ker}(L - \lambda I)^n = \mathrm{Ker}(L - \lambda I)$, that is, the geometric and algebraic multiplicities coincide. To do so it is sufficient to show that $\mathrm{Ker}(L - \lambda I)^2 = \mathrm{Ker}(L - \lambda I)$. If $u \in \mathrm{Ker}(L - \lambda I)^2$, then there is $v \in \mathrm{Ker}(L - \lambda I)$ such that $(L - \lambda I)u = v$. However, since L is self-adjoint, we obtain

$$\|v\|_H^2 = \langle (L - \lambda I)u, v\rangle_H = \langle u, (L - \lambda I)v\rangle_H = 0,$$

that is $v = 0$ and $u \in \mathrm{Ker}(L - \lambda I)$. If λ is a multiple eigenvalue, there is a basis of eigenvectors in $\mathrm{Ker}(L - \lambda I)$, which can be orthogonalized and normalized by the Gram–Schmidt orthogonalization procedure. $\qquad\square$

B.12 Sturm Comparison Theorem

Let us consider nonzero solutions $u(x; \lambda)$ of the second-order differential equation

$$u''(x) + (\lambda - V(x))u(x) = 0, \quad x \in \mathbb{R},$$

for any $V \in L^\infty(\mathbb{R})$. The Sturm Comparison Theorem allows to compare the distribution of zeros of the solution $u(x; \lambda)$ for two values of λ.

Theorem B.10 *Assume that $u(x; \lambda_1)$ has two consequent zeros at a and b, where*

$$-\infty < a < b < \infty.$$

If $\lambda_2 > \lambda_1$, then $u(x; \lambda_2)$ has a zero in (a, b).

Proof Let $u_{1,2}(x) = u(x; \lambda_{1,2})$ be two solutions of the differential equations such that $u_{1,2} \in C^2(a, b)$. Assume without loss of generality that $u_1(x) > 0$ on (a, b). It follows from the second-order differential equation that

$$\frac{d}{dx}(u_1'(x)u_2(x) - u_1(x)u_2'(x)) = (\lambda_2 - \lambda_1)u_1(x)u_2(x), \quad x \in [a, b].$$

By contradiction, let us assume that $u_2(x)$ is sign-definite on (a, b) and again let $u_2(x) > 0$ without loss of generality. Then, integrating the above equation on $[a, b]$ and using $u_1(a) = u_1(b) = 0$, we obtain

$$0 \geq u_1'(b)u_2(b) - u_1'(a)u_2(a) = (\lambda_2 - \lambda_1) \int_a^b u_1(x)u_2(x)dx > 0.$$

The contradiction proves that $u_2(x)$ must change sign (have a zero) on (a, b). $\quad\square$

The proof extends if $a = -\infty$ and/or $b = +\infty$ provided that $u_{1,2}(x)$ are bounded in the corresponding limits $x \to -\infty$ and/or $x \to +\infty$.

B.13 Unstable Manifold Theorem

Consider the initial-value problem for the system of differential equations

$$\begin{cases} \dot{x}(t) = f(x), & t \in \mathbb{R}, \\ x(0) = x_0, \end{cases}$$

where $x \in \mathbb{R}^n$ and $f(x) : \mathbb{R}^n \to \mathbb{R}^n$ is a C^1 function. Solutions of the initial-value problem determine the phase flow $\phi_t(x_0) : \mathbb{R}^n \to \mathbb{R}^n$. The Unstable Manifold Theorem characterizes an invariant manifold under the phase flow $\phi_t(x)$ in negative times $t \leq 0$.

Theorem B.11 *Let $f(0) = 0$ and assume that $A = D_x f(0)$ has $k \leq n$ eigenvalues with positive real part. Let $\delta > 0$ be sufficiently small. There exists a k-dimensional C^1 unstable manifold of $\dot{x} = f(x)$ such that*

$$U(0) = \left\{ x \in B_\delta(0) : \quad \phi_{t \leq 0}(x) \in B_\delta(0), \quad \lim_{t \to -\infty} \|\phi_t(x)\|_{\mathbb{R}^n} = 0 \right\},$$

where $B_\delta(0) \subset \mathbb{R}^n$ is a ball of radius $\delta > 0$ centered at 0. Moreover, the unstable manifold $U(0)$ is tangent to the unstable manifold of the linearized system $\dot{x} = Ax$.

Remark B.7 Similarly, one can formulate the Stable Manifold Theorem for the stable manifold,

$$S(0) = \left\{ x \in B_\delta(0) : \quad \phi_{t \geq 0}(x) \in B_\delta(0), \quad \lim_{t \to +\infty} \|\phi_t(x)\|_{\mathbb{R}^n} = 0 \right\},$$

which is invariant in positive times $t \geq 0$.

A modification of the Unstable Manifold Theorem is applied to the system of differential equations with time-periodic coefficients

$$\begin{cases} \dot{x}(t) = A(t)x + f(x), & t \in \mathbb{R}, \\ x(0) = x_0, \end{cases}$$

where $A(t) : \mathbb{R} \to \mathbb{R}^{n \times n}$ is continuous and T-periodic and $f(x) : \mathbb{R}^n \to \mathbb{R}^n$ is a C^1 function such that

$$\limsup_{\|x\|_{\mathbb{R}^n} \to 0} \frac{\|f(x)\|_{\mathbb{R}^n}}{\|x\|_{\mathbb{R}^n}} = 0.$$

The modified Unstable Manifold Theorem characterizes the unstable manifold for the T-periodic system of differential equations.

Theorem B.12   *Assume that the linearized system $\dot{x} = A(t)x$ has $k \leq n$ Floquet multipliers outside the unit circle. Let $\delta > 0$ be sufficiently small. There exists a k-dimensional C^1 unstable manifold of $\dot{x} = A(t)x + f(x)$ such that*

$$U(0) = \left\{ x \in B_\delta(0) : \quad \phi_{t \leq 0}(x) \in B_\delta(0), \quad \lim_{t \to -\infty} \|\phi_t(x)\|_{\mathbb{R}^n} = 0 \right\}.$$

Moreover, the unstable manifold $U(0)$ is tangent to the unstable manifold of the linearized system $\dot{x} = A(t)x$.

B.14   Weak Convergence Theorem

Let H be a Hilbert space with the inner product $\langle \cdot, \cdot \rangle_H$ and the induced norm $\|\cdot\|_H$. If a sequence in a Hilbert space H is bounded, it does not imply that there exists a subsequence, which converges to an element in H, because H is generally non-compact. The best that it implies is the existence of a weakly convergent subsequence in H. The following definition distinguishes between weak and strong convergence of sequences in H.

Definition B.6   We say that a sequence $\{u_k\}_{k \in \mathbb{N}} \in H$ converges strongly in H if there exists an element $u \in H$ such that

$$u_k \to u \text{ in } H \quad \Longleftrightarrow \quad \lim_{k \to \infty} \|u_k - u\|_H = 0.$$

We say that the sequence converges weakly in H if there exists an element $u \in H$ such that

$$u_{k_j} \rightharpoonup u \text{ in } H \quad \Longleftrightarrow \quad \forall v \in H : \quad \langle v, u_k - u \rangle_H = 0.$$

It is clear that the strong convergence implies the weak convergence thanks to the Cauchy–Schwarz inequality

$$|\langle v, u_k - u \rangle_H| \leq \|v\|_H \|u_k - u\|_H.$$

However, the converse statement is valid only in compact vector spaces such as $H = \mathbb{R}^n$.

The Weak Convergence Theorem guarantees the weak convergence of a bounded sequence in the Hilbert space H.

Theorem B.13 *If a sequence $\{u_k\}_{k\in\mathbb{N}} \in H$ is bounded, there exists a subsequence $\{u_{k_j}\}_{j\in\mathbb{N}} \subset \{u_k\}_{k\in\mathbb{N}}$ and an element $u \in H$ such that $u_{k_j} \rightharpoonup u$ in H.*

Proof For simplification, let us consider a *separable* Hilbert space H, that is, we assume that there exists a countable orthonormal basis $\{v_n\}_{n\in\mathbb{N}}$ in H. Since $\langle u_k, v_1\rangle_H \le \|u_k\|_H \|v_1\|_H$ for all $k \in \mathbb{N}$, if $\{u_k\}_{k\in\mathbb{N}}$ is a bounded sequence in H, then $\langle u_k, v_1\rangle_H$ is a bounded sequence of numbers. Therefore, there exists a subsequence $\{u_{k_j}^{(1)}\}_{j\in\mathbb{N}} \subset \{u_k\}_{k\in\mathbb{N}}$ such that $|a_1| < \infty$, where

$$a_1 := \lim_{j\to\infty} \langle u_{k_j}^{(1)}, v_1\rangle_H.$$

Continuing this process inductively, for each $n \ge 1$, there exists a subsequence $\{u_{k_j}^{(n+1)}\}_{j\in\mathbb{N}} \subset \{u_{k_j}^{(n)}\}_{j\in\mathbb{N}}$ such that $|a_n| < \infty$, where

$$a_n := \lim_{j\to\infty} \langle u_{k_j}^{(n)}, v_n\rangle_H, \quad n \in \mathbb{N}.$$

Let us denote $w_n := u_{k_n}^{(n)}$, $n \in \mathbb{N}$. Then, for each $k \in \mathbb{N}$,

$$\lim_{n\to\infty} \langle w_n, v_k\rangle_H = a_k.$$

Because the sequence $\{w_n\}_{n\in\mathbb{N}}$ is bounded in H and the set $\{v_n\}_{n\in\mathbb{N}}$ is dense in H, for each $v \in H$ and $\epsilon > 0$, there is $n_0(\epsilon) \ge 1$ such that

$$|\langle w_n - w_m, v\rangle_H| \le |\langle w_n - w_m, v_k\rangle| + |\langle w_n - w_m, v - v_k\rangle_H| < \epsilon,$$

for all $n, m \ge n_0(\epsilon)$. Therefore, as $n \to \infty$, $\langle w_n, v\rangle$ converges to a number $a(v)$ for each $v \in H$.

The map $v \mapsto a(v)$ is linear and bounded by

$$\forall v \in H: \quad |a(v)| \le \|v\|_H \sup_{n\in\mathbb{N}} \|w_n\|_H.$$

By the Riesz Representation Theorem, there is $u \in H$ such that $a(v) = \langle u, v\rangle_H$ for all $v \in H$. Therefore, $\langle w_n, v\rangle_H \to \langle u, v\rangle_H$ as $n \to \infty$ for all $v \in H$ and, consequently, $w_n \rightharpoonup u$ as $n \to \infty$. $\qquad\square$

B.15 Weyl Theorem

Let H be a Hilbert space with the inner product $\langle\cdot,\cdot\rangle_H$ and the induced norm $\|\cdot\|_H$. Let $L : \mathrm{Dom}(L) \subseteq H \to H$ be a self-adjoint linear operator. Recall the definition of the discrete $\sigma_d(L)$ and essential $\sigma_{\mathrm{ess}}(L)$ spectra of the self-adjoint operator L in Section B.11. The Weyl Theorem gives a rigorous method to find if $\lambda \in \sigma_{\mathrm{ess}}(L)$.

Theorem B.14 *Let L be a self-adjoint linear operator in Hilbert space H. Then, $\lambda \in \sigma_{\mathrm{ess}}(L)$ if and only if there exists a sequence $\{u_n\}_{n\ge 1} \in \mathrm{Dom}(L)$ such that $\|u_n\|_H = 1$ for all $n \ge 1$, $u_n \rightharpoonup 0$ in H as $n \to \infty$, and $\lim_{n\to\infty} \|(L-\lambda I)u_n\|_H = 0$.*

Proof Let $\lambda \in \sigma_{\text{ess}}(L)$ and assume that $\text{Ker}(L - \lambda I) = \{0\}$ for simplification of arguments. Since $\lambda \in \sigma(L)$, the inverse of $(L - \lambda I)$ is unbounded. Therefore, there is a sequence $\{v_n\}_{n \in \mathbb{N}} \in H$ such that $\|v_n\|_H = 1$ for all $n \in \mathbb{N}$ and $\|(L - \lambda I)^{-1} v_n\| \to \infty$ as $n \to \infty$. Let

$$u_n := \frac{(L - \lambda I)^{-1} v_n}{\|(L - \lambda I)^{-1} v_n\|_H}, \quad n \in \mathbb{N},$$

then we have $\|u_n\|_H = 1$ for all $n \in \mathbb{N}$ and

$$\|(L - \lambda I) u_n\|_H = \frac{\|v_n\|_H}{\|(L - \lambda I)^{-1} v_n\|_H} \to 0 \quad \text{as} \quad n \to \infty.$$

It remains to show that $\{u_n\}_{n \in \mathbb{N}}$ converges weakly to $0 \in H$ in the sense of Definition B.6. For all $f \in \text{Dom}(L - \lambda I)^{-1}$, we have

$$|\langle f, u_n \rangle_H| = \frac{|\langle (L - \lambda I)^{-1} f, v_n \rangle_H|}{\|(L - \lambda I)^{-1} v_n\|_H} \leq \frac{\|(L - \lambda I)^{-1} f\|_H}{\|(L - \lambda I)^{-1} v_n\|_H},$$

where the right-hand side converges to zero as $n \to \infty$. Since $\text{Dom}(L - \lambda I)^{-1}$ is dense in H, then $u_n \rightharpoonup 0$ in H as $n \to \infty$.

In the opposite direction, let $\{u_n\}_{n \geq 1} \in \text{Dom}(L)$ be a sequence such that $\|u_n\|_H = 1$ for all $n \geq 1$, $u_n \rightharpoonup 0$ in H as $n \to \infty$, and $\lim_{n \to \infty} \|(L - \lambda I) u_n\|_H = 0$. Therefore, $\lambda \in \sigma(L)$. It remains to show that λ is not an isolated eigenvalue of finite multiplicity. If $\text{Ker}(L - \lambda I) = \{0\}$, we are done, so let us assume that $\{\phi_j\}_{j=1}^k$ is a finite orthonormal basis for $\text{Ker}(L - \lambda I)$. Let P_λ and Q_λ be orthogonal projections onto $\text{Ker}(L - \lambda I)$ and $[\text{Ker}(L - \lambda I)]^\perp$, respectively. Since $u_n \rightharpoonup 0$ in H as $n \to \infty$, we have

$$\|P_\lambda u_n\|_H^2 = \sum_{j=1}^k \|\langle u_n, \phi_j \rangle_H|^2 \to 0 \quad \text{as} \quad n \to \infty.$$

Therefore, $\|Q_\lambda u_n\|_H^2 \to 1$ as $n \to \infty$, so that we can define

$$v_n := \frac{Q_\lambda u_n}{\|Q_\lambda u_n\|_H}, \quad n \in \mathbb{N}.$$

We have $\|v_n\|_H = 1$ for all $n \in \mathbb{N}$ and

$$\|(L - \lambda I) v_n\|_H = \frac{\|(L - \lambda I) u_n\|_H}{\|Q_\lambda u_n\|_H} \to 0 \quad \text{as} \quad n \to \infty.$$

The inverse of $Q_\lambda(A - \lambda I)Q_\lambda$ is unbounded, that is, $\lambda \in \sigma(Q_\lambda A Q_\lambda)$. However, $\text{Ker}(Q_\lambda(A - \lambda I)Q_\lambda) = \{0\}$, hence $\lambda \in \sigma_{\text{ess}}(A)$. $\square$

Let us use this theorem to establish the location of the continuous spectrum of the self-adjoint operator $L = -\partial_x^2 + c$ for any $c \in \mathbb{R}$ with $\text{Dom}(L) = H^2(\mathbb{R})$ and $H = L^2(\mathbb{R})$.

Lemma B.5 *If $L = -\partial_x^2 + c$ for any $c \in \mathbb{R}$, then $\sigma_{\text{ess}}(L) = [c, \infty)$.*

Proof We recall that

$$-u''(x) + cu(x) = \lambda u(x), \quad x \in \mathbb{R}$$

has no bounded solutions for $\lambda < c$ and has bounded solutions $u(x;k) = e^{ikx}$ for any $\lambda = c + k^2 \geq c$. Fix the point $\lambda = c + k_0^2$ and define the sequence $\{u_n\}_{n\geq 1} \in H^2(\mathbb{R})$ by

$$u_n(x) = \frac{a_n}{2\pi} \int_{k_0 - n^{-1}}^{k_0 + n^{-1}} e^{ikx}dk, \quad n \geq 1,$$

where a_n is found from the normalization $\|u_n\|_{L^2} = 1$. After explicit integration, we obtain

$$u_n(x) = \frac{a_n}{\pi x} \sin\left(\frac{x}{n}\right) e^{ik_0 x} \in H^2(\mathbb{R}), \quad n \geq 1,$$

thanks to analyticity of $u_n(x)$ at $x = 0$ and the sufficient decay of $u_n(x)$ to zero as $|x| \to \infty$. The normalization condition is satisfied if

$$\|u_n\|_{L^2}^2 = \frac{a_n^2}{\pi^2 n} \int_{\mathbb{R}} \frac{\sin^2(z)}{z^2} dz = 1, \quad n \geq 1,$$

which shows that $a_n \sim \sqrt{n}$ as $n \to \infty$. Therefore, we have

$$u_n(x) = \frac{a_n}{\pi n} \frac{\sin(x/n)}{x/n} e^{ik_0 x} \to 0 \quad \text{as} \quad n \to \infty.$$

Using the same computations, we obtain

$$(L - \lambda_0 I)u_n = \frac{a_n}{2\pi} \int_{k_0 - n^{-1}}^{k_0 + n^{-1}} (k^2 - k_0^2)e^{ikx}dk$$

$$= \frac{a_n}{\pi} e^{ik_0 x} \left(\frac{\sin(x/n)}{n^2 x} + \frac{2\cos(x/n)}{nx^2} - \frac{2\sin(x/n)}{x^3} - \frac{2ik_0 \cos(x/n)}{nx} + \frac{2ik_0 \sin(x/n)}{x^2} \right).$$

Again this function is analytic at $x = 0$ and has the sufficient decay to zero as $|x| \to \infty$ to be in $L^2(\mathbb{R})$. Integrating over x, we obtain

$$\|(L - \lambda_0 I)u_n\|_{L^2}^2$$

$$= \frac{a_n^2}{\pi^2} \int_{\mathbb{R}} \left[\frac{1}{n^5} \left(\frac{\sin(z)}{z} + \frac{2\cos(z)}{z^2} - \frac{2\sin(z)}{z^3} \right)^2 + \frac{4k_0^2}{n^3} \left(\frac{\cos(z)}{z} - \frac{\sin(z)}{z^2} \right)^2 \right] dz.$$

Since $a_n \sim \sqrt{n}$ as $n \to \infty$, we conclude that $\lim_{n\to\infty} \|(L - \lambda_0 I)u_n\|_{L^2} = 0$, which shows that $\lambda_0 = c + k_0^2 \in \sigma_c(L)$ for any $k_0 \in \mathbb{R}$. Therefore, $\sigma_{\text{ess}}(L) = [c, \infty)$. $\square$

Arguments in Lemma B.5 can be extended to the linear operators $L = -\partial_x^2 + c + V(x)$ if $V(x)$ decays to zero as $|x| \to \infty$ exponentially fast. The key ingredient in this proof is the construction of the eigenfunctions in $L^\infty(\mathbb{R})$ for any $\lambda \in \sigma_{\text{ess}}(L)$.

An alternative version of the Weyl Theorem states that if L_0 and L are two self-adjoint operators in a Hilbert space and $L - L_0$ is L_0-compact in the sense that $\text{Dom}(L_0) \subseteq \text{Dom}(L - L_0)$ and $(L - L_0)(L_0 - \lambda I)^{-1}$ is compact for some $\lambda \notin \sigma(L_0)$, then $\sigma_c(L) = \sigma_c(L_0)$. See the proof in Hislop & Sigal [89, Chapter 14]. The alternative version of the Weyl Theorem can also be used to establish that if $V \in L^\infty(\mathbb{R}) \cap L^2(\mathbb{R})$, then

$$\sigma_{\text{ess}}(-\partial_x^2 + c + V(x)) = \sigma_{\text{ess}}(-\partial_x^2 + c) = [c, \infty).$$

However, to use this theorem, a more delicate spectral analysis is needed to establish that

$$V(-\partial_x^2 + c - \lambda)^{-1}$$

is a compact operator for $\lambda \notin [c, \infty)$.

B.16 Wiener algebra

Let $W(\mathbb{R})$ be the space of all 2π-periodic functions whose Fourier series converge absolutely, that is, all functions $f(x) : \mathbb{R} \to \mathbb{C}$ whose Fourier series $f(x) = \sum_{n \in \mathbb{Z}} \hat{f}_n e^{inx}$ satisfy

$$\|f\|_W := \sum_{n \in \mathbb{Z}} |\hat{f}_n| < \infty.$$

If $\hat{f} \in l^1(\mathbb{Z})$, then $f \in C_{\mathrm{per}}(\mathbb{R})$. Therefore, space $W(\mathbb{R})$ is a commutative Banach algebra of continuous functions with respect to the pointwise multiplication (Appendix B.1), which is usually called the *Wiener algebra*.

The Wiener algebra of $W(\mathbb{R})$ with respect to pointwise multiplication is isomorphic to the Banach algebra of $l^1(\mathbb{Z})$ with respect to the convolution product since if

$$\forall f, g \in W(\mathbb{R}): \quad \widehat{(fg)}_n = \sum_{m \in \mathbb{Z}} \hat{f}_m \hat{g}_{n-m} = (\hat{f} \star \hat{g})_n, \quad n \in \mathbb{Z},$$

then

$$\forall \hat{f}, \hat{g} \in l^1(\mathbb{Z}): \quad \|\hat{f} \star \hat{g}\|_{l^1} \le \|\hat{f}\|_{l^1} \|\hat{g}\|_{l^1}.$$

Similarly, we can define the Wiener space $W^s(\mathbb{R})$ for any $s \ge 0$ to be the space of all 2π-periodic functions $f(x)$ with the norm

$$\|f\|_{W^s} := \|\hat{f}\|_{l_s^1}, \quad s \ge 0.$$

Recall the discrete embedding of $l_s^2(\mathbb{Z})$ to $l_r^1(\mathbb{Z})$ for any $r \ge 0$ and $s > \frac{1}{2} + r$ with the bound

$$\exists C_{s,r} > 0: \quad \forall \hat{f} \in l_s^2(\mathbb{Z}): \quad \|\hat{f}\|_{l_r^1} \le C_{s,r} \|\hat{f}\|_{l_s^2}, \quad r \ge 0, \quad s > \frac{1}{2} + r.$$

As a result, the discrete space $l_r^1(\mathbb{Z})$ is also a Banach algebra with respect to the convolution product for any $r \ge 0$ and the Wiener space $W^r(\mathbb{R})$ is a Banach algebra with respect to pointwise multiplication (Wiener algebra).

The concept of Wiener space and Wiener algebra extends to continuous functions on line $\mathbb{R}$ with the decay to zero at infinity, whose Fourier transform belongs to $L^1(\mathbb{R})$.

References

[1] Ablowitz, M.J.; Clarkson, P.A.; *Solitons, Nonlinear Evolution Equations and Inverse Scattering* (Cambridge University Press, Cambridge, 1991).

[2] Ablowitz, M.J.; Prinari, B.; Trubach, A.D.; *Discrete and Continuous Nonlinear Schrödinger Systems* (Cambridge University Press, Cambridge, 2004).

[3] Aftalion, A.; *Vortices in Bose–Einstein Condensates*, Progress in Nonlinear Differential Equations and Their Applications **67** (Birkhäuser, Boston, 2006).

[4] Aftalion, A.; Helffer, B.; "On mathematical models for Bose–Einstein condensates in optical lattices", Rev. Math. Phys. **21** (2009), 229–278.

[5] Agrawal, G.P.; Kivshar, Yu.S.; *Optical Solitons: From Fibers to Photonic Crystals* (Academic Press, San Diego, 2003).

[6] Agueev, D.; Pelinovsky, D.; "Modeling of wave resonances in low-contrast photonic crystals", SIAM J. Appl. Math. **65** (2005), 1101–1129.

[7] Akhmediev, N.; Ankiewicz, A.; *Solitons: Nonlinear Pulses and Beams* (Kluwer Academic, Dordrecht, 1997).

[8] Al Khawaja, U.; "A comparative analysis of Painlevé, Lax pair, and similarity transformation methods in obtaining the integrability conditions of nonlinear Schrödinger equations", J. Math. Phys. **51** (2010), 053506 (11 pp.).

[9] Alama, S.; Li, Y.Y.; "Existence of solutions for semilinear elliptic equations with indefinite linear part", J. Diff. Eqs. **96** (1992), 89–115.

[10] Alfimov, G.L.; Brazhnyi, V.A.; Konotop, V.V.; "On classification of intrinsic localized modes for the discrete nonlinear Schrödinger equation", Physica D **194** (2004), 127–150.

[11] Angulo Pava, J.; *Nonlinear Dispersive Equations (Existence and Stability of Solitary and Periodic Travelling Wave Solutions)* (AMS, Providence, RI, 2009).

[12] Azizov, T.Ya.; Iohvidov, I.S.; *Elements of the Theory of Linear Operators in Spaces with Indefinite Metric* (Nauka, Moscow, 1986) [in Russian].

[13] Bambusi, D.; Sacchetti, A.; "Exponential times in the one-dimensional Gross–Pitaevskii equation with multiple well potential", Comm. Math. Phys. **275** (2007), 1–36.

[14] Barashenkov, I.V.; Oxtoby, O.F.; Pelinovsky, D.E.; "Translationally invariant discrete kinks from one-dimensional maps", Phys. Rev. E **72** (2005), 035602(R) (4 pp.).

[15] Barashenkov, I.V.; van Heerden, T.C.; "Exceptional discretizations of the sine-Gordon equation", Phys. Rev. E **77** (2008), 036601 (9 pp.).

[16] Bao, W.; Du, Q.; "Computing the ground state solution of Bose–Einstein condensates by a normalized gradient flow", SIAM J. Sci. Comput. **25** (2004), 1674–1697.

[17] Belmonte-Beitia, J.; Pérez-García, V.M.; Vekslerchik, V.; Torres, P. J.; "Lie symmetries, qualitative analysis and exact solutions of nonlinear Schrödinger equations with inhomogeneous nonlinearities", Discr. Cont. Dyn. Syst. B **9** (2008), 221–233.

[18] Belmonte-Beitia, J.; Pelinovsky, D.E.; "Bifurcation of gap solitons in periodic potentials with a periodic sign-varying nonlinearity coefficient", Applic. Anal. **89** (2010), 1335–1350.

[19] Belmonte-Beitia, J.; Cuevas, J.; "Solitons for the cubic–quintic nonlinear Schrödinger equation with time- and space-modulated coefficients", J. Phys. A: Math. Theor. **42** (2009), 165201 (11 pp.).

[20] Bloch, F.; "Über die Quantenmechanik der Electronen in Kristallgittern", Z. Phys. **52** (1928), 555–600.

[21] Bluman, G.W.; Kumei, S.; *Symmetries and Differential Equations* (Springer, New York, 1989).

[22] Brugarino, T.; Sciacca, M.; "Integrability of an inhomogeneous nonlinear Schrödinger equation in Bose–Einstein condensates and fiber optics", J. Math. Phys. **51** (2010), 093503 (18 pp.).

[23] Buffoni, B.; Sere, E.; "A global condition for quasi-random behaviour in a class of conservative systems", Commun. Pure Appl. Math. **49** (1996), 285–305.

[24] Busch, K.; Schneider, G.; Tkeshelashvili, L.; Uecker, H.; "Justification of the nonlinear Schrödinger equation in spatially periodic media", Z. Angew. Math. Phys. **57** (2006), 905–939.

[25] Buslaev, V.S.; Perelman, G.S.; "Scattering for the nonlinear Schrödinger equation: states close to a soliton", St. Petersburg Math. J. **4** (1993), 1111–1142.

[26] Buslaev, V.S.; Perelman, G.S.; "On the stability of solitary waves for nonlinear Schrödinger equations", Amer. Math. Soc. Transl. **164** (1995), 75–98.

[27] Buslaev, V.S; Sulem, C.; "On asymptotic stability of solitary waves for nonlinear Schrödinger equations", Ann. Inst. H. Poincaré Anal. Non Linéaire **20** (2003), 419–475.

[28] Calvo, D.C.; Akylas, T.R.; "On the formation of bound states by interacting nonlocal solitary waves", Physica D **101** (1997), 270–288.

[29] Carretero-Gonzáles, R.; Talley, J.D.; Chong, C.; Malomed, B.A.; "Multistable solitons in the cubic–quintic discrete nonlinear Schrödinger equation", Physica D **216** (2006), 77–89.

[30] Casenawe, T.; *Semilinear Schrödinger Equations*, Courant Lecture Notes (Courant Institute, New York, 2003).

[31] Champneys, A.R.; Yang, J.; "A scalar nonlocal bifurcation of solitary waves for coupled nonlinear Schrödinger systems", Nonlinearity **15** (2002), 2165–2192.

[32] Chang, S.M.; Gustafson, S.; Nakanishi, K.; Tsai, T.P.; "Spectra of linearized operators for NLS solitary waves", SIAM J. Math. Anal. **39** (2007), 1070–1111.

[33] Chugunova, M.; Pelinovsky, D.; "Count of unstable eigenvalues in the generalized eigenvalue problem", J. Math. Phys. **51** (2010), 052901 (19 pp.).

[34] Chugunova, M.; Pelinovsky, D.; "Two-pulse solutions in the fifth-order KdV equation: rigorous theory and numerical approximations", Discr. Cont. Dyn. Syst. B **8** (2007), 773–800.

[35] Chugunova, M.; Pelinovsky, D.; "Block-diagonalization of the symmetric first-order coupled-mode system", SIAM J. Appl. Dyn. Syst. **5** (2006), 66–83.

[36] Chugunova, M.; Pelinovsky, D.; "Spectrum of a non-self-adjoint operator associated with the periodic heat equation", J. Math. Anal. Appl. **342** (2008), 970–988.

[37] Chugunova, M.; Pelinovsky, D.; "On quadratic eigenvalue problems arising in stability of discrete vortices", Lin. Alg. Appl. **431** (2009), 962–973.

[38] Colin, Th.; "Rigorous derivation of the nonlinear Schrödinger equation and Davey–Stewartson systems with quadratic hyperbolic systems", Asymptot. Anal. **31** (2002), 69–91.

[39] Colin, M.; Lannes, D.; "Short pulse approximations in dispersive media", SIAM J. Math. Anal. **41** (2009), 708–732.

[40] Comech, A.; Pelinovsky, D.; "Purely nonlinear instability of standing waves with minimal energy", Commun. Pure Appl. Math. **56** (2003), 1565–1607.

[41] Cuccagna, S.; "On asymptotic stability in energy space of ground states of NLS in 1D", J. Diff. Eqs. **245** (2008), 653–691.

[42] Cuccagna, S.; Pelinovsky, D.; "Bifurcations from the end points of the essential spectrum in the linearized NLS problem", J. Math. Phys. **46** (2005) 053520 (15 pp.).

[43] Cuccagna, S; Pelinovsky, D.; Vougalter, V.; "Spectra of positive and negative energies in the linearized NLS problem", Commun. Pure Appl. Math. **58** (2005), 1–29.

[44] Cuccagna, S.; Tarulli, M.; "On asymptotic stability of standing waves of discrete Schrödinger equation in $\mathbb{Z}$", SIAM J. Math. Anal. **41** (2009), 861–885.

[45] Demanet, L.; Schlag, W.; "Numerical verification of a gap condition for a linearized NLS equation", Nonlinearity **19** (2006), 829–852.

[46] Derks, G.; Gottwald, G.A.; "A robust numerical method to study oscillatory instability of gap solitary waves", SIAM J. Appl. Dyn. Syst. **4** (2005), 140–158.

[47] Dimassi, M.; Sjöstrand, J.; *Spectral Asymptotics in the Semi-classical Limit*, London Mathematical Society Lecture Notes **268** (Cambridge University Press, Cambridge, 1999).

[48] Dmitriev, S.V.; Kevrekidis, P.G.; Yoshikawa, N.; "Discrete Klein–Gordon models with static kinks free of the Peierls–Nabarro potential", J. Phys. A: Math. Gen. **38** (2005), 7617–7627.

[49] Dmitriev, S.V.; Kevrekidis, P.G.; Sukhorukov, A.A.; Yoshikawa, N.; Takeno, S.; "Discrete nonlinear Schrödinger equations free of the Peierls–Nabarro potential", Phys. Lett. A **356** (2006), 324–332.

[50] Dohnal, D.; Pelinovsky, D.; Schneider, G.; "Coupled-mode equations and gap solitons in a two-dimensional nonlinear elliptic problem with a separable periodic potential", J. Nonlin. Sci. **19** (2009), 95–131.

[51] Dohnal, D.; Uecker, H.; "Coupled-mode equations and gap solitons for the 2D Gross–Pitaevskii equation with a non-separable periodic potential", Physica D **238** (2009), 860–879.

[52] Eastham, M.S.; *The Spectral Theory of Periodic Differential Equations* (Scottish Academic Press, Edinburgh, 1973).

[53] Eckmann, J.-P.; Schneider, G.; "Nonlinear stability of modulated fronts for the Swift–Hohenberg equation", Comm. Math. Phys. **225** (2002), 361–397.

[54] Esteban, M.J.; Séré, É.; "Stationary states of the nonlinear Dirac equation: a variational approach", Comm. Math. Phys. **171** (1995), 323–350.

[55] Evans, L.C.; *Partial Differential Equations* (AMS, Providence, RI, 1998).

[56] Feckan, M.; *Topological Degree Approach to Bifurcation Problems* (Springer, Heidelberg, 2008).

[57] Flach, S.; Kladko, K.; "Moving discrete breathers?", Physica D **127** (1999), 61–72.

[58] Floer, A.; Weinstein, A.; "Nonspreading wave packets for the cubic Schrödinger equation with a bounded potential", J. Funct. Anal. **69** (1986), 397–408.

[59] Floquet, G.; "Sur les équations différentielles linéaires à coefficients périodique", Ann. École Norm. Sup. **12** (1883), 47–88.

[60] Friesecke, G.; Pego, R.L.; "Solitary waves on FPU lattices", Nonlinearity **12** (1999), 1601–1627; **15** (2002), 1343–1359; **17** (2004), 207–227; **17** (2004), 229–251.

[61] Gagnon, L.; Winternitz, P.; "Symmetry classes of variable coefficient nonlinear Schrödinger equations", J. Phys. A: Math. Gen. **26** (1993), 7061–7076.

[62] Gang, Z.; Sigal, I.M.; "Asymptotic stability of nonlinear Schrödinger equations with potential", Rev. Math. Phys. **17** (2005), 1143–1207.

[63] Gang, Z.; Sigal, I.M.; "Relaxation of solitons in nonlinear Schrödinger equations with potential", Adv. Math. **216** (2007), 443–490.

388   *References*

[64] García-Ripoll, J.J.; Pérez-García, V.M.; "Optimizing Schrödinger functionals using Sobolev gradients: applications to quantum mechanics and nonlinear optics", SIAM J. Sci. Comput. **23** (2001), 1315–1333.

[65] Georgieva, A.; Kriecherbauer, T.; Venakides, S.; "Wave propagation and resonance in a one-dimensional nonlinear discrete periodic medium", SIAM J. Appl. Math. **60** (1999), 272–294.

[66] Giannoulis, J.; Mielke, A.; "The nonlinear Schrödinger equation as a macroscopic limit for an oscillator chain with cubic nonlinearities", Nonlinearity **17** (2004), 551–565.

[67] Giannoulis, J.; Mielke, A.; Sparber C.; "Interaction of modulated pulses in the nonlinear Schrödinger equation with periodic potential", J. Diff. Eqs. **245** (2008), 939–963.

[68] Giannoulis, J.; Herrmann, M.; Mielke, A.; "Continuum descriptions for the dynamics in discrete lattices: derivation and justification", in *Analysis, Modeling and Simulation of Multiscale Problems* (Springer, Berlin, 2006), pp. 435–466.

[69] Glazman, I.M.; *Direct Methods of Qualitative Spectral Analysis of Singular Differential Operators* (Israel Program for Scientific Translation, Jerusalem, 1965).

[70] Gohberg I.G.; Krein, M.G.; *Introduction to the Theory of Linear Non-self-adjoint Operators*, Translations of Mathematical Monographs **18** (AMS, Providence, RI, 1969).

[71] Golubitsky, M.; Schaeffer, D.G.; *Singularities and Groups in Bifurcation Theory*, Volume **1** (Springer, Berlin, 1985).

[72] Goodman, R.H.; Weinstein, M.I.; Holmes, P.J.; "Nonlinear propagation of light in one-dimensional periodic structures", J. Nonlin. Sci. **11** (2001), 123–168.

[73] Gorshkov, K.A.; Ostrovsky, L.A.; "Interactions of solitons in nonintegrable systems: Direct perturbation method and applications", Physica D **3** (1981), 428–438.

[74] Grillakis, M.; Shatah, J.; Strauss, W.; "Stability theory of solitary waves in the presence of symmetry. I", J. Funct. Anal. **74** (1987), 160–197.

[75] Grillakis, M.; Shatah, J.; Strauss, W.; "Stability theory of solitary waves in the presence of symmetry. II", J. Funct. Anal. **94** (1990), 308–348.

[76] Grillakis, M.; "Linearized instability for nonlinear Schrödinger and Klein–Gordon equations", Commun. Pure Appl. Math. **41** (1988), 747–774.

[77] Grillakis, M.; "Analysis of the linearization around a critical point of an infinite dimensional Hamiltonian system", Commun. Pure Appl. Math. **43** (1990), 299–333.

[78] Grimshaw, R.; "Weakly nonlocal solitary waves in a singularly perturbed nonlinear Schrödinger equation", Stud. Appl. Math. **94** (1995), 257–270.

[79] Groves, M.D.; "Solitary-wave solutions to a class of fifth-order model equations", Nonlinearity **11** (1998), 341–353.

[80] Groves, M.D.; Schneider, G.; "Modulating pulse solutions for a class of nonlinear wave equations", Comm. Math. Phys. **219** (2001), 489–522.

[81] Groves, M.D.; Schneider, G.; "Modulating pulse solutions for quasilinear wave equations", J. Diff. Eqs. **219** (2005), 221–258.

[82] Groves, M.D.; Schneider, G.; "Modulating pulse solutions to quadratic quasilinear wave equations over exponentially long length scales", Comm. Math. Phys. **278** (2008), 567–625.

[83] Gross, E.P.; "Hydrodynamics of a superfluid concentrate", J. Math. Phys. **4** (1963), 195–207.

[84] Gurski, K.F.; Kollar, R.; Pego, R.L.; "Slow damping of internal waves in a stably stratified fluid", Proc. R. Soc. Lond. **460** (2004), 977–994.

[85] Haragus, M.; Kapitula, T.; "On the spectra of periodic waves for infinite-dimensional Hamiltonian systems", Physica D **237** (2008), 2649–2671.

[86] Helffer, B.; *Semi-classical Analysis for the Schrödinger Operator and Applications*, Lecture Notes in Mathematics **1336** (Springer, New York, 1988).

[87] Heinz, H.P.; Küpper, T.; Stuart, C.A.; "Existence and bifurcation of solutions for nonlinear perturbations of the periodic Schrödinger equation", J. Diff. Eqs. **100** (1992), 341–354.

[88] Hernández-Heredero, R.; Levi, D.; "The discrete nonlinear Schrödinger equation and its Lie symmetry reductions", J. Nonlin. Math. Phys. **10** (2003), 77–94.

[89] Hislop, P.D.; Sigal, I.M.; *Introduction to Spectral Theory with Applications to Schrödinger Operators* (Springer, New York, 1996).

[90] Ilan, B.; Weinstein, M. I.; "Band-edge solitons, nonlinear Schrödinger/Gross–Pitaevskii equations and effective media", Multiscale Model. Simul. **8** (2010), 1055–1101.

[91] Iohvidov, I.S.; Krein, M.G.; Langer, H.; *Introduction to the Spectral Theory of Operators in Spaces with an Indefinite Metric* (Mathematische Forschung, Berlin, 1982).

[92] Iooss, G.; "Travelling waves in the Fermi–Pasta–Ulam lattice", Nonlinearity **13** (2000), 849–866.

[93] Iooss, G.; Adelmeyer, M.; *Topics in Bifurcation Theory and Applications* (World Scientific, Singapore, 1998).

[94] Iooss, G.; Kirchgassner, K.; "Travelling waves in a chain of coupled nonlinear oscillators", Comm. Math. Phys. **211** (2000), 439–464.

[95] Iooss, G.; Pelinovsky, D.; "Normal form for travelling kinks in discrete Klein–Gordon lattices", Physica D **216** (2006), 327–345.

[96] James, G.; Sire, Y.; "Travelling breathers with exponentially small tails in a chain of nonlinear oscillators", Comm. Math. Phys. **257** (2005), 51–85.

[97] James, G.; Sire, Y.; "Numerical computation of travelling breathers in Klein–Gordon chains", Physica D **204** (2005), 15–40.

[98] JeanJean, L.; Tanaka, K.; "A remark on least energy solutions in $\mathbb{R}^{N}$", Proc. Amer. Math. Soc. **131** (2002), 2399–2408.

[99] Jones, C.K.R.T.; "An instability mechanism for radially symmetric standing waves of a nonlinear Schrödinger equation", J. Diff. Eqs. **71** (1988), 34–62.

[100] Jones, C.K.R.T.; "Instability of standing waves for nonlinear Schrödinger-type equations", Ergod. Theor. Dynam. Sys. **8** (1988), 119–138.

[101] Kapitula, T.; "Stability of waves in perturbed Hamiltonian systems", Physica D **156** (2001), 186–200.

[102] Kapitula, T.; Kevrekidis, P.G.; "Stability of waves in discrete systems", Nonlinearity **14** (2001), 533–566.

[103] Kapitula, T.; Kevrekidis, P.; "Bose–Einstein condensates in the presence of a magnetic trap and optical lattice: two-mode approximation", Nonlinearity **18** (2005), 2491–2512.

[104] Kapitula, T.; Kevrekidis, P.; Chen, Z.; "Three is a crowd: solitary waves in photorefractive media with three potential wells", SIAM J. Appl. Dyn. Syst. **5** (2006), 598–633.

[105] Kapitula, T.; Kevrekidis, P.; Sandstede, B.; "Counting eigenvalues via the Krein signature in infinite-dimensional Hamiltonian systems", Physica D **195** (2004), 263–282; Addendum: Physica D **201** (2005), 199–201.

[106] Kapitula, T.; Law, K.J.H.; Kevrekidis, P.G.; "Interaction of excited states in two-species Bose–Einstein condensates: a case study", SIAM J. Appl. Dyn. Syst. **9** (2010), 34–61.

[107] Kapitula, T.; Sandstede, B.; "Edge bifurcations for near integrable systems via Evans function techniques", SIAM J. Math. Anal. **33** (2002), 1117–1143.

[108] Kato, T.; *Perturbation Theory for Linear Operators* (Springer, New York, 1976).

[109] Kaup, D.J.; "Perturbation theory for solitons in optical fibers", Phys. Rev. A **42** (1990), 5689–5694.

[110] Kevrekidis, P.G.; *The Discrete Nonlinear Schrödinger Equation: Mathematical Analysis, Numerical Computations and Physical Perspectives*, Springer Tracts in Modern Physics **232** (Springer, New York, 2009).

[111] Kevrekidis, P.G.; "On a class of discretizations of Hamiltonian nonlinear partial differential equations", Physica D **183** (2003), 68–86.

[112] Kevrekidis, P.G.; Pelinovsky, D.E.; Stefanov, A.; "Asymptotic stability of small bound states in the discrete nonlinear Schrödinger equation in one dimension", SIAM J. Math. Anal. **41** (2009), 2010–2030.

[113] Kirrmann, P.; Schneider, G.; Mielke, A.; "The validity of modulation equations for extended systems with cubic nonlinearities", Proc. Roy. Soc. Edinburgh A **122** (1992), 85–91.

[114] Khare, A.; Dmitriev, S.V.; Saxena, A.; "Exact moving and stationary solutions of a generalized discrete nonlinear Schrödinger equation", J. Phys. A: Math. Gen. **40** (2007), 11301–11317.

[115] Klaus, M.; Pelinovsky, D.; Rothos, V.M.; "Evans function for Lax operators with algebraically decaying potentials", J. Nonlin. Sci. **16** (2006), 1–44.

[116] Klaus, M.; Shaw, K.; "On the eigenvalues of Zakharov–Shabat systems", SIAM J. Math. Anal. **34** (2003), 759–773.

[117] Kohn, W.; "Analytic properties of Bloch waves and Wannier functions", Phys. Rev. **115** (1959), 809–821.

[118] Kollar, R.; Pego, R.L.; "Spectral stability of vortices in two-dimensional Bose–Einstein condensates via the Evans function and Krein signature", Appl. Math. Res. Express **2011** (2011), in press (46 pp.).

[119] Kollar, R.; "Homotopy method for nonlinear eigenvalue pencils with applications", SIAM J. Math. Anal. **43** (2011), 612–633.

[120] Koukouloyannis, V.; Kevrekidis, P.G.; "On the stability of multibreathers in Klein–Gordon chains", Nonlinearity **22** (2009), 2269–2285.

[121] Kominis, Y.; Hizanidis, K.; "Power dependent soliton location and stability in complex photonic structures", Opt. Express **16** (2008), 12124–12138.

[122] Krieger, J.; Schlag, W.; "Stable manifolds for all monic supercritical focusing nonlinear Schrödinger equations in one dimension", J. Amer. Math. Soc. **19** (2006), 815–920.

[123] Kruskal, M.D.; Segur, H.; "Asymptotics beyond all orders in a model of crystal growth", Stud. Appl. Math. **85** (1991), 129–181.

[124] Küpper, T.; Stuart, C.A.; "Necessary and sufficient conditions for gap-bifurcation", Nonlin. Anal. **18** (1992), 893–903.

[125] Lakoba, T.; Yang, J.; "A generalized Petviashvili iteration method for scalar and vector Hamiltonian equations with arbitrary form of nonlinearity", J. Comput. Phys. **226** (2007), 1668–1692.

[126] Lakoba, T.I.; Yang, J.; "A mode elimination technique to improve convergence of iteration methods for finding solitary waves", J. Comput. Phys. **226** (2007), 1693–1709.

[127] Lakoba, T.I.; "Conjugate gradient method for finding fundamental solitary waves", Physica D **238** (2009), 2308–2330.

[128] Lafortune, S.; Lega, J.; "Spectral stability of local deformations of an elastic rod: Hamiltonian formalism", SIAM J. Math. Anal. **36** (2005), 1726–1741.

[129] Lannes, D.; "Dispersive effects for nonlinear geometrical optics with rectification", Asymptot. Anal. **18** (1998), 111–146.

[130] Lieb, E.H.; Loss, M.; *Analysis*, 2nd edn (AMS, Providence, RI, 2001).

[131] Lieb, E.H.; Seiringer, R.; Solovej, J.P.; Yngvason, J.; *The Mathematics of the Bose Gas and its Condensation*, Oberwolfach Seminars **34** (Birkhäuser, Basel, 2005).

[132] Lin, Z.; "Instability of nonlinear dispersive solitary waves", J. Funct. Anal. **255** (2008), 1191–1224.

[133] Linares, F.; Ponce, G.; *Introduction to Nonlinear Dispersive Equations* (Springer, New York, 2009).

[134] Lions, P.-L.; "The concentration compactness principle in the calculus of variations. The locally compact case", Ann. Inst. Henri Poincaré **1** (1984), 223–283.

[135] Lukas, M.; Pelinovsky, D.E.; Kevrekidis, P.G.; "Lyapunov–Schmidt reduction algorithm for three-dimensional discrete vortices", Physica D **237** (2008), 339–350.

[136] MacKay, R.S.; Aubry, S.; "Proof of existence of breathers for time-reversible or Hamiltonian networks of weakly coupled oscillators", Nonlinearity **7** (1994), 1623–1643.

[137] MacKay, R.S.; "Slow manifolds", in *Energy Localization and Transfer*, eds. T. Dauxois, A. Litvak-Hinenzon, R.S. MacKay, and A. Spanoudaki (World Scientific, Singapore, 2004), pp. 149–192.

[138] Mallet-Paret, J.; "The Fredholm alternative for functional differential equations of mixed type", J. Dyn. Diff. Eqs. **11** (1999), 1–47.

[139] Melvin, T.R.O.; Champneys, A.R.; Kevrekidis, P.G.; Cuevas, J.; "Travelling solitary waves in the discrete Schrödinger equation with saturable nonlinearity: existence, stability and dynamics", Physica D **237** (2008), 551–567.

[140] Melvin, T.R.O.; Champneys, A.R.; Pelinovsky, D.E.; "Discrete travelling solitons in the Salerno model", SIAM J. Appl. Dyn. Syst. **8** (2009), 689–709.

[141] Mizumachi, T.; "Asymptotic stability of small solitary waves to 1D nonlinear Schrödinger equations with potential", J. Math. Kyoto Univ. **48** (2008), 471–497.

[142] Morgante, A.M.; Johansson, M.; Kopidakis, G.; Aubry, S.; "Standing wave instabilities in a chain of nonlinear coupled oscillators", Physica D **162** (2002), 53–94.

[143] Newell, A.C.; Moloney, J.V.; *Nonlinear Optics* (Westview Press, Boulder, CO, 2003).

[144] Olver, P.J.; *Applications of Lie Groups to Differential Equations* (Springer, New York, 1993).

[145] Oxtoby, O.F.; Barashenkov, I.V.; "Moving solitons in the discrete nonlinear Schödinger equation", Phys. Rev. E **76** (2007), 036603 (18 pp.).

[146] Oxtoby, O.F.; Pelinovsky, D.E.; Barashenkov, I.V.; "Travelling kinks in discrete phi-4 models", Nonlinearity **19** (2006), 217–235.

[147] Pankov, A.; *Travelling Waves and Periodic Oscillations in Fermi–Pasta–Ulam Lattices* (Imperial College Press, London, 2005).

[148] Pankov, A.; "Periodic nonlinear Schrödinger equation with application to photonic crystals", Milan J. Math. **73** (2005), 259–287.

[149] Pankov, A.; "Gap solitons in periodic discrete nonlinear Schrödinger equations", Nonlinearity **19** (2006), 27–40.

[150] Pankov, A.; "Gap solitons in periodic discrete nonlinear Schrödinger equations II: A generalized Nehari manifold approach", Discr. Cont. Dyn. Syst. **19** (2007), 419–430.

[151] Pego, R.L.; Warchall, H.A.; "Spectrally stable encapsulated vortices for nonlinear Schrödinger equations", J. Nonlin. Sci. **12** (2002), 347–394.

[152] Pelinovsky, D.E.; "Asymptotic reductions of the Gross–Pitaevskii equation", in *Emergent Nonlinear Phenomena in Bose–Einstein Condensates*, eds. P.G. Kevrekidis, D.J. Franzeskakis, and R. Carretero-Gonzalez (Springer, New York, 2008), pp. 377–398.

[153] Pelinovsky, D.E.; "Inertia law for spectral stability of solitary waves in coupled nonlinear Schrödinger equations", Proc. Roy. Soc. Lond. A, **461** (2005), 783–812.

[154] Pelinovsky, D.E.; "Translationally invariant nonlinear Schrödinger lattices", Nonlinearity **19** (2006), 2695–2716.

[155] Pelinovsky, D.E.; Kevrekidis, P.G.; Frantzeskakis, D.J.; "Stability of discrete solitons in nonlinear Schrödinger lattices", Physica D **212** (2005), 1–19.

[156] Pelinovsky, D.E.; Kevrekidis, P.G.; Frantzeskakis, D.J.; "Persistence and stability of discrete vortices in nonlinear Schrödinger lattices", Physica D **212** (2005), 20–53.

[157] Pelinovsky, D.E.; Melvin, T.R.O.; Champneys, A.R.; "One-parameter localized traveling waves in nonlinear Schrödinger lattices", Physica D **236** (2007), 22–43.

[158] Pelinovsky, D.E.: Rothos, V.M.; "Bifurcations of travelling breathers in the discrete NLS equations", Physica D **202** (2005), 16–36.

[159] Pelinovsky, D.; Sakovich, A.; "Internal modes of discrete solitons near the anti-continuum limit of the dNLS equation", Physica D **240** (2011), 265–281.

[160] Pelinovsky, D.E.; Stepanyants, Yu.A.; "Convergence of Petviashvili's iteration method for numerical approximation of stationary solutions of nonlinear wave equations", SIAM J. Numer. Anal. **42** (2004), 1110–1127.

[161] Pelinovsky, D.E.; Sukhorukov, A.A.; Kivshar, Yu.S.; "Bifurcations and stability of gap solitons in periodic potentials", Phys. Rev. E **70** (2004), 036618 (17 pp.).

[162] Pelinovsky, D.; Schneider, G.; "Justification of the coupled-mode approximation for a nonlinear elliptic problem with a periodic potential", Applic. Anal. **86** (2007), 1017–1036.

[163] Pelinovsky, D.; Schneider, G.; "Moving gap solitons in periodic potentials", Math. Meth. Appl. Sci. **31** (2008), 1739–1760.

[164] Pelinovsky, D.; Schneider, G.; "Bounds on the tight-binding approximation for the Gross–Pitaevskii equation with a periodic potential", J. Diff. Eqs. **248** (2010), 837–849.

[165] Pelinovsky, D.; Schneider, G.; MacKay, R.; "Justification of the lattice equation for a nonlinear elliptic problem with a periodic potential", Comm. Math. Phys. **284** (2008), 803–831.

[166] Pelinovsky, D.E.; Yang, J.; "Instabilities of multihump vector solitons in coupled nonlinear Schrödinger equations", Stud. Appl. Math. **115** (2005), 109–137.

[167] Perelman, G.; "On the formation of singularities in solutions of the critical nonlinear Schrödinger equation", Ann. Henri Poincaré **2** (2001), 605–673.

[168] Petviashvili, V.I.; "Equation of an extraordinary soliton", Sov. J. Plasma Phys. **2** (1976), 257–258.

[169] Pillet, C.A.; Wayne, C.E.; "Invariant manifolds for a class of dispersive, Hamiltonian, partial differential equations", J. Diff. Eqs. **141** (1997), 310–326.

[170] Pitaevskii, L.P.; "Vortex lines in an imperfect Bose gas", Sov. Phys. JETP **13** (1961), 451–454.

[171] Pitaevskii, L.; Stringari, S.; *Bose–Einstein Condensation* (Oxford University Press, Oxford, 2003).

[172] Pomeau, Y.; Ramani, A.; Grammaticos, B.; "Structural stability of the Korteweg–de Vries solitons under a singular perturbation", Physica D **31** (1988), 127–134.

[173] Pontryagin, L.S.; "Hermitian operators in spaces with indefinite metric", Izv. Akad. Nauk SSSR Ser. Mat. **8** (1944), 243–280.

[174] Porter, M.A.; Chugunova, M.; Pelinovsky, D.E.; "Feshbach resonance management of Bose–Einstein condensates in optical lattices", Phys. Rev. E **74** (2006), 036610 (8 pp.).

[175] Qin, W.-X.; Xiao, X.; "Homoclinic orbits and localized solutions in nonlinear Schrödinger lattices", Nonlinearity **20** (2007), 2305–2317.

[176] Reed, M.; Simon, B.; *Methods of Modern Mathematical Physics. IV. Analysis of Operators* (Academic Press, New York, 1978).

[177] Salerno, M.; "Quantum deformations of the discrete nonlinear Schrödinger equation", Phys. Rev. A **46** (1992), 6856–6859.

[178] Sánchez, A.; Bishop, A.R.; "Collective coordinates and length-scale competition in spatially inhomogeneous soliton-bearing equations", SIAM Rev. **40** (1998), 579–615.

[179] Sandstede, B.; "Stability of multiple-pulse solutions", Trans. Amer. Math. Soc. **350** (1998), 429–472.

[180] Scheel, A.; Van Vleck, E.S.; "Lattice differential equations embedded into reaction-diffusion systems", Proc. Roy. Soc. Edinburgh A **139** (2009), 193–207.

[181] Schlag, W.; "Stable manifolds for an orbitally unstable NLS", Ann. Math. **169** (2009), 139–227.

[182] Schneider, G.; "Validity and limitation of the Newell–Whitehead equation", Math. Nachr. **176** (1995), 249–263.

[183] Schneider, G.; "The validity of generalized Ginzburg–Landau equations", Math. Meth. Appl. Sci. **19** (1996), 717–736.

[184] Schneider, G.; "Justification of modulated equations for hyperbolic systems via normal forms", Nonlin. Diff. Eqs. Appl. **5** (1995), 69–82.

[185] Schneider, G.; Uecker, H.; "Nonlinear coupled mode dynamics in hyperbolic and parabolic periodically structured spatially extended systems", Asymptot. Anal. **28** (2001), 163–180.

[186] Schneider, G.; Wayne, C.E.; "The long-wave limit for the water wave problem. I. The case of zero surface tension", Commun. Pure Appl. Math. **53** (2000), 1475–1535.

[187] Schneider, G.; Wayne, C.E.; "The rigorous approximation of long-wavelength capillary–gravity waves", Arch. Ration. Mech. Anal. **162** (2002), 247–285.

[188] Serkin, V.N.; Hasegawa, A.; Belyaeva, T.L.; "Nonautonomous solitons in external potentials", Phys. Rev. Lett. **98** (2007), 074102 (4 pp.).

[189] Shatah, J.; Strauss, W.; "Instability of nonlinear bound states", Comm. Math. Phys. **100** (1985), 173–190.

[190] Shi, H.; Zhang, H.; "Existence of gap solitons in periodic discrete nonlinear Schrödinger equations", J. Math. Anal. Appl. **361** (2010), 411–419.

[191] Sivan, Y.; Fibich, G.; Efremidis, N.K.; Bar-Ad, S.; "Analytic theory of narrow lattice solitons", Nonlinearity **21** (2008), 509–536.

[192] Skorobogatiy, M.; Yang, J.; *Fundamentals of Photonic Crystal Guiding* (Cambridge University Press, Cambridge, 2009).

[193] Soffer, A.; Weinstein, M.I.; "Multichannel nonlinear scattering theory for nonintegrable equations", Comm. Math. Phys. **133** (1990), 119–146.

[194] Soffer, A.; Weinstein, M.I.; "Multichannel nonlinear scattering theory for nonintegrable equations II: The case of anisotropic potentials and data", J. Diff. Eqs. **98** (1992), 376–390.

[195] Soffer, A.; Weinstein, M.I.; "Selection of the ground state for nonlinear Schrödinger equations", Rev. Math. Phys. **16** (2004), 977–1071.

[196] Speight, J.M.; "Topological discrete kinks", Nonlinearity **12** (1999), 1373–1387.

[197] de Sterke, C.M.; Sipe, J.E.; "Gap solitons", Prog. Opt. **33** (1994), 203–259.

[198] Stuart, C.A.; "Bifurcations into spectral gaps", Bull. Belg. Math. Soc. Simon Stevin, 1995, suppl. (59 pp.).

[199] Sulem, C.; Sulem, P.L.; *The Nonlinear Schrödinger Equation: Self-focusing and Wave Collapse* (Springer, New York, 1999).

[200] Sylvester, J.J.; "A demonstration of the theorem that every homogeneous quadratic polynomial is reducible by real orthogonal substitutions to the form of a sum of positive and negative squares", Philos. Mag. IV (1852), 138–142.

[201] Tao, T.; *Nonlinear Dispersive Equations: Local and Global Analysis*, CBMS Regional Conference Series in Mathematics **106** (AMS, Providence, RI, 2006).

[202] Tao, T.; "Why are solitons stable?", Bull. Amer. Math. Soc. **46** (2009), 1–33.

[203] Tovbis, A.; Tsuchiya, M.; Jaffe, C.; "Exponential asymptotic expansions and approximations of the unstable and stable manifolds of singularly perturbed systems with the Henon map as an example", Chaos **8** (1998), 665–681.

[204] Tovbis, A.; "On approximation of stable and unstable manifolds and the Stokes phenomenon", Contemp. Math. **255** (2000), 199–228.

[205] Tovbis, A.; "Breaking homoclinic connections for a singularly perturbed differential equation and the Stokes phenomenon", Stud. Appl. Math. **104** (2000), 353–386.

[206] Tovbis, A.; Pelinovsky, D.; "Exact conditions for existence of homoclinic orbits in the fifth-order KdV model", Nonlinearity **19** (2006), 2277–2312.

[207] Vakhitov, M.G.; Kolokolov, A.A.; "Stationary solutions of the wave equation in a medium with nonlinearity saturation", Radiophys. Quantum Electron. **16** (1973), 783–789.

[208] Vougalter V.; "On the negative index theorem for the linearized nonlinear Schrödinger problem", Canad. Math. Bull. **53** (2010), 737–745.

[209] Vougalter, V.; Pelinovsky, D.; "Eigenvalues of zero energy in the linearized NLS problem", J. Math. Phys. **47** (2006), 062701 (13 pp.).

[210] Weder, R.; "The $W_{k,p}$-continuity of the Schrödinger wave operators on the line", Comm. Math. Phys. **208** (1999), 507–520.

[211] Weinstein, M.I.; "Liapunov stability of ground states of nonlinear dispersive evolution equations", Commun. Pure Appl. Math. **39** (1986), 51–68.

[212] Weinstein, M.; "Excitation thresholds for nonlinear localized modes on lattices", Nonlinearity **12** (1999), 673–691.

[213] Yang, J.; "Classification of the solitary waves in coupled nonlinear Schrödinger equations", Physica D **108** (1997), 92–112.

[214] Yang, J.; Akylas, T.R.; "Continuous families of embedded solitons in the third-order nonlinear Schrödinger equation", Stud. Appl. Math. **111** (2003), 359–375.

[215] Yang, J.; Pelinovsky, D.E.; "Stable vortex and dipole vector solitons in a saturable nonlinear medium", Phys. Rev. E **67** (2003), 016608 (12 pp.).

[216] Yang, J.; Lakoba, T.; "Universally-convergent squared-operator iteration methods for solitary waves in general nonlinear wave equations", Stud. Appl. Math. **118** (2007), 153–197.

[217] Yang, J.; Lakoba, T.; "Accelerated imaginary-time evolution methods for computations of solitary waves", Stud. Appl. Math. **120** (2008), 265–292.

[218] Yau, H.T.; Tsai, T.P.; "Asymptotic dynamics of nonlinear Schrödinger equations: resonance dominated and radiation dominated solutions", Commun. Pure Appl. Math. **55** (2002), 1–64.

[219] Yau, H.T.; Tsai, T.P.; "Stable directions for excited states of nonlinear Schrödinger equations", Comm. Part. Diff. Eqs. **27** (2002), 2363–2402.

[220] Yau, H.T.; Tsai, T.P.; "Relaxation of excited states in nonlinear Schrödinger equations", Int. Math. Res. Not. **2002** (2002), 1629–1673.

[221] Zelik, S.; Mielke, A.; "Multi-pulse evolution and space-time chaos in dissipative systems", Mem. Amer. Math. Soc. **198** (2009), 1–97.

[222] Zhang, G.; "Breather solutions of the discrete nonlinear Schrödinger equations with unbounded potentials", J. Math. Phys. **50** (2009), 013505 (10 pp.).

[223] Zhou, Z.; Yu, J.; Chen, Y.; "On the existence of gap solitons in a periodic discrete nonlinear Schrödinger equation with saturable nonlinearity", Nonlinearity **23** (2010), 1727–1740.

Index

Printed in the United States
by Baker & Taylor Publisher Services